# Sex Differences in the
# CENTRAL NERVOUS SYSTEM

# Sex Differences in the
# CENTRAL NERVOUS SYSTEM

Edited by

**REBECCA M. SHANSKY**
Northeastern University, Boston, MA, USA

ELSEVIER

Amsterdam • Boston • Heidelberg • London • New York • Oxford
Paris • San Diego • San Francisco • Singapore • Sydney • Tokyo
Academic Press is an imprint of Elsevier

Academic Press is an imprint of Elsevier
125, London Wall, EC2Y 5AS, UK
525 B Street, Suite 1800, San Diego, CA 92101-4495, USA
225 Wyman Street, Waltham, MA 02451, USA
The Boulevard, Langford Lane, Kidlington, Oxford OX5 1GB, UK



**British Library Cataloguing-in-Publication Data**
A catalogue record for this book is available from the British Library

**Library of Congress Cataloging-in-Publication Data**
A catalog record for this book is available from the Library of Congress

ISBN: 978-0-12-802114-9

For information on all Academic Press publications
visit our website at http://store.elsevier.com/

*Publisher:* Mica Haley
*Acquisition Editor:* Mica Haley
*Editorial Project Manager:* Kathy Padilla
*Production Project Manager:* Caroline Johnson
*Designer:* Greg Harris

Typeset by Thomson Digital

# TABLE OF CONTENTS

# LIST OF CONTRIBUTORS

**Anthony P. Auger**
University of Wisconsin–Madison, Neuroscience Training Program, Departments of Zoology and Psychology, Madison, WI, USA

**Jill B. Becker**
University of Michigan, Molecular and Behavioral Neuroscience Institute; University of Michigan, Department of Psychiatry; University of Michigan, Department of Psychology, Ann Arbor, MI, USA

**Kristen M. Culbert**
University of Nevada Las Vegas, Department of Psychology, Las Vegas, NV, USA

**Kelly M. Dumais**
Boston College, Neurobiology of Social Behavior Laboratory, Department of Psychology, Chestnut Hill, MA, USA

**Yosefa Ehrlich**
Brooklyn College of the City University of New York, Department of Psychology, Brooklyn, NY, USA

**Liisa A.M. Galea**
University of British Columbia, Department of Psychology; University of British Columbia, Centre for Brain Health, Vancouver, BC, Canada

**Juan L. Gomez**
Oregon Health & Sciences University, Department of Behavioral Neuroscience, Portland, OR, USA

**Gian D. Greenberg**
Oregon Health Sciences University, Department of Behavioral Neuroscience, Portland, OR, USA

**Robert J. Handa**
University of Arizona College of Medicine, Department Basic Medical Sciences, Phoenix, AZ, USA

**Ashley A. Keiser**
University of Michigan, Department of Psychology, Ann Arbor, MI, USA

**Kelly L. Klump**
Michigan State University, Department of Psychology, East Lansing, MI, USA

**Victoria N. Luine**
Hunter College of CUNY, Department of Psychology, New York, NY, USA

**Lisa Y. Maeng**
Harvard Medical School, Department of Psychiatry, Boston, MA, USA

**Anna M. Malysz**
Baylor College of Medicine, Department of Molecular and Cellular Biology, Houston, TX, USA

**Shailaja K. Mani**
Baylor College of Medicine, Department of Neuroscience, Houston; Baylor College of Medicine, Department of Molecular and Cellular Biology, Houston, TX, USA

**Christian J. Merz**
Ruhr-University Bochum, Institute of Cognitive Neuroscience, Department of Cognitive Psychology, Bochum, Germany

**Mohammed R. Milad**
Massachusetts General Hospital, Department of Psychiatry, Charlestown, MA, USA

**Gretchen N. Neigh**
Emory University, Department of Physiology, Department of Psychiatry & Behavioral Sciences, Atlanta, GA, USA

**Christina L. Nemeth**
Emory University, Department of Physiology, Department of Psychiatry & Behavioral Sciences, Atlanta, GA, USA

**Mario G. Oyola**
Baylor College of Medicine, Department of Neuroscience, Houston, TX, USA

**Adam N. Perry**
University of Texas at El Paso, Department of Biology, El Paso, TX, USA

**Sarah E. Racine**
Ohio University, Department of Psychology, Athens, OH, USA

**Doodipala Samba Reddy**
TAMHSC College of Medicine, Department of Neuroscience and Experimental Therapeutics, Bryan, TX, USA

**Meighen Roes**
University of British Columbia, Department of Psychology, Vancouver, BC, Canada

**Sydney A. Rowson**
Emory University, Department of Physiology, Department of Psychiatry & Behavioral Sciences, Atlanta, GA, USA

**Jaclyn M. Schwarz**
University of Delaware, Department of Psychological and Brain Sciences, Newark, DE, USA

**Farida Sohrabji**
TAMHSC College of Medicine, Department of Neuroscience and Experimental Therapeutics, Bryan, TX, USA

**Brian C. Trainor**
University of California, Department of Psychology, Davis, CA, USA

**Natalie C. Tronson**
University of Michigan, Department of Psychology, Ann Arbor, MI, USA

**Alexa H. Veenema**
Boston College, Neurobiology of Social Behavior Laboratory, Department of Psychology, Chestnut Hill, MA, USA

**Deborah J. Walder**
Brooklyn College of the City University of New York, Department of Psychology, Brooklyn, NY, USA

**C. Jane Welsh**
Texas A&M University, Department of Veterinary Integrative Biosciences, College Station, TX, USA

**Christel Westenbroek**
University of Michigan, Molecular and Behavioral Neuroscience Institute, Ann Arbor, MI, USA

**Oliver T. Wolf**
Ruhr-University Bochum, Institute of Cognitive Neuroscience, Department of Cognitive Psychology, Bochum, Germany

**Beril Yaffe**
Brooklyn College of the City University of New York, Department of Psychology, Brooklyn, NY, USA

# CHAPTER 1

# Sex Differences in Immunity and Inflammation: Implications for Brain and Behavior

**Gretchen N. Neigh, Christina L. Nemeth, Sydney A. Rowson**
Emory University, Department of Physiology, Department of Psychiatry & Behavioral Sciences, Atlanta, GA, USA

## 1 INTRODUCTION

Are sex differences in the immune system evolutionary in nature? Sir Peter Medawar was the first to address the question of how a mother is able to immunologically tolerate her fetus (Trowsdale and Betz, 2006). This complex issue where a fetus, up to 50% immunologically foreign, is able to pass inert has been discussed since the 1950s (Abrams and Miller, 2011; Trowsdale and Betz, 2006). The necessity for a mother to have an immune system that can fluctuate in order to prevent the rejection of the foreign fetus is one explanation for the baseline differences we observe in the male and female immune systems (Van Lunzen and Altfeld, 2014; Abrams and Miller, 2011).

Sex differences in the immune response and susceptibility to immune-related diseases cannot be disputed. The exact nature of these differences and how these differences contribute to sickness and disease incidence is quite complex, depending on a multitude of factors including age, genetics, and environment. The evolutionary "need" brought on by these factors carries with it far reaching effects on both peripheral and central functions of the immune system. The sections within this chapter highlight the basic immune-related differences between men and women, citing studies of both human disease and model animal systems. Differences in disease incidence and baseline immune activity will be outlined, followed by potential mechanisms to explain these differences, and finally, sex-dependent immune effects on behavior and the manifestation of comorbid disease states will be discussed.

### 1.1 Immune mediators

A vast array of inflammatory cells mediates peripheral and central immune responses. The immune response is divided into two component systems, the innate and adaptive systems, which differ in their ability to recognize and remember specific pathogens and antigens. While the innate system mounts a generalized and nonspecific response, the adaptive system triggers a response that is both pathogen/antigen specific and based on an immunological memory of previous responses. Despite this major difference, both

*Sex Differences in the Central Nervous System.* http://dx.doi.org/10.1016/B978-0-12-802114-9.00001-9

the innate and adaptive immune responses are composed of cell-mediated and humoral components. It is important to note that while these immunological components may originate peripherally, the brain is not unaffected – or immune privileged. Macrophages and dendritic cells are located within the brain and respond to inflammatory stimuli (Dantzer et al., 2008), and furthermore, activation of microglial cells, the brain's resident macrophages, occurs readily following infection. In addition to resident immune cells in the brain, several different routes of immune-to-brain access are possible including humoral access through circumventricular organs (Rivest, 2009), primary afferent nerve activation (vagal and trigeminal nerves; Goehler et al., 2000; Dantzer et al., 2008), cytokine trafficking through increased permeability of the blood–brain barrier, and activation of macrophage or endothelial cell interleukin-1 (IL-1) receptors, which cause a local increase of cytokine and prostaglandin release (Kubera et al., 2011; Rivest, 2009; Dantzer et al., 2008).

Within the body, lymphoid tissues and immune-relevant organs house the immunological system. Four major organs of the immunological system are bone marrow, thymus, spleen, and lymph nodes. All cells of the body are derived from the bone marrow. Here, stem cells develop into mature red blood cells, platelets, lymphocytes, and granulocytes while some migrate out of the bone marrow to mature. The following is a brief review of inflammatory mediators to aid in the understanding of how basic sex differences affect these cell populations and the immune response (Figure 1.1).

*T-cells*: They derive from immature lymphocytes, mature in the thymus and are released into the blood stream. In action, T-cells have two very different functions. T helper (Th) cells coordinate the immune response and activate other necessary immune cells. There are two distinct types of Th cells, Th1 and Th2, which are differentiated by the inflammatory cytokines that they release. Th1 cells release inflammatory cytokines that promote phagocytosis while Th2 cells release cytokines that stimulate the production of antibodies. Cytotoxic T lymphocytes (Tc cells) are important for the downregulation or destruction of parasites, tumor cells, and virus-infected cells. Like many other immune cells, T-cell recruitment to the brain occurs following injury and in many immune-related disorders (Engelhardt and Ransohoff, 2012).

*B-cells*: They originate in the spleen and are important for the development of specific antigens against foreign bacteria, viruses, and tumor cells. B-cells respond to inflammation within the brain and are thought to be one of the main effector cells in multiple sclerosis pathogenesis, promoting inflammatory activity on both sides of the blood–brain barrier (Büdingen et al., 2012).

*Natural killer (NK) cells*: These cells, similar to T- and B-cells, derive from lymphocytes. NK cells are the most effective killer cells and are similar to Tc cells. NK cells destroy parasitic or infected foreign targets. Unlike Tc cells, NK cells do not

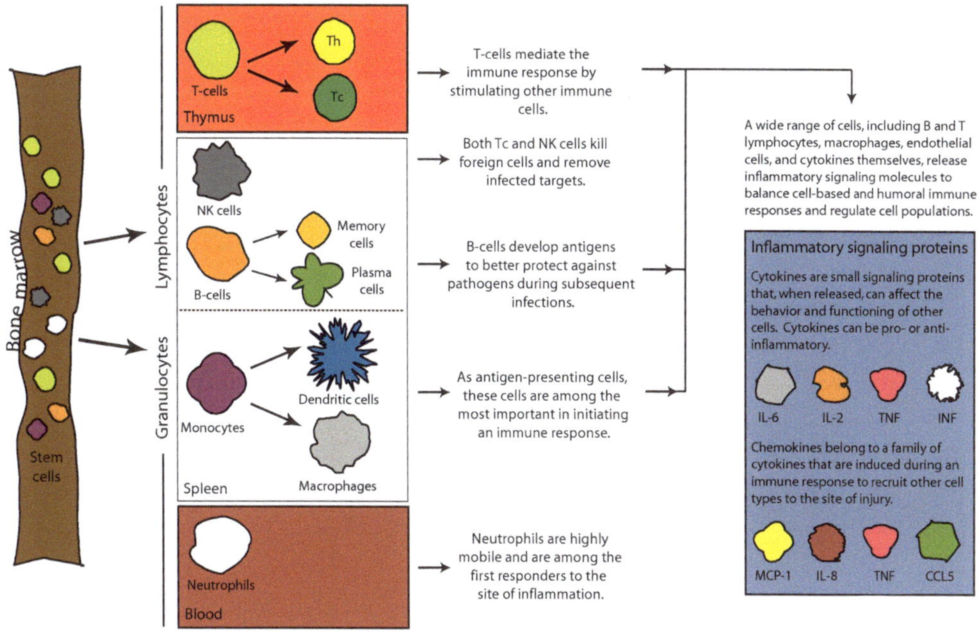


**Figure 1.1** ***Immune cells originate from hematopoietic stem cells in the bone marrow.*** A number of these cells then mature within the bone marrow while other cells migrate to other tissues to mature and further differentiate (left). Immune cells are important for the recognition of pathogens and the initiation of inflammatory processes. Once activated, many cells are capable of releasing cytokines and chemokines to further regulate cellular activity and the immune response (right).

require target recognition prior to killing infected cells, and therefore work more efficiently than Tc cells. Because NK cells act in a general fashion, they are one of the quickest responders to cerebral injury and are recruited to sites of injury within the brain. NK cells have been shown to have detrimental effects following injury, such as stroke (Gan et al., 2014).

*Neutrophils*: These are the most abundant white blood cells, and are produced in the bone marrow. Neutrophils are one of the first responders to the site of injury where they promote secretion of anti-inflammatory molecules while reducing cell death and the release of toxic substances. In the brain, neutrophils migrate to the site of injury within hours and can contribute to neurotoxicity in response to injury (Allen et al., 2012).

*Macrophages*: These are essential to the immune response. Activation of the immune system is triggered when macrophages and/or dendritic cells present antigens to T- or B-cells within the spleen. Macrophages also participate in phagocytosis and release cytokines that modulate the immunological response. Macrophages stem from perivascular monocytes, which are derived from blood-borne monocytes, and reside just outside of the basement membrane of the brain. These cells cycle in and out of blood vessels and play an important role in immune responses within the brain.

*Microglia*: They serve as the brain's resident macrophages and are highly involved in the central immune response and the release of inflammatory signaling proteins. These cells are derived from hematopoietic stem cells and enter the brain early in gestation to mature. Though similar in nature, recent evidence suggests microglia to be very different from macrophages in both origin and function (Prinz and Priller, 2014). Microglial involvement is implicated in a variety of immune-related disorders as discussed throughout this chapter.

*Dendritic cells:* Similar to macrophages, dendritic cells originate in the bone marrow and are capable of presenting antigens. Dendritic cells, due to their vast presence, are more efficient antigen-presenting cells than macrophages and work to reduce cytokine release while increasing cell death pathways. Derived from monocytes, dendritic cells also infiltrate to the brain and participate in localized increases in inflammation (Prinz and Priller, 2014).

## 2 SEX-DEPENDENT BASELINE DIFFERENCES IN IMMUNE FUNCTIONING AND RESPONSE

Given that immune cells of both peripheral and central origin have the capacity to alter function of the central nervous system, we will first review the relationship between sex steroids and the immune system with a focus on peripheral organ systems before honing in on specific interactions and implications within the central nervous system. Sex hormones play a pivotal role in the differences and fluctuations in immune activation between men and women (see Table 1.1). Estrogen exerts a biphasic effect on the immune system: low levels stimulate the immune system, while high levels suppress it. Similarly, progesterone, a highly cyclical hormone, reduces immune activity (Abrams and Miller, 2011). While baseline immune strength is higher for women, immune activity, including the presence and activity of immune cells, fluctuates with the menstrual cycle as well as with reproductive stage (Fish, 2008). Spiking levels of estrogen and progesterone during pregnancy serves to inhibit immune function and reduce cell-mediated immune activity (Abrams and Miller, 2011) leaving pregnant women more susceptible to illness, such as influenza. Furthermore, higher levels of estrogen and progesterone during pregnancy actively shift the helper T-cell ratio toward Th2 thereby decreasing Th1 activity (Fish, 2008), a pattern consistent with observations of reduced symptoms of certain autoimmune disorders (those related to Th2 cells) during pregnancy (Fish, 2008). Therefore, a balance must exist between reducing immunity so as not to attack the maturing fetus and stabilizing immunity to prevent infection, also a serious threat to a healthy fetus.

**Table 1.1** Quick reference guide of the sex hormone-induced differences and fluctuations in immune activation between males and females

**Effects of sex hormones on immune activity**

| Estrogen | Progesterone | Testosterone |
|---|---|---|
| • Increases dendritic cell production of IL-6 and IL-8<br>• Increases B lymphocyte production of IgG and IgM<br>• Increases number of regulatory T-cells<br>• Increases expression of *Cd22*, *Ptpn6*, *Bcl2*, and *Vcam1*<br>• Increases nitric oxide synthase<br>• Decreases monocyte secretion of IL-6 and IL-12<br>• Decreases natural killer cell activity<br>• Decreases monocyte levels<br>• Inhibits neutrophil chemotaxis<br>• Protects B-cells from apoptosis<br>• Favors Th1 over Th2 in helper T cell differentiation | • Favors Th2 over Th1 cell type in helper T cell differentiation<br>• Increases production of INF-α | • Decreases IgG and IgM<br>• Increases IL-12<br>• Increases IL-10 |

Full descriptions of these effects are provided in the text.

Regulatory T-cells are required to maintain immune tolerance of the body's cells, and deficiencies in regulatory T-cells have been implicated in sex-dependent differences and the development of autoimmune disorders (Gratz and Campbell, 2014). Regulatory T-cells are particularly susceptible to hormonal fluctuations during a woman's menstrual cycle. For example, when estrogen levels are lowered during the luteal phase, the number of regulatory T-cells decreases, potentially contributing to increased immune activity (Fish, 2008). Regulatory T-cell effects have been implicated in multiple sclerosis and rheumatoid arthritis and may contribute to the increased incidence rates in females in these two diseases (Fish, 2008). Estrogen effects on B-cell levels have also been proposed to contribute to the difference in rates of autoimmune diseases by increasing circulating levels of immunoglobulins G and M (IgG, IgM; Fish, 2008; Grimaldi et al., 2002; Lamason et al., 2006).

Later, in adulthood, expression of multiple signaling molecules differs between the sexes, which further influences immune functioning, and the susceptibility to disease. Gene expression of IL-10, its receptor (IL-10r), IL-16, IL1α, and toll-like receptor signaling protein as well as IL-1β protein expression are higher in females. Male rats, on the other hand, express higher levels of C–C motif chemokine 22 (CCL22) and C–C chemokine receptor type 4 (CCR4), its receptor. CCL22 and CCR4 are important players in regulatory T-cell recruitment, and specifically during allograft tolerance (Schwarz et al., 2012; Lee et al., 2005;Yoshie and Matsushima, 2014).

The immune response triggered by viral exposure differs distinctly between males and females. For example, following viral exposure, the production of chemokines and cytokines is higher in females than males. More specifically, lymphocyte activation, the production of antibodies, as well as CD4+, T-cell, and CD8+ cell activity become elevated in the female response compared with males. Furthermore, females sustain a higher level of immune activation after viral clearance – an effect that can be both beneficial and damaging. Females have a higher likelihood of resisting persistence of virus, but the increased period of heightened immune activation can also be detrimental to the female central nervous system (Nau and Brück, 2002; Klein, 2012). Unlike females, males avoid this immunopathology by returning to baseline immune activity more quickly after viral clearance, but males then risk the re-emergence of viral infection (Klein, 2012). Moreover, sustained immune activity in females can also contribute to comorbidities associated with increased immune activation that will be discussed in later sections (Dantzer and Kelley, 2007; Schwarz and Bilbo, 2012).

## 2.1 Sex-dependent development and maturation of immune cells, immune responses, and immune signaling pathways

Sex hormones play an important role in the development of the immune system. Estrogen in particular has been thoroughly investigated and found to impact the production of many important immune signaling molecules. In addition to regulating immune signaling, estrogen has a wide impact on the functioning of immune cells (see Table 1.2). Estrogen increases the production of multiple immune molecules and expression of genes associated with the adaptive immune system (Oertelt-Prigione, 2012; Fish, 2008). Furthermore, hormonal regulation of immune processes is not limited to estrogen, as progesterone, testosterone, and allopregnanolone all play roles in several immune mediated pathways (Piccinni et al., 1995;Trinchieri, 1995; Oertelt-Prigione, 2012;

**Table 1.2** Estrogen alone acts on a variety of immune markers

| **Effects of estrogen on immune markers** | |
|---|---|
| IL-6 | Increases production in dendritic cells<br>Decreases secretion from monocytes |
| IL-8 | Decreases production in dendritic cells |
| IL-12 | Increases secretion from monocytes |
| IgG | Increases production |
| IgM | Increases production |
| Nitric oxide synthase | Increases production |

Though generally anti-inflammatory in nature, estrogen induced effects may vary by host cell type.

Shiau et al., 2014). Differences in hormonal concentrations provide one explanation for differences in immune activity between males and females, but hormonal fluctuations over an individual's lifetime also play an important role in altering immune function. Many genes affected by sex hormones (especially those associated with the adaptive immune system) change in expression after the onset of puberty in females (Lamason et al., 2006). Along these lines, the onset of menopause alters levels of monocytes to concentrations comparable to what is observed in males (Oertelt-Prigione, 2012). Further details on the effects of sex hormones on immune molecules are discussed in Section 4.

Though hormones are important, they may not be solely responsible for sex differences in immune cell function. Without immune stimulation, girls as young as the age of 6 years still have higher levels of IgM than boys. It is therefore likely that additional sex differences, beyond the hormonal influence over immune activation, are at play in the larger picture of sex-dependent immune activity (Butterworth et al., 1967). Potential mechanisms that underlie these differences will be discussed later in the chapter.

## 2.2 Incidence of autoimmune disorders in men versus women

Autoimmune disorders are characterized by the body's immune system attacking healthy tissue to generate disease and dysfunction. Overall, the incidence of autoimmune disorders is higher in females than males, though the incidence ratio between males and females differs greatly among diseases and sometimes among geographic location (Ngo et al., 2014). Diseases such as systemic lupus erythematosus, Sjogren's syndrome, rheumatoid arthritis, Grave's disease, and primary biliary cirrhosis favor females while ankylosing spondylitis (a form of arthritis) is more common in males. Other diseases, such as ulcerative colitis and Crohn's disease have close to equal incidences in males and females (Ngo et al., 2014). One possible explanation for the increased incidence of many autoimmune diseases in women is due to the increased immune reactivity observed in women (Hewagama et al., 2009) though multiple potential mechanisms will be discussed later in this chapter.

## 2.3 Incidence of infectious diseases in men versus women

Due to differences in baseline and activated immune activity, the prevalence of infectious diseases varies between males and females. While behavioral differences that influence the likelihood of infection are partially responsible for differences in incidence rates of various infectious diseases, it is becoming increasingly clear that biological differences between the sexes are at least partially responsible for differing incidence rates. The mechanism of transmission of various infectious diseases differs between men and women, particularly in the case of sexually transmitted diseases; however, this does not account for the entirety of incidence rate differences.

The incidence of the human immunodeficiency virus (HIV) is significantly higher in females compared with males. The progression of HIV infection to acquired immune deficiency syndrome (AIDS) is faster in women, and females with HIV have higher immune activation and higher levels of helper T-cells. The increased activity of innate immunity in females has been proposed to be responsible for the differences in incidence rates (Addo and Altfeld, 2014); however, it is important to note that transmission mechanism may be partly responsible for differences in rates of various diseases, as is the case in HIV.

Males are more commonly affected with hepatitis C viral infection (HCV). Because HCV is transmitted both sexually and through blood, higher HCV incidence rates in males must be dependent on other sex-dependent mechanisms that, in this case, cause males to be more susceptible. In a study in the United States between 1999 and 2002, of those found to have anti-HCV antibodies, 89% of males developed a chronic HCV infection while only 63.4% of females experienced chronic infection. Furthermore, the progression of HCV is often more serious in males, and females are more likely to spontaneously clear the virus (Armstrong et al., 2006; Baden et al., 2014). Similar to other conditions, female hormones have been cited as a major contributor to the decreased prevalence in females. Additionally, throughout the lifespan of the disease, females have been found to have varying progression rates of HCV, with the highest rate of progression observed after menopause, further implicating the influence of female hormones over HCV incidence (Baden et al., 2014).

Viral diseases such as hepatitis B, dengue virus, West Nile virus, and hantavirus are all more prevalent in males. On the other hand, cytomegalovirus, herpes simplex virus type 2, influenza A, and measles are all more prevalent in females. Furthermore, while the prevalence of hantavirus is higher in males, mortality is greater in females. Because females have a tendency to exhibit a heightened innate and adaptive immune response, females may be better suited to fight off viral infection. In contrast, female elevations in the inflammatory response due to heightened immune activity may actually be detrimental in the consequences of disease exposure: enhanced inflammation may ultimately increase the severity of symptoms and subsequent damage, an effect commonly observed in influenza (Klein, 2012). Influenza infection results in higher female mortality compared with males, but in theory, enhanced immune activation should be better suited to protect the body from the virus. This difference could be explained by the fact that mortality correlates with levels of proinflammatory immune activity in severe strains of influenza. Heightened female immune activity in response to influenza viral infection may mediate higher female mortality rates due to complications from enhanced and sustained immune activation (Klein, 2012).

## 3 SEX-DEPENDENT DIFFERENCES WITHIN THE CENTRAL NERVOUS SYSTEM

As referenced in Section 1.1, microglia are the brain's resident macrophages and make up 5–12% of the cells in the central immune system (Lenz and McCarthy, 2014). Microglia are the predominant immune cell of the central nervous system and are essential for detection and clearance of pathogens (Lehnardt, 2010). Microglia are antigen-presenting cells capable of releasing cytokines, characteristic of immune-competent cells (Schwarz and Bilbo, 2012). When microglia are activated by an immune challenge, they change morphology and adopt what is classified as an amoeboid form, and perform a multitude of actions to protect the central nervous system. Microglia release growth factors, remove debris and microorganisms, and release signaling molecules such as IL-1β, IL-6, and tumor necrosis factor-α (TNF-α; Kreutzberg, 1996; Lenz and McCarthy, 2014). Interestingly, though baseline differences exist, microglia also differ widely in males and females during development and throughout life (see Section 4.2).

Similarly, astrocytes have a sexually dimorphic nature in the brain. At birth, astrocytes in the preoptic area and arcuate nucleus in males have more complex projections. The processes of male astrocytes are longer, more frequent, and have a greater complexity of branching than females (McCarthy et al., 2003). In the medial amygdala, female rats have simpler and fewer astrocytes (Johnson et al., 2008). Differences in astrocyte makeup can be, at least partially, attributed to the presence of sex hormones though the activity and influence of other signaling molecules within the CNS likely drive differences between the sexes (Conejo et al., 2005).

While female hormones are commonly examined for their role in increasing susceptibility to autoimmune disorders, male hormones may also play a protective role after disease onset. For example, testosterone has been found to reduce brain atrophy and improve cognitive performance in men with multiple sclerosis (MS; Sicotte et al., 2007), an autoimmune disease of the central nervous system. MS is characterized by an immune attack on oligodendrocytes, cells that produce myelin. Females are disproportionately diagnosed with MS at a significantly younger age than males, causing investigation into the mechanism of this sex bias (Arnold and Chen, 2009). In a rodent model of multiple sclerosis, termed experimental autoimmune encephalomyelitis (EAE), sex differences in EAE severity are observed, reminiscent of the sex bias in human MS. In the specific sex bias EAE model, supplemental treatment with androgens exerts a protective effect from EAE symptoms and furthermore, castrated males experienced symptoms of increased severity (Palaszynski et al., 2004). Another study found that supplementation with dihydrotestosterone (an androgen hormone) caused a decreased severity of EAE (Dalal et al., 1997). Further implicating sex hormones in a role in MS severity, and a protective role of androgens, 24% of human males in an MS patient cohort had low levels of testosterone (Wei and Lightman, 1997). Additionally, the microglial-produced immune molecules nitric oxide and TNF-α are toxic to oligodendrocytes, further implicating the neuroimmune system in the pathology of MS (Drew et al., 2003).

Although females are more likely to suffer from multiple sclerosis, the progression of neurodegeneration in MS is faster in men. In addition to the influence of sex hormones, male and female sex chromosomes are also a potential contributor to observed sex differences within the disease. Mouse bone marrow chimera models have been used to determine the effects of sex chromosomes on disease outcomes. For example, work with this model has demonstrated a greater degree of neurodegeneration in mice with XY chromosomes compared with those with XX chromosomes, implicating the male chromosomes in at least some aspects of MS severity (Du et al., 2014). The influence of sex chromosomes has also been implicated in the initiation and progression of systemic lupus erythematosus (Arnold and Chen, 2009). Additional information about the role of X-linked regulation of immune-related diseases can be found in Section 4.5, and additional mechanisms conferring immune differences between the sexes will be explored in detail in Section 4.

## 3.1 Neurodegenerative responses

The brain's response to injury differs greatly in males and females. During reproductive years, females experience improved clinical prognosis from stroke, traumatic brain injury (TBI), cerebral hypoxia, and subarachnoid hemorrhage (Xiong et al., 2007). Although the exact mechanisms behind this bias are unclear, a number of studies suggest a role for ovarian hormones. In stroke, a strong sex-by-age interaction exists; men outnumber women in incidence until age 65. Later, women experience similar incidence and mortality until around age 85, when women exceed men in stroke incidence and mortality (Haast et al., 2012; Ritzel et al., 2013). While such a profound bias occurs in ischemic stroke, no differences in mortality were detected in hemorrhagic stroke in males or females over age 65, though women were at a greater risk than men prior to that age (Bushnell et al., 2014).

Sex differences in the response to cerebral injury are also consistently reported in preclinical studies using model animal systems. In a rodent study of TBI, male and female adult mice exhibited the same extent of cortical tissue damage following TBI; however, only males manifested sensorimotor deficits (Xiong et al., 2007). Furthermore, females exhibited a higher degree of cellular proliferation in the dentate gyrus compared with males though rates of angiogenesis were similar between the two sexes. Similarly, in a rodent model of microvascular stroke, male and female adult rats had comparable microglial responses to injury; however, only males displayed increased anxiety-like behavior (Nemeth et al., 2014). In terms of ischemic injury, young female mice fare better than young male mice, and as expected, lesion sizes are comparable to males' when young female mice are ovariectomized (Liu et al., 2009).

Together clinical and preclinical findings support the differential effects of sex on injury and outcome from cerebral injury. Further details on how sex steroids may influence neurodegenerative responses may be found in Section 4.2.

# 4 MECHANISMS OF SEX-DEPENDENT IMMUNE ACTIVITY

## 4.1 Introduction

Sex differences in immune physiology and functioning begin during development as referenced earlier and in other chapters of this book. During development, the placenta serves as the interface between the mother and offspring, and it mediates sex-specific patterns of gene regulation. Placental gene expression shows a wide array of genes to be sexually dimorphic from across the genome, participating in gene transcription, cell growth, cell proliferation, and hormonal function (Buckberry et al., 2014). In humans, larger birth size of males compared with females suggests that the male placenta promotes growth and development over extra-embryonic tissue development (Buckberry et al., 2014). The dedication to growth at the expense of placental reserve in males is just one example of how early life programming differs between males and females and how such variations may give rise to sexually dimorphic disease susceptibilities later in life.

## 4.2 Sex steroids

The influence of sex steroids on sex-dependent differences in immune functioning has been alluded to within this chapter, but by what mechanisms do these differences manifest? Neuron–glial plasticity is heavily influenced by sex steroids and contributes to differential immune responses by sex (Berczi et al., 2010). Sex steroids regulate the transcription of genes relevant to the development and maturation of immune cells, immune responses, and immune signaling (Van Lunzen and Altfeld, 2014). As a result, the basic inflammatory response differs based on sex, and the control of the immune system by sex steroids is perhaps most evident during pregnancy when hormone concentrations are high and immune responses are generally suppressed to prevent rejection of the fetus (Van Lunzen and Altfeld, 2014; Gabriel and Arck, 2014).

In general, estrogens improve the antibody response while also increasing B- and T-cell availability, anti-inflammatory actions, and inflammatory cell recruitment via the increased availability of chemokine receptors (Gabriel and Arck, 2014). Estrogen increases the production of IgG and IgM in B lymphocytes, levels of nitric oxide synthase, and the production of IL-6 and IL-8 in dendritic cells (Fish, 2008; Oertelt-Prigione, 2012). Estrogen also decreases monocyte levels and monocyte secretion of IL-6 and IL-12 at low concentrations, which is further supported by the observation that both females during menopause and males have higher numbers of monocytes (Oertelt-Prigione, 2012). Similarly, numbers of regulatory T-cells increase in females during times of high estrogen and decrease when estrogen levels subside (Fish, 2008), which may have implications in multiple sclerosis and other autoimmune diseases (Viglietta et al., 2004; Fish, 2008). Dendrites increase stimulation of T lymphocytes in the presence of estrogen (Oertelt-Prigione, 2012), and estrogen also increases chemokine receptor expression, in particular CCR and CCR1 in helper T-cells (Fish, 2008). Furthermore, high levels of estrogen suppress NK cell activity (Oertelt-Prigione, 2012) and inhibit neutrophil chemotaxis demonstrating estrogen's tight regulation of immune responders (Shiau et al., 2014). Similarly, some evidence shows that low levels of estrogen tips the helper T-cell balance toward Th1 while high levels enhance Th2 activity (Fish, 2008). The ratio of Th1 to Th2 is dependent on concentration and shifts function from a protective immunity to a susceptibility toward immune-related illness including asthma, eczema, and allergic reactions. Furthermore, estrogens enhance the response of Th1 while testosterone has the opposite effect (Nhamoyebonde and Leslie, 2014; Ngo et al., 2014).

B-cell health is also affected by estrogen as B-cells express both estrogen receptors α and β, and stimulation of this receptor protects against cell-receptor mediated apoptosis and upregulates the expression of *Cd22*, *Ptpn6*, *Bcl2*, and *Vcam1*, genes implicated in B-cell survival (Grimaldi et al., 2002). Finally, and in addition to B-cell survival genes, after puberty, females have increased expression of multiple genes associated with the adaptive immune system and increased production of IgG1, IgG2a, IgG2b, and IgM (Lamason et al., 2006).

In contrast, progesterone enhances migration of neutrophils (enhancing the immune response); however, unlike the actions of estrogen, progesterone has detrimental consequences on the ability to mount a T-cell response, manage viral clearance, and maintain epithelial cell repair mechanisms (Gabriel and Arck, 2014). Progesterone participates in helper T-cell differentiation and treatment with progesterone can cause T-helper cells to favor Th2 cells and promote production of the cytokines IL-4 and IL-5 (Piccinni et al., 1995). In both men and women, the presence of progesterone correlates with the ability of dendritic cells to produce interferon-α (INF-α), and female peripheral blood lymphocytes produce increased levels of INF-α, an activator of dendritic cells, after stimulation of toll-like receptor 7 (TLR7). This results in increased activation of cytotoxic T-cells compared with males (Addo and Altfeld, 2014). On the other hand, androgens inhibit both B- and T-cell functioning, thereby reducing cytokine production, virus-neutralizing antibodies, and virus clearing capabilities. Furthermore, androgen-induced decreases of TLR4 expression suppress T-cell responses illustrating the wide reaching and heavy influence sex steroids have on several aspects of innate and adaptive immune responses (Gabriel and Arck, 2014).

Similarly, and in contrast to the increased production of IgG and IgM with estrogen, testosterone decreases the production of IgG and IgM but increases IL-12, which is essential in bridging the adaptive and innate immune systems in response to a lipopolysaccharide (LPS) challenge (Trinchieri, 1995; Oertelt-Prigione, 2012; Shiau et al., 2014). Testosterone also increases production of IL-10, which has been shown to be protective in disease models such as experimental autoimmune encephalomyelitis (EAE; Dalal et al., 1997).

Allopregnanolone is derived from locally-synthesized or peripheral progesterone, which, instead of binding to intracellular steroid receptors, is a positive allosteric modulator of the γ-aminobutyric acid $(GABA)_A$ receptor. Several inflammatory cells express functional $GABA_A$ receptors allowing for modulation by allopregnanolone. Macrophages, microglia, lymphocytes, and dendritic cells show decreases in inflammatory activity following $GABA_A$ activation, including the suppressed release of TNF and indoleamine 2,3-dioxygenase (Noorbakhsh et al., 2014). Furthermore, resulting responses by T-cells can be altered as a result of activated GABA receptors. Despite the effects on immune response mediators, concentrations of allopregnanolone in males and females do not wildly differ. Women experience increases in allopregnanolone during the luteal phase of the menstrual cycle as well as during pregnancy (Luisi et al., 2000; Genazzani et al., 1998). Fertile women and age-matched men have comparable circulating concentrations, and though levels of allopregnanolone in males decrease as they age, no differences are detectable in women as a function of age or menopause (Genazzani et al., 1998). With respect to inflammatory or autoimmune disease states, decreases in allopregnanolone are observed in women with multiple sclerosis (Noorbakhsh et al., 2014); however, more work is necessary to unravel the complete interaction of allopregnanolone with sex steroid and immune modulators that may contribute to immune-related disease states.

Females with autoimmune disorders mount an exacerbated inflammatory response to subsequent stimuli compared with males or healthy females (Ngo et al., 2014). Under these conditions, estrogen stimulates release of INF-γ, TNF, and IL-12, and inhibits production of the anti-inflammatory cytokine IL-10; testosterone, on the other hand, has the opposite effect (Nhamoyebonde and Leslie, 2014). Finally, and as previously mentioned, T-helper cells are also affected by steroid concentrations such that low levels of 17β-estradiol stimulate cytokine production and differentiation of Th1 and polarization of Th2 (Nhamoyebonde and Leslie, 2014; Ngo et al., 2014).

## 4.3 Microglia

As a major source of inflammatory activation following brain injury or infection, and during degenerative disorders, microglial cells play an important role in establishing the inflammatory tone of the brain. During development, the prevalence and activation state of microglia varies as a function of sex (Lenz and McCarthy, 2014), and deviations in activity during development may set the stage for a host of immune-related disorders.

Early in development, microglia, through the influence of steroid hormones (specifically estradiol), shape the cytoarchitecture to program brain sexual differentiation within the preoptic area of the hypothalamus (Lenz and McCarthy, 2014). In males, later in development and again at birth, a surge in testosterone from the testis is converted to estradiol in the brain via aromatase, which then upregulates prostaglandin E2, masculinizing the brain, as evidenced by masculinized microglia (greater density of dendritic spines) and male copulatory behavior in adulthood (Lenz et al., 2013; Schwarz and Bilbo, 2012). Conversely, the brain takes on a female phenotype in the absence of hormonal stimulation though many argue a role for estrogens and the protection against excessive estrogen exposure (via α-fetoprotein) in female development (Bakker and Baum, 2008). Interestingly, masculinization of the brain is dependent upon microglial activation as microglial inhibition at any stage during this critical period of development prevents any masculinizing effects (Lenz et al., 2013; Welberg, 2013).

The number and phenotype of microglia vary considerably as a function of sex and time. Before birth and sexual differentiation of the brain (embryonic day 17), rat pups exhibit no differences in the number or morphology of microglia (Schwarz et al., 2012). By postnatal day (PND) 0, genes potentially involved in microglial colonization and migration, CCL20 and CCL4, are higher in males. Shortly thereafter on PND 4, and following sexual differentiation of the brain, males show significantly increased numbers of microglia in several brain regions including the hippocampus, amygdala, dentate gyrus, and parietal cortex (Schwarz et al., 2012). Neonatal male rats also have increased amoeboid (activated morphology) microglia compared with females (Lenz et al., 2013). This pattern reverses later, with females exhibiting a higher number of microglia with an activated morphology in these same brain regions beginning by PND 30, and is maintained until at least PND 60 (Schwarz et al., 2012; Lenz et al., 2013). Also during this time, males express inflammatory markers, chemokine ligands (CCL) 4 and 20 at levels multi-fold higher than females (Schwarz and Bilbo, 2012).

Given the sex-dependent disparity in microglial number and morphology throughout development, it is reasonable to assume that neonatal and developmental susceptibilities to inflammatory disorders and disease, mediated by microglial activity, stem from these developmental origins. Males, who show a greater number and reactivity of microglia prior to birth, are more susceptible to developmental disorders, while in accordance, females, who show a greater number and reactivity of microglia later in life, are more susceptible to diseases with an inflammatory component that develop later in life, such as multiple sclerosis, anxiety, and depressive disorders (see Figure 1.2) (Schwarz et al., 2012; Ganguly and Brenhouse, 2014).

## 4.4 Neurotransmitters

The interactions of neurotransmitter systems with the immune system are plentiful and only beginning to be unraveled. Immune cells express classic neurotransmitter receptor types, including serotonin (5-hydroxytryptamine receptors), dopamine, acetylcholine, and glutamate receptors (Pacheco et al., 2009). Inflammatory cells may encounter neurotransmitters in various ways, including through blood plasma concentrations, through the release from inflammatory cells (by itself, or others), and in the brain, where immune cells may gain access to the brain through a permissive blood–brain barrier (Levite, 2008; Pacheco et al., 2009). Neurotransmitter effects on immune cell functioning are quite

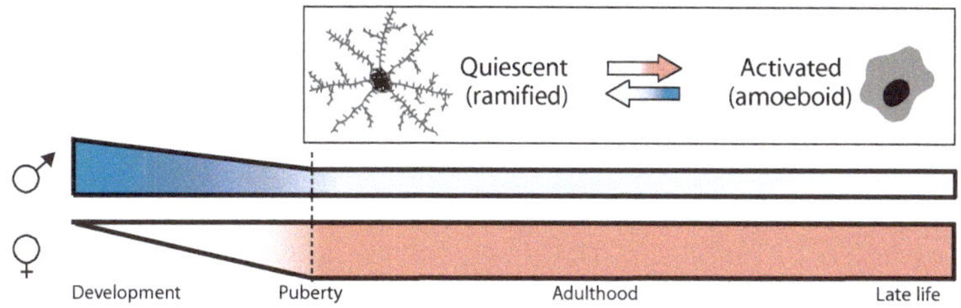


In males, microglia are more numerous and in an activated (amoeboid) state beginning early in development, a pattern that begins reversal by puberty. In contrast, females exhibit activated microglia beginning during puberty and this pattern is maintained throughout adulthood.

| Disease | Average age of onset | ♂ Prevalence | ♀ Prevalence | Citation |
|---|---|---|---|---|
| Hepatitis C | n/a | + | | Armstrong et al. (2006); Baden et al. (2014) |
| Human immunodeficiency virus | n/a | | + | Addo and Altfeld (2014) |
| Influenza A | n/a | | + | Klein (2012) |
| Autism | <1 | + | | Werling and Geschwind (2013) |
| Attention deficit hyperactivity disorder | 7 | + | | |
| Social phobia | 15 | | + | Schneier et al. (1992) |
| Crohn's disease | 15–30; 50+ | + Before puberty | + After puberty | Laass et al. (2014); Brant and Nguyen (2008) |
| Bipolar disorder | 20–25 | = | = | Weissman (1996) |
| Post-traumatic stress disorder | 23 | | + | Donner and Lowry (2013) |
| Panic disorders | 24 | | + | Kessler (2003) |
| Multiple sclerosis | 29 | | ++ | Ngo et al. (2014); Olek (2005) |
| Major depressive disorder | 25–30 | | + | Weissman (1996) |
| Lupus | 59 | | + | Ngo et al. (2014) |
| Heart disease | >60 | + | Greater mortality | Bushnell et al. (2014); Haast et al. (2012) |
| Alzheimer's disease | >60 | | + | Mielke et al. (2014) |
| Mild cognitive impairment | 83 | | + | |

**Figure 1.2** ***Sex-dependent differences in immune cell activation throughout development may account for differences in immune-related disease prevalence.*** While males express a higher degree of activated microglia during development, this pattern switches such that females express more activated microglia in later life (top). Although the exact role of microglial activation within immune-related diseases is unclear, the pattern of activation corresponds to sex differences in disease prevalence (bottom).

complex and context specific (Levite, 2008). Receptor subtype, immune cell type, activation state of the cell, and dose of neurotransmitter all contribute to differential response and functional outcome.

In rodent models, prenatal exposure to immune pathogens has been shown to alter the adult expression of various monoamines, their metabolites, as well as excitatory and inhibitory amino acids in a sex-specific fashion. In a study by Bitanihirwe et al. (2010), an immune challenge late in gestation decreased dopamine in the medial prefrontal cortex (mPFC) in both males and females while increasing serotonin (5-hydroxytryptamine) in both the amygdala and nucleus accumbens in females only. Similarly, excitatory glutamate was decreased in the mPFC of males only, and inhibitory GABA was decreased in the dorsal hippocampus of females only. Whether such effects translate to humans remains an unanswered question; however, it is plausible that prenatal immune exposure and subsequent neurochemical abnormalities may underlie sex-dependent immune responses and susceptibilities to immune-related diseases.

## 4.5 X-linked regulation of immunity

While the majority of sexually dimorphic genes are autosomal, sex chromosome-linked genes also regulate the differential expression of immune-related genes and immune disorders within the population. Of the 1100 genes in the human X chromosome and 100 genes in the human Y chromosome, only 54 are homologous between males and females (Dai and Ahmed, 2014). In females, random X-inactivation of one X chromosome prevents duplication of X chromosome genes. Variations in this process and escape from X inactivation occur in an estimated 15% of X chromosome-linked genes and may account for traits, including those related to immune function, to be different by sex (Buckberry et al., 2014; Dai and Ahmed, 2014). The X chromosome is home to several immune-related genes including *TLR7*, *TLR8*, *FOXP3* gene, CD40, IL-1, IκB kinase γ, and cytokine receptors (*IL9R*), to name a few (Dai and Ahmed, 2014; Van Lunzen and Altfeld, 2014). Mutations and varied expression of these genes are linked to disease susceptibility and presentation, as reviewed by Dai and Ahmed (2014).

Interestingly, the X chromosome is enriched in microRNAs (miRNAs) and suppression of several relevant target genes located on the X chromosome may occur through failure of X chromosome silencing (Sharma and Eghbali, 2014). Duplicate dosing of miRNA information within the X chromosomes may account for heightened immune responses in females.

## 4.6 MicroRNAs

MicroRNAs are small nonprotein-coding RNAs that regulate gene expression at the posttranscriptional level. Dysregulation of miRNAs are a factor in several disease states, including immune-related disorders (Dai and Ahmed, 2014; Dai et al., 2013; Sharma and Eghbali, 2014), and sex differences in miRNA expression relating to inflammatory signaling are observed in females at birth and males thereafter (Murphy et al., 2014).

Several studies have documented the sex-dependent expression of miRNAs in gonadal tissue as well as in the brain, liver, and lungs (Dai and Ahmed, 2014). Within the brain, differential expression of miRNAs was detected, particularly in the cortex, with the majority of upregulated sequences found in women during adolescence. Furthermore, differentially expressed miRNAs were related to Wnt signaling and transforming growth factor-β pathways – two pathways implicated in neurological disease (Ziats and Rennert, 2014).

In a rodent model of lupus (murine NZB/WF1 model), distinct sex differences in the expression of miRNAs are noted. Male NZB/WF1 mice administered estrogen experience increases in lupus symptoms as well as lupus-related miRNAs (Sharma and Eghbali, 2014). One such miRNA, miR-155, plays an important role in the regulation of adaptive and innate immune processes and has been shown to alter levels of inflammation within disease states such as experimental autoimmune encephalomyelitis and rheumatoid arthritis (Dai et al., 2013). Immune-related disorders appear to upregulate expression of miRNAs, and administration of sex hormones, such as estrogen, exacerbates immune responses and further miRNA dysregulation, suggesting an interaction of sex hormones and miRNA regulation in the susceptibility to, or progression of, immune-related disorders.

## 4.7 Microbiome

The intestinal biota play a role in the balance of the immune system and the response to immune challenges. Trillions of bacterial species found within the human gut interact with the host immune system to shape and maintain homeostasis. Further, new findings suggest that the immune system has evolved to accommodate such colonization by bacterial communities and that imbalances within this relationship may influence the immune response and the development of immune-related disorders (Cerf-Bensussan and Gaboriau-Routhiau, 2010). Following exposure to a pathogen (*Mycobacterium avium* subsp. *paratuberculosis)*, male and female mice mount a sex-dependent gut immune response such that levels of IL-1α, IL-1β, IL-17, IL-10, and IL-6 differ by sex (Karunasena et al., 2014). Further, females exposed to a pathogen, probiotics, or a combination of pathogen and probiotics exhibited increased levels of IL-6 and IL-1α, but decreased levels of IL-10 compared with exposed males and control mice illustrating an overall decrease of anti-inflammatory mediators (Karunasena et al., 2014). Within this study, increases in the ratios of specific disease-related bacterial species (*Stenotrophomonas* sp. and *Eubacterium contortum* related to ulcerative colitis, inflammatory bowel disease, and Crohn's disease) were detected in females compared with males, and maintenance on probiotics reversed this effect (Karunasena et al., 2014). Additional evidence for sex-dependent control over the gut microbiota is illustrated in nonobese diabetic (NOD) mice in which males and females have equal incidence of diabetes under germ-free conditions. In contrast, under specific pathogen-free conditions, females develop diabetes at twice the rate of male NOD mice, demonstrating a differential control of commensal microbes, which is often related to autoimmunity (Chervonsky, 2010). These findings are important given the interaction of the gut microbiota with the immune response and suggest that sex-dependent factors within the gut environment influence overall immune health and susceptibility.

## 5 CONSEQUENCES OF SEX DIFFERENCES WITHIN CENTRAL NERVOUS SYSTEM TRAUMA AND DISEASE

### 5.1 Introduction

Exposure to immunogens during gestation, or critical periods during development, may incur susceptibilities to certain systemic or neurological pathologies. Distinct sex differences are noted within the susceptibilities and incidences of these pathologies and the contribution of sex steroids to these disease states are of interest. Early immune activation serves as a risk factor for adult schizophrenia and autism, namely in males (Schwarz and Bilbo, 2012; Rana et al., 2012). Moreover, early immune activation of microglia has been shown to play a role in brain aging and increasing susceptibility to Alzheimer's disease (Bitanihirwe et al., 2010). Indeed, several rodent models have demonstrated the impact of early life immune challenges. Such consequences to the offspring include effects on behavior, cognition, learning, memory (Rana et al., 2012; Bilbo et al., 2006), cytoarchitecture, and neurochemical release (see Section 4.4; Rana et al., 2012; Madhusudan et al., 2013; Bitanihirwe et al., 2010). Important to keep in mind is that the factors that underlie these late-life susceptibilities are multifarious and highly dependent on both the timing and nature of the immune challenge as well as on the animal species and maternal environment. Furthermore, many effects of prenatal immune exposure are tissue or brain-region specific, adding another layer of complexity. Though model animals systems have highlighted that early immune activation leads to lifelong alterations to several bodily systems, much work is necessary to tease apart such effects.

### 5.2 Sex differences in behavior

Male and female behavioral responses to immune activation are also sex-dependent. Male offspring of pregnant rats exposed to an LPS-immune challenge show reduced social play behavior during PND 26 to 40 compared with control rats, while female rats were less susceptible to behavioral change (Taylor et al., 2012). Further, these sex-dependent responses to early life immune challenges can persist into adulthood. For example, in a mouse model of late prenatal immune exposure, the sex-dependent effects on adult conditioned avoidance learning were assessed. Males exposed to polyriboinosinic-polyribocytidylic acid (Poly-I:C) on gestational day 17 showed enhanced latent inhibition in adulthood compared with control mice of both sexes and Poly-I:C exposed females, demonstrating a delayed learning ability specific to male exposed offspring (Bitanihirwe et al., 2010). In a separate assessment, rats treated with LPS on PND 3 and 5 show altered sex-dependent behavior when exposed to a subsequent challenge in adulthood. Upon the second LPS challenge, males demonstrated reduced locomotor activity in a non-novel open field compared with males who did not undergo a neonatal immune challenge. In contrast, females did not exhibit differences in locomotor activity regardless of previous immune exposures (Tenk et al., 2008).

Early life immune challenges have been shown to further alter neurochemical responses in a sex-dependent manner. Although increased locomotor activity is commonly observed in rats after exposure to a dopaminergic agonist, females previously exposed to LPS exhibit a greater increase in locomotor activity following dopaminergic agonist treatment compared with males and saline treated controls (Tenk et al., 2007). Furthermore, males exposed to LPS on PND 3 and 5 experienced a greater degree of anxiety-like behaviors than saline treated controls while females did not exhibit this difference (Walker et al., 2009). Although this apparent sex bias for behavioral alterations could be attributed partially to lack of study in females, some studies, as discussed at the beginning of this section, do compare both sexes, and these findings underscore the potential increased susceptibility of males to immune activation at a young age.

The pattern of increased susceptibility to immune activation in young males is not specific to mice and rats. Juvenile immune activation impairs learning in a sex-specific manner in birds as well. For example, juvenile male zebra finches exposed to an LPS challenge had impaired performance in a novel foraging task; females did not experience the same deficiencies (Grindstaff et al., 2013). Similarly, when male white-crowned sparrows were exposed to LPS on long and short days, mimicking seasonal differences, males exposed to LPS on long days lost more weight than males exposed to LPS on short days, while females did not exhibit such differences (Owen-Ashley et al., 2006).

In contrast to the discussion above, exposure later in life seems to impact females more severely. Female rats that experience a repeated immune challenge in adulthood exhibit increased depressive-like behaviors compared with untreated females and immune-challenged males (Tonelli et al., 2008). Similarly, females (at or after PND 74) who experienced repeated LPS administration in the nasal cavity had reduced escape behavior and increased immobility compared with both males and females not treated with LPS (Tonelli et al., 2008). These data provide additional support for a sex-dependent variation of temporal susceptibility to immune activity. Consistently, and in a separate model, experimentally induced gastritis in both rats and mice caused increased depressive- and anxiety-like symptoms in females compared with males when tested in adulthood (Luo et al., 2013; Painsipp et al., 2007). This hypothesis of temporal sensitivity is further discussed in the review by Schwarz and Bilbo (2012).

Environmental factors also contribute to the apparent sex difference in immune response. Male rats exposed to LPS in adulthood who were single-housed reduced the number of times they visited the food hopper while LPS-group-housed and control males did not. Furthermore, LPS-exposed females who were group-housed decreased their food hopper visits while single-housed and control females did not (Yee and Prendergast, 2010).

Finally, in short-term measures of immune sickness behavior, females may fare better. Adult females exposed to mild LPS treatment fared better in the forced swim test than their male counterparts while other measures such as sucrose consumption and food intake were not different between males and females (Pitychoutis et al., 2009). These findings, though in contrast, may highlight an important distinction between males and females and their responses to immune activity in the long and short term.

## 5.3 Sex differences in other comorbidities

A growing body of evidence exists implicating the immune system in many diseases that initially were not believed to hold immune relevance. With better understanding of the impact of immune system activation in a multitude of different disorders, sex differences in immune function are becoming more widely recognized as a possible mechanism of the sex bias in prevalence of these diseases.

The discrepancy in cellular makeup (discussed in Sections 3 and 4.2) at different stages of life likely underlies differences in susceptibility to immune-activity-related comorbidities. Males may be more susceptible to immune events and diseases that develop during early life while females appear to be more susceptible to diseases that manifest in adulthood. As referenced earlier, immune challenges in young males lead to long-term memory and learning impairments with no effect to females (Schwarz and Bilbo, 2012). Conversely, immune challenges after adolescence may be more impactful in females due to increased cellular activity in the female brain during these later stages of life. For example, disorders such as autism and schizophrenia, which are more frequently observed in males, are thought to be promoted by early life immune stress, while diseases such as depression and anxiety, which manifest later in life, have a female sex bias promoted by susceptibilities to later life immune stress (Schwarz and Bilbo, 2012). Early life immune exposure and later risk for disease is well documented in studies of influenza during pregnancy. Women who experience influenza while pregnant run a fourfold increased risk of their offspring developing bipolar disorder, and similar risks are associated with schizophrenia (Parboosing et al., 2013). Patients with schizophrenia have higher circulating levels of IL-6 and soluble IL-6 receptor protein compared with healthy controls (Lin et al., 1998; Maes et al., 1994), and both depression and obsessive-compulsive disorder are similarly associated with immune activation and the presence of high levels of immune molecules such as IL-12, IL-1$\beta$, IL-8, and TNF-$\alpha$ (Ganguly and Brenhouse, 2014). Differing levels of immune molecules implicated in psychiatric disorders may stem from baseline sex differences or differential responses to immune stress previously discussed. In animal models, neonatal male rats challenged with LPS experienced increased levels of TNF-α in adulthood while females did not, highlighting the sex-dependent and long-term molecular changes that occur in response to immune insults. As these molecular effectors manifest in a sex-dependent manner, so does their impact on the differential incidences of psychiatric disorders (Walker et al., 2010).

An alternate theory has been proposed to account for increased psychiatric disease in women, highlighting the similarity between psychiatric diseases and immune-sickness behavior. It is suggested that in both of these conditions, the presence of inflammatory cytokines in the brain is responsible for the characteristic symptoms of disorders such as depression and anxiety. Symptoms such as lack of appetite, lack of interest in social activities, and fatigue are associated with immune system sickness, but are also characteristics of many psychiatric disorders. The similarity of symptoms suggests that increased sustained inflammatory cytokines signaling could be a mechanism of development of mental health disorders. It is possible, therefore, that sustained female neuroimmune activation throughout adulthood contributes to the increased frequency with which particular mental health disorders manifest in females (Schwarz and Bilbo, 2012; Dantzer and Kelley, 2007).

Autism spectrum disorders (ASD) have been associated with both increased cytokine activity as well as an increased density of microglial cells (Schaafsma and Pfaff, 2014). ASD affects approximately four males for every one female, and several theories have been put forth to explain the sex bias of ASD. While most of these theories depend on sex differences alone, immune activity may be involved. The influence of sex hormones over both immune and glial activity in patients with ASD has not gone unnoticed and it is likely that a balance between sex hormones and immune activity plays a role, at least in part, in the risk and severity of ASD (Werling and Geschwind, 2013).

Alzheimer's disease (AD) also shows a strong sex bias in incidence. Several genetic risk factors have been identified in the susceptibility to developing AD, and many of these carry profound sex differences. Expression of apolipoprotein ε (APOE) is one of the strongest known risk factors for AD and shown to confer a fourfold increased risk of AD in women, but only a slight increased risk in males (Mielke et al., 2014). Expression of APOE is also associated with an earlier age of onset for AD. Similarly, decreased expression of brain-derived neurotrophic factor (BDNF) is an important risk factor for the development of AD, and BDNF expression is tightly regulated by estrogen (Mielke et al., 2014). Decreases in estrogen in aged women may therefore increase the risk for AD via reductions in BDNF. Though more work needs to be done to establish the mechanisms of sex differences in AD, gonadal hormone regulation of various immune and regulatory factors likely underlies these discrepancies, at least in part. Other variables, including the increased lifespan of women versus men certainly contributes to increased prevalence of AD; however, overall incidence as well as differences in age of onset support a biological basis for sex differences within AD.

## 6 CONCLUSIONS

We began this chapter by positing the question of whether or not sex differences in immune function were evolutionary based. This is a question that cannot truly be answered, but it has provided a framework for us in which to consider the differences in peripheral and central immune function between men and women. Consideration of immune differences from the evolutionary prospective does not give clear explanation to many of the conditions that we considered in this chapter such as HIV and AD; however, it is important to recognize that neither of these challenges contributed to shaping males and females in the evolutionary sense. Sex differences likely arose from the main biological goal of each sex. For males, that goal is to gain access to resources and territory that best position the organism to deposit its genes in the next generation. For females, that goal is to successfully generate progeny to get its genes into the next generation. These goals, although similar in purpose, are distinctly different in application. The essential and highly specific role of the immune system for successful completion of each of these strategies is likely what shaped the male and female immune responses. Sex differences in immune function are not themselves pathological; they are adaptive, but every adaptation comes with an opportunity cost. A thorough understanding of sex-specific adaptations and the mechanisms that mediate them may ultimately guide treatment approaches in a sex-specific manner.

## ACKNOWLEDGMENT

This work was supported by the National Institutes of Health (T32-GM008602 to SAR); and the American Heart Association (13PRE14700010 to CLN).

## REFERENCES

Abrams, E.T., Miller, E.M., 2011. The roles of the immune system in women's reproduction: evolutionary constraints and life history trade-offs. Yearb. Phys. Anthropol. 146 (Suppl.), 134–154.

Addo, M.M., Altfeld, M., 2014. Sex-based differences in HIV type 1 pathogenesis. J. Infect. Dis. 209 (Suppl. 3), S86–S92.

Allen, C., et al., 2012. Neutrophil cerebrovascular transmigration triggers rapid neurotoxicity through release of proteases associated with decondensed DNA. J. Immunol. 189 (1), 381–392.

Armstrong, G.L., et al., 2006. The prevalence of Hepatitis C virus infection in the United States, 1999 through 2002. Ann. Int. Med. 144 (10), 705–714.

Arnold, A.P., Chen, X., 2009. What does the "four core genotypes" mouse model tell us about sex differences in the brain and other tissues? Front. Neuroendocrinol. 30 (1), 1–9.

Baden, R., Rockstroh, J.K., Buti, M., 2014. Natural history and management of hepatitis C: does sex play a role? J. Infect. Dis. 209 (Suppl. 3), S81–S85.

Bakker, J., Baum, M., 2008. Role for estradiol in female-typical brain and behavioral sexual differentiation. Front. Neuroendocrinol. 29 (1), 1–16.

Berczi, I., Laatikainen, R., Pulkkinen, J., 2010. Sex hormones and their analogues in neuroimmune biology. Immun. Endoc. Metab. Agents Med. Chem. 10 (3), 142–181.

Bilbo, S.D., et al., 2006. A behavioural characterization of neonatal infection-facilitated memory impairment in adult rats. Behav. Brain Res. 169 (1), 39–47.

Bitanihirwe, B.K.Y., et al., 2010. Late prenatal immune activation in mice leads to behavioral and neurochemical abnormalities relevant to the negative symptoms of schizophrenia. Neuropsychopharmacology 35 (12), 2462–2478.

Brant, S.R., Nguyen, G.C., 2008. Is there a gender difference in the prevalence of Crohn's disease or ulcerative colitis? Inflamm. Bowel Dis. 14 (Suppl. 2), S2–S3.

Buckberry, S., et al., 2014. Integrative transcriptome meta-analysis reveals widespread sex-biased gene expression at the human fetal-maternal interface. Mol. Hum. Reprod. 20 (8), 810–819.

Büdingen, H.,Von, et al., 2012. B cell exchange across the blood–brain barrier in multiple sclerosis. J. Clin. Invest. 122 (12), 4533–4543.

Bushnell, C., et al., 2014. Guidelines for the prevention of stroke in women: a statement for healthcare professionals from the American Heart Association/American Stroke Association. Stroke 45 (5), 1545–1588.

Butterworth, M., McClellan, B., Aklansmith, M., 1967. Influence of sex on immunoglobulin levels. Nature 214, 1224–1225.

Cerf-Bensussan, N., Gaboriau-Routhiau, V., 2010. The immune system and the gut microbiota: friends or foes? Nat. Rev. 10 (10), 735–744.

Chervonsky, A.V., 2010. Influence of microbial environment on autoimmunity. Nat. Immunol. 11 (1), 28–35.

Conejo, N.M., et al., 2005. Influence of gonadal steroids on the glial fibrillary acidic protein-immunoreactive astrocyte population in young rat hippocampus. J. Neurosci. Res. 79 (4), 488–494.

Dai, R., Ahmed, S.A., 2014. Sexual dimorphism of miRNA expression: a new perspective in understanding the sex bias of autoimmune diseases. Ther. Clin. Risk Manag. 10, 151–163.

Dai, R., et al., 2013. Sex differences in the expression of lupus-associated miRNAs in splenocytes from lupus-prone NZB/WF1 mice. Biol. Sex Differ. 4 (1), 19.

Dalal, M., Kim, S., Voskuh, R.R., 1997. Testosterone therapy ameliorates experimental autoimmune encephalomyelitis and induces a T helper 2 bias in the autoantigen-specific T lymphocyte response. J. Immunol. 159 (1), 3–6.

Dantzer, R., Kelley, K.W., 2007. Twenty years of research on cytokine-induced sickness behavior. Brain Behav. Immun. 21 (2), 153–160.

Dantzer, R., et al., 2008. From inflammation to sickness and depression: when the immune system subjugates the brain. Nat. Rev. Neurosci. 9 (1), 46–56.

Donner, N., Lowry, C., 2013. Sex differences in anxiety and emotional behavior. Eur. J. Physiol. 465 (5), 601–626.

Drew, P., Chavis, J., Bhatt, R., 2003. Sex steroid regulation of microglial cell activation: relevance to multiple sclerosis. Ann. N.Y. Acad. Sci. 5 (1007), 329–334.

Du, S., et al., 2014. XY sex chromosome complement, compared with XX, in the CNS confers greater neurodegeneration during experimental autoimmune encephalomyelitis. Proc. Natl. Acad. Sci. USA 111 (7), 2806–2811.

Engelhardt, B., Ransohoff, R.M., 2012. Capture, crawl, cross: the T cell code to breach the blood–brain barriers. Trends Immunol. 33 (12), 579–589.

Fish, E.N., 2008. The X-files in immunity: sex-based differences predispose immune responses. Nat. Rev. 8 (September), 737–744.

Gabriel, G., Arck, P.C., 2014. Sex, immunity and influenza. J. Infect. Dis. 209 (Suppl. 3), S93–S99.

Gan, Y., et al., 2014. Ischemic neurons recruit natural killer cells that accelerate brain infarction. Proc. Natl. Acad. Sci. USA 111 (7), 2704–2709.

Ganguly, P., Brenhouse, H.C., 2014. Broken or maladaptive? Altered trajectories in neuroinflammation and behavior after early life adversity. Develop. Cogn. Neurosci. 11, 18–30.

Genazzani, A., et al., 1998. Circulating levels of allopregnanolone in humans: gender, age, and endocrine influences. J. Clin. Endocrinol. Metab. 83 (6), 2099–2103.

Goehler, L.E., et al., 2000. Vagal immune-to-brain communication: a visceral chemosensory pathway. Autonomic. Neurosci. 85, 49–59.

Gratz, I.K., Campbell, D.J., 2014. Organ-specific and memory treg cells: specificity, development, function, and maintenance. Front. Immunol. 5 (July), 333.

Grimaldi, C.M., et al., 2002. Estrogen alters thresholds for B cell apoptosis and activation. J. Clin. Invest. 109 (12), 1625–1633.

Grindstaff, J., Hunsaker, V., Cox, S., 2013. Maternal and developmental immune challenges alter behavior and learning ability of offspring. Horm. Behav. 62 (3), 337–344.

Haast, R.A.M., Gustafson, D.R., Kiliaan, A.J., 2012. Sex differences in stroke. J. Cereb. Blood Flow. Metab. 32 (12), 2100–2107.

Hewagama, A., et al., 2009. Stronger inflammatory/cytotoxic T cell response in women identified by microarray analysis. Genes Immun. 10 (5), 509–516.

Johnson, R.T., Breedlove, S.M., Jordan, C.L., 2008. Sex differences and laterality in astrocyte number and complexity in the adult rat medial amygdala. J. Comp. Neurol. 511 (5), 599–609.

Karunasena, E., et al., 2014. Host response to the pathogen Mycobacterium avium subspecies paratuberculosis and beneficial microbes are directed through host sex-specificity. Appl. Environ. Microbiol. 80 (15), 4481–4490.

Kessler, R.C., et al., 2003. The epidemiology of major depressive disorder: results from the National Comorbidity Survery Replication. JAMA 289 (23), 3095–3105.

Klein, S.L., 2012. Sex influences immune responses to viruses, and efficacy of prophylaxis and therapeutic treatments for viral diseases. Bioessays 34 (12), 1050–1059.

Kreutzberg, G.W., 1996. Microglia: a sensor for pathological events in the CNS. Trends Neurosci. 19 (8), 312–318.

Kubera, M., et al., 2011. In animal models, psychosocial stress-induced (neuro)inflammation, apoptosis and reduced neurogenesis are associated to the onset of depression. Prog. Neuro-Psychopharmacol. Biol. Psychiat. 35 (3), 744–759.

Laass, M.W., Roggenbuck, D., Conrad, K., 2014. Diagnosis and classification of Crohn's disease. Autoimmun. Rev. 13 (4–5), 467–471.

Lamason, R., et al., 2006. Sexual dimorphism in immune response genes as a function of puberty. BMC Immunol. 7, 2.

Lee, I., et al., 2005. Recruitment of Foxp3+ T regulatory cells mediating allograft tolerance depends on the CCR4 chemokine receptor. J. Exp. Med. 201 (7), 1037–1044.

Lehnardt, S., 2010. Innate immunity and neuroinflammation in the CNS: the role of microglia in Toll-like receptor-mediated neuronal injury. Glia 58 (3), 253–263.

Lenz, K.M., McCarthy, M.M., 2014. A starring role for microglia in brain sex differences. Neuroscientist, May.

Lenz, K.M., et al., 2013. Microglia are essential to masculinization of brain and behavior. J. Neurosci. 33 (7), 2761–2772.

Levite, M., 2008. Neurotransmitters activate T-cells and elicit crucial functions via neurotransmitter receptors. Curr. Opin. Pharmacol. 8 (4), 460–471.

Lin, A., et al., 1998. The inflammatory response system in treatment-resistant schizophrenia: increased serum interleukin-6. Schizophren. Res. 32 (1), 9–15.

Liu, F., et al., 2009. Changes in experimental stroke outcome across the life span. J. Cereb. Blood Flow Metab. 29 (4), 792–802.

Luisi, S., et al., 2000. Serum allopregnanolone levels in pregnant women: changes during pregnancy, at delivery, and in hypertensive patients. J. Clin. Endocrinol. Metab. 85 (7), 2429–2433.

Luo, J., et al., 2013. Experimental gastritis leads to anxiety- and depression-like behaviors in female but not male rats. Behav. Brain Funct. 9, 46.

Madhusudan, A., Vogel, P., Knuesel, I., 2013. Impact of prenatal immune system disturbances on brain development. J. Neuroimmune. Pharmacol. 8 (1), 79–86.

Maes, M., Meltzer, H.Y., Bosmans, E., 1994. Immune-inflammatory markers in schizophrenia: comparison to normal controls and effects of clozapine. Acta Psychiatr. Scand. 89 (5), 346–351.

McCarthy, M., Todd, B., Amateau, S., 2003. Estradiol modulation of astrocytes and the establishment of sex differences in the brain. Ann. N.Y. Acad. Sci. 7 (1007), 283–297.

Mielke, M., Vemuri, P., Rocca, W., 2014. Clinical epidemiology of Alzheimer's disease: assessing sex and gender differences. Clin. Epidemiol. 6, 37–48.

Murphy, S.J., et al., 2014. Sex differences in microRNA expression during developmentin rat cortex. Neurochem. Int. 77 (Nov), 24–32.

Nau, R., Brück, W., 2002. Neuronal injury in bacterial meningitis: mechanisms and implications for therapy. Trends Neurosci. 25 (1), 38–45.

Nemeth, C.L., et al., 2014. Microglial activation occurs in the absence of anxiety-like behavior following microembolic stroke in female, but not male, rats. J. Neuroinflamm. 11 (1), 174.
Ngo, S.T., Steyn, F.J., McCombe, P.A., 2014. Gender differences in autoimmune disease. Front. Neuroendocrinol. 35 (3), 347–369.
Nhamoyebonde, S., Leslie, A., 2014. Biological differences between the sexes and susceptibility to tuberculosis. J. Infect. Dis. 209 (Suppl. 3), S100–S106.
Noorbakhsh, F., Baker, G.B., Power, C., 2014. Allopregnanolone and neuroinflammation: a focus on multiple sclerosis. Front. Cell Neurosci. 8 (June), 134.
Oertelt-Prigione, S., 2012. The influence of sex and gender on the immune response. Autoimmun. Rev. 11 (6–7), A479–A485.
Olek, M., 2005. Multiple Sclerosis: Etiology, Diagnosis, and New Treatment Strategies. Humana Press, Totowa, New Jersey.
Owen-Ashley, N.T., et al., 2006. Hormonal, behavioral, and thermoregulatory responses to bacterial lipopolysaccharide in captive and free-living white-crowned sparrows (*Zonotrichia leucophrys gambelii*). Horm. Behav. 49 (1), 15–29.
Pacheco, R., et al., 2009. Role of dopamine in the physiology of T-cells and dendritic cells. J. Neuroimmunol. 216 (1–2), 8–19.
Painsipp, E., et al., 2007. Experimental gastritis in mice enhances anxiety in a gender-related manner. Neuroscience 150 (3), 522–536.
Palaszynski, K.M., et al., 2004. Androgens are protective in experimental autoimmune encephalomyelitis: implications for multiple sclerosis. J. Neuroimmunol. 146 (1–2), 144–152.
Parboosing, R., et al., 2013. Gestational influenza and bipolar disorder in adult offspring. JAMA Psych. 70 (7), 677–685.
Piccinni, M.-P., et al., 1995. Progesterone favors the development of human T helper cells producing Th2-type cytokines and promotes both 1L-4 production and membrane CD30 expression in established Th1 cell clones. J. Immunol. 155 (1), 128–133.
Pitychoutis, P.M., et al., 2009. Neurochemical and behavioral alterations in an inflammatory model of depression: sex differences exposed. Am. J. Physi. Anthropol. 159 (4), 1216–1232.
Prinz, M., Priller, J., 2014. Microglia and brain macrophages in the molecular age: from origin to neuropsychiatric disease. Nat. Rev. Neurosci. 15 (5), 300–312.
Rana, S.A., Aavani, T., Pittman, Q.J., 2012. Sex effects on neurodevelopmental outcomes of innate immune activation during prenatal and neonatal life. Horm. Behav. 62 (3), 228–236.
Ritzel, R.M., Capozzi, L.A., McCullough, L.D., 2013. Sex, stroke, and inflammation: the potential for estrogen-mediated immunoprotection in stroke. Horm. Behav. 63 (2), 238–253.
Rivest, S., 2009. Regulation of innate immune responses in the brain. Nat. Rev. Immunol. 9 (6), 429–439.
Schaafsma, S.M., Pfaff, D.W., 2014. Etiologies underlying sex differences in Autism Spectrum Disorders. Front. Neuroendocrinol. 35 (3), 255–271.
Schneier, F.R., et al., 1992. Social phobia comorbidity and morbidity in an epidemiologic sample. Arch. Gen. Psychiat. 49, 282–288.
Schwarz, J.M., Bilbo, S.D., 2012. Sex, glia, and development: interactions in health and disease. Horm. Behav. 62 (3), 243–253.
Schwarz, J.M., Sholar, P.W., Bilbo, S.D., 2012. Sex differences in microglial colonization of the developing rat brain. J. Neurochem. 120 (6), 948–963.
Sharma, S., Eghbali, M., 2014. Influence of sex differences on microRNA gene regulation in disease. Biol. Sex Differ. 5 (1), 3.
Shiau, H.J., Aichelmann-Reidy, M.E., Reynolds, M.A., 2014. Influence of sex steroids on inflammation and bone metabolism. Periodontology 64 (1), 81–94, 2000.
Sicotte, N., et al., 2007. Testosterone treatment in multiple sclerosis: a pilot study. Arch. Neurol. 64, 683–688.
Taylor, P.V., et al., 2012. Sexually dimorphic effects of a prenatal immune challenge on social play and vasopressin expression in juvenile rats. Biol. Sex Differ. 3 (1), 15.
Tenk, C.M., et al., 2007. Neonatal immune system activation with lipopolysaccharide enhances behavioural sensitization to the dopamine agonist, quinpirole, in adult female but not male rats. Brain Behav. Immun. 21 (7), 935–945.

Tenk, C.M., Kavaliers, M., Ossenkopp, K.-P., 2008. Sexually dimorphic effects of neonatal immune system activation with lipopolysaccharide on the behavioural response to a homotypic adult immune challenge. Int. J. Devlop. Neurosci. 26 (3–4), 331–338.

Tonelli, L.H., Holmes, A., Postolache, T.T., 2008. Intranasal immune challenge induces sex-dependent depressive-like behavior and cytokine expression in the brain. Neuropsychopharmacology 33 (5), 1038–1048.

Trinchieri, G., 1995. Interleukin-12: a proinflammatory cytokine with immunoregulatory functions that bridge innate resistance and antigen-specific adaptive immunity. Ann. Rev. Immunol. 13, 521–576.

Trowsdale, J., Betz, A.G., 2006. Mother's little helpers: mechanisms of maternal-fetal tolerance. Nat. Immunol. 7 (3), 241–246.

Van Lunzen, J., Altfeld, M., 2014. Sex differences in infectious diseases-common but neglected. J. Infect. Dis. 209 (Suppl. 3), S79–S80.

Viglietta, V., et al., 2004. Loss of functional suppression by CD4+ CD25+ regulatory T cells in patients with multiple sclerosis. J. Exp. Med. 199 (7), 971–979.

Walker, A.K., et al., 2009. Neonatal lipopolysaccharide and adult stress exposure predisposes rats to anxiety-like behaviour and blunted corticosterone responses: implications for the double-hit hypothesis. Psychoneuroendocrinology 34 (10), 1515–1525.

Walker, A., Nakamura, T., Hodgson, D.M., 2010. Neonatal lipopolysaccharide exposure alters central cytokine responses to stress in adulthood in Wistar rats. Stress 13 (6), 506–515.

Wei, T., Lightman, S.L., 1997. The neuroendocrine axis in patients with multiple sclerosis. Brain. 120, 1067–1076.

Weissman, M.M., et al., 1996. Cross-national epidemiology of major depression and bipolar disorder. JAMA 276 (4), 293–299.

Welberg, L., 2013. Microglia maketh the male. Nat. Rev. Neurosci. 14 (4), 226.

Werling, D.M., Geschwind, D.H., 2013. Sex differences in autism spectrum disorders. Curr. Opin. Neurol. 26 (2), 146–153.

Xiong, Y., et al., 2007. Role of gender in outcome after traumatic brain injury and therapeutic effect of erythropoietin in mice. Brain Res. 1185, 301–312.

Yee, J.R., Prendergast, B.J., 2010. Sex-specific social regulation of inflammatory responses and sickness behaviors. Brain Behav. Immun. 24 (6), 942–951.

Yoshie, O., Matsushima, K., 2014. CCR4 and its ligands: from bench to bedside. Int. Immunol. 4 (3), 1–10.

Ziats, M.N., Rennert, O.M., 2014. Identification of differentially expressed microRNAs across the developing human brain. Mol. Psychiatr. 19 (7), 848–852.

# CHAPTER 2

# Molecular Mechanisms of Memory in Males and Females

**Ashley A. Keiser, Natalie C. Tronson**
University of Michigan, Department of Psychology, Ann Arbor, MI, USA

## 1 INTRODUCTION

There has been growing attention to sex differences in basic cognitive and neural functions, including learning, memory, and synaptic plasticity (Simpson and Kelly, 2012). Part of the impetus for this shift toward sex as an important variable is the observation of strikingly different susceptibilities to disorders of memory in men and women. For example, women are almost three times as likely than men to develop posttraumatic stress disorder (Kessler et al., 2012), approximately twice as likely to develop Alzheimer's disease (Seshadri et al., 1997; Lin and Doraiswamy, 2015), and more susceptible to learned aspects of addiction including cue-induced craving (Robbins et al., 1999) and habit formation (Quinn et al., 2007). In contrast, males are more susceptible to genetic disorders that cause intellectual disabilities, largely due to the prevalence of X-linked genes in the brain that mediate neural development and function (McCarthy and Arnold, 2011).

In the nondisordered population, males and females perform equally well across a wide variety of memory tasks. How these tasks are processed, however, often relies on different strategies and recruitment of different neural circuits in males and females. In spatial tasks, for example, males rely predominantly on distal cues, whereas females rely on landmarks or proximal cues (Rodríguez et al., 2011; Bettis and Jacobs, 2013; Keeley et al., 2013; Shah et al., 2013). In memory tests with an emotional component, men and women show differences in amygdala lateralization for emotion-related information (Gasbarri et al., 2007; Cahill, 2011), and females show increased recruitment of hippocampal circuitry for cues with an emotional component (Bellace et al., 2013). There are also several memory tasks in which consistent sex differences are observed. Females typically outperform males in autobiographic, verbal memory whereas males often outperform females in spatial memory tasks (Andreano and Cahill, 2009). Females also show enhanced extinction of fear conditioning compared with males (Milad et al., 2006; Lebrón-Milad et al., 2013; Rey et al., 2013; Cover et al., 2014). These findings are consistent across both human subjects (Chai and Jacobs, 2010; Lebrón-Milad et al., 2013) and animal models (Dalla and Shors, 2009; Lebrón-Milad et al., 2013; Shah et al., 2013), suggesting that these effects are not solely attributable to differences in environmental factors or gendered socialization of girls and boys. These differences in circuitry and

*Sex Differences in the Central Nervous System.* http://dx.doi.org/10.1016/B978-0-12-802114-9.00002-0

cognitive strategies strongly suggest that the information storage process is fundamentally different between males and females.

The goal of studying the intracellular mechanisms of memory formation is to understand how information is stored as a physical state of cells and circuits. To date, the majority of studies on sex differences in neural functions have focused on the role of sex hormones on development, cognition, and memory. In this chapter, we focus further downstream of these effects to describe the similarities and differences between males and females in the intracellular pathways activated by learning and required for memory formation. In developing a coherent view of molecular mechanisms of memory in females compared with males, we will begin to be able to understand sex differences in what information is stored, how it is differentially processed, where it is stored, and gain insight into the strategies used by males and females.

## 2 MOLECULAR MECHANISMS UNDERLYING MEMORY FORMATION

There are many coordinated processes that are required for synaptic plasticity and memory formation (Figure 2.1). These include signaling via neurotransmitter release and receptor activation, calcium and second messenger signaling, transcription of genes, de novo protein synthesis, and long-lasting histone modifications that alter subsequent gene expression. Receptors are trafficked in and out of the membrane, proteins are ubiquitinated and broken down, and scaffolding and cytoskeletal proteins are reorganized.

These basic categories of mechanisms are consistent across many forms of synaptic plasticity, from synapse development and pruning during development, to all types of memory formation; as such they are likely to be the same in males and females. Yet there is substantial redundancy in how each of these steps is instantiated. For example, gene transcription can be induced by many different transcription factors, and multiple signaling pathways converge to activate the same transcription factors. Whether the precise signaling mechanisms are the same in males and females is less clear.

Many of the specific signaling pathways required for memory formation, in particular the transcription factor cre-response element binding protein (CREB) activation via extracellular signaling regulated kinase 1 and 2 (ERK1/2), protein kinase A (PKA), calcium calmodulin kinases (CaMKII, CaMKIV), and mammalian target of rapomycin-AKT (mTOR-AKT) pathways, have been well defined in male rodents (Figure 2.2), and are described in detail elsewhere (Tronson and Taylor, 2007; Johansen et al., 2011). Briefly, during a learning experience, glutamatergic signaling at both α-amino-3-hydroxy-5-methyl-4-isoxazolepropionic acid (AMPA) and *N*-methyl-D-aspartate (NMDA) receptors triggers calcium influx that leads to calcium dependent signaling via calmodulin (CaM), CaMKII (Silva et al., 1992; Lucchesi et al., 2011), CaMKIV (Kang et al., 2001), ERK1/2 (Sananbenesi et al., 2002; Shalin et al., 2004), protein kinase C (PKC), and protein kinase M zeta (PKMζ) (Sacktor, 2011) activity. G-protein coupled receptors

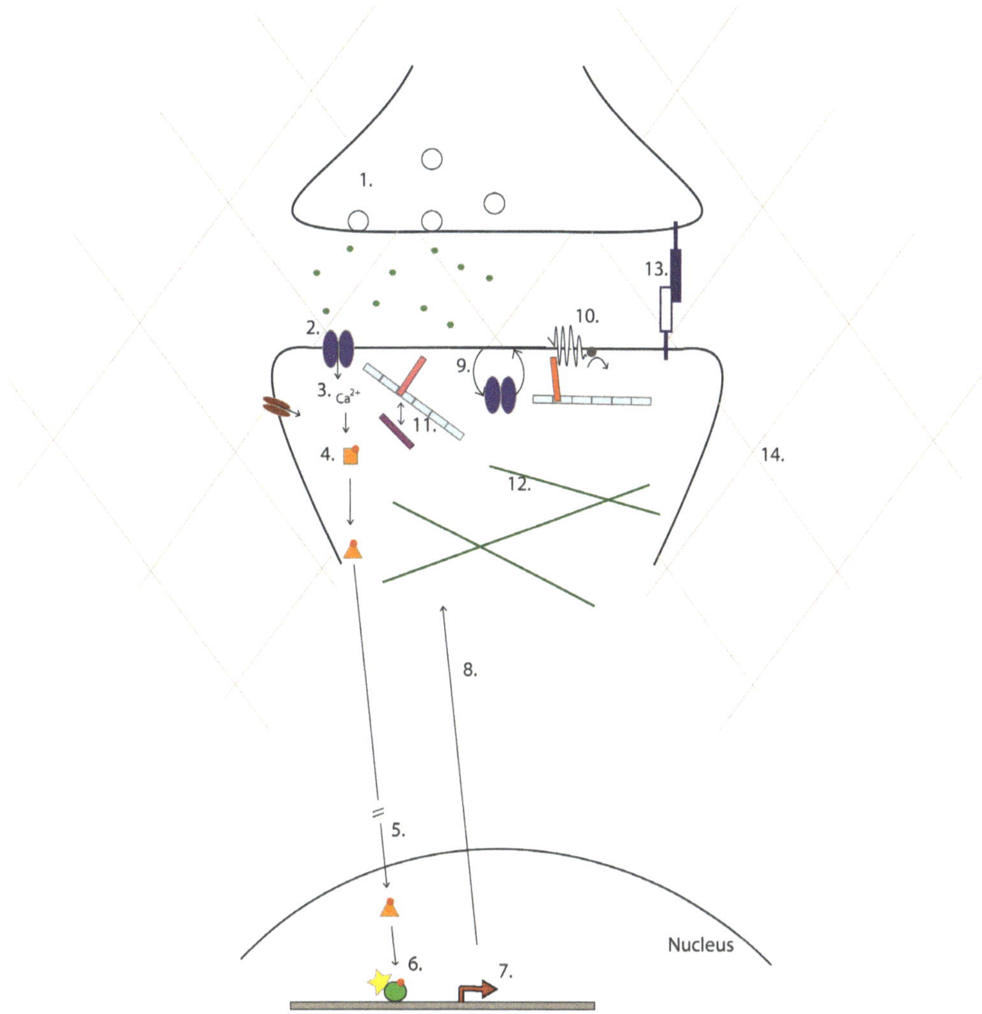


**Figure 2.1** ***Schematic of processes underlying synaptic plasticity.*** (1) Neurotransmitter release; (2) glutamate receptor activation; (3) calcium (and other ion) influx; (4) second messenger activation; (5) transduction to nucleus; (6) transcription factor activation and recruitment of coactivators; (7) gene transcription; (8) de novo protein synthesis and transport to synapse; (9) receptor trafficking; (10) G-protein receptor activation; (11) scaffold protein rearrangement; (12) actin rearrangement; (13) adhesion molecules; (14) perineuronal net rearrangement.

additionally result in activation of PKA (Abel et al., 1997) and the mTOR-AKT (Horwood et al., 2006; Jobim et al., 2012) pathways. These signaling cascades mediate ongoing receptor activity and trafficking (Malenka, 2003; Rumpel, 2005) at the synapse, and transduce the synaptic signals to the nucleus, activating transcription factors that include CREB (Kogan et al., 1997; Pittenger et al., 2002; Josselyn et al., 2004; Alberini, 2009)

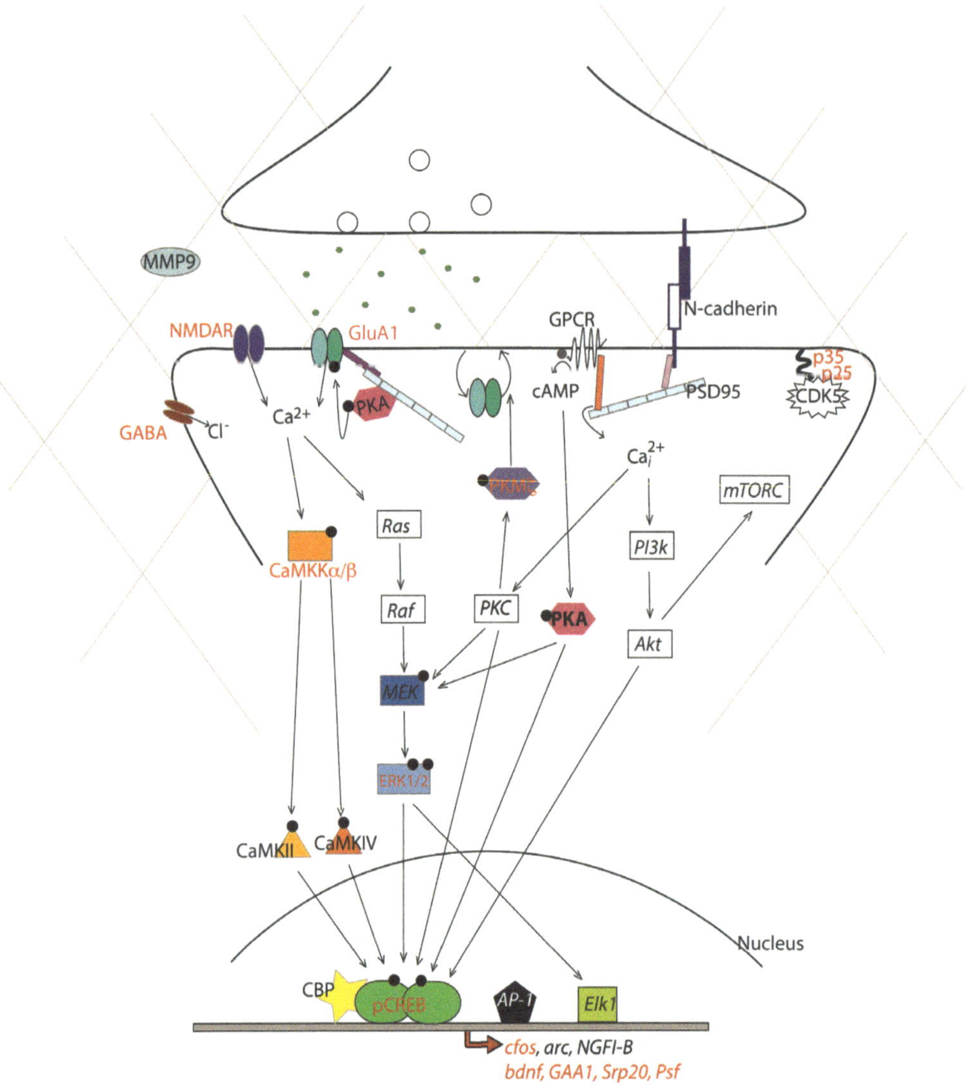


**Figure 2.2** ***Summary of signaling mechanisms that mediate molecular mechanisms of memory.*** Text in red represents proteins that are differentially involved in males and females. Key: Red dot, phosphorylation; arrows, demonstrated connections between signaling molecules; pink/purple rectangles, scaffolding proteins; gray grid represents extracellular matrix. NMDA, *N*-methyl-D-aspartate receptor; GluA1, glutamate receptor 1; GABAR, GABA receptor; GPCR, G-protein coupled receptor; $Ca^{2+}$, calcium, $Ca_i^{2+}$, calcium released from internal stores; $Cl^-$ chloride ions; AC, adenylyl cyclase; cAMP, cyclic AMP; PKA, protein kinase A; PKC, protein kinase C; Akt, AKT/protein kinase B; mTOR, mammalian target of rapamycin; MEK, mitogen activated protein kinase kinase; ERK, extracellular signal regulated kinase; PKMζ, protein kinase M zeta; PI3K, phosphoinositide 3 kinase; CDK5 cyclin dependent kinase 5; CaMKKα/β, calcium modulated kinase kinase α and β; CaMKII/CaMKIV, calcium modulated kinase II/ IV; Elk1; CPB, CREB binding protein; pCREB, phosphorylated cyclic AMP responsive element binding protein; AP-1, activator protein 1; MMP9, matrix metalloproteinase 9.

and cofactors such as CREB Binding Protein (CBP) and p300 (Alarcón et al., 2004; Maurice et al., 2008), resulting in transcription of immediate early genes (i.e.g.) such as cFos (Radulovic et al., 1998), early growth response protein 1 (Egr1, Zif268), activity-regulated cytoskeleton-associated protein (Arc, Arg3.1) (Guzowski et al., 2000), and other new proteins required for synaptic strengthening. Histone modifications, including changes in histone acetylation and methylation, are induced via activation of histone deacetylases, histone acetylases, and histone demethylases (Lubin and Sweatt, 2007; Tsai, 2009; Penney and Tsai, 2014).

Structural elements in and around the synapse are also modified so that spines and synapses can become plastic. ERK, PKA, and other kinases regulate destabilization and restabilization of dendritic spines and synapses by triggering changes in scaffolding protein interactions (Colledge et al., 2000; Moita et al., 2002; Gao et al., 2013; Fitzgerald et al., 2014), breakdown of proteins (Artinian et al., 2008; Jarome et al., 2011), modification of adhesion molecules that link the pre- and postsynapses (Schrick et al., 2007), loosening of the perineuronal net (Kaczmarek et al., 2002; Stawarski et al., 2014; Tsilibary et al., 2014), and cytoskeleton reorganization (Emes and Grant, 2012; Huang et al., 2013). These signaling, structural, and epigenetic alterations together mediate lasting stability of synapses and play an integral role in the cellular storage of memories.

The idea that sex differences in molecular mechanisms in memory formation exist runs counter to the widely held notion that these pathways represent fundamental processes, and small changes in the patterns of kinase activity result in engagement of alternative downstream mechanisms and can dramatically change the outcome. Yet, despite the gross similarities in the mechanisms underlying memory formation, there also exist striking sex differences at all levels, from receptor involvement, kinase signaling, transcription factors, and gene expression (Table 2.1). It remains possible, therefore, that alternative signal transduction pathways may result in broadly similar memory in males and females. In the subsequent section we describe the sex differences in molecular mechanisms of memory that have been identified to date.

## 2.1 Receptors and neural transmission

### *2.1.1 AMPA receptors*

Glutamate is a crucial neurotransmitter in learning and memory, and AMPA receptors fulfill several unique roles in synaptic plasticity underlying memory formation. First, activation of AMPA receptors is required for initial depolarization of the postsynapse and subsequent activation of NMDA receptors. In addition, AMPA receptor trafficking into the postsynaptic membrane is required for increased synaptic strength after learning (Rumpel, 2005) or long-term potentiation (LTP) (Malenka, 2003). AMPA receptors are composed of heterodimers, most commonly GluA1/GluA2 or GluA2/GluA3, although other combinations and at least one additional subunit (GluA4) exist. GluA1, in particular, is believed to be essential for several forms of memory, including spatial memory and

**Table 2.1** Differences between male (M) and female (F) protein expression and activation in memory formation

| Pathway | Protein | Expression/activation | Pharmacological manipulations | Transgenic manipulation |
|---|---|---|---|---|
| NMDA receptors | GluA1 | F > M d.h., mPFC, amyg. (Katsouli et al., 2014) | | M impaired CFC<br>F normal CFC (Dachtler et al., 2011) |
| | GluA2 | M > F d.h. mPFC (Katsouli et al., 2014) | | |
| | GluA4 | M > F d.h., mPFC (Katsouli et al., 2014) | | |
| | GluN1 | | M enhanced (Waddell, 2010) | |
| | GluN2A | F > M d.h. (Monfort and Felipo, 2007) | | LTP F > M (Monfort and Felipo, 2007) |
| | mGluR8 | | | M and F impaired [NLR]<br>F impaired [MWM] (Duvoisin et al., 2010; Iñiguez et al., 2012; Chen et al., 2010; Mendez-Lopez et al., 2009; Shalin et al., 2006; Ter Horst, 2012) |
| GABA receptors | δ-GABA$_A$ | | | F enhanced [TFC] (Wiltgen et al., 2005) |
| | α4-GABA$_A$ | | | F enhanced [Context post TFC]<br>M enhanced [Context post TFC] (Moore et al., 2010) |
| | α5-GABA$_A$ | | | F enhanced [TFC] (Yee et al., 2004) |
| | γ2-GABA$_A$ | F > M (Nani et al., 2013) | | F enhanced [IPSP] (Nani et al., 2013) |
| | PKA | F > M d.h. [MWM] (Iñiguez et al., 2012) | | M and F impaired [MWM, NOR, CFC] (Abel et al., 1997) |
| | ERK1/2 | M > F, v.h. [CFC] (Gresack et al., 2009; Antunes-Martins et al., 2007) | | |
| | CaMKKα | | | M impaired FC, CFC, and MWM (Mizuno et al., 2006) |
| | CaMKKβ | | | M impaired MWM; M impaired LTP (Mizuno et al., 2007) |

| | | | | |
|---|---|---|---|---|
| | CBP | | | M = F [MWM, CFC, NOR] (Chen et al., 2010) |
| | CREB | M > F CFC, d.h (Kudo et al., 2004) | | M and F impaired [MWM, NOR] (Pittenger et al., 2002)<br>M and F impaired (Kogan et al., 1997)<br>F impaired (Hebda-Bauer et al., 2007) |
| | cFos | F > M [MWM] IL, Ca1, Ca3 (Mendez-Lopez et al., 2009) | | |
| | BDNF | M > F [CFC, HPC] (Mizuno et al., 2006) | | |
| | *srp20* | M > F [CFC,MWM] (Antunes-Martins et al., 2007) | | |
| | *psf* | M = F [MWM] (Antunes-Martins et al., 2007) | | |
| | GAA1 | M > F d.h. [MWM, CFC] (Mizuno et al., 2007) | | |
| | PKMζ | M > F synaptic localization (Sebastian et al., 2013) | | |
| | NOS | | | M impaired [LTP] (Dachtler et al., 2012)<br>M and F impaired [CFC] (Kelley et al., 2009) |
| CDK5 | p25 | | | F enhanced [MWM]<br>M and F enhanced [FC]<br>F enhanced [LTP] (Ris et al., 2005) |
| | p35 | | | F impaired [MWM] (Engmann et al., 2011) |
| | KSR1 | | | M and F impaired [CFC, PA] (Shalin et al., 2006) |
| | MR-1 | | | F enhanced CFC; F impaired FC extinction (Ter Horst, 2012) |

Abbreviations: d.h., dorsal hippocamps; v.h., ventral hippocampus; amyg., amygdala; mPFC, medial prefrontal cortex; FC, fear conditioning (cued); CFC, context fear conditioning; TFC, trace fear conditioning; MWM, Morris water maze; LTP, long-term potentiation; NLR, novel location recognition; IPSP, inhibitory postsynaptic potential; nd, no sex difference; MR-1, mineralocorticoid receptor; KSR1, kinase suppressor of Ras 1; NOS, nitric oxide signaling; NOR, novel object recognition; PA, passive avoidance.

context fear conditioning (Rumpel, 2005). GluA1 trafficking into synaptic membranes is activity-dependent (Shi et al., 1999) and leads to increased calcium permeability of the synapse, as well as stabilization of enlarged synapses and persistence of LTP (Kopec et al., 2007).

Given the importance of GluA1 in memory formation and plasticity, it is surprising that this AMPA receptor subunit plays a different role in memory formation between the sexes. In GluA1 constitutive knockout mice, male animals, but not females, exhibit impaired context fear conditioning (Dachtler et al., 2011). This effect was neither due to locomotor or nociceptive effects, nor learning deficits as male and female GluR1 knockout (KO) animals showed similar freezing levels immediately after footshock during fear conditioning. Thus, GluA1 has a sex-specific requirement in context fear conditioning (Dachtler et al., 2011). Supporting a differential role of GluA1 in males and females, differential patterns of activation of cFos in response to novelty in GluA1 KO mice have been observed. Here, hippocampal cFos is markedly more pronounced in males, whereas lateral septum cFos is more pronounced in females (Procaccini et al., 2013). These results suggest that not only does GluA1 play different roles in learning and memory, but also supports differential recruitment of neural circuits in males and females during memory formation.

Further supporting sex differences in AMPA receptor function, GluA1, 2, and 4 subunits are differentially expressed in memory-related brain regions. Here, female mice have higher expression of GluA1 in dorsal hippocampus, amygdala, and medial prefrontal cortex, and lower GluA2 and GluA4 expression in dorsal hippocampus and medial prefrontal cortex compared with males (Katsouli et al., 2014). It is surprising that females show more GluA1 expression in hippocampus (Katsouli et al., 2014) but fewer impairments of hippocampal-dependent memory when deleted (Dachtler et al., 2011). It remains possible that compensatory mechanisms, and not the role of GluA1 per se, mediate the differential effect of GluA1 knockout in male and female mice.

There is clear evidence that AMPA receptor expression and function show some sex-specific effects in learning and memory processes. This suggests that intracellular signaling mechanisms activated by AMPA receptors will also show marked differences between males and females. AMPA receptors and consequent downstream signaling are not sufficient, however, and additional glutamatergic receptors, in particular NMDA receptors, are required for successful memory formation.

### *2.1.2 NMDA receptors*

NMDA receptors (NMDARs) are ligand gated calcium channels composed of GluN1 and at least one other (GluN2A, GluN2B, GluN3, GluN4) subunit. NMDARs become active with glutamate binding and prior depolarization of the neuron, and as such NMDARs are a central mechanism for the acquisition of memory and induction of LTP (Tsien et al., 1996; Morris, 2013). It is therefore surprising that NMDA receptors also have differential effects on learning and memory in males and females. Sex differences in memory as a consequence of NMDA receptor manipulation have been observed in both humans and animal models. In general, NMDA receptors are critical for learning in both male and female animals; however, there remains differential activation and performance in NDMA-dependent tasks. For example, men are more vulnerable to amnestic effects of ketamine than women, despite an overall lower behavioral sensitivity to NMDA antagonists (Morgan et al., 2006). In rodents, context fear conditioning and NMDA-dependent LTP is lower in females than males, suggesting limited activation of NMDARs in females (Maren and Fanselow, 1994). Similarly, enhancing GluN1 subunit function with D-cycloserine enhances trace eyeblink conditioning in male, but not female, animals (Waddell, 2010).

The differential role of NMDARs in male and female synaptic plasticity is also evident during aging, where females exhibit less of an increase in adulthood compared with males (Maren and Fanselow, 1994) and LTP in female animals remains intact during aging, whereas declines are observed in males, correlating with a decrease in NR2A subunits (Monfort and Felipo, 2007). Together, these effects may reflect differential levels of hippocampal GluN1 subunits in the hippocampus of female and male animals (Monfort and Felipo, 2007), or differential efficacy of intracellular signaling downstream of NMDARs.

There is additional evidence that modulation of NMDARs has very different outcomes on neuronal plasticity in males and females. Notably, stress causes an NMDA-dependent decrease in dendritic spines in males, and an NMDA-dependent increase in females (Shors, 2004). Although this effect is mediated, in part, by sex differences in hypothalamic–pituitary–adrenal axis signaling it also suggests that the downstream effectors of NMDAR activity exert very different effects in males and females.

### 2.1.3 GABA receptors

Concomitant with the differences in glutamatergic excitatory transmission, sex differences have also been observed in inhibitory transmission, specifically gamma-aminobutyric acid (GABA) receptors. GABA receptor activation inhibits memory formation, and pharmacological inhibition (Brioni et al., 1989) or genetic knockdown (Collinson et al., 2002) of GABA receptors results in increased memory formation. There are striking sex differences in the role of δ-, α4, α5, and γ2 $GABA_A$ subunits in fear conditioning. Genetic deletion of δ-$GABA_A$ receptor subunits profoundly enhances trace fear conditioning in female, but not male, mice, but does not affect delay conditioning or context conditioning in either males or females (Wiltgen et al., 2005).

In contrast, both male and female α4-$GABA_A$ KO mice show enhanced fear conditioning. Male and female α4-$GABA_A$ KO mice also show opposite patterns of freezing to context after delay and trace tone fear conditioning, where males exhibit increased freezing to context in delay but not trace tone conditioning, and females have increased context freezing after trace, but not delay conditioning (Moore et al., 2010). Similarly, loss of α5-$GABA_A$ subunits in the hippocampus leads to a female-specific enhancement of trace fear conditioning. In these α5-$GABA_A$ H105R mutant animals, delay conditioning was intact and extinction impaired in both sexes (Yee et al., 2004).

Consistent with differential roles of $GABA_A$ receptors in males and females, constitutive phosphorylation of γ2-$GABA_A$ receptor subunit causes sex-specific changes in expression of α4- and δ-$GABA_A$ subunits and increased tonic currents in the hippocampus only in female mice (Nani et al., 2013). GABAergic modulation of memory therefore requires sex-specific recruitment of GABA receptor subunits. This suggests that females require differential activation of signaling mechanisms downstream of GABA receptors during memory formation. Consistent with this possibility, previous work has demonstrated sex differences in subunit composition and pharmacology (Gulinello and Smith, 2003) and the role of androgens (Nuñez and McCarthy, 2008), progesterone (Andrade et al., 2012), and estrogens (Nuñez and McCarthy, 2009) on GABAergic functions. The divergent roles of glutamatergic and GABAergic receptors in memory formation supports the hypothesis that memory processes are differentially regulated in males and females. Furthermore, this suggests that downstream effectors including endocannabinoids (Huang and Woolley, 2012), kinase signaling cascades, and transcriptional regulation will be markedly different in males and females during memory formation.

## 2.2 Intracellular signaling mechanisms

As described previously (see Figure 2.2), signal transduction pathways and transcriptional mechanisms of memory consolidation have been well described in male animals. Many of the targets were initially identified in knockout and transgenic animals, when female and male animals were routinely tested together. These studies, from drosophila to mice, were the first indication of sex differences in molecular mechanisms of memory (previously summarized by Jazin and Cahill (2010)). The majority of studies in knockout mice observe no sex differences on the effects on memory. For example, genetic deletion of PKA, one of the key second messengers in memory formation, eliminates memory in male and female mice. This suggests that there are fundamental pathways or proteins that must be activated in order for memory processes to be stored.

Nevertheless, a growing literature on molecular mechanisms of memory in female animals demonstrates that many pathways and proteins thought to be key for memory processes are differentially activated by learning and required for consolidation in females and males. To date, the majority of sex differences in memory have been observed in calcium signaling and pathways that culminate in CREB-dependent transcription.

### 2.2.1 Calcium-dependent signaling

NMDA receptors and GluA1 play key roles in memory consolidation, in part due to their calcium permeability, and calcium influx appears to have differential roles in memory formation in males and females. In addition to sex differences in NMDAR and GluA1 manipulations, males are more sensitive than females to enhancement of memory by nonspecific blockade of calcium channels (Wilmott and Thompson, 2013), suggesting that calcium signaling is central to memory formation in males, but that this relationship is less clear in females.

Consistent with the suggestion that calcium signaling has sex-specific roles in learning and memory, calcium calmodulin kinase kinase α and β (CaMKKα and CaMKKβ) have been shown to be required for context fear conditioning (Blaeser et al., 2006; Mizuno et al., 2006) and spatial or novel object recognition memory (Peters et al., 2003; Mizuno et al., 2007; Bachstetter et al., 2014), respectively. Neither CaMKKα nor β are required for memory formation in female animals (Mizuno et al., 2006, 2007). Therefore CaMKKα and β are the first identified male-specific mechanisms of memory formation.

The transcriptional targets of CaMKKα/β also differ between males and females. In males, CaMKKβ knockout results in decreased *Srp20* and increased *Psf* messenger RNA expression in males, but not in females (Antunes-Martins et al., 2007). This suggests that the mechanisms controlling transcription of these genes differ between the sexes. Moreover, upregulation of *Srp20* after context fear conditioning or Morris water maze training is male-specific, and upregulation of *Psf* is stronger in males than females (Antunes-Martins et al., 2007). Thus, calcium signaling via CaMKKα and β, leading to transcriptional regulation of splicing factors *Srp20* and *Psf*, is required for hippocampal-dependent memory formation in males, but not females. Similarly, glycosylphosphatidylinositol anchor attachment protein (GAA1) is unregulated after spatial and contextual fear conditioning, but only in male animals (Mizuno et al., 2007). The importance of *Srp20*, *Psf*, and GAA1 in memory, and their contribution to sex differences in mechanisms of plasticity, is not yet known.

In contrast, there is little evidence for sex differences in other components of the CaMKKα/β signaling pathway. The immediate downstream substrates of CaMKKα and β are CaMKI and CaMKIV, respectively, both of which play key roles in memory consolidation, with no evidence for a differential role in males and females (Takao et al., 2010). This leaves the question of how CaMKKα/β differentially regulates transcription. One possibility is that calcium-dependent signaling, and CaMKKα/β specifically, is coupled to different downstream effectors. Indeed, CaMKKα/β also couples to the adenosine monophosphate-activated kinase (AMPK) signaling pathway (Birnbaum, 2005; Hawley et al., 2005; Hurley et al., 2005), and CaMKKα phosphorylates and activates Akt (protein kinase B, PKB) (Yano et al., 1998). Alternate signaling pathways may thus be recruited in males compared with females as a consequence of CaMKKα/β activation during memory formation.

Calcium-dependent signaling culminates in the activation of CREB via CaMKK and CaMKIV, and via CaMKII. Sex differences in CaMKK signaling and transcription additionally suggest a differential role of CREB in males and females. In particular, CaMKKβ is required for CREB activation in the hippocampus in male mice (Peters et al., 2003). Therefore, the failure of CaMKKβ deletion to impair memory in females strongly implicates CaMKKβ-CREB as a male-specific mechanism for memory formation. The signaling pathways that mediate this role in females are yet to be identified.

### 2.2.2 CREB

There is mixed evidence for a diverging role of CREB in males and females. Loss-of-function CREB transgenic manipulations lead to memory impairments in both male and female mice (Kogan et al., 1997; Pittenger et al., 2002). In contrast, direct and indirect evidence suggests a differential role of the transcription factor CREB in memory formation in males and females. After context fear conditioning, males show higher levels of CREB phosphorylation (Kudo et al., 2004) in the CA1 area of the hippocampus, but not the dentate gyrus or CA3. This suggests that CREB is more readily induced in males compared with females, and further, CREB may be more important in memory processes in males. In contrast, in CREB-deficient mutants ($CREB^{\alpha\delta-}$), females show greater spatial memory impairments compared with males during aging (Hebda-Bauer et al., 2007). Given that the latter study observed sex differences only in aged mice, it is possible that these results reflect more general susceptibility to memory impairments in females compared with males.

Indirect evidence for sex differences in the role of CREB in memory processes comes from analysis of upstream kinases and downstream CREB-dependent transcription during memory formation in male and female animals. As described previously, CaMKKα/β activation is a critical part of calcium signaling that culminates in CREB-mediated transcription in males, but not females. Alternative pathways for CREB activation include the extracellular regulated kinase/mitogen activated protein kinase (ERK/MAPK) pathway and PKA. Although PKA appears to be required for fear memory formation in both males and females (Abel et al., 1997), ERK is differentially activated after context fear conditioning, with males but not females showing significant ERK phosphorylation in ventral hippocampus (Gresack et al., 2009). The roles of these kinases vary depending on memory task. Substantial sex differences in PKA activity are observed during reward-related tasks (Becker et al., 2007; Iñiguez et al., 2012) and after retrieval of a spatial memory (Iñiguez et al., 2012), whereas ERK is similarly activated in both male and female animals in these tasks (Nygard et al., 2013). It is not sufficient, therefore, to think of signaling proteins or pathways as male- or female-specific. Rather, the pathways activated reflect the information processing in that task, suggesting differential information processing in males and females may drive differential recruitment of signal transduction pathways.

CREB-dependent gene expression also shows differential patterns in male and female animals. Brain-derived neurotrophic factor is more highly expressed in males compared with females whereas other transcripts, including nerve growth factor IB, show similar expression after a learning experience (Mizuno et al., 2006). There are many mechanisms by which transcription may differ despite activation of a transcription factor. Co-activators (e.g., CBP and P300) (Maurice et al., 2008) and transcriptional repressors (e.g., activator protein 1, AP-1) are key components of transcriptional regulation (Guedea et al., 2011). There is evidence for sex differences in AP-1 binding to DNA (Zhu and Pfaff, 1998), suggesting that beyond transcription factors, regulators of transcription may be important targets for sex differences research. At this time, the contribution of AP-1 and other transcriptional regulators to differential CREB-related transcription in males and females remains unknown.

The convergence of PKA, ERK, PKC, calcium-dependent signaling, and AKT, among other pathways, on CREB (see Figure 2.2) suggests that this transcription factor may be a hub for integrating input from multiple sources. Such hubs are of great importance as they may allow different patterns of upstream activity to converge into the same output, or may provide diverging output depending on specific patterns of kinase activity.

### 2.2.3 Other signaling pathways

There are several other kinases that exert significant sex differences in memory. Nitric oxide synthase 1 (NOS) has a male-specific role in cortical plasticity (Dachtler et al., 2012), with cortical plasticity impaired only in male, but not female, αNOS1 knockout mice. In contrast, both male and female neuronal NOS knockouts show similar impairments in fear conditioning (Kelley et al., 2009). The interactions between GluA1 and NO signaling in memory formation, and compensation between these pathways, may explain, in part, the sex differences in GluA1 KO in memory (Dachtler et al., 2011). PKMζ plays a male-specific role in remote memory (Sebastian et al., 2013) but remains important during recent memory processes in females. This kinase regulates memory by maintaining AMPA receptor trafficking (Sacktor, 2011); however, whether the sex-dependent effects of PKMζ are mediated via GluA1 remains unknown.

Independent of GluA1, cyclin-dependent kinase 5 (Cdk5) plays several critical roles in learning and memory, including regulation of NMDA receptors, calmodulation, PKA, and synaptic spine proliferation (Norrholm et al., 2003; Shah and Lahiri, 2014). Overexpression of p25, an activator of Cdk5 signaling pathways, leads to enhanced context and impaired cue-dependent memory in female, but not male, mice (Ris et al., 2005). Additional evidence for sex-specific roles of Cdk5 in memory processes is the differential role of the p35 small activator of Cdk5 in spatial working memory, where female p35 heterozygous knockout ($p35^{+/-}$) mice were impaired, but male animals were not (Engmann et al., 2011). Perhaps more importantly, male and female $p35^{+/-}$ mice show different downstream regulation of histone modifications and expression of spine-related proteins (Engmann et al., 2011). Although Cdk5 is also required for memory formation in males (Fischer et al., 2002), its specific roles and mechanisms of action may differ between males and females.

There are many additional signaling pathways, transcription factors, and proteins identified as important for memory formation in males, but whether or not the same mechanisms mediate memory in females has yet to be studied. Several classes of signaling and synaptic reorganization may be of particular interest in understanding sex differences in signal transduction. Adhesion molecules and scaffolding proteins are crucial for both synaptic reorganization, and for holding kinases and other proteins in close proximity to their targets, thereby potentially mediating differential recruitment of kinase pathways during consolidation in males and females.

In addition, both adhesion molecules and scaffold proteins show sex differences in expression. For example, adhesion proteins such as N-cadherin and neuroligins show sex differences in expression and roles in synapse formation (Gokce and Südhof, 2013). In addition, scaffold proteins are critically involved in determining the signaling mechanisms coupled to receptors thereby changing receptor function (Mao et al., 2005; Kammermeier and Worley, 2007; Tronson et al., 2010; Gao et al., 2011). Whether adhesion and scaffold proteins, their interactions with receptors, or their functional relevance to memory differ in males and females is yet to be studied.

### *2.2.4 Summary*

Our current understanding of sex differences in molecular mechanisms of memory is very limited. Despite our extensive knowledge of signal transduction, protein interactions, and transcriptional regulation in male animals across an array of memory tasks, we know very little about the similarities and differences in females. At this stage, however, there is strong evidence for diverging roles of several pathways that are central to memory formation (at least in males), notably glutamate and GABA receptors, calcium-dependent signaling, and CREB.

There are several issues that deserve additional attention. First, there is some contradiction between studies using knockout mice that demonstrate no sex differences in mechanisms of memory, whereas studies of direct manipulations or observations of post-translational modifications of the same protein demonstrate differential effects on memory in males and females. Forebrain deletion of CREB, for example, eliminates memory in both males and females (Kogan et al., 1997; Pittenger et al., 2002), yet females are both more sensitive to more subtle manipulations (Hebda-Bauer et al., 2007) and show less CREB activation after learning (Kudo et al., 2004). Knockout studies, therefore, may miss subtleties or be confounded by differential compensatory mechanism in males and females. Despite these limitations, converging evidence from studies utilizing transgenic (including knockout) animals, pharmacological manipulations, and observing changes in gene expression or protein provides strong – and growing – evidence for sex differences in key mechanisms of memory.

## 2.3 Hormonal influences in memory mechanisms

The influence of the hormonal milieu on acute signaling, connectivity of brain regions, and strategies used to solve tasks likely contributes to differential recruitment of signaling mechanisms in males and females. Sex hormones critically mediate synaptic plasticity (Woolley et al., 1990; Zamani et al., 2000; Srivastava et al., 2008; Barha and Galea, 2010; Vierk et al., 2014) and gene expression during development (Tsai, 2009; McCarthy and Arnold, 2011; Bowers et al., 2014). Resultant differences in neural circuitry in males and females may lead to sex-specific strategies to solve a task and differences in activation of specific brain regions, both of which would likely result in a quantitative difference between intracellular signaling between males and females.

Multiple studies have demonstrated sex-specific roles for estrogen in receptor activity and kinase signaling in the brain. Metabotropic glutamate receptors (mGluRs) are differentially regulated by estradiol in males and females. In particular, plasma membrane estrogen receptor activates MAPK/ERK signaling and CREB-dependent transcription via mGluR1/5 in the absence of depolarization, or inhibits CREB activity during depolarization via mGluR2/3 (Boulware et al., 2005). These bidirectional effects on CREB may be mediated by altered interactions of mGluRs with scaffolding proteins that can change the kinase signaling pathways coupled to the receptor (Mao et al., 2005; Tronson et al., 2010). Regulation of GABA by mGluRs is also modulated by estrogen in females, suggesting that mGluRs and their downstream MAPK/ERK signaling may play a sex-specific role in memory processes (Huang and Woolley, 2012). This possibility is particularly intriguing given roles of mGluR in the induction of long-term depression (Huber et al., 2001) and metaplasticity (Tronson et al., 2010).

Estradiol regulates many intracellular signaling cascades, for example, AMPK's role in energy metabolism is, in part, regulated by estrogen (Briski et al., 2014); however, its direct role in memory processes has yet to be studied. mTOR-Akt signaling (Fortress et al., 2013), together with ERK (Tuscher et al., 2014), are activated by estrogen in the hippocampus and mediate the enhancing effect of estradiol on memory consolidation. Furthermore, estrogens (Woolley et al., 1990; Zamani et al., 2000; Srivastava et al., 2008; Barha and Galea, 2010; Vierk et al., 2014) and androgens (Matsumoto, 1991; Ooishi et al., 2012; Hajszan et al., 2013; Scharfman and MacLusky, 2014) play direct roles in synaptic plasticity; however, the roles of gonadal hormones in mediating sex differences in molecular mechanisms of memory remains unclear (Anagnostaras et al., 1998). Regardless, differences in hormonal milieu are sufficient to alter the state of neuron, which in turn could qualitatively alter the signaling cascades recruited to orchestrate the large array of processes required for synaptic plasticity and memory formation.

## 2.4 Chromosomal complement

In concert with hormonal differences between males and females, there are also constitutive differences in the chromosomal complement. That is, females have two X-chromosomes and males have one X and one Y, resulting in differences in development that are independent of hormonal environment (Xu et al., 2002; Arnold and Burgoyne, 2004). In consequence, males are more susceptible to X-linked genetic disorders. Furthermore, females have higher levels of expression of a subset of X-linked genes compared with males (Xu and Disteche, 2006). This differential expression of X-linked genes occurs because although most genes on one X-allele are silenced by X-inactivation, approximately 15% of genes in humans (and 3% of genes in mice) escape X-inactivation (Xu and Disteche, 2006; Prothero et al., 2009; Berletch et al., 2011). In addition to differences of increased X-linked gene dosing in females, the presence of specific Y-linked genes in males also potentially mediates differences during development and in adulthood (Xu et al., 2002; Arnold and Burgoyne, 2004; McCarthy and Arnold, 2011; Armoskus et al., 2014).

One example of differential gene expression due to chromosomal complement is the X-specific demethylase *UTX* and the Y-specific analog *UTY*, which are differentially expressed regardless of hormonal state (Xu et al., 2008b). In the amygdala, *UTX* in females is higher than *UTY* in males, whereas the opposite pattern is seen in hypothalamus (Xu et al., 2008b). *UTX* and *UTY* are particularly interesting for their histone demethylation at the histone 3 lysine 27 site (Hong et al., 2007; Lan et al., 2007). Another X-linked demethylase gene, *Jarid1c*, is more highly expressed in adult female than male hippocampus and cerebellum, both key areas for memory formation (Iwase et al., 2007; Xu et al., 2008a).

Given the role of epigenetic modifications in the regulation of transcription and suppression of gene expression, the differential expression of *UTX* and *UTY* during development results in lasting differences in the role of the amygdala: for example, in memory later in life (Xu et al., 2008b; McCarthy, 2009). Similarly, greater *Jarid1c* expression may lead to increased gene expression at the H3K4 site in females during memory formation (Xu et al., 2008a). There are regional differences in sex-dependent gene expression (Xu et al., 2002; Armoskus et al., 2014); however, the direct effects of these differences in gene expression and gene dosing on molecular mechanisms of memory remain unknown.

## 2.5 Interpretations of sex differences in mechanisms of memory

It is well known that males and females differ at chromosomal and hormonal levels, and it is well established that sex differences during development result in structural and circuit-level differences in the brain (McCarthy and Arnold, 2011; Gruene et al., 2014). Yet the possibility that the basic intracellular mechanisms of fundamental neural processes might differ – especially when task performance is the same – is more difficult to accept.

There are several fundamental difficulties with this concept. One is that a central assumption for studying molecular mechanisms of memory is that we can reduce memory formation to a series of molecular steps that can be understood in terms of their precise roles in the process. The suggestion that these molecular steps might do something different in female compared with male animals is inconsistent with this strong reductionist view. On the other hand, we know that different memory tasks also require different cognitive strategies, neural circuits, and molecular mechanisms, despite the same underlying process (storage of information). If, as other studies suggest, males and females are using different strategies and neural circuits, then it follows that there should be significant differences in molecular mechanisms as well. Delineating sex differences in molecular mechanisms of memory will therefore provide insight into how and what males and females are learning in the same memory task.

A second problem is that we do not yet know precisely which signaling mechanisms are required in memory formation in both males and females, and which play a sex-specific role. To date, studies have demonstrated surprising differences, where molecules that are considered central to memory processes in males are not required in females (e.g., GluA1), and others have shown robust conservation of signaling (e.g., CamKIV) despite differences in upstream kinases (e.g., CaMKKβ). Many questions remain unanswered: what mechanisms fulfill the role in females that GluA1 does in males? What kinases are upstream of CaMKIV in females? At this stage, the pathways targeted in females are limited to those that we know are important for memory formation in males. A broader approach (for example, proteomics (Yang et al., 2013)) is needed to determine female-specific analogs of the male-specific pathways we already know.

Another limitation in our view of intracellular signaling and memory is conceptual. We typically envision signaling pathways as linear cascades of receptors, ions, second messengers, and transcription factors. Here, changing or deleting any one protein causes elimination of that pathway and disruption of the delicate orchestration of kinase-dependent signal transduction and protein rearrangement required for memory formation (e.g., Figure 2.2). Thus, any differences in mechanism in males and females should result in very different memory performance in males and females.

An alternative is to conceptualize postsynaptic signaling as a networked model of interactions within multiprotein complexes (Grant, 2003) (Figure 2.3). This model suggests that individual proteins might matter less than the network property of each cluster, or hub, thereby building in significant redundancy between signaling pathways. From the data so far, CREB appears to act as a major hub, integrating multiple upstream signals, and critically determining efficacy of memory storage.

The concept of networks of proteins acting via interacting hubs has several advantages for understanding sex differences in mechanisms of memory. First, it allows for differential activation of specific proteins within a similar network pattern of kinase activity (see Figure 2.3). Second, this and similar models will help us move toward more targeted, hypothesis-driven search for female-specific molecular mechanisms of memory. Data on similarities and differences between males and females in molecular mechanism of memory will feed forward to inform these (and other) more complex models of synaptic function.

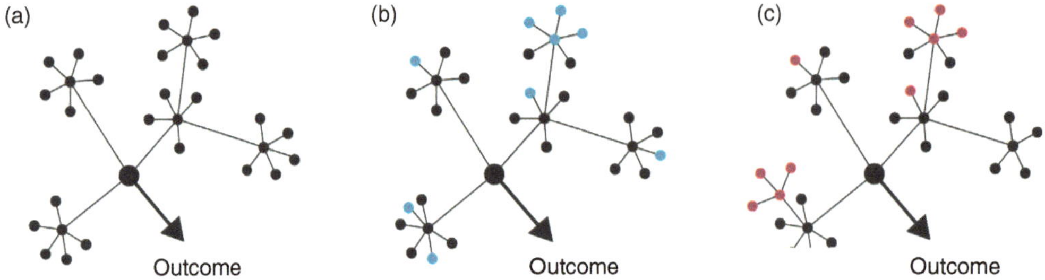


**Figure 2.3** ***Networked interactions between kinases as an organizational principle of complex signaling pathways.*** (a) Postsynaptic signal transduction may be organized as multiprotein complexes where the hub is more important than each specific protein. (b, c) This framework allows the basic signaling structure to be maintained despite sex-specific patterns of signaling in (b) males (blue) versus (c) females (pink). *Modified from Grant (2003).*

## 3 CONCLUSIONS

There are significant sex differences in molecular mechanisms underlying memory formation at every level of intracellular signaling, from receptors, to second messengers, transcription factors, gene expression, and histone modifications. At this stage, we have only small snapshots of individual proteins, and sometimes pathways. Despite these current limitations, several things are clear: males and females utilize different cognitive strategies, neural circuits, and molecular mechanisms during memory tasks. Detailing these differences in function and biological mechanism will be essential for understanding susceptibility to disorders of memory, and developing targeted treatments for males and females. Analysis of sex differences in molecular mechanisms of memory also provides the unique opportunity to study network properties of synaptic function. By developing new models to understand intracellular signaling, with all the redundancies and interaction between signaling pathways, we will gain new insights into the molecular machinery of synaptic plasticity and memory.

## REFERENCES

Abel, T., Nguyen, P.V., Barad, M., Deuel, T.A.S., Kandel, E.R., Bourtchouladze, R., 1997. Genetic demonstration of a role for PKA in the late phase of LTP and in hippocampus-based long-term memory. Cell 88, 615–626.

Alarcón, J.M., Malleret, G., Touzani, K., Vronskaya, S., Ishii, S., Kandel, E.R., Barco, A., 2004. Chromatin acetylation, memory, and LTP are impaired in CBP+/− mice. Neuron 42, 947–959.

Alberini, C.M., 2009. Transcription factors in long-term memory and synaptic plasticity. Physiol. Rev. 89, 121–145.

Anagnostaras, S.G., Maren, S., DeCola, J.P., Lane, N.I., Gale, G.D., Schlinger, B.A., Fanselow, M.S., 1998. Testicular hormones do not regulate sexually dimorphic Pavlovian fear conditioning or perforant-path long-term potentiation in adult male rats. Behav. Brain Res. 92, 1–9.

Andrade, S., Arbo, B.D., Batista, B.A.M., Neves, A.M., Branchini, G., Brum, I.S., Barros, H.M.T., Gomez, R., Ribeiro, M.F.M., 2012. Effect of progesterone on the expression of GABAA receptor subunits in the prefrontal cortex of rats: implications of sex differences and brain hemisphere. Cell Biochem. Funct. 30, 696–700.
Andreano, J.M., Cahill, L., 2009. Sex influences on the neurobiology of learning and memory. Learn. Mem. 16, 248–266.
Antunes-Martins, A., Mizuno, K., Irvine, E.E., Lepicard, E.M., Giese, K.P., 2007. Sex-dependent up-regulation of two splicing factors, Psf and Srp20, during hippocampal memory formation. Learn. Mem. 14, 693–702.
Armoskus, C., Moreira, D., Bollinger, K., Jimenez, O., Taniguchi, S., Tsai, H.-W., 2014. Identification of sexually dimorphic genes in the neonatal mouse cortex and hippocampus. Brain Res. 1562, 23–38.
Arnold, A.P., Burgoyne, P.S., 2004. Are XX and XY brain cells intrinsically different? Trends Endocrinol. Metab. 15, 6–11.
Artinian, J., McGauran A-MT, De Jaeger, X., Mouledous, L., Frances, B., Roullet, P., 2008. Protein degradation, as with protein synthesis, is required during not only long-term spatial memory consolidation but also reconsolidation. Eur. J. Neurosci. 27, 3009–3019.
Bachstetter, A.D., Webster, S.J., Tu, T., Goulding, D.S., Haiech, J., Watterson, D.M., Van Eldik, L.J., 2014. Generation and behavior characterization of CaMKIIβ knockout mice. PLOS One 9, e105191.
Barha, C.K., Galea, L.A.M., 2010. Influence of different estrogens on neuroplasticity and cognition in the hippocampus. Biochim. Biophys. Acta 1800, 1056–1067.
Becker, J.B., Monteggia, L.M., Perrot-Sinal, T.S., Romeo, R.D., Taylor, J.R., Yehuda, R., 2007. Stress and disease: is being female a predisposing factor? J. Neurosci. 27, 11851–11855.
Bellace, M., Williams, J.M., Mohamed, F.B., Faro, S.H., 2013. An fMRI study of the activation of the hippocampus by emotional memory. Int. J. Neurosci. 123, 121–127.
Berletch, J.B., Yang, F., Xu, J., Carrel, L., Disteche, C.M., 2011. Genes that escape from X inactivation. Hum. Genet. 130, 237–245.
Bettis, T.J., Jacobs, L.F., 2013. Sex differences in memory for landmark arrays in C57BL/J6 mice. Anim. Cogn 16, 873–882.
Birnbaum, M.J., 2005. Activating AMP-activated protein kinase without AMP. Mol. Cell 19, 289–290.
Blaeser, F., Sanders, M.J., Truong, N., Ko, S., Wu, L.J., Wozniak, D.F., Fanselow, M.S., Zhuo, M., Chatila, T.A., 2006. Long-term memory deficits in Pavlovian fear conditioning in Ca2+/calmodulin kinase kinase alpha-deficient mice. Mol. Cell. Biol. 26, 9105–9115.
Boulware, M.I., Weick, J.P., Becklund, B.R., Kuo, S.P., Groth, R.D., Mermelstein, P.G., 2005. Estradiol activates group I and II metabotropic glutamate receptor signaling, leading to opposing influences on cAMP response element-binding protein. J. Neurosci. 25, 5066–5078.
Bowers, J.M., Perez-Pouchoulen, M., Roby, C.R., Ryan, T.E., McCarthy, M.M., 2014. Androgen modulation of Foxp1 and Foxp2 in the developing rat brain: impact on sex specific vocalization. Endocrinology 155, 4881–4894.
Brioni, J.D., Nagahara, A.H., McGaugh, J.L., 1989. Involvement of the amygdala GABAergic system in the modulation of memory storage. Brain Res. 487, 105–112.
Briski, K.P., Ibrahim, B.A., Tamrakar, P., 2014. Energy metabolism and hindbrain AMPK: regulation by estradiol. Horm. Mol. Biol. Clin. Invest. 17 (3), 129–136.
Cahill, L., 2011. Sex influences on brain and emotional memory: the burden of proof has shifted. In: Savic, I. (Ed.), Progress in Brain Research Sex Differences in the Human Brain, their Underpinnings and Implications, vol. 29.
Chai, X.J., Jacobs, L.F., 2010. Effects of cue types on sex differences in human spatial memory. Behav. Brain Res. 208, 336–342.
Chen, G., Zou, X., Watanabe, H., van Deursen, J.M., Shen, J., 2010. CREB binding protein is required for both short-term and long-term memory formation. J. Neurosci. 30 (39), 13066–13077.
Colledge, M., Dean, R.A., Scott, G.K., Langeberg, L.K., Huganir, R.L., Scott, J.D., 2000. Targeting of PKA to glutamate receptors through a MAGUK-AKAP complex. Neuron 27, 107–119.
Collinson, N., Kuenzi, F.M., Jarolimek, W., Maubach, K.A., Cothliff, R., Sur, C., Smith, A., Otu, F.M., Howell, O., Atack, J.R., McKernan, R.M., Seabrook, G.R., Dawson, G.R., Whiting, P.J., Rosahl, T.W.,

2002. Enhanced learning and memory and altered GABAergic synaptic transmission in mice lacking the alpha 5 subunit of the GABAA receptor. J. Neurosci. 22, 5572–5580.
Cover, K.K., Maeng, L.Y., Lebrón-Milad, K., Milad, M.R., 2014. Mechanisms of estradiol in fear circuitry: implications for sex differences in psychopathology. Transl. Psychiat. 4, e422.
Dachtler, J., Fox, K.D., Good, M.A., 2011. Gender specific requirement of GluR1 receptors in contextual conditioning but not spatial learning. Neurobiol. Learn. Mem. 96, 461–467.
Dachtler, J., Hardingham, N.R., Fox, K., 2012. The role of nitric oxide synthase in cortical plasticity is sex specific. J. Neurosci. 32, 14994–14999.
Dalla, C., Shors, T.J., 2009. Sex differences in learning processes of classical and operant conditioning. Physiol. Behav. 97, 229–238.
Duvoisin, R.M., Villasana, L., Pfankuch, T., Raber, J., 2010. Sex-dependent cognitive phenotype of mice lacking mGluR8. Behav. Brain Res. 209 (1), 21–26.
Emes, R.D., Grant, S.G.N., 2012. Evolution of synapse complexity and diversity. Ann. Rev. Neurosci. 35, 111–131.
Engmann, O., Hortobágyi, T., Pidsley, R., Troakes, C., Bernstein, H.-G., Kreutz, M.R., Mill, J., Nikolic, M., Giese, K.P., 2011. Schizophrenia is associated with dysregulation of a Cdk5 activator that regulates synaptic protein expression and cognition. Brain 134, 2408–2421.
Fischer, A., Sananbenesi, F., Schrick, C., Spiess, J., Radulovic, J., 2002. Cyclin-dependent kinase 5 is required for associative learning. J. Neurosci. 22, 3700–3707.
Fitzgerald, P.J., Pinard, C.R., Camp, M.C., Feyder, M., Sah, A., Bergstrom, H.C., Graybeal, C., Liu, Y., Schlüter, O.M., Grant, S.G., Singewald, N., Xu, W., Holmes, A., 2014. Durable fear memories require PSD-95. Mol. Psychiat.doi: 10.1038/mp.2014.161, [Epub ahead of print].
Fortress, A.M., Fan, L., Orr, P.T., Frick, K.M., 2013. Estradiol-induced object recognition memory consolidation is dependent on activation of mTOR signaling in the dorsal hippocampus. Learn. Mem. 20, 147–155.
Gao, C., Frausto, S.F., Guedea, A.L., Tronson, N.C., Radulovic, J., 2011. IQGAP1 regulates NR2A signaling, spine density, and cognitive processes. J. Neurosci. 31 (23), 8533–8542.
Gao, C., Tronson, N.C., Radulovic, J., 2013. Modulation of behavior by scaffolding proteins of the postsynaptic density. Neurobiol. Learn. Mem. 105, 3–12.
Gasbarri, A., Arnone, B., Pompili, A., Pacitti, F., Pacitti, C., Cahill, L., 2007. Sex-related hemispheric lateralization of electrical potentials evoked by arousing negative stimuli. Brain Res. 1138, 178–186.
Gokce, O., Südhof, T.C., 2013. Membrane-tethered monomeric neurexin LNS-domain triggers synapse formation. J. Neurosci. 33, 14617–14628.
Grant, S.G.N., 2003. Synapse signalling complexes and networks: machines underlying cognition. Bioessays 25, 1229–1235.
Gresack, J.E., Schafe, G.E., Orr, P.T., Frick, K.M., 2009. Sex differences in contextual fear conditioning are associated with differential ventral hippocampal extracellular signal-regulated kinase activation. Neuroscience 159, 451–467.
Gruene, T.M., Roberts, E., Thomas, V., Ronzio, A., Shansky, R.M., 2014. Sex-specific neuroanatomical correlates of fear expression in prefrontal-amygdala circuits. Biol. Psychiat. [Epub ahead of Print].
Guedea, A.L., Schrick, C., Guzman, Y.F., Leaderbrand, K., Jovasevic, V., Corcoran, K.A., Tronson, N.C., Radulovic, J., 2011. ERK-associated changes of AP-1 proteins during fear extinction. Mol. Cell. Neurosci. 47, 137–144.
Gulinello, M., Smith, S.S., 2003. Anxiogenic effects of neurosteroid exposure: sex differences and altered GABAA receptor pharmacology in adult rats. J. Pharmacol. Exp. Ther. 305, 541–548.
Guzowski, J.F., Lyford, G.L., Stevenson, G.D., Houston, F.P., McGaugh, J.L., Worley, P.F., Barnes, C.A., 2000. Inhibition of activity-dependent arc protein expression in the rat hippocampus impairs the maintenance of long-term potentiation and the consolidation of long-term memory. J. Neurosci. 20, 3993–4001.
Hajszan, T., MacLusky, N.J., Johansen, J.A., Jordan, C.L., Leranth, C., 2013. Effects of androgens and estradiol on spine synapse formation in the prefrontal cortex of normal and testicular feminization mutant male rats. Endocrinology 148, 1963–1967.
Hawley, S.A., Pan, D.A., Mustard, K.J., Ross, L., Bain, J., Edelman, A.M., Frenguelli, B.G., Hardie, D.G., 2005. Calmodulin-dependent protein kinase kinase-beta is an alternative upstream kinase for AMP-activated protein kinase. Cell Metab. 2, 9–19.

Hebda-Bauer, E.K., Luo, J., Watson, S.J., Akil, H., 2007. Female CREBalphadelta-deficient mice show earlier age-related cognitive deficits than males. Neuroscience 150, 260–272.

Hong, S., Cho, Y.-W., Yu, L.-R., Yu, H., Veenstra, T.D., Ge, K., 2007. Identification of JmjC domain-containing UTX and JMJD3 as histone H3 lysine 27 demethylases. Proc. Natl. Acad. Sci. 104, 18439–18444.

Horwood, J.M., Dufour, F., Laroche, S., Davis, S., 2006. Signalling mechanisms mediated by the phosphoinositide 3-kinase/Akt cascade in synaptic plasticity and memory in the rat. Eur. J. Neurosci. 23, 3375–3384.

Huang, G.Z., Woolley, C.S., 2012. Estradiol acutely suppresses inhibition in the hippocampus through a sex-specific endocannabinoid and mGluR-dependent mechanism. Neuron 74, 801–808.

Huang, W., Zhu, P.J., Zhang, S., Zhou, H., Stoica, L., Galiano, M., Krnjević, K., Roman, G., Costa-Mattioli, M., 2013. mTORC2 controls actin polymerization required for consolidation of long-term memory. Nat. Neurosci. 16, 441–448.

Huber, K.M., Roder, J.C., Bear, M.F., 2001. Chemical induction of mGluR5- and protein synthesis-dependent long-term depression in hippocampal area CA1. J. Neurophysiol. 86, 321–325.

Hurley, R.L., Anderson, K.A., Franzone, J.A., Kemp, B.E., Means, A.R., Witters, L.A., 2005. The Ca2+/calmodulin-dependent protein kinase kinases are AMP-activated protein kinase kinases. J. Biol. Chem. 280, 29060–29066.

Iñiguez, S.D., Charntikov, S., Baella, S.A., Herbert, M.S., Bolaños-Guzmán, C.A., Crawford, C.A., 2012. Post-training cocaine exposure facilitates spatial memory consolidation in C57BL/6 mice. Hippocampus 22, 802–813.

Iwase, S., Lan, F., Bayliss, P., la Torre-Ubieta de, L., Huarte, M., Qi, H.H., Whetstine, J.R., Bonni, A., Roberts, T.M., Shi, Y., 2007. The X-linked mental retardation gene SMCX/JARID1C defines a family of histone H3 lysine 4 demethylases. Cell 128, 1077–1088.

Jarome, T.J., Kwapis, J.L., Werner, C.T., Parsons, R.G., Gafford, G.M., Helmstetter, F.J., 2011. Activity dependent protein degradation is critical for the formation and stability of fear memory in the amygdala. Learn. Mem. 6, e24349.

Jazin, E., Cahill, L., 2010. Sex differences in molecular neuroscience: from fruit flies to humans. Nat. Rev. Neurosci. 11, 9–17.

Jobim, P.F.C., Pedroso, T.R., Christoff, R.R., Werenicz, A., Maurmann, N., Reolon, G.K., Roesler, R., 2012. Inhibition of mTOR by rapamycin in the amygdala or hippocampus impairs formation and reconsolidation of inhibitory avoidance memory. Neurobiol. Learn. Mem. 97, 105–112.

Johansen, J.P., Cain, C.K., Ostroff, L.E., LeDoux, J.E., 2011. Molecular mechanisms of fear learning and memory. Cell 147, 509–524.

Josselyn, S.A., Kida, S., Silva, A.J., 2004. Inducible repression of CREB function disrupts amygdala-dependent memory. Neurobiol. Learn. Mem. 82, 159–163.

Kaczmarek, L., Lapinska-Dzwonek, J., Szymczak, S., 2002. Matrix metalloproteinases in the adult brain physiology: a link between c-Fos, AP-1 and remodeling of neuronal connections? EMBO J. 21, 6643–6648.

Kammermeier, P.J., Worley, P.F., 2007. Homer 1a uncouples metabotropic glutamate receptor 5 from postsynaptic effectors. Proc. Natl. Acad. Sci. USA 104, 6055–6060.

Kang, H., Sun, L.D., Atkins, C.M., Soderling, T.R., Wilson, M.A., Tonegawa, S., 2001. An important role of neural activity-dependent CaMKIV signaling in the consolidation of long-term memory. Cell 106, 771–783.

Katsouli, S., Stamatakis, A., Giompres, P., Kouvelas, E.D., Stylianopoulou, F., Mitsacos, A., 2014. Sexually dimorphic long-term effects of an early life experience on AMPA receptor subunit expression in rat brain. Neuroscience 257, 49–64.

Keeley, R.J., Tyndall, A.V., Scott, G.A., Saucier, D.M., 2013. Sex difference in cue strategy in a modified version of the morris water task: correlations between brain and behaviour. Nat. Neurosci. 8, e69727.

Kelley, J.B., Balda, M.A., Anderson, K.L., Itzhak, Y., 2009. Impairments in fear conditioning in mice lacking the nNOS gene. Learn. Mem. 16, 371–378.

Kessler, R.C., Petukhova, M., Sampson, N.A., Zaslavsky, A.M., Wittchen, H.-U., 2012. Twelve-month and lifetime prevalence and lifetime morbid risk of anxiety and mood disorders in the United States. Int. J. Methods Psychiatr. Res. 21, 169–184.

Kogan, J.H., Frankland, P.W., Blendy, J.A., Coblentz, J., Marowitz, Z., Schütz, G., Silva, A.J., 1997. Spaced training induces normal long-term memory in CREB mutant mice. Curr. Biol. 7, 1–11.

Kopec, C.D., Real, E., Kessels, H.W., Malinow, R., 2007. GluR1 links structural and functional plasticity at excitatory synapses. J. Neurosci. 27, 13706–13718.

Kudo, K., Qiao, C.-X., Kanba, S., Arita, J., 2004. A selective increase in phosphorylation of cyclic AMP response element-binding protein in hippocampal CA1 region of male, but not female, rats following contextual fear and passive avoidance conditioning. Brain Res. 1024, 233–243.

Lan, F., Bayliss, P.E., Rinn, J.L., Whetstine, J.R., Wang, J.K., Chen, S., Iwase, S., Alpatov, R., Issaeva, I., Canaani, E., Roberts, T.M., Chang, H.Y., Shi, Y., 2007. A histone H3 lysine 27 demethylase regulates animal posterior development. Nature 449, 689–694.

Lebrón-Milad, K., Tsareva, A., Ahmed, N., Milad, M.R., 2013. Sex differences and estrous cycle in female rats interact with the effects of fluoxetine treatment on fear extinction. Behav. Brain Res. 253, 217–222.

Lin, K.A., Doraiswamy, P.M., 2015. When Mars versus Venus is not a cliché: gender differences in the neurobiology of Alzheimer's disease. Front. Neurol. 5, 288.

Lubin, F.D., Sweatt, J.D., 2007. The IκB kinase regulates chromatin structure during reconsolidation of conditioned fear memories. Neuron 55, 942–957.

Lucchesi, W., Mizuno, K., Giese, K.P., 2011. Novel insights into CaMKII function and regulation during memory formation. Brain Res. Bull. 85, 2–8.

Malenka, R.C., 2003. Synaptic plasticity and AMPA receptor trafficking. Ann. NY Acad. Sci. 1003, 1–11.

Mao, L., Yang, L., Tang, Q., Samdani, S., Zhang, G., Wang, J.Q., 2005. The scaffold protein homer1b/c links metabotropic glutamate receptor 5 to extracellular signal-regulated protein kinase cascades in neurons. J. Neurosci. 25, 2741–2752.

Maren, S., Fanselow, M.S., 1994. Sex differences in hippocampal long-term potentiation (LTP) and Pavlovian fear conditioning in rats: positive correlation between LTP and contextual learning. Brain Res. 661, 25–34.

Matsumoto, A., 1991. Synaptogenic action of sex steroids in developing and adult neuroendocrine brain. Psychoneuroendocrinology 16, 25–40.

Maurice, T., Duclot, F., Meunier, J., Naert, G., Givalois, L., Meffre, J., Célérier, A., Jacquet, C., Copois, V., Mechti, N., Ozato, K., Gongora, C., 2008. Altered memory capacities and response to stress in p300/CBP-associated factor (PCAF) histone acetylase knockout mice. Neuropsychopharmacology 33, 1584–1602.

McCarthy, M.M., 2009. The epigenetics of sex differences in the brain. J. Neurosci 29, 12815–12823.

McCarthy, M.M., Arnold, A.P., 2011. Reframing sexual differentiation of the brain. Nat. Neurosci. 14, 677–683.

Méndez-López, M., Méndez, M., López, L., Arias, J.L., 2009. Sexually dimorphic c-Fos expression following spatial working memory in young and adult rats. Physiol. Behav. 98 (3), 307–317.

Milad, M.R., Goldstein, J.M., Orr, S.P., Wedig, M.M., Klibanski, A., Pitman, R.K., Rauch, S.L., 2006. Fear conditioning and extinction: influence of sex and menstrual cycle in healthy humans. Behav. Neurosci. 120, 1196–1203.

Mizuno, K., Ris, L., Sanchez-Capelo, A., Godaux, E., Giese, K.P., 2006. Ca2+/calmodulin kinase kinase is dispensable for brain development but is required for distinct memories in male, though not in female, mice. Mol. Cell. Biol. 26, 9094–9104.

Mizuno, K., Ris, L., Godaux, E., Giese, K.P., 2007. Calcium/calmodulin kinase kinase β has a male-specific role in memory formation. Neuroscience 145, 393–402.

Moita, M.A.P., Lamprecht, R., Nader, K., LeDoux, J.E., 2002. A-kinase anchoring proteins in amygdala are involved in auditory fear memory. Nat. Neurosci. 5, 837–838.

Monfort, P., Felipo, V., 2007. Hippocampal long-term potentiation is reduced in mature compared to young male rats but not in female rats. Neuroscience 146, 504–508.

Moore, M.D., Cushman, J., Chandra, D., Homanics, G.E., Olsen, R.W., Fanselow, M.S., 2010. Trace and contextual fear conditioning is enhanced in mice lacking the α4 subunit of the GABAA receptor. Neurobiol. Learn. Mem. 93, 383–387.

Morgan, C.J.A., Perry, E.B., Cho, H.-S., Krystal, J.H., D'Souza, D.C., 2006. Greater vulnerability to the amnestic effects of ketamine in males. Psychopharmacology 187, 405–414.

Morris, R.G.M., 2013. NMDA receptors and memory encoding. Neuropharmacology 74, 32–40.

Nani, F., Bright, D.P., Revilla-Sanchez, R., Tretter, V., Moss, S.J., Smart, T.G., 2013. Tyrosine phosphorylation of GABAA receptor γ2-subunit regulates tonic and phasic inhibition in the thalamus. J. Neurosci. 33, 12718–12727.

Norrholm, S.D., Bibb, J.A., Nestler, E.J., Ouimet, C.C., Taylor, J.R., Greengard, P., 2003. Cocaine-induced proliferation of dendritic spines in nucleus accumbens is dependent on the activity of cyclin-dependent kinase-5. Neuroscience 116, 19–22.

Nuñez, J.L., McCarthy, M.M., 2008. Androgens predispose males to GABAA-mediated excitotoxicity in the developing hippocampus. Exp. Neurol. 210, 699–708.

Nuñez, J.L., McCarthy, M.M., 2009. Resting intracellular calcium concentration, depolarizing gamma-aminobutyric acid and possible role of local estradiol synthesis in the developing male and female hippocampus. Neuroscience 158, 623–634.

Nygard, S.K., Klambatsen, A., Hazim, R., Eltareb, M.H., Blank, J.C., Chang, A.J., Quinones-Jenab, V., Jenab, S., 2013. Sexually dimorphic intracellular responses after cocaine-induced conditioned place preference expression. Brain Res. 1520, 121–133.

Ooishi, Y., Kawato, S., Hojo, Y., Hatanaka, Y., Higo, S., Murakami, G., Komatsuzaki, Y., Ogiue-Ikeda, M., Kimoto, T., Mukai, H., 2012. Modulation of synaptic plasticity in the hippocampus by hippocampus-derived estrogen and androgen. J. Steroid Biochem. Mol. Biol. 131, 37–51.

Penney, J., Tsai, L.-H., 2014. Histone deacetylases in memory and cognition. Sci. Signal 7, re12–re112.

Peters, M., Mizuno, K., Ris, L., Angelo, M., Godaux, E., Giese, K.P., 2003. Loss of Ca2+/calmodulin kinase kinase β affects the formation of some, but not all, types of hippocampus-dependent long-term memory. J. Neurosci. 23, 9752–9760.

Pittenger, C., Huang, Y.Y., Paletzki, R.F., Bourtchouladze, R., Scanlin, H., Vronskaya, S., Kandel, E.R., 2002. Reversible inhibition of CREB/ATF transcription factors in region CA1 of the dorsal hippocampus disrupts hippocampus-dependent spatial memory. Neuron 34, 447–462.

Procaccini, C., Maksimovic, M., Aitta-aho, T., Korpi, E.R., Linden, A.M., 2013. Reversal of novelty-induced hyperlocomotion and hippocampal c-Fos expression in GluA1 knockout male mice by the mGluR2/3 agonist LY354740. Neuroscience 250, 189–200.

Prothero, K.E., Stahl, J.M., Carrel, L., 2009. Dosage compensation and gene expression on the mammalian X chromosome: one plus one does not always equal two. Chromosome Res. 17, 637–648.

Quinn, J.J., Arnold, A.P., Taylor, J.R., 2007. Sex chromosome complement regulates habit formation. Nat. Neurosci. 10, 1398–1400.

Radulovic, J., Kammermeier, J., Spiess, J., 1998. Relationship between fos production and classical fear conditioning: effects of novelty, latent inhibition, and unconditioned stimulus preexposure. J. Neurosci. 18, 7452–7461.

Rey, C.D., Lipps, J., Shansky, R.M., 2013. Dopamine D1-receptor activation rescues extinction impairments in low-estrogen female rats and induces cortical layer-specific activation changes in prefrontal-amygdala circuits. Neuropsychopharmacology 39 (5), 1282–1289.

Ris, L., Angelo, M., Plattner, F., Capron, B., Errington, M.L., Bliss, T.V.P., Godaux, E., Giese, K.P., 2005. Sexual dimorphisms in the effect of low-level p25 expression on synaptic plasticity and memory. Eur. J. Neurosci. 21, 3023–3033.

Robbins, S.J., Ehrman, R.N., Childress, A.R., O'Brien, C.P., 1999. Comparing levels of cocaine cue reactivity in male and female outpatients. Drug Alcohol Depend. 53, 223–230.

Rodríguez, C.A., Chamizo, V.D., Mackintosh, N.J., 2011. Overshadowing and blocking between landmark learning and shape learning: the importance of sex differences. Learn. Behav. 39, 324–335.

Rumpel, S., 2005. Postsynaptic receptor trafficking underlying a form of associative learning. Science 308, 83–88.

Sacktor, T.C., 2011. How does PKMζ maintain long-term memory? Nat. Rev. Neurosci. 12, 9–15.

Sananbenesi, F., Fischer, A., Schrick, C., Spiess, J., Radulovic, J., 2002. Phosphorylation of hippocampal Erk-1/2, Elk-1, and p90-Rsk-1 during contextual fear conditioning: interactions between Erk-1/2 and Elk-1. Mol. Cell. Neurosci. 21, 463–476.

Scharfman, H.E., MacLusky, N.J., 2014. Differential regulation of BDNF, synaptic plasticity and sprouting in the hippocampal mossy fiber pathway of male and female rats. Neuropharmacology 76 Pt C, 696–708.

Schrick, C., Fischer, A., Srivastava, D.P., Tronson, N.C., Penzes, P., Radulovic, J., 2007. N-Cadherin regulates cytoskeletally associated IQGAP1/ERK signaling and memory formation. Neuron 55, 786–798.

Sebastian, V., Vergel, T., Baig, R., Schrott, L.M., Serrano, P.A., 2013. PKMζ differentially utilized between sexes for remote long-term spatial memory. PLOS One 8, e81121.

Seshadri, S., Wolf, P.A., Beiser, A., Au, R., McNulty, K., White, R., D'Agostino, R.B., 1997. Lifetime risk of dementia and Alzheimer's disease. The impact of mortality on risk estimates in the Framingham Study. Neurology 49, 1498–1504.

Shah, K., Lahiri, D.K., 2014. Cdk5 activity in the brain – multiple paths of regulation. J. Cell Sci. 127, 2391–2400.

Shah, D.S., Prados, J., Gamble, J., De Lillo, C., Gibson, C.L., 2013. Sex differences in spatial memory using serial and search tasks. Behav. Brain Res. 257, 90–99.

Shalin, S.C., Zirrgiebel, U., Honsa, K.J., Julien, J.-P., Miller, F.D., Kaplan, D.R., Sweatt, J.D., 2004. Neuronal MEK is important for normal fear conditioning in mice. J. Neurosci. Res. 75, 760–770.

Shalin, S.C., Hernandez, C.M., Dougherty, M.K., Morrison, D.K., Sweatt, J.D., 2006. Kinase suppressor of Ras1 compartmentalizes hippocampal signal transduction and subserves synaptic plasticity and memory formation. Neuron 50 (5), 765–779.

Shi, S.H., Hayashi, Y., Petralia, R.S., Zaman, S.H., Wenthold, R.J., Svoboda, K., Malinow, R., 1999. Rapid spine delivery and redistribution of AMPA receptors after synaptic NMDA receptor activation. Science 284, 1811–1816.

Shors, T.J., 2004. The opposite effects of stress on dendritic spines in male vs. female rats are NMDA receptor-dependent. Eur. J. Neurosci. 19, 145–150.

Silva, A.J., Paylor, R., Wehner, J.M., Tonegawa, S., 1992. Impaired spatial learning in alpha-calcium-calmodulin kinase II mutant mice. Science 257, 206–211.

Simpson, J., Kelly, J.P., 2012. An investigation of whether there are sex differences in certain behavioural and neurochemical parameters in the rat. Behav. Brain Res. 229, 289–300.

Srivastava, D.P., Woolfrey, K.M., Woolfrey, K., Jones, K.A., Shum, C.Y., Lash, L.L., Swanson, G.T., Penzes, P., 2008. Rapid enhancement of two-step wiring plasticity by estrogen and NMDA receptor activity. Proc. Natl. Acad. Sci. 105, 14650–14655.

Stawarski, M., Stefaniuk, M., Wlodarczyk, J., 2014. Matrix metalloproteinase-9 involvement in the structural plasticity of dendritic spines. Front. Neuroanat. 8, 68.

Takao, K., Tanda, K., Nakamura, K., Kasahara, J., Nakao, K., Katsuki, M., Nakanishi, K., Yamasaki, N., Toyama, K., Adachi, M., Umeda, M., Araki, T., Fukunaga, K., Kondo, H., Sakagami, H., Miyakawa, T., 2010. Comprehensive behavioral analysis of calcium/calmodulin-dependent protein kinase IV knockout mice. PLOS One 5, e9460.

Ter Horst, J.P., Carobrez, A.P., van der Mark, M.H., de Kloet, E.R., Oitzl, M.S., 2012. Sex differences in fear memory and extinction of mice with forebrain-specific disruption of the mineralocorticoid receptor. Eur. J. Neurosci. 36 (8), 3096–3102.

Tronson, N.C., Taylor, J.R., 2007. Molecular mechanisms of memory reconsolidation. Nat. Rev. Neurosci. 8, 262–275.

Tronson, N.C., Guzman, Y.F., Guedea, A.L., Huh, K.H., Gao, C., Schwarz, M.K., Radulovic, J., 2010. Metabotropic glutamate receptor 5/homer interactions underlie stress effects on fear. Biol. Psychiat. 68, 1007–1015.

Tsai, H.-W., 2009. Sex differences in histone modifications in the neonatal mouse brain. Epigenetics 4, 47–53.

Tsien, J.Z., Huerta, P.T., Tonegawa, S., 1996. The essential role of hippocampal CA1 NMDA receptor-dependent synaptic plasticity in spatial memory. Cell 87, 1327–1338.

Tsilibary, E., Tzinia, A., Radenovic, L., Stamenkovic, V., Lebitko, T., Mucha, M., Pawlak, R., Frischknecht, R., Kaczmarek, L., 2014. Neural ECM proteases in learning and synaptic plasticity. Preface. Brain Extracellular Matrix in Health and Disease. Elsevier, pp. 135–157.

Tuscher, J.J., Fortress, A.M., Kim, J., Frick, K.M., 2014. Regulation of object recognition and object placement by ovarian sex steroid hormones. Behav. Brain Res. 285, 140–157.

Vierk, R., Brandt, N., Rune, G.M., 2014. Hippocampal estradiol synthesis and its significance for hippocampal synaptic stability in male and female animals. Neuroscience 274, 24–32.

Waddell, J., 2010. D-cycloserine reverses the detrimental effects of stress on learning in females and enhances retention in males. Neurobiol. Learn. Mem. 93, 31–36.
Wilmott, L.A., Thompson, L.T., 2013. Sex- and dose-dependent effects of post-trial calcium channel blockade by magnesium chloride on memory for inhibitory avoidance conditioning. Behav. Brain Res. 257, 49–53.
Wiltgen, B.J., Sanders, M.J., Ferguson, C., Homanics, G.E., Fanselow, M.S., 2005. Trace fear conditioning is enhanced in mice lacking the delta subunit of the GABAA receptor. Learn. Mem. 12, 327–333.
Woolley, C.S., Gould, E., Frankfurt, M., McEwen, B.S., 1990. Naturally occurring fluctuation in dendritic spine density on adult hippocampal pyramidal neurons. J. Neurosci. 10, 4035–4039.
Xu, J., Disteche, C.M., 2006. Sex differences in brain expression of X- and Y-linked genes. Brain Res. 1126, 50–55.
Xu, J., Burgoyne, P.S., Arnold, A.P., 2002. Sex differences in sex chromosome gene expression in mouse brain. Hum. Mol. Genet. 11, 1409–1419.
Xu, J., Deng, X., Disteche, C.M., 2008a. Sex-specific expression of the X-linked histone demethylase gene Jarid1c in brain. PLOS One 3, e2553.
Xu, J., Deng, X., Watkins, R., Disteche, C.M., 2008b. Sex-specific differences in expression of histone demethylases Utx and Uty in mouse brain and neurons. J. Neurosci. 28, 4521–4527.
Yang, J., Hu, L.-L., Liu, L.-Y., Zhao, L.-Y., Hou, N., Ni, L., Li, Z.-F., Wang, A.-Y., Song, T.-S., Huang, C., 2013. Proteomics reveals intersexual differences in the rat brain hippocampus. Anat. Rec. 296, 462–469.
Yano, S., Tokumitsu, H., Soderling, T.R., 1998. Calcium promotes cell survival through CaM-K kinase activation of the protein-kinase-B pathway. Nature 396, 584–587.
Yee, B.K., Möhler, H., Feldon, J., 2004. GABAA receptors containing the α5 subunit mediate the trace effect in aversive and appetitive conditioning and extinction of conditioned fear. Eur. J. Neurosci. 20, 1928–1936.
Zamani, M.R., Desmond, N.L., Levy, W.B., 2000. Estradiol modulates long-term synaptic depression in female rat hippocampus. J. Neurophysiol. 84, 1800–1808.
Zhu, Y.S., Pfaff, D.W., 1998. Differential regulation of AP-1 DNA binding activity in rat hypothalamus and pituitary by estrogen. Brain Res. Mol. Brain Res. 55, 115–125.

# CHAPTER 3

# Sex Differences in Anxiety Disorders: Gonadal Hormone Interactions with Pathophysiology, Neurobiology, and Treatment

**Lisa Y. Maeng*, Mohammed R. Milad****
*Harvard Medical School, Department of Psychiatry, Boston, MA, USA
**Massachusetts General Hospital, Department of Psychiatry, Charlestown, MA, USA

## 1 INTRODUCTION

Women have twice as high a risk as men of developing anxiety and stress-related psychiatric disorders (Tolin and Foa, 2006). This sex bias suggests that women may have a greater sensitivity to stressful and traumatic life experiences. To support this, a substantial number of studies have identified sex differences in the response to stress as well as sexual dimorphisms in the critical neurocircuitry pertaining to stress and arousal (Goldstein et al., 2010; Kogler et al., 2014; Maeng et al., 2010). Despite these findings, the mechanisms that mediate sex differences in stress and anxiety disorders remain unclear.

Many of these psychopathologies significantly disrupt cognitive processes, and the resulting dysfunction within these systems can lead to debilitating symptoms that manifest as impairments in learning and memory. Indeed, individuals suffering from anxiety often have difficulty with learning and memory tasks. Gonadal hormones, such as estrogen and testosterone, are known to affect learning and memory processes. In this chapter, we focus on processes of fear regulation (acquisition and extinction) to examine the contributions of gonadal hormones, estrogen in particular, to sex differences in anxiety disorders. By describing the influence of gonadal hormones on fear-related learning and memory modulation, we hope to elucidate the underlying mechanisms of sex differences in anxiety disorders. In this discussion, we aim to promote a better understanding of the vulnerability factors in anxiety disorders to inform future treatment options to improve outcomes.

## 2 SEX DIFFERENCES IN ANXIETY DISORDERS

Anxiety disorders are the most commonly occurring class of mental disorders in the general population, with an estimated lifetime prevalence rate of nearly 16% globally and 20% in just the United States (Kessler et al., 2005, 2009). As mentioned previously,

*Sex Differences in the Central Nervous System.* http://dx.doi.org/10.1016/B978-0-12-802114-9.00003-2

women are at about a twofold higher risk for any anxiety-related disorder than men according to epidemiological reports (Breslau et al., 1997; Foa and Street, 2001; Kessler et al., 1994, 1995; 2005; McLean et al., 2011). This higher prevalence in women is observed across all anxiety and fear-related disorders, including social anxiety disorder, phobias, generalized anxiety disorder, panic disorder, and posttraumatic stress disorder (PTSD) (Breslau et al., 1997, 2009; Kessler et al., 1994, 1995). Although PTSD and obsessive-compulsive disorder (OCD) are no longer defined as anxiety disorders in the DSM-V, we will include it in our discussion given that they share critical aspects of anxiety disorders.

Aside from prevalence, men and women differ with regard to the manner or degree to which they suffer from these disorders. For instance, women appear to experience anxiety symptoms to a greater degree (Altemus et al., 2014). Women also comprise more than half of the population with generalized anxiety disorder and have a greater vulnerability to comorbid mental disorders that persists later in life (60 years of age and older; American Psychiatric Association, 2013; Bakish, 1999; Van der Veen et al., 2014). Moreover, women also experience more severe, debilitating, and persistent symptoms. In panic disorder, this increased severity of symptoms is demonstrated by women experiencing a higher frequency of panic attacks than men (Kessler et al., 2006; Reed and Wittchen, 1998). These differences contribute to an overall worse quality of life for women with panic disorder as compared to men (Breslau et al., 1998, 2002; Frans et al., 2005; Holbrook et al., 2002; Kilpatrick et al., 2013; Perrin et al., 2014; Seedat et al., 2005). Men and women also differentially express the characteristics and symptoms of anxiety disorders. For instance, women are more likely to show OCD symptoms in the contamination/cleaning domain, whereas men exhibit more obsessive behaviors related to the sexual or religious dimension of OCD (Labad et al., 2008).

Epidemiological studies suggest that women may be at a greater risk for developing anxiety disorders, or exacerbating their present symptoms, during different phases of their reproductive lives (e.g., puberty, menses, pregnancy, postpartum, and menopause). These periods of elevated risk coincide with times of drastic hormonal fluctuations, implicating a role for gonadal hormones in the onset, maintenance, and persistence of anxiety disorders in women. This risk for developing anxiety and related disorders may be due to an inability to downregulate negative emotional responses to stress and fear during these times (Campbell-Sills et al., 2006; Mennin et al., 2005; Nolen-Hoeksema, 1991).

## 3 NEUROBIOLOGY OF FEAR EXTINCTION

In the laboratory, Pavlovian fear conditioning is commonly used to investigate fear learning and memory processes. In fear conditioning, the subject learns that a neutral conditioned stimulus (CS), such as a light or tone, predicts an aversive unconditioned stimulus (US), such as a mild footshock, and expresses conditioned responses (CR) to CS presentations that follow. The behavior that is frequently measured to assess fear is conditioned freezing in rodents and skin conductance response (SCR) or fear-potentiated startle in humans. During fear extinction, the subject learns that the CS no longer predicts the aversive US and exhibits a reduction in CR to the CS. The neurobiology of fear extinction has been studied extensively and implicated in the etiology of anxiety and stress-related disorders (Dias et al., 2013; Hofmann, 2008). Here, we focus on the key nodes of the fear extinction network in both rodents and humans, which include regions of the prefrontal cortex (PFC), amygdala, and hippocampus.

## 3.1 Rodent fear circuitry

The neural circuitry that modulates fear extinction behavior has been identified and well studied in male rodents. In the circuit driving fear, the prelimbic (PL) medial prefrontal cortex (mPFC) sends excitatory input to the basolateral amygdala (BLA), which activates the central amygdala (CeA) for enhanced conditioned fear expression (Likhtik et al., 2005; Sierra-Mercado et al., 2011; Sotres-Bayon et al., 2004). Studies that examined fear extinction, or the inhibition of fear, highlight the infralimbic (IL) area of the mPFC. During fear extinction, the IL activates inhibitory intercalated cells in the amygdala. These cells inhibit CeA activity and thus reduce fear output (Likhtik et al., 2005; Quirk et al., 2003; Quirk and Mueller, 2008; Sierra-Mercado et al., 2011; Sotres-Bayon et al., 2004). The hippocampus responds to contextual cues and communicates with the amygdala and hippocampus to then either drive or suppress fear expression depending on the context; in the extinction context, the hippocampus sends inputs that induce IL activity to suppress fear (Sotres-Bayon et al., 2012; Corcoran and Maren, 2001).

## 3.2 Human fear circuitry

Similar to rodents, human neuroimaging studies have identified brain regions that are associated with increased and reduced fear expression, which appear to be functionally homologous (Milad and Quirk, 2012). Brain activations during fear conditioning and extinction suggest that the human dorsal anterior cingulate cortex (dACC) and ventromedial prefrontal cortex (vmPFC) are homologous with the rodent PL and IL, respectively (Linnman et al., 2012). Increased activation of the vmPFC that is positively associated with reduced SCR occurs during presentation of the extinguished CS during extinction recall (Kalisch et al., 2006; Milad et al., 2007; Phelps et al., 2004). In addition, the amygdala and hippocampus are also activated during these tasks, demonstrating similar functional roles as in rodents (Kalisch et al., 2006).

## 3.3 Relevance to anxiety disorders

Exaggerated and uncontrollable fear is a signature characteristic of anxiety- and stress-related disorders. This is especially true for individuals suffering from PTSD, who cannot control their fear to stimuli associated with past traumatic events. The fear extinction network can inform research on anxiety disorders not only because fear extinction processes model some of the core behavioral features of anxiety disorders, but also because of their shared neurobiology (for a review, see Graham and Milad, 2011). These shared characteristics allow findings from the rodent fear extinction model to be easily translated and applied clinically. Indeed, patients with anxiety-related disorders, such as social anxiety disorder, and patients with PTSD, exhibit decreased hippocampal and prefrontal volume and amygdala hyperactivity (Bremner et al., 1997; Coffey et al., 1993; Lebron-Milad et al., 2012; Machado-de-Sousa et al., 2014; Shin et al., 2006). PTSD patients exhibit poor extinction recall with disruptions within the fear extinction network, e.g., hyperactivity within the dACC and the amygdala and hypoactivity within the vmPFC (Milad et al., 2009a; Pitman et al., 2012). Similarly, these impairments in extinction recall and altered brain function are also observed in OCD patients (Milad et al., 2013). These findings provide some evidence that the fear extinction model can be an effective transdiagnostic tool that can be used in both humans and rodents to detect susceptibility to anxiety disorders after a stressful life event. In addition to identifying biomarkers of anxiety disorders, this model may also help evaluate treatment efficacy. Prolonged exposure therapy is based on fear extinction processes and is also one of the most effective forms of cognitive behavioral therapy for the treatment of anxiety disorders (Foa, 2000, 2011; McLean and Foa, 2013). Therefore, fear extinction may be a good experimental model for investigating the neural mechanisms that are impacted by these treatments and how changes in these systems might influence how an individual responds to therapy.

## 4 SEX DIFFERENCES IN FEAR EXTINCTION

Interestingly, in both human and nonhuman animals, the brain regions that subserve cognition and emotional processing are also differentially activated between the sexes during stress and fear learning (Bangasser and Shors, 2010; Baran et al., 2010; Goldstein et al., 2010; Lebron-Milad et al., 2012). Given the marked sex differences observed in psychopathology, it is not surprising that the regions comprising the fear extinction network also exhibit sexual dimorphisms. The sexually dimorphic nature of the brain regions that respond to stress and fear learning (e.g., mPFC, amygdala, and hippocampus) may underlie the sex differences observed in fear conditioning and extinction, although the direction of these behavioral sex differences does not appear to be consistent.

### 4.1 Rodents

Studies report that female rodents do not perform as well as males in fear conditioning and extinction learning (Baker-Andresen et al., 2013; Baran et al., 2009, 2010; Fenton et al., 2014). Male rats fear condition faster than females – they exhibit more freezing to the CS in fewer trials as compared to females (Aguilar et al., 2003; Baran et al., 2009, 2010; Daviu et al., 2014; Maren et al., 1994; Pryce et al., 1999; Ribeiro et al., 2010). Moreover, in another task in which female and male rats were trained to avoid an aversive arm (bright light and loud noise) in a plus maze, the males made more entries into the arm when the aversive stimuli were no longer present during extinction; in contrast, the female rats decreased aversive arm exploration, indicating a failure to extinguish the behavior (Ribeiro et al., 2010). In a separate study, males also performed better in a contextual fear conditioning task; however, contrary to the aforementioned study, females appeared to extinguish better, expressing less fear and exhibiting a higher extinction rate than males (Daviu et al., 2014). Baran et al. (2009) found no differences in fear acquisition between males and females acquired, but reported that the female rats did not extinguish as well as males.

## 4.2 Humans

The mixed findings observed in the rodent studies are also observed in human studies. Milad et al. (2006) demonstrated in humans that although men exhibited a greater conditioned fear response, measured by skin conductance, compared to women during fear acquisition, they did not differ during extinction learning. In contrast, women exhibited greater fear conditioning to a CS paired with viewing mock panic attacks, with increased electrodermal responding and distress ratings compared to men in this observational fear conditioning procedure (Kelly and Forsyth, 2007). Although the literature in both humans and rodents appears inconsistent in these findings, this may be due to differences in fear paradigm, strain, or species studied as well as not precisely controlling for menstrual cycle phase or oral contraceptive use.

## 4.3 Sexually dimorphic circuitry

Differences in the neural mechanisms modulating fear may underlie these sex differences in fear learning and extinction (Baran et al., 2010; Ter Horst et al., 2012). In a study examining the role of the mPFC, rats were fear conditioned to a paired tone-footshock. During extinction training 1 h later, males with and without mPFC lesions and sham-lesioned females successfully reduced freezing, whereas lesioned mPFC female rats continued to freeze to the tone and failed to extinguish (Baran et al., 2010). This effect on extinction learning persisted even after a 24-h delay in the mPFC-lesioned females. Medial PFC lesions impaired extinction recall in males but, like the sham females, these males were able to reacquire extinction. Thus, the mPFC was necessary for extinction recall in males and for extinction acquisition in females (Baran et al., 2010). Others have reported no behavioral differences between males and females during fear conditioning and extinction (Milad et al., 2009a). Despite a lack of observed sex differences in performance, males and females may engage different brain regions to execute the same behavior. In a functional magnetic resonance imaging study, men and women exhibit similar SCR during fear conditioning and extinction, but show differential activations within fear circuitry (Lebron-Milad et al., 2012). Women show greater activations than men within the amygdala as well as the dorsal and rostral anterior cingulate cortices during fear conditioning. During extinction recall, men exhibit greater activation in the rostral anterior cingulate cortex signal (Lebron-Milad et al., 2012).

The sex differences observed across the various fear paradigms may have important clinical implications, as reports of sex differences in fear conditioning and extinction in individuals with PTSD are beginning to emerge (Felmingham et al., 2010; Inslicht et al., 2013; Shvil et al., 2014). In one such study, enhanced fear conditioning, as indicated by higher conditioned SCR, was observed in women as compared to men with PTSD (Inslicht et al., 2013). With regard to fear extinction, neural activations and psychophysiological responses that were associated with poor extinction memory recall were exhibited in men with PTSD but not in women with PTSD (Shvil et al., 2014). Because of the potential clinical relevance of sex differences in fear behaviors and circuitry, it is imperative that studies take sex into account when considering the neurobiological mechanisms underlying the pathophysiology of anxiety and related disorders.

# 5 ESTROGEN

## 5.1 Role of estrogen in fear – rodents

Sex differences in fear conditioning and extinction processes may be partly explained by the influence of gonadal hormones. Sex hormones, estrogen in particular, can modulate various learning and memory processes, and the modulation of these processes, which include fear extinction learning and memory, is evident in both rodents and humans. In rodents, hormonal changes are typically experienced in relation to the estrous cycle. The rodent estrous cycle consists of 4–5 days with four distinctive phases: proestrus, estrus, metestrus, and diestrus. During the proestrus phase, estrogen and progesterone levels naturally peak. Ovarian hormone levels decrease during estrus and metestrus before rising again as the cycle progresses to late diestrus and proestrus. Unfortunately, many rodent fear extinction studies do not control for estrous phase, overlooking the potential effects that naturally varying ovarian hormone levels may be having on extinction learning and memory. Not accounting for estrous phase can mask underlying sex differences and distort the understanding of important aspects of stress- and fear-related behaviors. For instance, Milad et al. (2009a) did not observe sex differences in auditory cued fear conditioning, extinction, and recall in male and female rats until the estrous phase was considered. When the females were divided into groups based on estrous phase during extinction training, sex differences in fear expression were observed in extinction recall. The females that underwent extinction learning in the metestrus phase (low ovarian hormones) exhibited more freezing during recall compared to both females extinguished in the proestrus phase (high ovarian hormones) and male rats (Milad et al., 2009a; Rey et al., 2014). Moreover, increasing estrogen levels in low-estrogen metestrus females via exogenous administration of estrogen prior to and within 4 h after extinction training enhanced fear extinction memory and reduced freezing during recall (Zeidan et al., 2011). The facilitatory role of estrogen has also been described in the extinction of other tasks. Estrogen administration enhances extinction in a passive avoidance task in male Wistar rats (Rivas-Arancibia and Vazquez-Pereyra, 1994), as well as conditioned taste aversion in both gonadectomized males and females (Yuan and Chambers, 1999). Intrahippocampal infusions of estrogen in ovariectomized female rats also enhance contextual fear extinction (Chang et al., 2009). Together, these data demonstrate that estrogen is a potent neuromodulator of fear extinction learning and memory mechanisms. The molecular mechanisms underlying the effect of estrogen on extinction processes have been discussed in a review (Cover et al., 2014).

Interestingly, estrogen appears to also enhance fear acquisition. Ovariectomized female mice given estrogen exhibited enhanced fear conditioning (Jasnow et al., 2006; Morgan and Pfaff, 2001), which was associated with increased corticotropin-releasing hormone mRNA expression in the CeA (Jasnow et al., 2006). Similarly, fear-potentiated startle was facilitated by estrogen administration in ovariectomized female rats (Hiroi and Neumaier, 2006). In contrast, estrogen replacement in ovariectomized female rats reduced contextual fear conditioning (Gupta et al., 2001). Although some of these findings appear to challenge estrogen's enhancement of fear extinction processes, it is important to note that these studies only examined the role of estrogen in males and ovariectomized females, neither of which experiences steady variation in hormone levels. This makes it somewhat challenging to directly compare these findings with those that study naturally cycling females. Moreover, given the overall learning and memory enhancement provided by estrogen administration, findings that report estrogen-facilitated fear conditioning should not be wholly unexpected. While these findings appear counterintuitive considering estrogen's beneficial effects, they may simply be additional support for estrogen's ability to increase learning capacity. Furthermore, estrogen depletion via ovariectomy may alter receptor density and function, producing varying dose-dependent effects of estrogen replacement and making interpretation of the results difficult. This may be due to the inverted U-shaped dose effect relationship of estrogen administration, which result in some optimal doses producing protective effects while other doses that are too low or high may induce impairments (Barha et al., 2010; Inagaki et al., 2010). In our studies, we have shown that naturally cycling high- and low-estrogen females exhibit no significant differences in fear conditioning, suggesting that high estrogen at normal physiological levels enhances extinction recall but does not increase fear expression during conditioning and extinction (Milad et al., 2009a).

## 5.2 Role of estrogen in fear – humans

Similar to what is observed in female rodents, the hormonal milieu in women is constantly changing throughout their lifespan depending on reproductive stage. The first significant hormonal event that females experience after birth takes place during puberty and the onset of menses. During this time, estrogen levels are dramatically increased. Following the onset of menses, a woman's 28-day (on average) menstrual cycle fluctuates between phases of low and high ovarian hormone levels. These levels are low during the early follicular phase and high during the luteal phase. During pregnancy, estrogen and progesterone levels are high and dramatically decrease soon after parturition. As women age, estrogen levels dramatically decrease and begin to cease cycling. Interestingly, sex differences and vulnerability to anxiety disorders appear to be associated with these periods of low and fluctuating levels of estrogen. Higher prevalence of anxiety- and stress-related mental illness typically begins during puberty and lasts until menopause (Hyde et al., 2008; Kessler et al., 1993; Nolen-Hoeksema and Girgus, 1994; Silberg et al., 1999; Sonnenberg et al., 2000; Kaltiala-Heino et al., 2003; Piccinelli and Wilkinson, 2000). Moreover, menstrual cycle-related effects on cognition and emotional memory have been reported with anxiolytic effects associated with the high-estrogen luteal phase (Bayer et al., 2014; Sundström Poromaa and Gingnell, 2014). The influence of estrogen on extinction retention in women follows the same pattern as in the rodent model as well, with low-estrogen status associated with impaired extinction memory (Glover et al., 2013; Milad et al., 2010). Women that underwent extinction training during the high-estrogen luteal phase had comparable extinction recall to men, with both of these groups showing significantly better extinction memory than women trained during their low-estrogen early follicular phase (Milad et al., 2010). Compared with women with higher levels of estrogen, women with low estrogen not only exhibited more fear responses during extinction, but they also showed stronger intrusive memories (Wegerer et al., 2014). Suppression of endogenous levels of estrogen in women taking hormonal contraceptives also impairs fear extinction, an effect that was produced in female rodents as well (Graham and Milad, 2013). These alterations in extinction behavior in women taking hormonal contraceptives are associated with differences in activity of neural substrates within fear circuitry, e.g., increased differential activation to paired CS compared with unpaired CS in the amygdala, thalamus, anterior cingulate, and vmPFC during extinction (Merz et al., 2012).

## 5.3 Estrogen and fear circuitry

Neural activity within the fear network differs across phases of the menstrual cycle in women and also across estrous cycle phases in rodents (Goldstein et al., 2005; Milad et al., 2006; Protopopescu et al., 2005). In studying how estrogen influences synaptic plasticity in learning and memory, the hippocampus, a site of estrogen's local synthesis, has been a prominent focus (Hojo et al., 2008). However, the estrogen-induced facilitation of fear extinction in female rodents is associated with enhanced neuronal activity within the IL and evident specifically during the retrieval of the extinction memory one day after training (Zeidan et al., 2011). Endogenous estrogen also facilitates synaptic potentiation within the IL, possibly contributing to strengthened extinction circuitry and eliciting better extinction recall (Galvin and Ninan, 2014). Thus, it appears that estrogen modulates the consolidation of extinction memory by increasing IL activity. In women, estrogen appears to enhance the functional activation of the vmPFC, the human homolog of IL, which is a critical structure in fear extinction (Protopopescu et al., 2005; Zeidan et al., 2011). Women in the late follicular–midcycle menstrual phase (high estrogen and progesterone) not only have better extinction retention than women in the early follicular menstrual phase (low estrogen and progesterone), but they also have significantly increased activation of the vmPFC during extinction learning and recall; vmPFC activation positively correlated with estrogen levels (Zeidan et al., 2011). Interestingly, low-estrogen women with PTSD exhibited higher fear-potentiated startle during extinction compared with low-estrogen women in the trauma-exposed control group, but this differentiation was not observed in high-estrogen women (Glover et al., 2012). Together, these findings and those mentioned previously indicating that individuals with PTSD exhibit impaired extinction recall (which is associated with dysfunction in the vmPFC and possibly related to activity during low-estrogen states in women) highlight the vmPFC as a brain region that not only modulates fear, but is also affected by estrogen levels and appears to be disrupted in anxiety and stress disorders.

The nodes of the fear extinction network are also modulated by stress and influenced differently between the sexes, suggesting an interaction between stress and estrogen within these critical brain regions. These interactions should be further examined, as traumatic life experiences can lead to the development of anxiety disorders, and subsequent stressors can exacerbate preexisting symptoms (Finlay-Jones and Brown, 1981; Francis et al., 2012; McLaughlin and Hatzenbuehler, 2009; Moitra et al., 2011). Sex differences have been reported in the effects of stress exposure on fear conditioning and extinction (Farrell et al., 2013; Zorawski et al., 2005, 2006). In humans, the Trier social stress test, a psychosocial stress task, modifies conditioned responding differently in men and in women taking oral contraceptives during fear conditioning. Stressed men display SCR reduction during fear conditioning compared to unstressed men, which correlates with impaired fear responses in the anterior cingulate cortex and amygdala, whereas stressed women showed enhanced fear learning compared with unstressed women, which correlated with facilitated differential responses within these same brain regions (Merz et al., 2010; Merz et al., 2013). Fear extinction and recall were not assessed in this study. One study that did assess fear extinction found no differences. Although stress exposure (cold pressor test) following fear acquisition reduced conditioned fear memory retrieval in men compared to women, there were no differences in extinction (Bentz et al., 2013). In rodents, sex differences in both fear conditioning and extinction have been observed in response to a stressor. Chronic exposure to restraint stress impairs extinction recall in males, but it facilitates fear extinction recall and impairs fear acquisition in female rats (Baran et al., 2009). The effect of stress on fear conditioning/extinction learning appears to be mixed and may also be attributed to species differences, type and intensity of stressors, and/or fear paradigms used. Despite these discrepancies, the differential effects of stress on males and females are consistently observed. One contributing factor to these variable effects may be the lack of controlling for menstrual/estrous phase in these studies.

In fact, estrogen has been implicated to play a role in the modulation of fear responding by stress via differences observed across the menstrual cycle in naturally cycling women. A study examined the interaction of menstrual phase, stress, and fear during exposure to a psychosocial stressor prior to fear conditioning. The data demonstrated that there were no differential effects on fear acquisition, but there was enhanced extinction recall in women stressed during the high-estrogen phase of their menstrual cycles. The opposite was seen in women exposed to the stressor during their low-estrogen phase (Antov and Stockhorst, 2014). These data suggest that experiencing trauma in a low-estrogen state may contribute to resistance to extinction and greater fear recovery during extinction recall. Because of their interconnected relationship with stress and fear mechanisms, ovarian hormones such as estrogen have been implicated in the etiology of anxiety disorders in women. These data suggest that low levels or fluctuations in estrogen that occur in cycling women at the time of a traumatic event, in addition to their hormone status during exposure therapy, may help predict treatment outcomes.

The mPFC is a critical brain region in the fear extinction network that is sexually dimorphic (Baran et al., 2010; Goldstein et al., 2005; Maroun, 2013; Quirk et al., 2000), sensitive to estrogen levels (Merz et al., 2012; Zeidan et al., 2011), and responsive to stress (Arnsten, 2009; Garrett and Wellman, 2009; Maeng and Shors, 2013; Maeng et al., 2010; Shansky and Lipps, 2013; Shansky and Morrison, 2009; Cook and Wellman, 2004; De Quervain et al., 1998; Luine et al., 1994; McEwen, 2005; Watanabe et al., 1992). Stress exposure increases the number and length of apical dendritic branches in the mPFC, but this process is prevented by ovariectomy, suggesting that it may be dependent on the presence of estrogen (Garrett and Wellman, 2009). Exposure to acute stress disrupts classical eyeblink conditioning in females, but enhances performance in male rats (Shors et al., 1998). These effects of stress on associative learning are dependent on mPFC activity in low-estrogen female rats but not in male rats (Maeng et al., 2010). Moreover, differential contributions of the PL and IL subregions of the mPFC have been identified in stress and fear (Akirav and Maroun, 2007; Fenton et al., 2014; Laurent and Westbrook, 2009; Maeng and Shors, 2013). Inactivation of the PL during acute forced swim stress exposure prevents the subsequent learning impairment, whereas the inactivation of the IL does not (Maeng and Shors, 2013). This suggests that the PL is engaged by stress to induce the negative effect of stress on associative learning. This is consistent with the lasting PL activity observed in females that express high contextual fear during extinction and recall, although estrous phase was not accounted for (Fenton et al., 2014). Together, these data suggest that the PL may be a brain target that is especially sensitive to stress and involved in impairments in fear extinction in female rodents. The IL–BLA pathway, on the other hand, has been indicated as being sensitive to both stress and estrogen, with chronic restraint stress inducing increases in dendritic branching in IL neurons that project to the BLA in estrogen-treated ovariectomized females (Shansky et al., 2010). This estrogen-enhanced IL plasticity may suggest increased pathway connectivity or functioning. Although fear learning and extinction were not assessed in this particular study, this effect on IL-BLA connectivity may be involved in the stress-enhanced extinction recall that was observed in women who experienced stress during their high-estrogen phase (Antov and Stockhorst, 2014).

The robust connections between the PFC and the amygdala have been studied extensively in stress and fear modulation in males (Garcia et al., 1999; Laurent and Westbrook, 2008; Maroun and Richter-Levin, 2003; Quirk et al., 2003; Sotres-Bayon et al., 2004). For instance, acute inescapable forced swim stress induces dendritic retraction in IL neurons and impairs fear extinction, but it does not impair fear conditioning in male mice (Izquierdo et al., 2006). The structure and function of the amygdala are also altered by stress (LeDoux, 2000; McGaugh, 2002). Acute immobilization stress, as well as a single administration of glucocorticoids, induces neuronal hypertrophy within the BLA and heightened anxiety 12 days after treatment (Kim et al., 2014; Mitra and Sapolsky, 2008; Mitra et al., 2005). However, the mPFC–BLA circuit may be especially critical for females (Maeng et al., 2010; Shansky et al., 2010). The amygdala contains estrogen receptors (Jasnow et al., 2006; Shughrue et al., 1998), and estrogen infusion influences amygdala function in fear and emotional responses (Frye and Walf, 2004). Disconnecting the mPFC and amygdala via contralateral lesions of these structures prevents stress-induced impairment in eyeblink conditioning, indicating that the mPFC communicates with the amygdala to modify learning after stress in females (Maeng et al., 2010). These data suggest that estrogen may be altering structure and function of the mPFC (via connections to the amygdala specifically in females) to modulate its role in stress and fear circuitry and the interactions between both systems. In humans, changes in stress sensitivity also manifest in altered neural activity. During the low-estrogen phase, healthy women presented with negative valence/high arousal images showed increased activation of brain regions within stress response circuitry, i.e., PFC, amygdala, hippocampus, paraventricular nucleus, and brainstem, compared with women in their high-estrogen phase (Goldstein et al., 2005). Together, these findings suggest that fluctuations in ovarian hormone levels may interact with the effects of stress in female fear circuitry, and that the alterations in neural structure and function can impact the ability to regulate fear, making females more susceptible to developing anxiety and stress-related mental disorders compared with males.

## 5.4 Estrogen and treatment for anxiety disorders

Understanding the mechanisms underlying estrogen modulation of fear processes can significantly impact future treatment of anxiety disorders. More sex-specific, hormone-sensitive treatments may not only minimize the sex bias in prevalence of these psychopathologies, but they could also: (1) eliminate or at least reduce symptom severity to levels that allow proper life functioning, and (2) increase the duration of improved treatment outcomes to make them longer lasting or permanent. More studies are necessary to establish where, how, and when estrogen can modify neural function to elicit its facilitation of extinction memory recall. Antidepressants are commonly used to treat anxiety disorders, with selective serotonin reuptake inhibitors (SSRIs) being the most commonly used. Given that there are known interactions between estrogen and serotonin and that estrogen has influences on fear processes, it may be important to increase our understanding of these interactions for administering pharmacological treatments and cognitive behavioral therapies. Sex differences have been reported in the response to antidepressants, including their influence on stress, associative learning, fear extinction, and pharmacokinetics (Dalla et al., 2010; Damoiseaux et al., 2014; Keers and Aitchison, 2010; Kokras et al., 2011; Lebrón-Milad et al., 2013; Leuner et al., 2004). Interestingly, some of these effects have been associated with naturally cycling female sex hormones, which may be influencing the differences in therapeutic response to antidepressant pharmacotherapy. For instance, sex and hormonal status may influence the pharmacokinetics of antidepressants, with similar doses producing higher plasma concentrations of medication in women than men and higher concentrations in low endogenous estrogen women (e.g., oral contraceptive users) than high-estrogen women (e.g., during pregnancy; Frackiewicz et al., 2000). Moreover, studies suggest that females respond better to SSRIs, a response that may be attributed to the known interactions between estrogen and serotonin. Chronic fluoxetine, for example, prevented the effect of stress on learning in females, but not in males (Leuner et al., 2004). Fluoxetine also improved extinction recall in females but not in males and improved extinction learning in low-estrogen rats while having no effect on high-estrogen rats (Lebrón-Milad et al., 2013). Together, these data highlight the need to consider the impact that sex hormones may have on treatment options. There is a need for regimens that are scheduled around menstrual cycle phase, account for oral contraceptive use, and/or include the measurement of hormone levels prior to the start of treatment.

Research aimed at localizing and identifying cellular, molecular, and genetic mechanisms by which estrogen modulates fear extinction and anxiety can guide our treatment targets to ultimately improve the efficacy of clinical applications. Of growing interest, pituitary adenylate cyclase-activating polypeptide, a stress- and fear-related peptide, was found to be associated with the diagnosis and symptoms of PTSD in women. This peptide appears to be modulated in the bed nucleus of the stria terminalis by estrogen administration in ovariectomized female rats (Ressler et al., 2011). Further advances could enable us to use the extinction-strengthening properties of estrogen (perhaps targeting critical regions such as the mPFC) without the harmful side effects that typically accompany it to enhance the benefits of pharmacotherapy and/or prolonged exposure therapy. To this end, we must improve our understanding of the complex processes that underlie stress and fear responses and interact with sex hormones that contribute to the etiology of anxiety disorders.

# 6 OTHER GONADAL HORMONES

## 6.1 Progesterone

Despite the plethora of literature on the role of estrogen in stress and fear learning and memory, progesterone is also an important female sex hormone involved in the menstrual/estrous cycle and pregnancy. High progesterone levels have been associated with enhanced memory for emotionally arousing stimuli in women (Ertman et al., 2011). Endogenous progesterone levels also interact with stress exposure to regulate emotional memory. In a free recall test, women in a high progesterone state of their menstrual cycle had increased stress hormone cortisol levels and better memory recall for negative images that were paired with the cold pressor stress task (Felmingham et al., 2010). Progesterone and estrogen have been reported to produce effects on arousal circuitry that interact or even oppose each other. In female rats, progesterone has been shown to prevent estrogen-enhanced performance in a spatial memory task (Bimonte-Nelson et al., 2006). Increased amygdala activity during presentation of negative images was observed in women in the luteal phase when both estrogen and progesterone levels are high, despite previous reports of reduced activity in the amygdala during high-estrogen states (Andreano and Cahill, 2010; Ferree et al., 2011).

Progesterone may also influence estrogen's role in extinction memory consolidation. Exogenous administration of progesterone facilitates extinction recall in female rats, similar to the effect of estrogen administration (Milad et al., 2009a). However, progesterone did not have this effect in women (Milad et al., 2010; Wegerer et al., 2014). This may be due to species differences; however, progesterone has been shown to suppress hypothalamic–pituitary–adrenal axis activity via conversion to its metabolite allopregnanolone (Biggio et al., 2014). In fact, a number of studies have reported anxiolytic properties of allopregnanolone (Pibiri et al., 2008; Toufexis et al., 2004). Human imaging studies have shown that allopregnanolone is associated with reduced amygdala responsivity to aversive stimuli, further supporting the anxiolytic role of this hormone (Sripada et al., 2013, 2014). Allopregnanolone modulates $GABA_A$ receptor activity and is of clinical relevance; women with PTSD have altered $GABA_A$ receptor sensitivity as well as reduced cerebrospinal fluid levels of allopregnanolone (Möller et al., 2014; Rasmusson et al., 2006). Administration of ganaxolone, a synthetic analog of allopregnanolone, was shown to reduce anxiety-like behavior in the elevated plus maze, reduce high fear expression during contextual fear conditioning, and enhance fear extinction retention in mice (Pinna and Rasmusson, 2014). Given the evidence that progesterone and its neuroactive steroid metabolite allopregnanolone are involved in emotional memory formation and mood disorders (Bäckström et al., 2014) and have significant interactions with estrogen, it is important to further examine their roles in fear extinction mechanisms as potential therapeutic agents for anxiety disorders.

### 6.2 Testosterone

Testosterone and its metabolites have also been shown to possess anxiolytic properties, reducing anxiety behaviors and enhancing cognition in male rodents (Frye et al., 2008; Hodosy et al., 2012; McDermott et al., 2012). The role of testosterone in fear conditioning and extinction is not yet clear. Some studies demonstrate that testosterone plays no role in contextual fear conditioning (Anagnostaras et al., 1998), while others suggest a role of testosterone in cued fear conditioning (Chen et al., 2014). Testosterone also has a relationship with estrogen; it is converted to estrogen in the brain via the enzyme aromatase. Thus, aromatase inhibitors, such as fadrozole, prevent estrogen synthesis, without affecting testosterone production. Interestingly, we have found that administration of fadrozole prior to extinction training impairs fear extinction recall in male rats, which can be rescued by estrogen administration (Graham and Milad, 2014). As the production of estrogen depends on the presence of testosterone in males, it may indirectly modulate fear extinction processes via this conversion. Extinction learning and extinction memory recall have been shown to be better in men with elevated testosterone to cortisol ratios (Pace-Schott et al., 2013), further implicating this hormone in fear extinction.

Men with low testosterone conditions have an increased risk for developing anxiety- and stress-related disorders (DiBlasio et al., 2008; Shores et al., 2004; Zarrouf et al., 2009). This can be reversed with testosterone administration, improving effect and reducing anxiety and depression (Kanayama et al., 2007; Pope et al., 2003; Wang et al., 1996; Zarrouf et al., 2009). Moreover, in castrated male rodents, testosterone replacement ameliorates anxiety behaviors (Hodosy et al., 2012; Khakpai, 2014). Although there are mixed reports on the role of testosterone in PTSD, there is evidence suggesting that a single administration of testosterone in women may reduce the stress response and startle reflex, which is typically heightened in PTSD (Hermans et al., 2006, 2007). Future studies will be necessary to examine the influence of testosterone on the mechanisms mediating fear extinction and its interactions with estrogen and other sex hormones.

## 7 CONCLUSIONS

Studies on stress, fear, and sex hormones indicate that heightened sensitivity to stress and impaired extinction memory consolidation are associated with low-estrogen states in females. This may be related to vulnerability to psychopathology. Stressful life experiences can disrupt fear extinction, a behavioral process that models the psychopathology of PTSD and anxiety disorders. This negative effect of stress can be further amplified during low levels of, or fluctuations in, estrogen. Because stress is a major contributor to the development of anxiety disorders, stress and fear mechanisms should be studied together and not in isolation.

More studies are needed to understand the roles and interactions these hormones may have with the effects of estrogen in fear extinction memory consolidation. In particular, the role of progesterone in fear extinction appears to be especially critical to investigate because of its cycle-dependent relationship with (and potentially opposing effects on) estrogen. Perhaps the ratio of estrogen to progesterone levels during these different phases may be critical in responses to stress and fear. It is important to emphasize that although low estrogen levels are linked with poor extinction memory recall, the elevated risk for anxiety may be more dependent on fluctuations and less on the absolute "low" versus "high" levels of estrogen. Studies examining ovariectomized females may provide some initial information about the role of estrogen on the specific behavior being tested; however, conducting experiments using naturally cycling females may provide insight on the many interactions that take place in normal animals and thus offer more clinical relevance to the affected population, who are typically of child-bearing age. Moreover, the effects of estrogen can be dose-dependent and follow an inverted U-shape in their interactions with stress that affect fear circuitry and mechanisms to influence subsequent learning and memory processes; estrogen levels that are too low or too high can have negative consequences. This may explain some of the inconsistencies in the effects of estrogen in the literature, and should also illustrate why it is imperative to study natural fluctuations, as they can determine whether estrogen treatment will have beneficial or harmful effects. These data describe the roles of sex hormones, especially estrogen, and the influence of stress in fear to elucidate their roles in the increased vulnerability of females to anxiety-related disorders. Based on the clinical implications of these data, it is critical that these interactions be further examined and considered when treating men and women with fear- and anxiety-related disorders.

## ACKNOWLEDGMENTS

We would like to thank the members of the Behavioral Neuroscience Program in the Department of Psychiatry at Massachusetts General Hospital for their helpful and thoughtful comments on this manuscript. MRM is supported by 1R01MH097880-001.

## REFERENCES

Aguilar, R., Gil, L., Gray, J.A., Driscoll, P., Flint, J., Dawson, G.R., Tobeña, A.AT Fearfulness and sex in F2 Roman rats: males display more fear though both sexes share the same fearfulness traits, 2003. Physiol. Behav. 78 (4–5), 723–732.

Akirav, I., Maroun, M., 2007. The role of the medial prefrontal cortex-amygdala circuit in stress effects on the extinction of fear. Neural Plast. 2007, 30873.

Altemus, M., Sarvaiya, N., Neill Epperson, C., 2014. Sex differences in anxiety and depression clinical perspectives. Front. Neuroendocrinol. 35 (3), 320–330.

American Psychiatric Association, 2013. Diagnostic and Statistical Manual of Mental Disorders, fifth ed. Author, Washington, D.C.

Anagnostaras, S.G., Maren, S., DeCola, J.P., Lane, N.I., Gale, G.D., Schlinger, B.A., Fanselow, M.S., 1998. Testicular hormones do not regulate sexually dimorphic Pavlovian fear conditioning or perforant-path long-term potentiation in adult male rats. Behav. Brain Res 92 (1), 1–9.

Andreano, J.M., Cahill, L., 2010. Menstrual cycle modulation of medial temporal activity evoked by negative emotion. Neuroimage 53 (4), 1286–1293.

Antov, M.I., Stockhorst, U., 2014. Stress exposure prior to fear acquisition interacts with estradiol status to alter recall of fear extinction in humans. Psychoneuroendocrinology 49, 106–118.

Arnsten, A.F.T., 2009. Stress signalling pathways that impair prefrontal cortex structure and function. Nat. Rev. Neurosci. 10 (6), 410–422.

Bäckström, T., Bixo, M., Johansson, M., Nyberg, S., Ossewaarde, L., Ragagnin, G., van Wingen, G., 2014. Allopregnanolone and mood disorders. Prog. Neurobiol. 113, 88–94.

Baker-Andresen, D., Flavell, C.R., Li, X., Bredy, T.W., 2013. Activation of BDNF signaling prevents the return of fear in female mice. Learn. Memory 20 (5), 237–240.

Bakish, D., 1999. The patient with comorbid depression and anxiety: the unmet need. J. Clin. Psychiat. 60 (Suppl. 6), 20–24.

Bangasser, D.A., Shors, T.J., 2010. Critical brain circuits at the intersection between stress and learning. Neurosci. Biobehav. Rev. 34 (8), 1223–1233.

Baran, S.E., Armstrong, C.E., Niren, D.C., Hanna, J.J., Conrad, C.D., 2009. Chronic stress and sex differences on the recall of fear conditioning and extinction. Neurobiol. Learn. Mem. 91 (3), 323–332.

Baran, S.E., Armstrong, C.E., Niren, D.C., Conrad, C.D., 2010. Prefrontal cortex lesions and sex differences in fear extinction and perseveration. Learn. Memory 17 (5), 267–278.

Barha, C.K., Dalton, G.L., Galea, L.A., 2010. Low doses of 17alpha-estradiol and 17beta-estradiol facilitate, whereas higher doses of estrone and 17alpha- and 17beta-estradiol impair, contextual fear conditioning in adult female rats. Neuropsychopharmacology 35 (2), 547–559.

Bayer, J., Schultz, H., Gamer, M., Sommer, T., 2014. Menstrual-cycle dependent fluctuations in ovarian hormones affect emotional memory. Neurobiol. Learn. Mem. 110, 55–63.

Bentz, D., Michael, T., Wilhelm, F.H., Hartmann, F.R., Kunz, S., von Rohr, I.R.R., de Quervain, D.J.-F., 2013. Influence of stress on fear memory processes in an aversive differential conditioning paradigm in humans. Psychoneuroendocrinology 38 (7), 1186–1197.

Biggio, G., Pisu, M.G., Biggio, F., Serra, M., 2014. Allopregnanolone modulation of HPA axis function in the adult rat. Psychopharmacology 231 (17), 3437–3444.

Bimonte-Nelson, H.A., Francis, K.R., Umphlet, C.D., Granholm, A.-C., 2006. Progesterone reverses the spatial memory enhancements initiated by tonic and cyclic oestrogen therapy in middle-aged ovariectomized female rats. Eur. J. Neurosci. 24 (1), 229–242.

Bremner, J.D., Randall, P., Vermetten, E., Staib, L., Bronen, R.A., Mazure, C., Charney, D.S., 1997. Magnetic resonance imaging-based measurement of hippocampal volume in posttraumatic stress disorder related to childhood physical and sexual abuse – a preliminary report. Biol. Psychiat. 41 (1), 23–32.

Breslau, N., 2002. Gender differences in trauma and posttraumatic stress disorder. JGSM 5 (1), 34–40.

Breslau, N., 2009. The epidemiology of trauma, PTSD, and other posttrauma disorders. Trauma Violence Abuse 10 (3), 198–210.

Breslau, N., Davis, G.C., Andreski, P., Peterson, E.L., Schultz, L.R., 1997. Sex differences in posttraumatic stress disorder. Arch. Gen. Psychiat. 54 (11), 1044–1048.

Breslau, N., Kessler, R.C., Chilcoat, H.D., Schultz, L.R., Davis, G.C., Andreski, P., 1998. Trauma and posttraumatic stress disorder in the community: the 1996 Detroit Area Survey of Trauma. Arch. Gen. Psychiat. 55 (7), 626–632.

Campbell-Sills, L., Barlow, D.H., Brown, T.A., Hofmann, S.G., 2006. Acceptability and suppression of negative emotion in anxiety and mood disorders. Emotion 6 (4), 587–595.

Chang, Y.-J., Yang, C.-H., Liang, Y.-C., Yeh, C.-M., Huang, C.-C., Hsu, K.-S., 2009. Estrogen modulates sexually dimorphic contextual fear extinction in rats through estrogen receptor beta. Hippocampus 19 (11), 1142–1150.

Chen, L.-S., Tzeng, W.-Y., Chuang, J.-Y., Cherng, C.G., Gean, P.-W., Yu, L., 2014. Roles of testosterone and amygdaloid LTP induction in determining sex differences in fear memory magnitude. Horm. Behav. 66 (3), 498–508.
Coffey, C.E., Wilkinson, W.E., Weiner, R.D., Parashos, I.A., Djang, W.T., Webb, M.C., Spritzer, C.E., 1993. Quantitative cerebral anatomy in depression. A controlled magnetic resonance imaging study. Arch. Gen. Psychiat. 50 (1), 7–16.
Cook, S.C., Wellman, C.L., 2004. Chronic stress alters dendritic morphology in rat medial prefrontal cortex. J. Neurobiol. 60 (2), 236–248.
Corcoran, K.A., Maren, S., 2001. Hippocampal inactivation disrupts contextual retrieval of fear memory after extinction. J. Neurosci. 21 (5), 1720–1726.
Cover, K.K., Maeng, L.Y., Lebrón-Milad, K., Milad, M.R., 2014. Mechanisms of estradiol in fear circuitry: implications for sex differences in psychopathology. Transl. Psychiat. 4, e422.
Dalla, C., Pitychoutis, P.M., Kokras, N., Papadopoulou-Daifoti, Z., 2010. Sex differences in animal models of depression and antidepressant response. Basic Clin. Pharmacol. Toxicol. 106 (3), 226–233.
Damoiseaux, V.A., Proost, J.H., Jiawan, V.C.R., Melgert, B.N., 2014. Sex differences in the pharmacokinetics of antidepressants: influence of female sex hormones and oral contraceptives. Clin. Pharmacokinet. 53 (6), 509–519.
Daviu, N., Andero, R., Armario, A., Nadal, R., 2014. Sex differences in the behavioural and hypothalamic-pituitary-adrenal response to contextual fear conditioning in rats. Horm. Behav. 66 (5), 713–723.
De Quervain, D.J., Roozendaal, B., McGaugh, J.L., 1998. Stress and glucocorticoids impair retrieval of long-term spatial memory. Nature 394 (6695), 787–790.
Dias, B.G., Banerjee, S.B., Goodman, J.V., Ressler, K.J., 2013. Towards new approaches to disorders of fear and anxiety. Curr. Opin. Neurobiol. 23 (3), 346–352.
DiBlasio, C.J., Hammett, J., Malcolm, J.B., Judge, B.A., Womack, J.H., Kincade, M.C., Derweesh, I.H., 2008. Prevalence and predictive factors for the development of de novo psychiatric illness in patients receiving androgen deprivation therapy for prostate cancer. Can. J. Urol. 15 (5), 4249–4256, discussion.
Ertman, N., Andreano, J.M., Cahill, L., 2011. Progesterone at encoding predicts subsequent emotional memory. Learn. Memory 18 (12), 759–763.
Farrell, M.R., Sengelaub, D.R., Wellman, C.L., 2013. Sex differences and chronic stress effects on the neural circuitry underlying fear conditioning and extinction. Physiol. Behav. 122, 208–215.
Felmingham, K., Williams, L.M., Kemp, A.H., Liddell, B., Falconer, E., Peduto, A., Bryant, R., 2010. Neural responses to masked fear faces: sex differences and trauma exposure in posttraumatic stress disorder. J. Abnorm. Psychol. 119 (1), 241–247.
Fenton, G.E., Pollard, A.K., Halliday, D.M., Mason, R., Bredy, T.W., Stevenson, C.W., 2014. Persistent prelimbic cortex activity contributes to enhanced learned fear expression in females. Learn. Memory 21 (2), 55–60.
Ferree, N.K., Kamat, R., Cahill, L., 2011. Influences of menstrual cycle position and sex hormone levels on spontaneous intrusive recollections following emotional stimuli. Conscious. Cogn. 20 (4), 1154–1162.
Finlay-Jones, R., Brown, G.W., 1981. Types of stressful life event and the onset of anxiety and depressive disorders. Psychol. Med. 11 (04), 803–815.
Foa, E.B., 2000. Psychosocial treatment of posttraumatic stress disorder. J. Clin. Psychiat. 61 (Suppl. 5), 43–48, discussion 49–51.
Foa, E.B., 2011. Prolonged exposure therapy: past, present, and future. Depress. Anxiety 28 (12), 1043–1047.
Foa, E.B., Street, G.P., 2001. Women and traumatic events. J. Clin. Psychiat. 62 (Suppl. 17), 29–34.
Frackiewicz, E.J., Sramek, J.J., Cutler, N.R., 2000. Gender differences in depression and antidepressant pharmacokinetics and adverse events. Ann. Pharmacother. 34 (1), 80–88.
Francis, J.L., Moitra, E., Dyck, I., Keller, M.B., 2012. The impact of stressful life events on relapse of generalized anxiety disorder. Depress. Anxiety 29 (5), 386–391.
Frans, O., Rimmö, P.-A., Aberg, L., Fredrikson, M., 2005. Trauma exposure and post-traumatic stress disorder in the general population. Acta Psychiatr. Scand. 111 (4), 291–299.
Frye, C.A., Walf, A.A., 2004. Estrogen and/or progesterone administered systemically or to the amygdala can have anxiety-, fear-, and pain-reducing effects in ovariectomized rats. Behav. Neurosci. 118 (2), 306–313.
Frye, C.A., Koonce, C.J., Edinger, K.L., Osborne, D.M., Walf, A.A., 2008. Androgens with activity at estrogen receptor beta have anxiolytic and cognitive-enhancing effects in male rats and mice. Horm. Behav. 54 (5), 726–734.

Galvin, C., Ninan, I., 2014. Regulation of the mouse medial prefrontal cortical synapses by endogenous estradiol. Neuropsychopharmacology 39 (9), 2086–2094.

Garcia, R., Vouimba, R.M., Baudry, M., Thompson, R.F., 1999. The amygdala modulates prefrontal cortex activity relative to conditioned fear. Nature 402 (6759), 294–296.

Garrett, J.E., Wellman, C.L., 2009. Chronic stress effects on dendritic morphology in medial prefrontal cortex: sex differences and estrogen dependence. Neuroscience 162 (1), 195–207.

Glover, E.M., Jovanovic, T., Mercer, K.B., Kerley, K., Bradley, B., Ressler, K.J., Norrholm, S.D., 2012. Estrogen levels are associated with extinction deficits in women with posttraumatic stress disorder. Biol. Psychiat. 72 (1), 19–24.

Glover, E.M., Mercer, K.B., Norrholm, S.D., Davis, M., Duncan, E., Bradley, B., Jovanovic, T., 2013. Inhibition of fear is differentially associated with cycling estrogen levels in women. J. Psychiat. Neurosci. 38 (5), 341–348.

Goldstein, J.M., Jerram, M., Poldrack, R., Ahern, T., Kennedy, D.N., Seidman, L.J., Makris, N., 2005. Hormonal cycle modulates arousal circuitry in women using functional magnetic resonance imaging. J. Neurosci. 25 (40), 9309–9316.

Goldstein, J.M., Jerram, M., Abbs, B., Whitfield-Gabrieli, S., Makris, N., 2010. Sex differences in stress response circuitry activation dependent on female hormonal cycle. J. Neurosci. 30 (2), 431–438.

Graham, B.M., Milad, M.R., 2011. The study of fear extinction: implications for anxiety disorders. Am. J. Psychiat. 168 (12).

Graham, B.M., Milad, M.R., 2013. Blockade of estrogen by hormonal contraceptives impairs fear extinction in female rats and women. Biol. Psychiat. 73 (4), 371–378.

Graham, B.M., Milad, M.R., 2014. Inhibition of estradiol synthesis impairs fear extinction in male rats. Learn. Memory 21 (7), 347–350.

Gupta, R.R., Sen, S., Diepenhorst, L.L., Rudick, C.N., Maren, S., 2001. Estrogen modulates sexually dimorphic contextual fear conditioning and hippocampal long-term potentiation (LTP) in rats(1). Brain Res. 888 (2), 356–365.

Hermans, E.J., Putman, P., Baas, J.M., Koppeschaar, H.P., van Honk, J., 2006. A single administration of testosterone reduces fear-potentiated startle in humans. Biol. Psychiat. 59 (9), 872–874.

Hermans, E.J., Putman, P., Baas, J.M., Gecks, N.M., Kenemans, J.L., van Honk, J., 2007. Exogenous testosterone attenuates the integrated central stress response in healthy young women. Psychoneuroendocrinology 32 (8–10), 1052–1061.

Hiroi, R., Neumaier, J.F., 2006. Differential effects of ovarian steroids on anxiety versus fear as measured by open field test and fear-potentiated startle. Behav. Brain Res. 166 (1), 93–100.

Hodosy, J., Zelmanová, D., Majzúnová, M., Filová, B., Malinová, M., Ostatníková, D., Celec, P., 2012. The anxiolytic effect of testosterone in the rat is mediated via the androgen receptor. Pharmacol. Biochem. Behav. 102 (2), 191–195.

Hofmann, S.G., 2008. Cognitive processes during fear acquisition and extinction in animals and humans: implications for exposure therapy of anxiety disorders. Clin. Psychol. Rev. 28 (2), 199–210.

Hojo, Y., Murakami, G., Mukai, H., Higo, S., Hatanaka, Y., Ogiue-Ikeda, M., Kawato, S., 2008. Estrogen synthesis in the brain – role in synaptic plasticity and memory. Mol. Cell. Endocrinol. 290 (1–2), 31–43.

Holbrook, T.L., Hoyt, D.B., Stein, M.B., Sieber, W.J., 2002. Gender differences in long-term posttraumatic stress disorder outcomes after major trauma: women are at higher risk of adverse outcomes than men. J. Trauma 53 (5), 882–888.

Hyde, J.S., Mezulis, A.H., Abramson, L.Y., 2008. The ABCs of depression: integrating affective, biological, and cognitive models to explain the emergence of the gender difference in depression. Psychol. Rev. 115 (2), 291–313.

Inagaki, T., Gautreaux, C., Luine, V., 2010. Acute estrogen treatment facilitates recognition memory consolidation and alters monoamine levels in memory-related brain areas. Horm. Behav. 58 (3), 415–426.

Inslicht, S.S., Metzler, T.J., Garcia, N.M., Pineles, S.L., Milad, M.R., Orr, S.P., Neylan, T.C., 2013. Sex differences in fear conditioning in posttraumatic stress disorder. J. Psychiatr. Res. 47 (1), 64–71.

Izquierdo, A., Wellman, C.L., Holmes, A., 2006. Brief uncontrollable stress causes dendritic retraction in infralimbic cortex and resistance to fear extinction in mice. J. Neurosci. 26 (21), 5733–5738.

Jasnow, A.M., Schulkin, J., Pfaff, D.W., 2006. Estrogen facilitates fear conditioning and increases corticotropin-releasing hormone mRNA expression in the central amygdala in female mice. Horm. Behav. 49 (2), 197–205.

Kalisch, R., Korenfeld, E., Stephan, K.E., Weiskopf, N., Seymour, B., Dolan, R.J., 2006. Context-dependent human extinction memory is mediated by a ventromedial prefrontal and hippocampal network. J. Neurosci. 26 (37), 9503–9511.

Kaltiala-Heino, R., Kosunen, E., Rimpelä, M., 2003. Pubertal timing, sexual behaviour and self-reported depression in middle adolescence. J. Adolesc. 26 (5), 531–545.

Kanayama, G., Amiaz, R., Seidman, S., Pope, H.G., 2007. Testosterone supplementation for depressed men: current research and suggested treatment guidelines. Exp. Clin. Psychopharmacol. 15 (6), 529–538.

Keers, R., Aitchison, K.J., 2010. Gender differences in antidepressant drug response. Int. Rev. Psychiat. 22 (5), 485–500.

Kelly, M.M., Forsyth, J.P., 2007. Sex differences in response to an observational fear conditioning procedure. Behav. Ther. 38 (4), 340–349.

Kessler, R.C., McGonagle, K.A., Swartz, M., Blazer, D.G., Nelson, C.B., 1993. Sex and depression in the National Comorbidity Survey. I: lifetime prevalence, chronicity and recurrence. J. Affect. Disord. 29 (2–3), 85–96.

Kessler, R.C., McGonagle, K.A., Zhao, S., Nelson, C.B., Hughes, M., Eshleman, S., Kendler, K.S., 1994. Lifetime and 12-month prevalence of DSM-III-R psychiatric disorders in the United States. Results from the National Comorbidity Survey. Arch. Gen. Psychiat. 51 (1), 8–19.

Kessler, R.C., Sonnega, A., Bromet, E., Hughes, M., Nelson, C.B., 1995. Posttraumatic stress disorder in the National Comorbidity Survey. Arch. Gen. Psychiat. 52 (12), 1048–1060.

Kessler, R.C., Chiu, W.T., Demler, O., Merikangas, K.R., Walters, E.E., 2005. Prevalence, severity, and comorbidity of 12-month DSM-IV disorders in the National Comorbidity Survey Replication. Arch. Gen. Psychiat. 62 (6), 617–627.

Kessler, R.C., Chiu, W.T., Jin, R., Ruscio, A.M., Shear, K., Walters, E.E., 2006. The epidemiology of panic attacks, panic disorder, and agoraphobia in the National Comorbidity Survey Replication. Arch. Gen. Psychiat. 63 (4), 415–424.

Kessler, R.C., Aguilar-Gaxiola, S., Alonso, J., Chatterji, S., Lee, S., Ormel, J., Wang, P.S., 2009. The global burden of mental disorders: an update from the WHO World Mental Health (WMH) surveys. Epidemiol. Psychiatr. Soc. 18 (1), 23–33.

Khakpai, F., 2014. The effect of opiodergic system and testosterone on anxiety behavior in gonadectomized rats. Behav. Brain Res. 263, 9–15.

Kilpatrick, D.G., Resnick, H.S., Milanak, M.E., Miller, M.W., Keyes, K.M., Friedman, M.J., 2013. National estimates of exposure to traumatic events and PTSD prevalence using DSM-IV and DSM-5 criteria. J. Trauma. Stress 26 (5), 537–547.

Kim, H., Yi, J.H., Choi, K., Hong, S., Shin, K.S., Kang, S.J., 2014. Regional differences in acute corticosterone-induced dendritic remodeling in the rat brain and their behavioral consequences. BMC Neurosci. 15, 65.

Kogler, L., Gur, R.C., Derntl, B., 2014. Sex differences in cognitive regulation of psychosocial achievement stress: Brain and behavior. Hum. Brain Mapp. 36 (3), 1028–1042.

Kokras, N., Dalla, C., Papadopoulou-Daifoti, Z., 2011. Sex differences in pharmacokinetics of antidepressants. Exp. Opin. Drug Metab. Toxicol. 7 (2), 213–226.

Labad, J., Menchon, J.M., Alonso, P., Segalas, C., Jimenez, S., Jaurrieta, N., Vallejo, J., 2008. Gender differences in obsessive-compulsive symptom dimensions. Depress. Anxiety 25 (10), 832–838.

Laurent, V., Westbrook, R.F., 2008. Distinct contributions of the basolateral amygdala and the medial prefrontal cortex to learning and relearning extinction of context conditioned fear. Learn. Memory 15 (9), 657–666.

Laurent, V., Westbrook, R.F., 2009. Inactivation of the infralimbic but not the prelimbic cortex impairs consolidation and retrieval of fear extinction. Learn. Memory 16 (9), 520–529.

Lebron-Milad, K., Abbs, B., Milad, M.R., Linnman, C., Rougemount-Bücking, A., Zeidan, M.A., Goldstein, J.M., 2012. Sex differences in the neurobiology of fear conditioning and extinction: a preliminary fMRI study of shared sex differences with stress-arousal circuitry. Biol. Mood. Anxiety Disord. 2 (1), 7.

Lebrón-Milad, K., Tsareva, A., Ahmed, N., Milad, M.R., 2013. Sex differences and estrous cycle in female rats interact with the effects of fluoxetine treatment on fear extinction. Behav. Brain Res. 253, 217–222.

LeDoux, J.E., 2000. Emotion circuits in the brain. Annu. Rev. Neurosci. 23, 155–184.

Leuner, B., Mendolia-Loffredo, S., Shors, T.J., 2004. Males and females respond differently to controllability and antidepressant treatment. Biol. Psychiat. 56 (12), 964–970.

Likhtik, E., Pelletier, J.G., Paz, R., Paré, D., 2005. Prefrontal control of the amygdala. J. Neurosci. 25 (32), 7429–7437.

Linnman, C., Zeidan, M.A., Furtak, S.C., Pitman, R.K., Quirk, G.J., Milad, M.R., 2012. Resting amygdala and medial prefrontal metabolism predicts functional activation of the fear extinction circuit. Am. J. Psychiat. 169 (4), 415–423.

Luine, V., Villegas, M., Martinez, C., McEwen, B.S., 1994. Repeated stress causes reversible impairments of spatial memory performance. Brain Res. 639 (1), 167–170.

Machado de Sousa, J.P., Osório, F., de, L., Jackowski, A.P., Bressan, R.A., Chagas, M.H.N., Torro-Alves, N., Hallak, J.E.C., 2014. Increased amygdalar and hippocampal volumes in young adults with social anxiety. PloS One 9 (2), e88523.

Maeng, L.Y., Shors, T.J., 2013. The stressed female brain: neuronal activity in the prelimbic but not infralimbic region of the medial prefrontal cortex suppresses learning after acute stress. Front. Neural. Circuits. 7, 198.

Maeng, L.Y., Waddell, J., Shors, T.J., 2010. The prefrontal cortex communicates with the amygdala to impair learning after acute stress in females but not in males. J. Neurosci. 30 (48), 16188–16196.

Maren, S., De Oca, B., Fanselow, M.S., 1994. Sex differences in hippocampal long-term potentiation (LTP) and Pavlovian fear conditioning in rats: positive correlation between LTP and contextual learning. Brain Res. 661 (1–2), 25–34.

Maroun, M., 2013. Medial prefrontal cortex: multiple roles in fear and extinction. Neuroscientist 19 (4), 370–383.

Maroun, M., Richter-Levin, G., 2003. Exposure to acute stress blocks the induction of long-term potentiation of the amygdala-prefrontal cortex pathway in vivo. J. Neurosci. 23 (11), 4406–4409.

McDermott, C.M., Liu, D., Schrader, L.A., 2012. Role of gonadal hormones in anxiety and fear memory formation and inhibition in male mice. Physiol. Behav. 105 (5), 1168–1174.

McEwen, B.S., 2005. Glucocorticoids, depression, and mood disorders: structural remodeling in the brain. Metabolism 54 (5 Suppl. 1), 20–23.

McGaugh, J.L., 2002. Memory consolidation and the amygdala: a systems perspective. Trends Neurosci. 25 (9), 456.

McLaughlin, K.A., Hatzenbuehler, M.L., 2009. Stressful life events, anxiety sensitivity, and internalizing symptoms in adolescents. J. Abnorm. Psychol. 118 (3), 659–669.

McLean, C.P., Foa, E.B., 2013. Dissemination and implementation of prolonged exposure therapy for posttraumatic stress disorder. J. Anxiety Disord. 27 (8), 788–792.

McLean, C.P., Asnaani, A., Litz, B.T., Hofmann, S.G., 2011. Gender differences in anxiety disorders: prevalence, course of illness, comorbidity and burden of illness. J. Psychiatr. Res. 45 (8), 1027–1035.

Mennin, D.S., Heimberg, R.G., Turk, C.L., Fresco, D.M., 2005. Preliminary evidence for an emotion dysregulation model of generalized anxiety disorder. Behav. Res. Ther. 43 (10), 1281–1310.

Merz, C.J., Tabbert, K., Schweckendiek, J., Klucken, T., Vaitl, D., Stark, R., Wolf, O.T., 2010. Investigating the impact of sex and cortisol on implicit fear conditioning with fMRI. Psychoneuroendocrinology 35 (1), 33–46.

Merz, C.J., Tabbert, K., Schweckendiek, J., Klucken, T., Vaitl, D., Stark, R., Wolf, O.T., 2012. Neuronal correlates of extinction learning are modulated by sex hormones. Soc. Cogn. Affect. Neurosci. 7 (7), 819–830.

Merz, C.J., Wolf, O.T., Schweckendiek, J., Klucken, T., Vaitl, D., Stark, R., 2013. Stress differentially affects fear conditioning in men and women. Psychoneuroendocrinology 38 (11), 2529–2541.

Milad, M.R., Quirk, G.J., 2012. Fear extinction as a model for translational neuroscience: ten years of progress. Ann. Rev. Psychol. 63, 129–151.

Milad, M.R., Goldstein, J.M., Orr, S.P., Wedig, M.M., Klibanski, A., Pitman, R.K., Rauch, S.L., 2006. Fear conditioning and extinction: influence of sex and menstrual cycle in healthy humans. Behav. Neurosci. 120 (6), 1196–1203.

Milad, M.R., Wright, C.I., Orr, S.P., Pitman, R.K., Quirk, G.J., Rauch, S.L., 2007. Recall of fear extinction in humans activates the ventromedial prefrontal cortex and hippocampus in concert. Biol. Psychiat. 62 (5), 446–454.
Milad, M.R., Igoe, S.A., Lebron-Milad, K., Novales, J.E., 2009a. Estrous cycle phase and gonadal hormones influence conditioned fear extinction. Neuroscience 164 (3), 887–895.
Milad, M.R., Pitman, R.K., Ellis, C.B., Gold, A.L., Shin, L.M., Lasko, N.B., Rauch, S.L., 2009b. Neurobiological basis of failure to recall extinction memory in posttraumatic stress disorder. Biol. Psychiat. 66 (12), 1075–1082.
Milad, M.R., Zeidan, M.A., Contero, A., Pitman, R.K., Klibanski, A., Rauch, S.L., Goldstein, J.M., 2010. The influence of gonadal hormones on conditioned fear extinction in healthy humans. Neuroscience 168 (3), 652–658.
Milad, M.R., Furtak, S.C., Greenberg, J.L., Keshaviah, A., Im, J.J., Falkenstein, M.J., Wilhelm, S., 2013. Deficits in conditioned fear extinction in obsessive-compulsive disorder and neurobiological changes in the fear circuit. JAMA Psychiat. 70 (6), 608–618.
Mitra, R., Sapolsky, R.M., 2008. Acute corticosterone treatment is sufficient to induce anxiety and amygdaloid dendritic hypertrophy. Proc. Natl. Acad. Sci. USA 105 (14), 5573–5578.
Mitra, R., Jadhav, S., McEwen, B.S., Vyas, A., Chattarji, S., 2005. Stress duration modulates the spatiotemporal patterns of spine formation in the basolateral amygdala. Proc. Natl. Acad. Sci. USA 102 (26), 9371–9376.
Moitra, E., Dyck, I., Beard, C., Bjornsson, A.S., Sibrava, N.J., Weisberg, R.B., Keller, M.B., 2011. Impact of stressful life events on the course of panic disorder in adults. J. Affect. Disord. 134 (1–3), 373–376.
Möller, A.T., Bäckström, T., Nyberg, S., Söndergaard, H.P., Helström, L., 2014. Women with PTSD have a changed sensitivity to GABA-A receptor active substances. Psychopharmacology. [Epub ahead of print]
Morgan, M.A., Pfaff, D.W., 2001. Effects of estrogen on activity and fear-related behaviors in mice. Horm. Behav. 40 (4), 472–482.
Nolen-Hoeksema, S., 1991. Responses to depression and their effects on the duration of depressive episodes. J. Abnorm. Psychol. 100 (4), 569–582.
Nolen-Hoeksema, S., Girgus, J.S., 1994. The emergence of gender differences in depression during adolescence. Psychol. Bull. 115 (3), 424–443.
Pace-Schott, E.F., Spencer, R.M.C., Vijayakumar, S., Ahmed, N.A.K., Verga, P.W., Orr, S.P., Milad, M.R., 2013. Extinction of conditioned fear is better learned and recalled in the morning than in the evening. J. Psychiatr. Res. 47 (11), 1776–1784.
Perrin, M., Vandeleur, C.L., Castelao, E., Rothen, S., Glaus, J., Vollenweider, P., Preisig, M., 2014. Determinants of the development of post-traumatic stress disorder, in the general population. Soc. Psychiat. Psychiatr. Epidemiol. 49 (3), 447–457.
Phelps, E.A., Delgado, M.R., Nearing, K.I., LeDoux, J.E., 2004. Extinction learning in humans: role of the amygdala and vmPFC. Neuron 43 (6), 897–905.
Pibiri, F., Nelson, M., Guidotti, A., Costa, E., Pinna, G., 2008. Decreased corticolimbic allopregnanolone expression during social isolation enhances contextual fear: A model relevant for posttraumatic stress disorder. Proc. Natl. Acad. Sci. USA 105 (14), 5567–5572.
Piccinelli, M., Wilkinson, G., 2000. Gender differences in depression. Br. J. Psychiat. 177, 486–492.
Pinna, G., Rasmusson, A.M., 2014. Ganaxolone improves behavioral deficits in a mouse model of post-traumatic stress disorder. Front. Cell. Neurosci. 8, 256.
Pitman, R.K., Rasmusson, A.M., Koenen, K.C., Shin, L.M., Orr, S.P., Gilbertson, M.W., Liberzon, I., 2012. Biological studies of post-traumatic stress disorder. Nat. Rev. Neurosci. 13 (11), 769–787.
Pope, H.G., Cohane, G.H., Kanayama, G., Siegel, A.J., Hudson, J.I., 2003. Testosterone gel supplementation for men with refractory depression: a randomized, placebo-controlled trial. Am. J. Psychiat. 160 (1), 105–111.
Protopopescu, X., Pan, H., Altemus, M., Tuescher, O., Polanecsky, M., McEwen, B., Stern, E., 2005. Orbitofrontal cortex activity related to emotional processing changes across the menstrual cycle. Proc. Natl. Acad. Sci. USA 102 (44), 16060–16065.
Pryce, C.R., Lehmann, J., Feldon, J., 1999. Effect of sex on fear conditioning is similar for context and discrete CS in Wistar, Lewis and Fischer rat strains. Pharmacol. Biochem. Behav. 64 (4), 753–759.
Quirk, G.J., Mueller, D., 2008. Neural mechanisms of extinction learning and retrieval. Neuropsychopharmacology 33 (1), 56–72.

Quirk, G.J., Russo, G.K., Barron, J.L., Lebron, K., 2000. The role of ventromedial prefrontal cortex in the recovery of extinguished fear. J. Neurosci. 20 (16), 6225–6231.
Quirk, G.J., Likhtik, E., Pelletier, J.G., Paré, D., 2003. Stimulation of medial prefrontal cortex decreases the responsiveness of central amygdala output neurons. J. Neurosci. 23 (25), 8800–8807.
Rasmusson, A.M., Pinna, G., Paliwal, P., Weisman, D., Gottschalk, C., Charney, D., Guidotti, A., 2006. Decreased cerebrospinal fluid allopregnanolone levels in women with posttraumatic stress disorder. Biol. Psychiat. 60 (7), 704–713.
Reed, V., Wittchen, H.U., 1998. DSM-IV panic attacks and panic disorder in a community sample of adolescents and young adults: how specific are panic attacks? J. Psychiatr. Res. 32 (6), 335–345.
Ressler, K.J., Mercer, K.B., Bradley, B., Jovanovic, T., Mahan, A., Kerley, K., May, V., 2011. Post-traumatic stress disorder is associated with PACAP and the PAC1 receptor. Nature 470 (7335), 492–497.
Rey, C.D., Lipps, J., Shansky, R.M., 2014. Dopamine D1 receptor activation rescues extinction impairments in low-estrogen female rats and induces cortical layer-specific activation changes in prefrontal-amygdala circuits. Neuropsychopharmacology 39 (5), 1282–1289.
Ribeiro, A.M., Barbosa, F.F., Godinho, M.R., Fernandes, V.S., Munguba, H., Melo, T.G., Silva, R.H., 2010. Sex differences in aversive memory in rats: possible role of extinction and reactive emotional factors. Brain Cogn. 74 (2), 145–151.
Rivas-Arancibia, S., Vazquez-Pereyra, F., 1994. Hormonal modulation of extinction responses induced by sexual steroid hormones in rats. Life Sci. 54 (21), PL363–367.
Seedat, S., Stein, D.J., Carey, P.D., 2005. Post-traumatic stress disorder in women: epidemiological and treatment issues. CNS Drugs 19 (5), 411–427.
Shansky, R.M., Lipps, J., 2013. Stress-induced cognitive dysfunction: hormone-neurotransmitter interactions in the prefrontal cortex. Front. Hum. Neurosci. 7, 123.
Shansky, R.M., Morrison, J.H., 2009. Stress-induced dendritic remodeling in the medial prefrontal cortex: effects of circuit, hormones and rest. Brain Res. 1293, 108–113.
Shansky, R.M., Hamo, C., Hof, P.R., Lou, W., McEwen, B.S., Morrison, J.H., 2010. Estrogen promotes stress sensitivity in a prefrontal cortex-amygdala pathway. Cereb. Cortex 20 (11), 2560–2567.
Shin, L.M., Rauch, S.L., Pitman, R.K., 2006. Amygdala, medial prefrontal cortex, and hippocampal function in PTSD. Ann. NY Acad. Sci. 1071, 67–79.
Shores, M.M., Sloan, K.L., Matsumoto, A.M., Moceri, V.M., Felker, B., Kivlahan, D.R., 2004. Increased incidence of diagnosed depressive illness in hypogonadal older men. Arch. Gen. Psychiat. 61 (2), 162–167.
Shors, T.J., Lewczyk, C., Pacynski, M., Mathew, P.R., Pickett, J., 1998. Stages of estrous mediate the stress-induced impairment of associative learning in the female rat. Neuroreport 9 (3), 419–423.
Shughrue, P.J., Scrimo, P.J., Merchenthaler, I., 1998. Evidence for the colocalization of estrogen receptor-beta mRNA and estrogen receptor-alpha immunoreactivity in neurons of the rat forebrain. Endocrinology 139 (12), 5267–5270.
Shvil, E., Sullivan, G.M., Schafer, S., Markowitz, J.C., Campeas, M., Wager, T.D., Neria, Y., 2014. Sex differences in extinction recall in posttraumatic stress disorder: a pilot fMRI study. Neurobiol. Learn. Mem. 113, 101–108.
Sierra-Mercado, D., Padilla-Coreano, N., Quirk, G.J., 2011. Dissociable roles of prelimbic and infralimbic cortices, ventral hippocampus, and basolateral amygdala in the expression and extinction of conditioned fear. Neuropsychopharmacology 36 (2), 529–538.
Silberg, J., Pickles, A., Rutter, M., Hewitt, J., Simonoff, E., Maes, H., Eaves, L., 1999. The influence of genetic factors and life stress on depression among adolescent girls. Arch. Gen. Psychiat. 56 (3), 225–232.
Sonnenberg, C.M., Beekman, A.T., Deeg, D.J., van Tilburg, W., 2000. Sex differences in late-life depression. Acta Psychiatr. Scand. 101 (4), 286–292.
Sotres-Bayon, F., Bush, D.E.A., LeDoux, J.E., 2004. Emotional perseveration: an update on prefrontal-amygdala interactions in fear extinction. Learn. Memory 11 (5), 525–535.
Sotres-Bayon, F., Sierra-Mercado, D., Pardilla-Delgado, E., Quirk, G.J., 2012. Gating of fear in prelimbic cortex by hippocampal and amygdala inputs. Neuron 76 (4), 804–812.
Sripada, R.K., Marx, C.E., King, A.P., Rampton, J.C., Ho, S.S., Liberzon, I., 2013. Allopregnanolone elevations following pregnenolone administration are associated with enhanced activation of emotion regulation neurocircuits. Biol. Psychiat. 73 (11), 1045–1053.

Sripada, R.K., Welsh, R.C., Marx, C.E., Liberzon, I., 2014. The neurosteroids allopregnanolone and dehydroepiandrosterone modulate resting-state amygdala connectivity. Hum. Brain Mapp. 35 (7), 3249–3261.
Sundström Poromaa, I., Gingnell, M., 2014. Menstrual cycle influence on cognitive function and emotion processing-from a reproductive perspective. Front. Neurosci. 8, 380.
Ter Horst, J.P., Carobrez, A.P., van der Mark, M.H., de Kloet, E.R., Oitzl, M.S., 2012. Sex differences in fear memory and extinction of mice with forebrain-specific disruption of the mineralocorticoid receptor. Eur. J. Neurosci. 36 (8), 3096–3102.
Tolin, D.F., Foa, E.B., 2006. Sex differences in trauma and posttraumatic stress disorder: a quantitative review of 25 years of research. Psychol. Bull. 132 (6), 959–992.
Toufexis, D.J., Davis, C., Hammond, A., Davis, M., 2004. Progesterone attenuates corticotropin-releasing factor-enhanced but not fear-potentiated startle via the activity of its neuroactive metabolite, allopregnanolone. J. Neurosci. 24 (45), 10280–10287.
Van der Veen, D.C., van Zelst, W.H., Schoevers, R.A., Comijs, H.C., Voshaar, R.C.O., 2014. Comorbid anxiety disorders in late-life depression: results of a cohort study. Int. Psychogeriatr. 1–9.
Wang, C., Alexander, G., Berman, N., Salehian, B., Davidson, T., McDonald, V., Swerdloff, R.S., 1996. Testosterone replacement therapy improves mood in hypogonadal men – a clinical research center study. J. Clin. Endocrinol. Metab. 81 (10), 3578–3583.
Watanabe, Y., Gould, E., McEwen, B.S., 1992. Stress induces atrophy of apical dendrites of hippocampal CA3 pyramidal neurons. Brain Res. 588 (2), 341–345.
Wegerer, M., Kerschbaum, H., Blechert, J., Wilhelm, F.H., 2014. Low levels of estradiol are associated with elevated conditioned responding during fear extinction and intrusive memories in daily life. Neurobiol. Learn. Mem. 116C, 145–154.
Yuan, D.L., Chambers, K.C., 1999. Estradiol accelerates extinction of a conditioned taste aversion in female and male rats. Horm. Behav. 36 (1), 1–16.
Zarrouf, F.A., Artz, S., Griffith, J., Sirbu, C., Kommor, M., 2009. Testosterone and depression: systematic review and meta-analysis. J. Psychiatr. Pract. 15 (4), 289–305.
Zeidan, M.A., Igoe, S.A., Linnman, C., Vitalo, A., Levine, J.B., Klibanski, A., Milad, M.R., 2011. Estradiol modulates medial prefrontal cortex and amygdala activity during fear extinction in women and female rats. Biol. Psychiat. 70 (10), 920–927.
Zorawski, M., Cook, C.A., Kuhn, C.M., LaBar, K.S., 2005. Sex, stress, and fear: individual differences in conditioned learning. Cogn. Affect. Behav. Neurosci. 5 (2), 191–201.
Zorawski, M., Blanding, N.Q., Kuhn, C.M., LaBar, K.S., 2006. Effects of stress and sex on acquisition and consolidation of human fear conditioning. Learn. Memory 13 (4), 441–450.

# CHAPTER 4

# Sex Differences in the Social Behavior Network and Mesolimbic Dopamine System

**Gian D. Greenberg*, Brian C. Trainor****
*Oregon Health Sciences University, Department of Behavioral Neuroscience, Portland, OR, USA
**University of California, Department of Psychology, Davis, CA, USA

## 1 INTRODUCTION TO SEX DIFFERENCES IN SOCIAL BEHAVIOR AND THE IMPORTANCE OF SEX-SPECIFIC RESPONSES IN NEURAL CIRCUITRY

Social interactions are a critical component of life for most species of vertebrates. Although there is considerable diversity in social behavior across species, certain features of social behavior are strongly conserved. This may explain why there is an evolutionarily conserved network of hypothalamic and limbic nuclei that regulate social behaviors. Based on lesion and immediate early gene expression studies, Sarah Winans Newman (1999) proposed that these nuclei form a "social behavior network" that controls social behaviors such as mating, parental behavior, and aggression. Over time this hypothesis has been supported with results from a variety of perspectives ranging from the comparative method (Goodson, 2005; O'Connell and Hofmann, 2012) to optogenetics (Lin et al., 2011). There is a growing appreciation for the importance of sex differences in social behaviors. Sex differences can range from subtle to more extreme (Shuster and Wade, 2003). In many fish, the actual act of spawning consists of motor patterns that are more similar in males and females whereas mating in rodents is mediated by motor patterns that typically are relatively unique to either males or females. The presence of sex differences in behavior has sparked an interest in identifying the underlying mechanisms. As might be expected, important neuroanatomical and neurochemical sex differences have been identified at several nodes within the social behavior network. Intriguingly, data also indicate that motivational systems, particularly the mesolimbic dopamine system, can also generate important sex differences in behavior (Becker, 2009).

Motivational systems play a key role in determining how individuals engage in social interactions. Dopamine neurons in the ventral tegmental area (VTA) project to forebrain areas such as the nucleus accumbens (NAc) and the frontal cortex. In the VTA, similar patterns of gene expression are observed in fish, amphibians, birds, and mammals suggesting that this small collection of neurons has a highly conserved function (O'Connell

*Sex Differences in the Central Nervous System.* http://dx.doi.org/10.1016/B978-0-12-802114-9.00004-4

and Hofmann, 2012). Although anatomical sexual dimorphism in the mesolimbic dopamine system is either subtle (Forlano and Woolley, 2010; Wissman et al., 2012) or absent (Campi et al., 2013), there is growing evidence for sex differences in how activity in this circuit contributes to sex differences in behavior. Furthermore, the mesolimbic dopamine system has direct and indirect connections with the social behavior network (O'Connell and Hofmann, 2011). The bed nucleus of the stria terminalis (BNST) in particular has emerged as a particularly important integrative node facilitating cross-talk between the social behavior network and mesolimbic dopamine system.

In this chapter, we review sex differences in structure and function across both the social behavior network and mesolimbic dopamine system. In some cases structures that are not obviously sexually dimorphic mediate very important sex differences in behavior. Context is also a key factor influencing behavior. Factors such as using more naturalistic behavior testing conditions can reveal key sex differences in brain function and behavior. Here, studying species with different social organizations can provide key insights into context-dependent function of brain circuits regulating social behavior. Compared with the social behavior network, the mesolimbic dopamine system is not particularly sexually dimorphic. However, at a functional level, dopaminergic signaling in this pathway is an important mediator of sex differences in behavior in both appetitive and aversive contexts. Finally, we review evidence for connectivity between the social behavior network and the mesolimbic dopamine system. The use of new optogenetic tools has provided novel insights into how connections between the BNST and VTA modulate affective states. Progress in understanding how these networks control behavior should continue to expand, increasing its focus on including both males and females in neuroscience research (Cahill, 2006; Beery and Zucker, 2011).

## 2 THE SOCIAL BEHAVIOR NETWORK AND SEX DIFFERENCES IN SOCIAL BEHAVIOR

The social behavior network consists of an interconnected network of hypothalamic and limbic nuclei that modulate social behaviors, such as reproduction, aggression, and affiliative behavior (Newman, 1999; Goodson, 2005; Goodson and Kingsbury, 2013). Most of these nuclei contain sex steroid hormone receptors such as androgen and estrogen receptors (ARs and ERs, respectively). These receptors have important effects on behavior both during development and in the adult brain (i.e., organizational versus activational effects (Pak and Handa, 2007). Steroid hormones can play a crucial role for organizing sexual dimorphisms in neuroanatomy (Arnold and Breedlove, 1985), but they can also influence more subtle sex differences in chemoarchitecture (Patisaul et al., 2003). Steroid hormones can also induce important patterns of sex- and region-specific gene expression (Xu et al., 2012). Steroid hormone levels are dynamic and are affected by interactions with the physical and social environment. Long-term changes in photoperiod (Prendergast et al., 2009) or rapid responses to social challenges (Gleason et al., 2009) have important effects on the levels of gonadal hormones that activate the social behavior network. We will discuss influences of steroid hormones within sexually dimorphic social behavior network nodes over various social behaviors, beginning with the most anatomically dimorphic regions and moving through its connections to less anatomically dimorphic structures. When discussing each region, we will emphasize the importance of connectivity between nodes that can allow for nonanatomically dimorphic subdivisions to have sex-specific effects on behavior. Finally, we will address how each node has direct or indirect connections to the mesolimbic dopamine system that regulate sex differences in motivation to engage in social behaviors.

## 2.1 Sexual dimorphism in the medial preoptic area (MPOA)

The sexually dimorphic nucleus of the preoptic area (SDN-POA) was discovered in rats (Gorski et al., 1978), and has been found to be larger in males in at least nine different species including humans (Hofman and Swaab, 1989), quail (Viglietti-Panzica et al., 1986), and the monogamous California mouse (Figure 4.1; Campi et al., 2013). The mechanisms for this sex difference are largely driven by the effects of estradiol early during the postnatal period (Gorski, 1986). In males, aromatase expressed in the SDN-POA converts circulating testosterone to estradiol, which in turn inhibits apoptosis in the SDN-POA resulting in a larger volume in the adult (Arai et al., 1996). Indeed, regulation of apoptosis during development, rather than neurogenesis, is observed in other sexually dimorphic regions of brain as well (Forger and de Vries, 2010). Effects of steroid hormones on the SDN-POA are not limited to development. Gonadectomy in the adult also reduces the size of the SDN-POA and this effect is blocked if testosterone replacement is added (Commins and Yahr, 1984a; Bloch and Gorski, 1988).

Sex differences in the MPOA are also observed outside of the SDN-POA. Across the entire MPOA, aromatase expression and aromatase activity (the rate at which androgens are converted to estrogens) is increased via androgen receptors (Roselli et al., 1997; Resko et al., 2000). Not surprisingly, aromatase activity in the MPOA is much higher in males than in females in many species (Roselli et al., 1985; Schumacher and Balthazart, 1986). In contrast, there is little evidence for sex differences in estrogen receptor expression

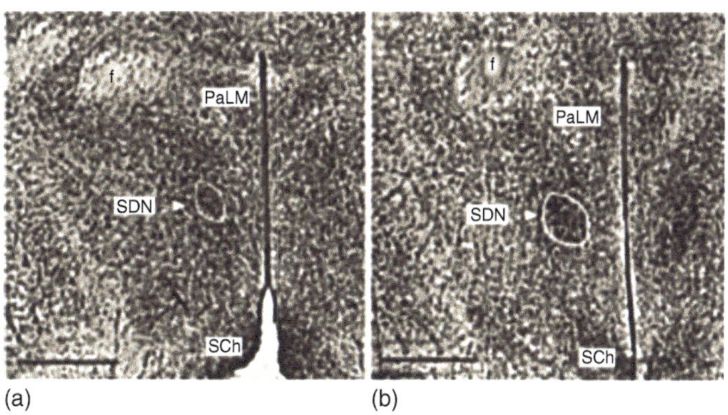


**Figure 4.1** ***Sexually dimorphic nucleus in female (a) and male (b) California mice.*** Thin solid lines indicate the boundary of the SDN-POA. Scale bars = 500 μm. *(From Campi et al. (2013).)*

in the MPOA (Cintra et al., 1986; Scott et al., 2000), although there do appear to be subtle differences in the cell types that express estrogen receptors (Herbison and Theodosis, 1992).

## 2.2 Sex differences in the MPOA on social behavior

Curiously, the region of the brain that is consistently larger in males versus females, the SDN-POA, has an uncertain role in behavior. In quail, testosterone implants placed in the preoptic area increased male sexual behavior and lesions reduced sexual behavior (Balthazart and Surlemont, 1990). However, in rodents, the SDN-MPOA in particular does not appear necessary for the performance of male sexual behavior (Hart and Leedy, 1985; Cherry et al., 1990). There is some evidence in mammals that the SDN-MPOA modulates mating preferences (reviewed by Campi et al., 2013). For example, rams that preferred to mount other rams had smaller SDN-POA volumes than rams that preferred to mount ewes (Roselli et al., 2004). Another intriguing case is the parthenogenic whiptail lizard *Cnemidophorus uniparens*. In this all-female species, some females engage in male-like courtship behavior directed at other females (Crews and Fitzgerald, 1980). The anterior hypothalamus-preoptic area (AH-POA), which includes the SDN-POA in mammals, is larger in males versus females in the sexually reproducing species *Cnemidophorus inoratus*. However, the size of the AH-POA in female *C. uniparens* is more similar to female *C. inoratus* rather than male (Grassman and Crews, 1990; Wade et al., 1993). It is possible that more subtle differences in the structure or function of the AH-POA may impact mating preferences in *C. uniparens*. However, it is clear that there is a great deal of diversity in how sexual dimorphism in the SDN impacts behavior.

In males, inactivation of the MPOA induces major deficits in male reproductive behavior (Hull and Dominguez, 2007). Initial lesion studies found that removing the entire MPOA inhibited copulatory behavior in male rats (Heimer and Larsson, 1967), while electrical stimulation of the MPOA facilitates mating behavior (Malsbury, 1971; Van Dis and Larsson, 1971). Indeed, the MPOA has been found to be a critical region for the modulation of sexual behavior in males of almost all mammalian species studied (Hart and Leedy, 1985), including mice (Bean et al., 1981), ferrets (Panzica et al., 1995), guinea pigs (Phoenix, 1961), and gerbils (Commins and Yahr, 1984b). Both androgens and estrogens act in the MPOA to facilitate male sexual behavior (Cornil et al., 2012). In adult rats aromatase activity in these nuclei is elevated in males compared with females (Roselli et al., 1996), and estrogens facilitate male sexual behavior (Scordalakes et al., 2002). Interestingly, careful study has shown that the motivation to engage in sexual behavior can be separated from mating itself, and that different neural circuits are involved. Although inactivation of the MPOA greatly reduces male copulatory behavior, sexual motivation appears to be less severely impacted (Everitt, 1990, but see Hull and Dominguez, 2007).

In females, the role of the MPOA is less clear with some studies reporting no effect of MPOA inactivation on lordosis (Malsbury et al., 1977; Gray et al., 1978) and other studies reporting evidence for an inhibitory effect of the MPOA on lordosis (Powers and Valenstein, 1972; Takeo et al., 1993). It appears that the MPOA has more important effects on proceptive behaviors that occur before mating, and can only be observed if investigators use more complex testing arenas (Pfaus, 1999). Under these conditions, lesions of the MPOA reduce hopping and darting (Whitney, 1986). This effect can also be induced by increased D1 dopamine receptor signaling or reduced D2 receptor signaling in the MPOA (Graham and Pfaus, 2010). Whereas the MPOA appears to have sex-specific roles for mating behaviors, its role appears to be more similar for males and females in the context of parental behaviors.

In many species reproduction is associated with an extended period of parental care, and the MPOA has been shown to play a very important role in this behavior. In most species of mammals, only females provide parental care. Thus, parental care frequently represents an extreme case of a sexually differentiated behavior. The MPOA is one of the most important nodes in the social behavior network controlling parental behavior (Stolzenberg and Numan, 2011). Lesions of the MPOA induce severe disruptions in maternal behavior (Numan, 1974, 1988). The MPOA also is an important mediator of parental motivation, as MPOA lesions will decrease operant bar pressing to gain access to pups (Lee et al., 1999). Intriguingly, the MPOA is the only brain region that shows a greater increase in c-fos immunoreactivity to pup-associated cues compared with cues associated with cocaine (Mattson and Morrell, 2005). Both hormones and experience have important effects on the MPOA.

Estradiol acts directly in the MPOA to prime virgin female rats to exhibit maternal care (Numan et al., 1977). Maternal behavior can also be induced by exposure to pups (Fleming and Rosenblatt, 1974). Virgin female mice exposed to pups have fewer neurons expressing c-fos in the MPOA compared with females with prior pup experience (Tsuneoka et al., 2013). However, virgin females exposed to pups over a 4-day period show significant increases in cyclic adenosine monophosphate (cAMP) response element-binding protein (CREB), ERβ, and oxytocin within the MPOA (Stolzenberg et al., 2012, 2014). Intriguingly, these changes can be mimicked in half the time if females are cotreated with a histone deactylase inhibitor, sodium butyrate. These results suggest that experience with pups induces epigenetic changes in the MPOA that facilitate maternal behavior (Dobolyi et al., 2014). Although almost all of our knowledge of the mechanisms controlling parental care is derived from females, a handful of studies from monogamous species have provided insights into neural mechanisms of paternal behavior. Overall it appears that some mechanisms of parental care are similar in males and females. For example, electrolytic lesions of the MPOA reduce male parental behavior in the monogamous California mouse (Lee and Brown, 2002, 2007). In California mice, estradiol promotes male parental care (Trainor and Marler, 2002) and California mouse fathers have more aromatase activity than virgin males (Trainor et al., 2003). Intriguingly, c-fos expression in the MPOA following a resident-intruder aggression test is significantly higher in parental males compared with virgin males (Trainor et al., 2008a). In general, the MPOA is considered less important for male-male aggression (Newman, 1999; Delville et al., 2000). However, the MPOA appears to play a much more important role in facilitating maternal aggression (Gammie, 2005; Arrati et al., 2006). Thus, in contrast to sexual behavior, the MPOA appears to have more similar effects on parental behavior in males and females.

## 2.3 Anatomical dimorphism in the medial amygdala

The medial amygdala (MeA) is an important node for processing social stimuli, especially social odor cues. Like the SDN-POA, the MeA is sexually dimorphic at an anatomical level. The volume of the posterior dorsal subdivision of the medial amygdala (MeAPD) in rats is larger in males than females (Hines et al., 1992), and this difference is associated with greater neuronal soma size in males than in females (Cooke et al., 1999; Morris et al., 2008). The difference in MeAPD size has been attributed to high concentrations of ARs and ERs in MeA neurons (Sheridan, 1979; Simerly et al., 1990), because the enlarged volumes and cell sizes are absent in androgen-insensitive rats (Cooke et al., 1999). The developmental mechanisms controlling sexual dimorphism in the MeA, such as cell size and chemoarchitecture, have interesting parallels with the MPOA. Like the MPOA, hormones during the perinatal period have crucial organizational influences on synaptic connectivity and cellular differentiation in the MeAPD (Cooke and Woolley, 2005). However, the maintenance of this sexual dimorphism appears to be more complicated. Interestingly, in adults, sex differences in MeAPD volume and neuronal soma size appear to be more dependent on circulating androgens (Cooke et al., 1999), whereas sex differences in total neuron number is independent of circulating androgens (Morris et al., 2008). Similar to the MPOA, more subtle sex differences derived from sex steroids are observed in MeA chemoarchitecture. The MeAPD contains some of the densest concentrations of the neuropeptide cholecystokinin in the male rat forebrain, and castration reduces cholecystokinin immunoreactivity (Simerly and Swanson, 1987). Intriguingly, sex differences in MeA chemoarchitecture are also observed in eusocial animals, such as the naked mole-rat. Males have more AR-positive nuclei than females, even though there are no sex differences in region size or cell numbers (Holmes et al., 2008). The importance of these differences in AR expression in MeA for behavior is still being studied.

Sex differences in chemoarchitecture of the MeA have been linked to sex differences in social behavior, particularly mating and aggression. Male mice have more aromatase-positive cells in the MeA and BNST compared with female mice, and the fibers from these neurons are more dense in males compared to females (Wu et al., 2009). In addition, both mating and the presentation of conspecific olfactory stimuli increase fos expression in AR-immunoreactive (ir) cells in the MeA of male Syrian and golden hamsters, respectively (Wood and Newman, 1993; Blake and Meredith, 2011). In other species, there is evidence that this relationship is not sex-specific. AR expression in the avian medial amygdala is positively associated with aggressive behavior in male and female juncos (Rosvall et al., 2012). However, it is important to note that within the MeA, there are discrete subpopulations of neurons that respond to different components of sexual behavior (Heeb and Yahr, 1996; Coolen et al., 1997; Kollack-Walker and Newman, 1997), and androgens and estrogens appear to work differently in different subdivisions of MeA.

The effect of prenatal androgen exposure on amygdala function in humans was tested using positron emission tomography of women diagnosed with congenital adrenal hyperplasia (CAH). Women diagnosed with CAH are exposed to elevated androgen levels during fetal development (Merke and Bornstein, 2005). This condition is typically diagnosed at birth and treatment normalizes androgen levels. However, girls with CAH typically engage in more male-typical play than their unaffected sister (Berenbaum, 1999). Although functional connectivity in amygdala and hypothalamus differed between unaffected men and women, women diagnosed with CAH did not differ from unaffected women (Ciumas et al., 2009). Although it is possible that prenatal androgen exposure may not affect brain function to the same extent in humans versus rodents, an alternative possibility is that androgens may need to be present in the adult to induce sex differences in connectivity.

## 2.4 Sex differences in effects of MeA on social behavior

As would be expected based on neuroanatomical sex differences in the MeA, there is also evidence for sex differences in the effects of the MeA on behavior. Infusion of testosterone into the MeAPD can restore sexual behavior in castrated male rats (Wood and Newman, 1995). In contrast, estradiol, but not dihydrotestosterone, implants placed in the MeAPD increased sexual behavior (Wood, 1996). Many neurons in the MeAPD project primarily to posterior-medial subregions of the BNST (Canteras et al., 1995; Coolen and Wood, 1998), including the principal nucleus of the BNST. In contrast, neurons in more anterior subregions of the MeA project to more lateral subregions of the posterior BNST. This differential circuitry may contribute to the different effects of the anterior MeA on behavior. Selective lesions to anterior MeA abolish both appetitive and consummatory components of male sexual behavior in hamsters (Lehman et al., 1980; Lehman and Winans, 1982). In contrast, male hamsters with lesions including the MeAPD still display mating behavior, but have deficits in their timing of appetitive sexual behavior (i.e., anogenital sniffing) (Lehman et al., 1983). In female rodents, the MeA appears to have a less important role in mediating sexual behaviors. Male but not female hamsters exposed to opposite sex olfactory cues had increased fos expression specifically in the MeAPD (Newman, 1999). Consistent with this observation MeA lesions did not block lordosis behavior, although it did reduce preferences for odors of intact males versus castrated males (Kondo and Sakuma, 2005). Studies using functional magnetic resonance imaging on humans are largely consistent with these data. Sexually arousing images increased activity in the left amygdala to a greater extent in men versus women (Hamann et al., 2004).

The MeA also modulates aggressive behavior, as lesions reduce aggression in male rodents (Shibata et al., 1982; Wang et al., 2013). Interestingly the impact of the MeA on aggression is stronger in animals with prior experience winning aggressive encounters (Vochteloo and Koolhaas, 1987). Winning aggressive encounters increases androgen receptor expression in the MeA of male California mice (Fuxjager et al., 2010). The experience of winning also increases the probability that an individual will win in the future, independently of intrinsic competitive ability (Oyegible and Marler, 2005; Hsu et al., 2007). Long-lasting changes in androgen sensitivity in the MeA may contribute to this effect. Future studies should further characterize which cell types in the MeA express ARs. For example, selective activation of gamma-aminobutyric acid (GABA) neurons within the MeA was found to increase aggressive behavior whereas activation of glutamate neurons in the MeA had the opposite effect (Hong et al., 2014).

Less is known about the effects of the MeA on female aggression, largely because female-female aggression levels in most rodent species are relatively low. However, Syrian hamsters are solitary (Gattermann et al., 2001) and females are extremely aggressive toward both males and females (Wise, 1974). Similarly, in the monogamous California mouse, females defend territories with males (Ribble and Salvioni, 1990) and are aggressive toward other females (Davis and Marler, 2003). In Syrian hamsters MeA lesions reduced female-female aggression (Takahashi and Gladstone, 1988) and female-female aggressive encounters in California mice induce increased c-fos (Davis and Marler, 2004) and phosphorylated extracellular signal regulated kinase (pERK) in the MeA (Silva et al., 2010). Strong increases in immediate early gene expression such as c-fos (Hasen and Gammie, 2005) and Egr-1 (Hasen and Gammie, 2006) are observed in the MeA following maternal aggression as well. In general, there has been little investigation of the neuroendocrine mechanisms in the MeA controlling aggression in males and females. In males, estrogens generally increase aggression (Laredo et al., 2014), and ERα cell counts in the MeA are positively correlated with aggression in males (Trainor et al., 2006). In contrast, experimental knockdown of ERα in the MeA of female rats increased agonistic behaviors directed toward juveniles (Spiteri et al., 2010). However, without a direct comparison to males it is unclear whether the inhibitor effects of ERα on aggression are context-dependent (e.g., aggression toward juveniles) or a genuine sex difference in how ERα in the MEA regulates aggression.

## 2.5 Sexual dimorphism in the bed nucleus of the stria terminalis

The BNST is a highly complex nucleus extending just posterior from the NAc and extending through to the hypothalamus (Moga et al., 1989). The posterior subregions of the BNST have been studied more intensely as they exhibit strong sexual dimorphism on both anatomical and neurochemical levels. Posterior divisions of BNST, including the principal nucleus of the BNST, are larger in males than females in a diverse group of species (Hines et al., 1985; del Abril et al., 1987; Allen and Gorski 1990; Campi et al., 2013). As in the MPOA, inhibition of apoptosis through aromatization of androgens during fetal development is an important process contributing to the masculinization of posterior BNST subregions (Chung et al., 2000). The posterior subdivisions of the BNST have dense concentrations of gonadal steroid receptors (Commins and Yahr, 1985; Simerly et al., 1990; Chen and Tu, 1992; Shah et al., 2004). Neuroanatomical tracers have revealed substantial sex differences in projections originating from the oval nucleus of the BNST (Gu et al., 2003). Most terminal fields had stronger projection in males versus females, and these differences were particularly strong in the MPOA and MEA.

In contrast to the posterior BNST, anterior subregions of the BNST (which includes the anteromedial BNST, BNSTam) generally lack anatomical sexual dimorphism (del Abril et al., 1987; Campi et al., 2013). Likewise, the concentration of estrogen and androgen receptors is quite low compared to posterior subregions of the BNST. Despite the lack of obvious sex differences in anatomy, evidence suggests that anterior subregions of the BNST respond to stressors in a sex-specific manner.

## 2.6 Sex differences in effects of anterior BNST on social behavior

The anterior BNST is highly responsive to social stress, as increased c-fos is observed in this area in male hamsters that lose aggressive encounters (Kollack-Walker et al., 1997). A series of experiments on California mice determined that the anterior BNST has an important role in mediating sex-specific responses to social defeat stress (Greenberg et al., 2014b). As the California mouse is one of the few species in which females are aggressive toward other females, this provides a unique opportunity to study both males and females exposed to an equivalent intensity of defeat stress (Trainor et al., 2013). The long-term effects of defeat stress are sex-specific and generally consistent with reactive and proactive coping strategies described by Koolhaas and colleagues (Koolhaas et al., 1999). In females, the long-term effects of social defeat are more consistent with reactive coping strategies such as social withdrawal (Greenberg et al., 2014b), reduced aggression (Steinman et al., 2015), and behavioral flexibility (Laredo et al., 2015). In contrast, stressed males adopt more proactive coping strategies such as social approach, aggressive behavior, and behavioral inflexibility. Sex differences in social withdrawal are mediated in part by sex differences in neurotrophin function in anterior subregions of the BNST.

First, immunoblot studies demonstrated that defeat stress increased protein expression of brain derived neurotrophic factor (BDNF) in anterior, but not posterior, micropunch samples of the BNST (Figure 4.2a; Greenberg et al., 2014b). Although BDNF is often considered to exert antidepressant effects (Duman and Monteggia, 2006), these effects appear to be mediated by tyrosine-related kinase B receptor (TrkB) activation in

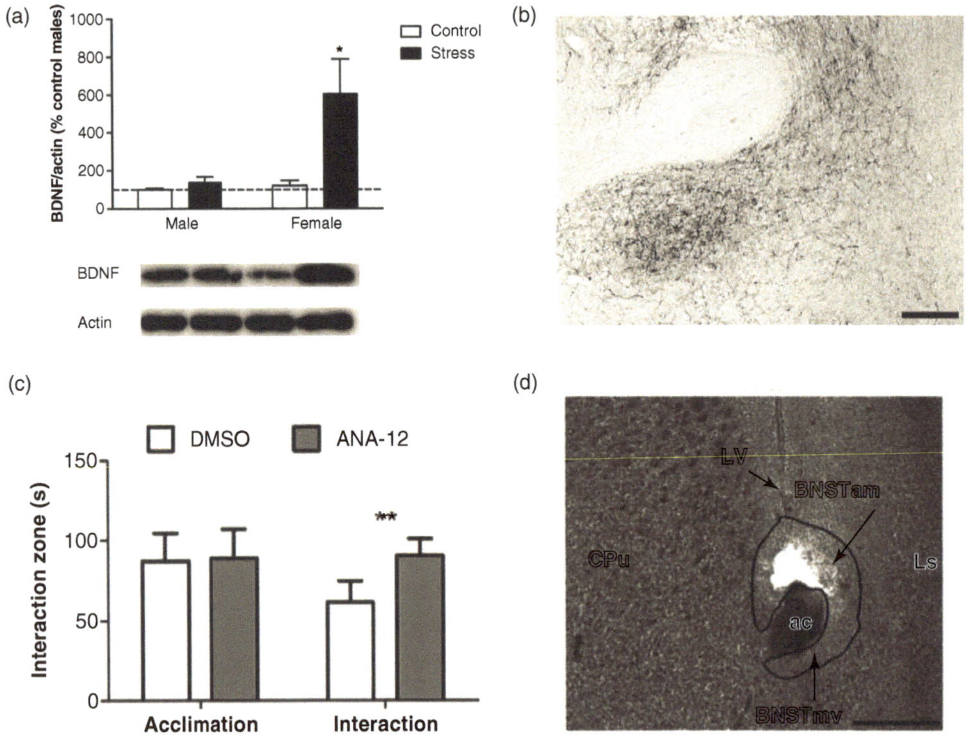


**Figure 4.2** ***Brain derived neurotrophic factor (BDNF) mediates sex differences in stress-induced social withdrawal.*** In punch samples of the anterior BNST, defeat stress increased BDNF protein expression in females but not males (a). Immunohistochemistry showed increased BDNF immunoreactivity in ventromedial subregions of the anterior BNST (b, scale bar = 200 μm). Infusion of a TrkB antagonist into the anterior BNST blocked stress-induced social withdrawal in females (c). Photomicrograph showing needle track in anterior BNST (d, scale bar = 500 μm). Structures caudate-putamen (CPu), anterior commissure (ac), anterior-medial bed nucleus of the stria terminalis (BNSTam), ventromedial bed nucleus of the stria terminalis (BNSTmv), and lateral septum (LS) shown for reference. $^*p < 0.05$, $^{**}p < 0.01$ versus control. *(From Greenberg et al. (2014b).)*

the hippocampus (Malberg and Duman, 2003). Successful antidepressant treatment increases expression of BDNF in the hippocampus (Nibuya et al., 1995; Autry et al., 2011) and bilateral infusion of mature BDNF into the dentate gyrus has antidepressant effects (Shirayama et al., 2002). However, BDNF action in the NAc had prodepressive effects (Krishnan, 2007). Results from female California mice indicate that BDNF acting in the anterior BNST also exerts prodepressive effects.

Immunohistochemistry was used to more accurately identify specific subregions of BNST expressing BDNF protein in California mice exposed to defeat stress (Figure 4.2b). Females but not males exposed to defeat had increased BDNF-ir within the ventral portion of the anterior BNST. Analyses of BDNF in the adjacent NAc revealed no sex differences or effects of stress on expression. Increased BDNF expression was reversed with chronic administration of a low dose of the selective serotonin reuptake inhibitor sertraline, which also blocked defeat-induced social withdrawal. Anterior-ventral BNST neurons receive serotonergic input (Phelix et al., 1992), which is one possible mechanism that could directly mediate the inhibitory effects of sertraline on BDNF. While the mechanism through which sertraline normalized BDNF levels is unresolved, another study demonstrated that the depressive-like effects of BDNF in the anterior BNST on social interaction are mediated by activation of TrkB.

Site-specific infusions of a TrkB selective antagonist into the anterior BNST blocked the expression of social aversion in stressed female mice (Figure 4.2c) while infusions of the TrkB antagonist that missed the anterior BNST had no effect on behavior (Greenberg et al., 2014b). Furthermore, TrkB antagonist infusions into the anterior BNST had no effect on behavior in females naïve to defeat. Intriguingly, the Val66Met point mutation in the human *Bdnf* gene impairs release of mature BDNF in the brain (Egan et al., 2003; Chen et al., 2006), yet this mutation is linked to reduced risk of depression (Frustaci et al., 2008; Matsuo et al., 2009; Grabe et al., 2012). This suggests that BDNF action in the NAc and BNST may have an underappreciated role in governing the effects of stress on psychological health.

To this point we have considered the impact of experience in the adult on BDNF expression. However, developmental experience also exerts critical effects on brain function and behavior (Lu et al., 2005). The developing brain is considerably plastic and influenced by environmental experience, and, interestingly, the social environment also has important effects on BDNF expression in the BNST. In prairie voles, juveniles that are raised with younger siblings have increased anxiety-like behavior as adults (Greenberg et al., 2012). Intriguingly, this increased anxiety-like behavior corresponds with increased BDNF expression in the BNST. In this study, anterior and posterior subregions of BNST were not examined individually. However, it seems likely that effects of early life experience on behavior and BDNF are mediated by anterior subregions of BNST.

## 2.7 Sex differences in effects of posterior BNST on social behavior

Some subcircuits from the posterior BNST are responsive to both sexual and aggressive behavior, and this activity is not selective to males. The anterior-dorsal MeA and posterior BNST have coordinated fos production both following mating and inter-male aggression in Syrian hamsters (Kollack-Walker and Newman, 1995). Interestingly, similar patterns of fos expression in these nodes was observed in female hamsters after aggression (Potegal et al., 1996) or mating (Joppa et al., 1995). Syrian hamster females are as aggressive (and sometimes more aggressive) as males. Thus, even though anterior-dorsal MeA and BNST circuits are sexually dimorphic, they respond similarly in the context of aggression, and a similar phenotype is observed. These observations support the hypothesis that in some cases sexually dimorphic brain circuits counter-intuitively facilitate similar behavior patterns in males and females (De Vries and Boyle, 1998).

Posterior subdivisions of the BNST also have important effects on reproductive behaviors. Lesions of posterior BNST severely disrupt male copulatory behavior (Powers et al., 1987). Selective lesions of the oval nucleus of the BNST also disrupt mating, but only in virgin males (Claro et al., 1995). Testosterone implants placed directly into the posterior BNST of castrated male hamsters greatly increased sexual behavior (Wood and Newman, 1995). This is consistent with the high concentrations of androgen and estrogen receptors in the posterior BNST. In female hamsters, electrolytic lesions of the BNST and lateral septum reduced ultrasonic vocalization but had no effect on lordosis (Kirn and Floody, 1985). In rats, not only did excitotoxic BNST lesions have no effect on lordosis, they did not interfere with paced mating either (Guarraci et al., 2004). These data indicate that the posterior BNST has a much stronger effect on mating behaviors in males versus females.

Despite the strong sex differences in how the posterior BNST regulates mating behavior, a consistent observation is that engaging in sexual behavior induces a strong increase in c-fos expression in the posterior BNST. Intriguingly, this effect is not dependent on steroid hormones. Although castration also reduces mating behavior in male rats, this behavior is not completely eliminated. When mating-induced c-fos immunoreactivity in the posterior BNST was examined in castrated rats implanted with testosterone, dihydrotestosterone, or estradiol capsules, there were no differences between these groups (Baum and Wesinger, 1993). Similar results have been observed in females (Flanagan-Cato and McEwen, 1995). In females, mating also induces increased c-fos immunoreactivity in the posterior BNST. Although gonadal hormones play an important role in promoting lordosis, the c-fos response in the posterior BNST appears less dependent on hormones and more dependent on vaginal-cervical stimulation (Tetel et al., 1993). Vaginal-cervical stimulation in ovariectomized females induced an equivalent c-fos response in posterior BNST to that observed in hormone primed rats that mated with a male (Pfaus et al., 1993). Curiously, although mating-induced c-fos in males and females is not dependent on gonadal hormones, it is clear that the affected cells are hormone sensitive. In male hamsters, nearly half of the neurons in posterior-medial BNST expressing mating-induced c-fos were also positive for androgen receptors (Wood and Newman, 1993). Similar results were observed in male rats with about 50% of mating-induced c-fos positive cells coexpressing ERα and 90% coexpressing androgen receptor (Greco et al., 1998). This organization of neural circuitry within the BNST may explain why mated males are more aggressive than virgin males (Flannelly and Lore, 1977). It would be interesting to determine whether mating-induced activation of hormone-sensitive neurons in the posterior BNST results in long-term changes in aggression. In females mating-induced c-fos in the MPOA is observed in ERα expressing cells in the MPOA and MeA (Greco et al., 2003); however, the BNST was not quantified in this study. Estrogen receptors in the BNST are closely associated with male aggressive behavior.

In several species, the number of ERα-ir positive cells in the BNST is positively correlated with aggressive behavior (Trainor et al., 2006, 2007a). This relationship extends beyond the standard resident-intruder aggression test, as ERα gene expression is positively correlated with aggressive behavior even in the home cage of *Mus musculus* (Greenberg et al., 2014a). Photoperiod appears to regulate aggression by modulating the mechanisms through which estrogens control behavior. During long, summer-like days, estrogens modulate aggression via slower, probably genomic mechanisms, whereas during short, winter-like days, estrogens act rapidly to modulate aggression via nongenomic mechanisms (Trainor et al., 2007b, 2008b; Laredo et al., 2014). Currently, it is unclear whether estrogens modulate seasonal variation in aggressive behavior in females. One of the pathways that can mediate rapid effects of estrogens on behavior is extracellular signal regulated kinase (ERK). Both male and female California mice display increased protein expression of another indirect marker of neural activity, phosphorylated ERK (pERK), in the posterior BNST following aggressive interactions (Silva et al., 2010; Trainor et al., 2010). Similar results in female California mice were observed using c-fos immunohistochemistry (Davis and Marler, 2004). Interestingly, there can be more subtle sex-specific environmental interactions that influence the aggression/pERK relationship in California mice. Increased pERK expression in posterior BNST occurs only during short day photoperiods in male mice, but it is observed in females regardless of photoperiod. While neural activity in the posterior BNST is activated in the context of aggressive behavior in both males and females, it is also linked to sex-specific reproductive behaviors.

## 3 THE MESOLIMBIC DOPAMINE SYSTEM

For a system that has been studied so intensively, there is still considerable debate over the exact function of the dopamine neurons within the VTA (Hyman et al., 2006; Niv and Schoenbaum, 2008; Berridge, 2012). However, it is clear that dopamine released within the mesolimbic dopamine system has important effects outside the context of reward. For example, both electric shocks and social stressors induce increased dopamine release in the mesolimbic system (reviewed in Miczek et al., 2008; Trainor, 2011). Similar to anterior subregions of the BNST, components of the mesolimbic dopamine system are characterized by a lack of striking sexual dimorphism in vertebrates. However, important differences in mesolimbic dopamine function are observed in males and females.

### 3.1 Sex differences in the neuroanatomy of the mesolimbic dopamine system

Although there are some reports in rats that the VTA volume is larger in females than in males (McArthur et al., 2007), this difference is not always found (Creutz and

Kritzer, 2002; Campi et al., 2013). This may be due to ambiguity in the boundaries of the VTA, as well as technical factors. For example, VTA dopamine neurons have an anterior to posterior orientation, so horizontal sections of the brain maximize the visibility of these cells (Margolis et al., 2006). Similarly, there is little evidence for sexual dimorphism in the size of the NAc. However, females have a higher density of dendritic spines on medium spiny neurons in the NAc than male rats (Forlano and Woolley, 2010; Wissman et al., 2012). The size of the prefrontal cortex (PFC) also does not appear to exhibit consistent sex differences, but has been linked instead to social organization. Females had larger PFC size in two species of voles but not two species of field mice (*Peromyscus*) (Kingsbury et al., 2012). However, socially monogamous prairie voles and California mice had significantly smaller PFC volumes than polygamous montane voles and white-footed mice. Examination of steroid receptor expression in the mesolimbic dopamine system has revealed some interesting differences between males and females.

Within the mesolimbic dopamine system, only the VTA shows abundant nuclear expression of estrogen receptors and androgen receptors. Nuclear localization of steroid receptors reflects the long-term action of these receptors on gene expression (Laredo et al., 2014). In rats, a greater percentage ofVTA dopamine neurons coexpress androgen receptors in males versus females, and are expressed on dopamine neurons projecting to the PFC (Kritzer and Creutz, 2008). This may explain why castration reduces dopamine release in the PFC (Aubele and Kritzer, 2012) and inhibits performance in behavioral tasks dependent upon PFC function, such as working memory (Kritzer et al., 2007). In contrast, ERα immunoreactivity was limited primarily to nondopaminergic cells and was more abundant in females than males (Kritzer and Creutz, 2008). Within the NAc and PFC, nuclear expression of estrogen and androgen receptors is generally absent. However, estrogens infused into the NAc act rapidly to increase dopamine release (Thompson and Moss, 1994). Estrogens also modulate PFC function during fear extinction learning. During proestrus, when estradiol levels are high, c-fos expression in PFC neurons projecting to the basolateral amygdala is positively associated with extinction learning (Rey et al., 2014). However, when estradiol levels are low, this relationship is reversed. There is growing evidence that posttranslational modifications of estrogen receptors can allow for insertion into the plasma membrane (Levin and Pietras, 2008), which can facilitate rapid action of estrogens independent from direct regulation of gene expression. Intriguingly, ERβ messenger RNA is detectable in the NAc (Shughrue et al., 1998) raising the possibility that ERβ expressed in the membrane could mediate rapid effects of estrogens in the NAc. The lack of a reliable ERβ antibody (Snyder et al., 2010) has hindered empirical evaluation of this hypothesis.

## 3.2 Sex differences in the mesolimbic dopamine system: appetitive social contexts

While sexual dimorphism at the anatomical levels is relatively subtle in the mesolimbic dopamine system, functional sex differences can be similar in magnitude as observed in the BNST and MPOA. For example, the extent to which appetitive stimuli induce dopamine release is much stronger in females versus males. This has been best described for psychostimulants such as amphetamine and cocaine, which induce stronger release of dopamine in females versus males (Becker, 2009; Gilles et al., 2014) and are more likely to be abused by women than men (Anker and Carroll, 2011). In anesthetized rats electrical stimulation of the medial forebrain bundle, which includes axons traveling from the VTA to the NAc, leads to greater release of dopamine in the ventral striatum and faster reuptake rate in females in males (Walker et al., 1999). Sex differences in dopamine function are observed in social contexts as well.

During reproductive behavior, dopamine can be released in the NAc in both males and females. However, there are important differences between males and females in the exact conditions required to induce dopamine release. For male rats, dopamine release is induced both by a sexually receptive female behind a screen and by mating with a female (Pfaus et al., 1990; Damsma et al., 1992). Functionally, dopaminergic neurotransmission in the NAc appears to be especially important in the anticipation of mating opportunities in male rats (Pfaus and Phillips, 1991). In contrast, dopamine release in the NAc in female rats is only observed in specific contexts. Under more naturalistic conditions, female rats control the pace of mating through solicitation displays such as hopping or darting (Erskine, 1989). If the testing arena has insufficient space for these proceptive behaviors to be expressed, dopamine release in the NAc is not observed during mating. However, if the female is able to control the pace of copulations large increases in dopamine release are observed (Mermelstein and Becker, 1995; Becker et al., 2001).

In rats, mating behavior is a short-term social interaction. In contrast, in socially monogamous prairie voles it can mark the beginning of a long-term social relationship known as a pair bond (Carter et al., 1995). In female prairie voles, activation of D2 dopamine receptors in the NAc contributes to the formation of pair bonds (Gingrich et al., 2000; Liu and Wang, 2003). In males, activation of D2 receptors in the NAc is also necessary for pair bond formation (Aragona et al., 2006). Interestingly, after mating there is a substantial upregulation of D1 receptor expression in the NAc of male prairie voles, which in turn stimulates aggression toward unfamiliar females. This pattern of behavior is very different in rats, in which D1 receptors in the NAc increase male sexual motivation (Everitt et al., 1989; Bialy et al., 2010).

Outside the context of mating, evidence from several sources suggests that dopamine neurons in the VTA are more reactive during social contexts in females versus males. A study in California mice used c-fos/tyrosine hydroxylase (TH) immunohistochemistry to examine the activity of dopamine neurons following a social interaction test (Greenberg et al., in review). Tyrosine hydroxylase is an essential enzyme for the synthesis of dopamine and the use of TH immunohistochemistry is considered a "gold standard" for identifying dopamine neurons in the VTA (Margolis et al., 2006). Females tested in a social interaction test had more c-fos/TH positive neurons across the entire VTA compared with males. There were no sex differences in c-fos/TH positive neurons in mice that were not exposed to a social context, suggesting that the sex difference was specific to social contexts. Consistent with this finding, a study on juvenile rats engaging in social play induced significant increases in TH/c-fos cells in the VTA of females but not males (Northcutt and Nhuyen, 2014). Interestingly, a study used *in vivo* calcium imaging to demonstrate increased activity of VTA dopamine neurons in female mice during social interactions with unfamiliar females (Gunaydin et al., 2014). Although these studies suggest that VTA dopamine neurons are more sensitive to social interactions in female, only a few species have been considered. It will be important in the future to determine whether this pattern generalizes to species with different social systems.

## 3.3 Sex differences in the mesolimbic dopamine system: aversive social contexts

Similar to appetitive social contexts, there is growing evidence that there are important sex differences in how the mesolimbic dopamine system responds to aversive social contexts. Here, too, the California mouse has proved to be a useful species for examining the effects of social stressors in both males and females. Results from these studies indicate that the mesolimbic dopamine system is more sensitive to social defeat in females compared with males. Intriguingly, after a single episode of defeat, no changes in c-fos/TH immunoreactivity are observed within the VTA in either males or females (Greenberg et al., in review). However, after a third episode of defeat an increase in c-fos/TH expressing cells was observed within the ventral subregion of the VTA of females but not males. This observation is interesting from two perspectives. First, an electrophysiology study on anesthetized rats found that footshocks preferentially activated ventral VTA neurons (Brischoux et al., 2009). This suggests that there is topographical organization to the VTA and that aversive contexts preferentially activate ventral dopamine neurons. Second, increased fos/TH activation was only observed in animals experiencing a third episode of defeat. Behavioral observations show that anxiety-like behaviors are elevated in the minutes immediately preceding a third episode of defeat (Greenberg et al., in review), indicating a level of anticipation that is not present after an initial episode of defeat. This result is consistent with the hypothesis that VTA dopamine neurons do not respond to rewards or punishment per se, but are activated by highly salient stimuli (Berridge, 2007).

Sex differences in the long-term effects of defeat stress have been identified in the NAc. An initial study used immunohistochemistry for phosphorylated CREB (pCREB) as an indirect marker for cellular activity (Trainor et al., 2011). Females exposed to social defeat had more pCREB positive cells in the NAc shell and core than control females (Figure 4.3a). While there was no effect of defeat on pCREB immunoreactivity in males, control males had more pCREB positive cells than control females in both NAc shell and core. Activation of D1 receptors increases cAMP production (Kebabian et al., 1972), which in turn increases phosphorylation of CREB. These results suggested that defeat stress might increase the activity of D1 receptors to a greater extent in females versus males. Total

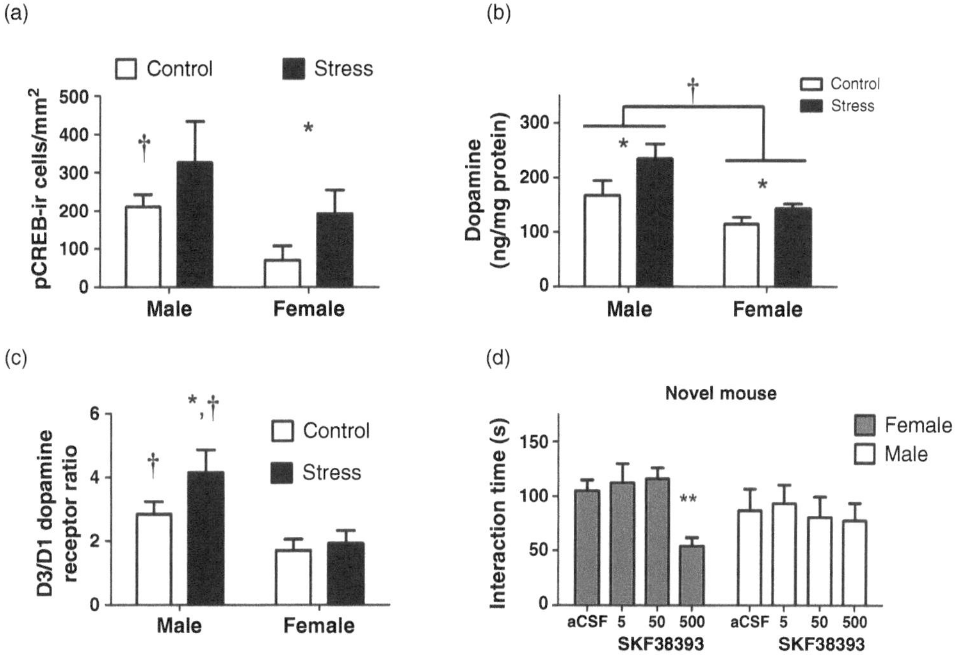


**Figure 4.3** ***Effects of defeat stress on the nucleus accumbens in male and female California mice.*** Defeat stress increases the number of pCREB positive neurons in the NAc shell in females but not males (a). In contrast, defeat stress increases total dopamine content of the NAc in both males and females (b). Control males have a higher ratio of D3/D1 receptors than control females and defeat stress increases this ratio further in males (c). This may explain increased sensitivity to D1 receptors in females. Infusion of 500 ng of the D1 agonist SKF3839 infused into the NAc shell reduced social interaction in females (d) but not males. $^{\dagger}p < 0.05$ sex difference, $^{*}p < 0.05$ effect of defeat, $^{**}p < 0.01$ versus aCSF. *(From Trainor et al. (2011); Campi et al. (2014).)*

dopamine (Figure 4.3b), dihydroxyphenylacetic acid, and homovanilic acid were measured in NAc punch samples collected during the inactive phase to assess baseline dopaminergic activity (Campi et al., 2014). Social defeat stress increased expression of both dopamine and its metabolites in both males and females. Although this result is consistent with previous studies in male mice and rats exposed to defeat (Krishnan et al., 2007; Razzoli et al., 2011), it does not explain sex differences in the effect of defeat on pCREB immunoreactivity or behavior. When real-time polymerase chain reaction was used to measure the expression of D1, D2, D3, and D5 receptors in the NAc, no effects of defeat stress were observed (Campi et al., 2014). However, defeat stress did have a more subtle effect on the ratio of the two most abundant receptor subtypes: D1 and D3. The D3 receptor is in the D2-like family of receptors and its activation decreases cAMP production. Intriguingly, control males had a higher ratio of D3/D1 receptors than females and defeat stress further increased this ratio of D3/D1 receptors in males but not females (Figure 4.3c). This result suggested that males might be less sensitive than females to D1 activation and that males exposed to defeat may compensate for increased dopaminergic activity by adjusting the expression of different receptor subtypes in the NAc. This hypothesis is supported by observations that a dose of the D1 agonist SKF38393 infused into the NAc, which induces social withdrawal in females, has no effect in males (Campi et al., 2014). Consistent with these data, infusion of the D1 antagonist SCH23390 into the NAc shell increased social interaction behavior in stressed females but not females naïve to defeat (Figure 4.3d). The results in California mice correspond well with the effects of chronic mild stress on behavior and NAc gene expression in male and female mice (LaPlant et al., 2009). Chronic mild stress induced more behavioral despair in the forced swim test in females versus males. However, males exposed to stress showed more changes in gene expression in the NAc than females, suggesting that some of the gene expression changes in males may confer resilience to stress. Overall, these data show that sex differences in the effects of defeat stress on dopaminergic signaling in the NAc have important effects on behavior.

## 4 CONNECTIVITY BETWEEN SOCIAL BEHAVIOR NETWORK AND THE MESOLIMBIC DOPAMINE SYSTEM

So far, the effects of the social behavior network and mesolimbic dopamine system have been considered in isolation. However, there are several nodes through which these two networks can communicate. One of the most important nodes appears to be the BNST (O'Connell and Hofmann, 2012). Posterior subregions of the BNST, which include the sexually dimorphic oval nucleus, receive dopaminergic input from the VTA (Swanson, 1982; Hasue and Shammah-Lagnado, 2002). The strongest evidence for functional connections between the BNST and VTA comes from research programs considering the behavioral processes that contribute to drug addiction (Jalabert et al., 2009). Electrodes implanted into the posterior mesencephalon (PM) support self-administered electrical stimulation (Rompré and Boye, 1989), similar to the medial forebrain bundle. Self-stimulation of the PM increased fos-ir in the VTA, NAc shell, and posterior BNST (Marcangione and Rompré, 2008). This coordinated activity between BNST and VTA plays an especially important role in animal models of drug relapse. A primary animal model of relapse is reinstatement of drug seeking behavior (Shaham et al., 2003), which is the recovery of a previously learned response after a period of extinction (Epstein et al., 2006). Reinstatement of drug seeking can be induced by psychosocial stress, and BNST-VTA connectivity plays a key role. Inactivation of the posterior BNST using sodium channel blockers (Erb and Stewart, 1999), GABA receptor agonists (McFarland et al., 2004), or norepinephrine receptor antagonists (Leri et al., 2002) can prevent footshock stress from reinstating cocaine self-administration. When a retrograde tracer was used to identify BNST neurons projecting to the VTA, swim stress induced a significant increase in the number of VTA projecting neurons coexpressing c-fos in the BNST (Briand et al., 2010). Anterior subregions of the

BNST also impact the VTA. Electrical stimulation of the anterior ventral BNST increased the activity of VTA neurons (Georges and Aston-Jones, 2001). The BNST-VTA circuit also appears to impact other aspects of reinforcement, as pharmacological disconnection of this circuit disrupts the formation of place preferences for cocaine (Sartor and Aston-Jones, 2012). A study using optogenetic tools provided insights into how BNST-VTA connectivity could impact behavior in different contexts.

Both excitatory and inhibitory projections from the anterior ventral BNST to the VTA were identified using a combination of optogenetic stimulation and electrophysiology (Jennings et al., 2013). Mice exposed to footshocks induced activation of the excitatory BNST-VTA pathway and inhibition of the inhibitory pathway. Experimental activation of excitatory neurons in the ventral BNST produced excitatory currents in the VTA, induced aversion, and reduced reward seeking. In contrast, activation of inhibitory neurons in the ventral BNST produced inhibitory currents in the VTA, induced place preferences, and enhanced reward seeking. Optogenetic stimulation of excitatory neurons in the BNST suppressed the activity of VTA GABA neurons, which in turn released dopamine neurons from inhibition. A different group, also using optogenetic approaches, reported slightly different results (Kim et al., 2013). Here activation of a projection from the oval nucleus of the BNST (which is sexually dimorphic) to the VTA was found to be anxiogenic whereas activating a projection from the anterodorsal BNST (which is not sexually dimorphic) was found to be anxiolytic. Clearly, knowledge of how BNST-VTA interactions impact behavioral states is moving quickly. A notable gap in this literature is that all of the studies reviewed in this section have been conducted on male rodents. This is surprising given the importance of posterior subregions of the BNST, which has some of the strongest and most evolutionarily conserved sexual dimorphisms in the brain.

There are also important interactions between the MPOA and mesolimbic dopamine circuitry for the motivation to engage in sex-typical social behaviors. These interactions have been extensively studied for maternal behavior in particular, which has been reviewed previously (Numan and Stolzenberg, 2009; Stolzenberg and Numan, 2011). The MPOA sends dense projections to the VTA, and these MPOA neurons are primed by the hormonal events of pregnancy to activate the VTA. Lesion studies show that the MPOA and mesolimbic structures are functionally associated. MPOA lesions, pharmacological inhibition of VTA neurons, and D1 receptor blockade in NAc all have similar effects on maternal behavior

Direct connections between the MeA and mesolimbic nodes are lacking. Tracer studies indicate few MeA neurons project to either the NAc (McDonald, 1991; Canteras et al., 1995) or the VTA (Geisler and Zahm, 2005). However, the MeA sends strong projections to the BNST. Interestingly, dorsal anterior MeA projects to both anterior and posterior subregions of the BNST, but more posterior subregions of the MEA project primarily to posterior subregions of the BNST but not anterior subregions of the BNST

(Gomez and Newman, 1992; Canteras et al., 1995; Coolen and Wood, 1998). This pathway is significant because there is growing evidence that the BNST acts as a critical node connecting the social behavior network with the mesolimbic dopamine system. This suggests that interactions between the MeA and mesolimbic dopamine system are likely mediated by the BNST.

## 5 CONCLUSIONS

In this chapter, we examined how sex differences in structure and function of the social behavior network and mesolimbic dopamine system impact social behavior in both males and females. Interestingly, sexual dimorphism within the social behavior network is strongly conserved across a diverse range mammalian species with different social systems. In some cases, such as the SDN-POA, we still do not fully understand the impact of these sex differences on behavior. Other nodes such as the BNST exert very different effects on behavior in males and females. An important observation is that sex differences in how the BNST impacts behavior are not limited to the sexually dimorphic posterior subregions. This theme carries over to the mesolimbic dopamine system, which also lacks sexual dimorphism at an anatomical level. Focus on connectivity between the BNST and VTA has provided insights into how these two systems communicate with one another. However, this circuit has not been studied in females and so it is unclear exactly how results identified in males will translate to females. Finally, we reviewed evidence that social experiences such as social stress or parental experience induce long-term changes in the function of both the social behavior network and mesolimbic dopamine system. Further study of how experience impacts the effects that these networks have in both males and females will be critical for understanding the neurobiological control of social behaviors.

## ACKNOWLEDGMENT

BCT was supported by MH097714 during the preparation of this manuscript.

## REFERENCES

Anker, J.J., Carroll, M.E., 2011. Females are more vulnerable to drug abuse than males: evidence from preclinical studies and the role of ovarian hormones. Curr. Top. Behav. Neurosci. 8, 73–96.

Aragona, B.J., Liu, Y., Yu, Y.J., Curtis, J.T., Detwiler, J.M., Insel, T.R., Wang, Z., 2006. Nucleus accumbens dopamine differentially mediates the formation and maintenance of monogamous pair bonds. Nat. Neurosci. 9, 133–139.

Arai, Y., Sekine, Y., Murakami, S., 1996. Estrogen and apoptosis in the developing sexually dimorphic preoptic area in female rats. Neurosci. Res. 25, 403–407.

Arnold, A.P., Breedlove, S.M., 1985. Organizational and activational effects of sex steroids on brain and behavior: a reanalysis. Horm. Behav. 19, 469–498.

Arrati, P.G., Carmona, C., Dominguez, G., Beyer, C., Rosenblatt, J.S., 2006. GABA receptor agonists in the medial preoptic area and maternal behavior in lactating rats. Physiol. Behav. 87, 51–65.

Aubele, T., Kritzer, M.F., 2012. Gonadectomy and hormone replacement affects *in vivo* basal extracellular dopamine levels in prefrontal cortex but not motor cortex of adult male rats. Cereb. Cortex 21, 222–232.

Autry, A.E., Adachi, M., Nosyreva, E., Na, E.S., Los, M.F., Cheng, P.-F., Kavalali, E.T., Monteggia, L.M., 2011. NMDA receptor blockade at rest triggers rapid behavioural antidepressant responses. Nature 475, 91–95.

Balthazart, J., Surlemont, C., 1990. Copulatory behavior is controlled by the sexually dimorphic nucleus of the quail. Brain Res. Bull. 25, 7–14.

Baum, M.J., Wesinger, S.R., 1993. Equivalent levels of mating-induced neural c-fos immunoreactivity in castrated male rats given androgen, estrogen, or no steroid replacement. Biol. Reprod. 48, 1341–1347.

Bean, N.J., Nunez, A.A., Conner, R., 1981. Effects of medial preoptic lesions on male mouse ultrasonic vocalizations and copulatory behavior. Brain Res. Bull. 6, 109–112.

Becker, J.B., 2009. Sexual differentiation of motivation: a novel mechanism? Horm. Behav. 55, 646–654.

Becker, J.B., Rudick, C.N., Jenkins, W.J., 2001. The role of dopamine in the nucleus accumbens and striatum during sexual behavior in the female rat. J. Neurosci. 21, 3236–3241.

Beery, A.K., Zucker, I., 2011. Sex bias in neuroscience and biomedical research. Neurosci. Biobehav. Rev. 35, 565–572.

Berenbaum, S.A., 1999. Effects of early androgens on sex-typed activities and interests in adolescents with congenetial adrenal hyperplasia. Horm. Behav. 35, 102–110.

Berridge, K.C., 2007. The debate over dopamine's role in reward: the case for incentive salience. Psychopharmacology 191, 391–431.

Berridge, K.C., 2012. From prediction error to incentive salience: mesolimbic computation of reward motivation. Eur. J. Neurosci. 35, 1124–1143.

Bialy, M., Kalata, U., Nikolaev-Diak, A., Nikolaev, E., 2010. D1 receptors involved in the acquisition of sexual experience in male rats. Behav. Brain Res. 206, 166–176.

Blake, C.B., Meredith, M., 2011. Change in number and activation of androgen receptor-immunoreactive cells in the medial amygdala in response to chemosensory input. Neuroscience 190, 228–238.

Bloch, G.J., Gorski, R.A., 1988. Cytoarchitectonis analysis of the SDN-POA of the intact and gonadectomized rat. J. Comp. Neurol. 275, 604–612.

Briand, L.A., Vassoler, F.M., Pierce, R.C., Valentino, R.J., Blendy, J.A., 2010. Ventral tegmental afferents in stress-induced reinstatement: the role of cAMP response element-binding protein. J. Neurosci. 30, 16149–16159.

Brischoux, F., Chakraborty, S., Brierley, D.I., Ungless, M.A., 2009. Phasic excitation of dopamine neurons in ventral VTA by noxious stimuli. Proc. Natl. Acad. Sci. USA 106, 4894–4899.

Cahill, L., 2006. Why sex matters for neuroscience. Nat. Rev. Neurosci. 7, 477–484.

Campi, K.L., Jameson, C.E., Trainor, B.C., 2013. Sexual dimorphism in the brain of the monogamous California mouse (*Peromyscus californicus*). Brain Behav. Evol. 81, 236–249.

Campi, K.L., Greenberg, G.D., Kapoor, A., Ziegler, T.E., Trainor, B.C., 2014. Sex differences in effects of dopamine D1 receptors on social withdrawal. Neuropharmacology 77, 208–216.

Canteras, N.S., Simerly, R.B., Swanson, L.W., 1995. Organization of projections from the medial nucleus of the amygdala: a PHAL study in the rat. J. Comp. Neurol. 360, 213–245.

Carter, C.S., DeVries, A.C., Getz, L.L., 1995. Physiological substrates of mammalian monogamy: the prairie vole model. Neurosci. Biobehav. Rev. 19, 303–314.

Chen, T.J., Tu, W.W., 1992. Sex differences in estrogen and androgen receptors in hamster brain. Life Sci. 50, 1639–1647.

Chen, Z.-Y., Jing, D., Bath, K.G., Ieraci, A., Khan, T., Siao, C.-J., Herrera, D.G., Toth, M., Yang, C., McEwen, B.S., Hempstead, B.L., Lee, F.S., 2006. Genetic variant BDNF (Val66Met) polymorphism alters anxiety-related behavior. Science 314, 140–143.

Cherry, J.A., Basham, M.E., Weaver, C.E., Krohmer, R.W., Baum, M.J., 1990. Ontogeny of the sexually dimorphic male nucleus in the preoptic/anterior hypothalamus of ferrets and its manipulation by gonadal steroids. J. Neurobiol. 21, 844–857.

Chung, W.C.J., Swaab, D.F., De Vries, G.J., 2000. Apoptosis during sexual differentiation of the bed nucleus of the stria terminalis in the rat brain. J. Neurobiol. 43, 234–243.

Cintra, A., Fuxe, K., Harfstrand, A., Agnati, L.F., Miller, L.S., Greene, J.L., Gustafsson, J.Å., 1986. Rapid important paper on the cellular localization and distribution of estrogen receptors in the rat tel- and diencephalon using monoclonal antibodies to human estrogen receptor. Neurochem. Int. 8, 587–595.

Ciumas, C., Hirschberg, A.L., Savic, I., 2009. High fetal testosterone and sexually dimorphic cerebral networks in females. Cereb. Cortex 19, 1167–1174.

Claro, F., Segovia, S., Guilamon, A., Del Abril, A., 1995. Lesions of the medial posterior region of the BST impair sexual behavior in sexually experienced and inexperienced male rats. Brain Res. Bull. 36, 1–10.
Commins, D., Yahr, P., 1984a. Adult testosterone levels influence the morphology of a sexually dimorphic area in the Mongolian gerbil brain. J. Comp. Neurol. 224, 132–140.
Commins, D., Yahr, P., 1984b. Lesions of the sexually dimorphic area disrupt mating and marking in male gerbils. Brain Res. Bull. 13, 185–193.
Commins, D., Yahr, P., 1985. Autoradiographic localization of estrogen and androgen receptors in the sexually dimorphic area and other regions of the gerbil brain. J. Comp. Neurol. 231, 473–489.
Cooke, B.M., Woolley, C.S., 2005. Sexually dimorphic synaptic organization of the medial amygdala. J. Neurosci. 25, 10759–10767.
Cooke, B.M., Tabibnia, G., Breedlove, S.M., 1999. A brain sexual dimorphism controlled by adult circulating androgens. Proc. Natl. Acad. Sci. 96, 7538–7540.
Coolen, L.M., Wood, R.I., 1998. Bidirectional connections of the medial amygdaloid nucleus in the Syrian hamster brain: simultaneous anterograde and retrograde tract tracing. J. Comp. Neurol. 399, 189–209.
Coolen, L.M., Olivier, B., Peters, H.J., Veening, J.G., 1997. Demonstration of ejaculation-induced neural activity in the male rat brain using 5-$HT_{1A}$ agonist 8-OH-DPAT. Physiol. Behav. 62, 881–891.
Cornil, C.A., Ball, G.F., Balthazart, J., 2012. Rapid control of male typical behaviors by brain-derived estrogens. Front. Neuroendocrinol. 33, 425–446.
Creutz, L.M., Kritzer, M.F., 2002. Estrogen receptor-beta immunoreactivity in the midbrain of adult rats: regional, subregional, and cellular localization in the A10, A9, and A8 dopamine cell groups. J. Comp. Neurol. 446, 288–300.
Crews, D., Fitzgerald, K.T., 1980. "Sexual" behavior in parthenogenetic lizards (*Cnemidophorus*). Proc. Natl. Acad. Sci. USA 77, 499–502.
Damsma, G., Pfaus, J.G., Wenkstern, D., Phillips, A.G., Fibiger, H.C., 1992. Sexual behavior increases dopamine transmission in the nucleus accumbens and striatum of male rats: comparison with novelty and locomotion. Behav. Neurosci. 106, 181–191.
Davis, E.S., Marler, C.A., 2003. The progesterone challenge: steroid hormone changes following a simulated territorial intrusion in female *Peromyscus californicus*. Horm. Behav. 44, 189–198.
Davis, E.S., Marler, C.A., 2004. C-fos changes following an aggressive encounter in female California mice: a synthesis of behavior, hormone changes and neural activity. Neuroscience 127, 611–624.
DeVries, G.J., Boyle, P.A., 1998. Double duty for sex differences in the brain. Behav. Brain Res. 92, 205–213.
del Abril, A., Segovia, S., Guillamón, A., 1987. The bed nucleus of the stria terminalis in the rat: regional sex differences controlled by gonadal steroids early after birth. Develop. Brain Res. 32, 295–300.
Delville, Y., DeVries, G.J., Ferris, C.F., 2000. Neural connections of the anterior hypothalamus and agonistic behavior in golden hamsters. Brain Behav. Evol. 55, 53–76.
Dobolyi, A., Grattan, D.R., Stolzenberg, D.S., 2014. Preoptic inputs and mechanisms that regulate maternal responsiveness. J. Neuroendocrinol. 26, 627–640.
Duman, R.S., Monteggia, L.M., 2006. A neurotrophic model for stress-related mood disorders. Biol. Psychiat. 59, 1116–1127.
Egan, M.F., Kojima, M., Callicott, J.H., Goldberg, T.E., Kolachana, B.S., Bertolino, A., Zaitsev, E., Gold, B., Goldman, D., Dean, M., Lu, B., Weinberger, D.R., 2003. The BDNF val66met polymorphism affects activity-dependent secretion of BDNF and human memory and hippocampal function. Cell 112, 257–269.
Epstein, D.H., Preston, K.L., Stewart, J., Shaham, Y., 2006. Towards a model of drug relapse: an assessment of the validity of the reinstatement procedure. Psychopharmacology 189, 1–16.
Erb, S., Stewart, J., 1999. A role for the bed nucleus of the stria terminalis, but not the amygdala, in the effects of corticotropin-releasing factor on stress-induced reinstatement of cocaine seeking. J. Neurosci. 19, RC35, 1–6.
Erskine, M.S., 1989. Solicitation behavior in the estrous female rat: a review. Horm. Behav. 23, 473–502.
Everitt, B.J., 1990. Sexual motivation: a neural and behavioral analysis of the mechanisms underlying appetitive and copulatory responses of male rats. Neurosci. Biobehav. Rev. 14, 217–232.
Everitt, B.J., Cador, M., Robbins, T.W., 1989. Interactions between the amygdala and ventral striatum in stimulus-reward associations: studies using a second-order schedule of sexual reinforcement. Neuroscience 30, 63–75.

Flanagan-Cato, L.M., McEwen, B.S., 1995. Pattern of Fos and Jun expression in the female rat forebrain after sexual behavior. Brain Res. 673, 53–60.

Flannelly, K., Lore, R., 1977. The influence of females upon aggression in domesticated male rats (*Rattus norvegicus*). Anim. Behav. 25, 654–659.

Fleming, A.S., Rosenblatt, J.S., 1974. Olfactory regulation of maternal behavior in rats: I effects of olfactory bulb removal in experienced and inexperienced lactating and cycling females. J. Comp. Physiol. Psychol. 86, 221–232.

Forger, N.G., de Vries, G.J., 2010. Cell death and sexual differentiation of behavior: worms, flies, and mammals. Curr. Opin. Neurobiol. 20, 776–783.

Forlano, P.M., Woolley, C.S., 2010. Quantitative analysis of pre- and postsynaptic sex differences in the nucleus accumbens. J. Comp. Neurol. 518, 1330–1348.

Frustaci, A., Pozzi, G., Gianfagna, F., Manzoli, L., Boccia, S., 2008. Meta-analysis of the brain-derived neurotrophic factor gene (BDNF) Val66Met polymorphism in anxiety disorders and anxiety-related personality traits. Neuropyschobiology 58, 163–170.

Fuxjager, M.J., Forbes-Lorman, R.M., Coss, D.J., Auger, C.J., Auger, A.P., Marler, C.A., 2010. Winning territorial disputes selectively enhances androgen sensitivity in neural pathways related to motivation and social aggression. Proc. Natl. Acad. Sci. USA 107, 12393–12398.

Gammie, S.C., 2005. Current models and future directions for understanding the neural circuitries of maternal behaviors in rodents. Behav. Cogn. Neurosci. Rev. 4, 119–135.

Gattermann, R., Fritzsche, P., Neumann, K., Al-Hussein, I., Kayser, A., Abiad, M., Yakti, R., 2001. Notes on the current distribution and the ecology of wild golden hamsters (*Mesocricetus auratus*). J. Zool. 254, 359–365.

Geisler, S., Zahm, D.S., 2005. Afferents of the ventral tegmental area in the rat-anatomical substratum for integrative functions. J. Comp. Neurol. 490, 270–294.

Georges, F., Aston-Jones, G., 2001. Potent regulation of midbrain dopamine neurons by the bed nucleus of the stria terminalis. J. Neurosci. 21, RC160–166.

Gilles, G.E., Virdee, K., McArthur, S., Dalley, J.W., 2014. Sex-dependent diversity in ventral tegmental dopaminergic neurons and developmental programing: a molecular, cellular and behavioral analysis. Neuroscience 282, 69–85.

Gingrich, B., Liu, Y., Wang, Z., Insel, T.R., 2000. Dopamine D2 receptors in the nucleus accumbens are important for social attachment in female prairie vole (*Microtus ochrogaster*). Behav. Neurosci. 114, 173–183.

Gleason, E.D., Fuxjager, M.J., Oyegbile, T.O., Marler, C.A., 2009. Testosterone release and social context: when it occurs and why. Front. Neuroendocrinol. 30, 460–469.

Gomez, D.M., Newman, S.W., 1992. Differential projections of the anterior and posterior regions of the medial amygdaloid nucleus in the Syrian hamster. J. Comp. Neurol. 317, 195–218.

Goodson, J.L., 2005. The vertebrate social behavior network: evolutionary themes and variations. Horm. Behav. 48, 11–22.

Goodson, J.L., Kingsbury, M.A., 2013. What's in a name? Considerations of homologies and nomenclature for vertebrate social behavior networks. Horm. Behav. 64, 103–112.

Gorski, R.A., 1986. Sexual differentiation of the brain: a model for drug-induced alterations of the reproductive system. Environ. Health Perspect. 70, 163.

Gorski, R.A., Gordon, J.H., Shryne, J.E., Southam, A.M., 1978. Evidence for a morphological sex difference within the medial preoptic area of the rat brain. Brain Res. 148, 333–346.

Grabe, H.J., Schwahn, C., Mahler, J., Appel, K., Schulz, A., Spitzer, C., Fenske, K., Barnow, S., Freyberger, H.J., Teumer, A., Petersmann, A., Biffar, R., Rosskopf, D., John, U., Völzke, H., 2012. Genetic epistasis between the brain-derived neurotrophic factor Val66Met polymorphism and the 5-HTT promoter polymorphism moderates the susceptibility to depressive disorders after childhood abuse. Prog. Neuro-psychopharmacol. Biol. Psychiat. 36, 264–270.

Graham, M.D., Pfaus, J.G., 2010. Differential regulation of female sexual behaviour by dopamine agonists in the medial preoptic area. Pharm. Biochem. Behav. 97, 284–292.

Grassman, M., Crews, D., 1990. Ovarian and adrenal function in the parthenogenetic whiptail lizard *Cnemidophorus uniparens* in the field and laboratory. Gen. Comp. Endocrinol. 76, 444–450.

Gray, G.D., Sodersten, P., Tallentire, D., Davidson, J.M., 1978. Effects of lesions in various structures of the suprachiasmatic-preoptic region on LH regulation and sexual behavior in female rats. Neuroendocrinology 25, 174–191.

Greco, B., Edwards, D.A., Zumpe, D., Michael, R.P., Clancy, A.N., 1998. Fos induced by mating or non-contact sociosexual interaction is colocalized with androgen receptors in neurons within the forebrain, midbrain, and lumbosacral spinal cord of male rats. Horm. Behav. 33, 125–138.

Greco, B., Blasberg, M.E., Kosinski, E.C., Blaustein, J.D., 2003. Response of ERα-IR and ERβ-IR cells in the forebrain of female rats to mating stimuli. Horm. Behav. 43, 444–453.

Greenberg, G.D., van Westerhuyzen, J.A., Bales, K.L., Trainor, B.C., 2012. Is it all in the family? The effects of early social structure on neural–behavioral systems of prairie voles (*Microtus ochrogaster*). Neuroscience 216, 46–56.

Greenberg, G.D., Howerton, C.L., Trainor, B.C., 2014a. Fighting in the home cage: agonistic encounters and effects on neurobiological markers within the social decision-making network of house mice (*Mus musculus*). Neurosci. Let. 566, 151–155.

Greenberg, G.D., Laman-Maharg, A., Campi, K.L., Voigt, H., Orr, V.N., Schaal, L., Trainor, B.C., 2014b. Sex differences in stress-induced social withdrawal: role of brain derived neurotrophic factor in the bed nucleus of the stria terminalis. Front. Behav. Neurosci. 7, 223.

Greenberg, G.D., Steinman, M.Q., Doig, I.E., Trainor, B.C. (in review). Topographical organization of ventral tegmental area dopamine neurons in response to aversive social contexts in California mice (*Peromyscus californicus*).

Gu, G., Cornea, A., Simerly, R.B., 2003. Sexual differentiation of projections from the principal nucleus of the bed nuclei of the stria terminalis. J. Comp. Neurol. 460, 543–562.

Guarraci, F.A., Megroz, A.B., Clark, A.S., 2004. Paced mating behavior in the female rat following lesions of three regions responsive to vaginocervical stimulation. Brain Res. 999, 40–52.

Gunaydin, L.A., Grosenick, L., Finkelstein, J.C., Kauvar, I.V., Fenno, L.E., Adhikari, A., Lammel, S., Mirzabekov, J.J., Airan, R.D., Zalocusky, K.A., Tye, K.M., Anikeeva, P., Malenka, R.C., Deisseroth, K., 2014. Natural neural projection dynamics underlying social behavior. Cell 157, 1535–1551.

Hamann, S., Herman, R.A., Nolan, C.L., Wallen, K., 2004. Men and women differ in amygdala response to visual sexual stimuli. Nat. Neurosci. 7, 411–416.

Hart, B., Leedy, M., 1985. Neurological bases of male sexual behavior. In: Adler, N., Pfaff, D., Goy, R. (Eds.), Reproduction. Springer, US, pp. 373–422, 7.

Hasen, N.S., Gammie, S.C., 2005. Differential fos activation in virgin and lactating mice in response to an intruder. Physiol. Behav. 84, 681–695.

Hasen, N.S., Gammie, S.C., 2006. Maternal aggression: new insights from Egr-1. Brain Res. 1108, 147–156.

Hasue, R.H., Shammah-Lagnado, S.J., 2002. Origin of the dopaminergic innervation of the central extended amygdala and accumbens shell: a combined retrograde tracing and immunohistochemical study in the rat. J. Comp. Neurol. 454, 15–33.

Heeb, M., Yahr, P., 1996. c-Fos immunoreactivity in the sexually dimorphic area of the hypothalamus and related brain regions of male gerbils after exposure to sex-related stimuli or performance of specific sexual behaviors. Neuroscience 72, 1049–1071.

Heimer, L., Larsson, K., 1967. Impairment of mating behavior in male rats following lesions in the preoptic-anterior hypothalamic continuum. Brain Res. 3, 248–263.

Herbison, A.E., Theodosis, D.T., 1992. Localization of oestrogen receptors in preoptic neurons containing neurotensin but not tyrosine hydroxylase, cholecystokinin or luteinizing hormone-releasing hormone in the male and female rat. Neuroscience 50, 283–298.

Hines, M., Davis, F.C., Coquelin, A., Goy, R.W., Gorski, R.A., 1985. Sexually dimorphic regions in the medial preoptic area and the bed nucleus of the stria terminalis of the guinea pig brain: a description and an investigation of their relationship to gonadal steroids in adulthood. J. Neurosci. 5, 40–47.

Hines, M., Allen, L.S., Gorski, R.A., 1992. Sex differences in subregions of the medial nucleus of the amygdala and the bed nucleus of the stria terminalis of the rat. Brain Res. 579, 321–326.

Hofman, M.A., Swaab, D.F., 1989. The sexually dimorphic nucleus of the preoptic area in the human brain: a comparative morphometric study. J. Anat. 164, 55.

Holmes, M.M., Goldman, B.D., Forger, N.G., 2008. Social status and sex independently influence androgen receptor expression in the eusocial naked mole-rat brain. Horm. Behav. 54, 278–285.

Hong, W., Kim, D.-W., Anderson, D.J., 2014. Antagonistic control of social versus repetitive self-grooming behaviors by separable amygdala neuronal subsets. Cell 158, 1348–1361.

Hsu, Y., Earley, R.L., Wolf, L.L., 2007. Modulation of aggressive behaviour by fighting experience: mechanism and contest outcomes. Biol. Rev. 81, 337–374.
Hull, E.M., Dominguez, J.M., 2007. Sexual behavior in male rodents. Horm. Behav. 52, 45–55.
Hyman, S.E., Malenka, R.C., Nestler, E.J., 2006. Addiction: the role of reward-related learning and memory. Ann. Rev. Neurosci. 29, 565–598.
Jalabert, M., Aston-Jones, G., Herzog, E., Manzoni, O., Georges, F., 2009. Role of the bed nucleus of the stria terminalis in the control of ventral tegmental area dopamine neurons. Prog. Neuropsychopharmacol. Biol. Psychiat. 33, 1336–1346.
Jennings, J.H., Sparta, D.R., Stamatakis, A.M., Ung, R.L., Pleil, K.E., Kash, T.L., Stuber, G.D., 2013. Distinct extended amygdala circuits for divergent motivational states. Nature 496, 224–228.
Joppa, M., Meisel, R., Garber, M., 1995. c-Fos expression in female hamster brain following sexual and aggressive behaviors. Neuroscience 68, 783–792.
Kebabian, J.W., Petzold, G.L., Greengard, P., 1972. Dopamine-sensitive adenylate cyclase in caudate nucleus of rat brain, and its similarity to the "dopamine receptor". Proc. Natl. Acad. Sci. USA 69, 2145–2149.
Kim, S.-Y., Adhikari, A., Lee, S.Y., Marshel, J.H., Kim, C.K., Mallory, C.S., Lo, M., Pak, S., Mattis, J., Lim, B.K., Malenka, R.C., Warden, M.R., Neve, R., Tye, K.M., Deisseroth, K., 2013. Diverging neural pathways assemble a behavioural state from separable features in anxiety. Nature 496, 219–223.
Kingsbury, M.A., Gleason, E.D., Ophir, A.G., Phelps, S.M., Young, L.J., Marler, C.A., 2012. Monogamous and promiscuous rodent species exhibit discrete variation in the size of the medial prefrontal cortex. Brain Behav. Evol. 80, 4–14.
Kirn, J., Floody, O.R., 1985. Differential effects of lesions in three limbic areas on ultrasound production and lordosis by female hamsters. Behav. Neurosci. 99, 1142–1152.
Kollack-Walker, S., Newman, S., 1995. Mating and agonistic behavior produce different patterns of Fos immunolabeling in the male Syrian hamster brain. Neuroscience 66, 721–736.
Kollack-Walker, S., Newman, S.W., 1997. Mating-induced expression of c-fos in the male Syrian hamster brain: role of experience, pheromones, and ejaculations. Neuroscience 66, 721–736.
Kollack-Walker, S., Watson, S.J., Akil, H., 1997. Social stress in Hamsters: defeat activates specific neurocircuits within the Brain. J. Neurosci. 17, 8842–8855.
Kondo, Y., Sakuma, Y., 2005. The medial amygdala controls the coital access of female rats: a possible involvement of emotional responsiveness. Jpn. J. Physiol. 55, 345–353.
Koolhaas, J., Korte, S., De Boer, S., Van Der Vegt, B., Van Reenen, C., Hopster, H., De Jong, I., Ruis, M., Blokhuis, H., 1999. Coping styles in animals: current status in behavior and stress-physiology. Neurosci. Biobehav. Rev. 23, 925–935.
Krishnan, V., Han, M.H., Graham, D.L., Berton, O., Renthal, W., Russo, S.J., LaPlant, Q., Graham, A., Lutter, M., Lagace, D.C., Ghose, S., Reister, R., Tannous, P., Green, T.A., Neve, R.L., Chakravarty, S., Kumar, A., Eisch, A.J., Self, D.W., Lee, F.S., Tamminga, C.A., Cooper, D.C., Gershenfeld, H.K., Nestler, E.J., 2007. Molecular adaptations underlying susceptibility and resistance to social defeat in brain reward regions. Cell 131, 391–404.
Kritzer, M.F., Creutz, L.M., 2008. Region and sex differences in constituent dopamine neurons and immunoreactivity for intracellular estrogen and androgen receptors in mesocortical projections in rats. J. Neurosci. 28, 9525–9535.
Kritzer, M.F., Brewer, A., Montalmant, F., Daveport, M., Robinson, J.K., 2007. Effects of gonadectomy on performance in operant tasks measuring prefrontal cortical function in adult male rats. Horm. Behav. 51, 183–194.
LaPlant, Q., Chakravarty, S., Vialou, V., Mukherjee, S., Koo, J.W., Kalahasti, G., Bradbury, K.R., Taylor, S.V., Maze, I., Kumar, A., Graham, A., Birnbaum, S.G., Krishnan, V., Truong, H.-T., Neve, R.L., Nestler, E.J., Russo, S.J., 2009. Role of nuclear factor κB in ovarian hormone-mediated stress hypersensitivity in female mice. Biol. Psychiatry 65, 874–880.
Laredo, S.A., Villalon Landeros, R., Trainor, B.C., 2014. Rapid effects of estrogens on behavior: environmental modulation and molecular mechanisms. Front. Neuroendocrinol. 35, 447–458.
Laredo, S.A., Steinman, M.Q., Robles, C.F., Ferrer, E., Ragen, B.J., Trainor, B.C., 2015. The mu-opioid receptor mediates sex differences in cognitive flexibility following social defeat stress. Eur. J. Neurosci. 41, 434–441.

Lee, A.W., Brown, R.E., 2002. Medial preoptic lesions disrupt parental behavior in both male and female California mice (*Peromyscus californicus*). Behav. Neurosci. 116, 968.

Lee, A.W., Brown, R.E., 2007. Comparison of medial preoptic, amygdala, and nucleus accumbens lesions on parental behavior in California mice (*Peromyscus californicus*). Physiol. Behav. 92, 617–628.

Lee, A., Clancy, S., Fleming, A.S., 1999. Mother rats bar-press for pups: effects of lesions of the mpoa and limbic sites on maternal behavior and operant responding for pup-reinforcement. Behav. Brain Res. 100, 15–31.

Lehman, M.N., Winans, S.S., 1982. Vomeronasal and olfactory pathways to the amygdala controlling male hamster sexual behavior: autoradiographic and behavioral analyses. Brain Res. 240, 27–41.

Lehman, M.N., Winans, S.S., Powers, J.B., 1980. Medial nucleus of the amygdala mediates chemosensory control of male hamster sexual behavior. Science 210, 557–560.

Lehman, M.N., Powers, J.B., Winans, S.S., 1983. Stria terminalis lesions alter the temporal pattern of copulatory behavior in the male golden hamster. Behav. Brain Res. 8, 109–128.

Leri, F., Flores, J., Rodaros, D., Stewart, J., 2002. Blockade of stress-induced but not cocaine-induced reinstatement by infusion of noradrenergic antagonists into the bed nucleus of the stria terminalis or the central nucleus of the amygdala. J. Neurosci. 22, 5713–5718.

Levin, E.R., Pietras, R.J., 2008. Estrogen receptors outside the nucleus in breast cancer. Breast Cancer Res. Treat. 108, 351–361.

Lin, D.Y., Boyle, M.P., Dollar, P., Lee, H., Lein, E.S., Perona, P., Anderson, D.J., 2011. Functional identification of an aggressive locus in the mouse hypothalamus. Nature 470, 221–226.

Liu, Y., Wang, Z.X., 2003. Nucleus accumbens oxytocin and dopamine interact to regulate pair bond formation in female prairie voles. Neuroscience 121, 537–544.

Lu, B., Pang, P.T., Woo, N.H., 2005. The yin and yang of neurotrophin action. Nat. Rev. Neurosci. 6, 603–614.

Malberg, J.E., Duman, R.S., 2003. Cell proliferation in adult hippocampus is decreased by inescapable stress: reversal by fluoxetine treatment. Neuropsychopharmacology 28, 1562–1571.

Malsbury, C.W., 1971. Facilitation of male rat copulatory behavior by electrical stimulation of the medial preoptic area. Physiol. Behav. 7, 797–805.

Malsbury, C.W., Kow, L.-M., Pfaff, D.W., 1977. Effects of medial hypothalamic lesions on the lordosis response and other behaviors in female golden hamsters. Physiol. Behav. 19, 223–237.

Marcangione, C., Rompré, P.-P., 2008. Topographical Fos induction within the ventral midbrain and projection sites following self-stimulation of the posterior mesencephalon. Neuroscience 154, 1227–1241.

Margolis, E.B., Lock, H., Hjelmstad, G.O., Fields, H.L., 2006. The ventral tegmental area revisited: is there an electrophysiological marker for dopaminergic neurons? J. Physiol. 577, 907–924.

Matsuo, K., Walss-Bass, C., Nery, F.G., Nicoletti, M.A., Hatch, J.P., Frey, B.N., Monkul, E.S., Zunta-Soares, G.B., Bowden, C.L., Escamilla, M.A., Soares, J.C., 2009. Neuronal correlates of brain-derived neurotrophic factor Val66Met polymorphism and morphometric abnormalities in bipolar disorder. Neuropsychopharmacology 34, 1904–1913.

Mattson, B.J., Morrell, J.I., 2005. Preference for cocaine- versus pup-associated cues differentially activates neurons expressing either Fos or cocaine- and amphetamine-regulated transcript in lactating, maternal rodents. Neuroscience 135, 315–328.

McArthur, S., McHale, E., Gillies, G.E., 2007. The size and distribution of midbrain dopaminergic populations are permanently altered by perinatal glucocorticoid exposure in a sex-region-and time-specific manner. Neuropsychopharmacology 32, 1462–1476.

McDonald, A., 1991. Topographical organization of amygdaloid projections to the caudatoputamen, nucleus accumbens, and related striatal-like areas of the rat brain. Neuroscience 44, 15–33.

McFarland, K., Davidge, S.B., Lapish, C.C., Kalivas, P.W., 2004. Limbic and motor circuitry underlying footshock-induced reinstatement of cocaine-seeking behavior. J. Neurosci. 24, 1551–1560.

Merke, D.P., Bornstein, S.R., 2005. Congenital adrenal hyperplasia. Lancet 365, 2125–2136.

Mermelstein, P.G., Becker, J.B., 1995. Increased extracellular dopamine in the nucleus-accumbens and striatum of the female rat during paced copulatory-behavior. Behav. Neurosci. 109, 354–365.

Miczek, K.A., Yap, J.J., Covington, H.E., 2008. Social stress, therapeutics and drug abuse: preclinical models of escalated and depressed intake. Pharmacol. Ther. 120, 102–128.

Moga, M.M., Saper, C.B., Gray, T.S., 1989. Bed nucleus of the stria terminalis: cytoarchitecture, immunohistochemistry, and projection to the parabrachial nucleus in the rat. J. Comp. Neurol. 283, 315–332.
Morris, J.A., Jordan, C.L., Breedlove, S.M., 2008. Sexual dimorphism in neuronal number of the posterodorsal medial amygdala is independent of circulating androgens and regional volume in adult rats. J. Comp. Neurol. 506, 851–859.
Newman, S.W., 1999. The medial extended amygdala in male reproductive behavior: a node in the mammalian social behavior network. Ann. NY Acad. Sci. 877, 242–257.
Nibuya, M., Morinobu, S., Duman, R.S., 1995. Regulation of BDNF and trkB messenger-RNA in rat-brain by chronic electroconvulsive seizure and antidepressant drug treatments. J. Neurosci. 15, 7539–7547.
Niv, Y., Schoenbaum, G., 2008. Dialogues on prediction errors. Trends Cogn. Sci. 12, 265–272.
Northcutt, K.V., Nhuyen, J.M., 2014. Female juvenile play elicits Fos expression in dopaminergic neurons of the VTA. Behav. Neurosci. 128, 178–186.
Numan, M., 1974. Medial preoptic area and maternal behavior in the female rat. J. Comp. Physiol. Psychol. 87, 746.
Numan, M., 1988. Neural basis of maternal behavior in the rat. Psychoneuroendocrinology 13, 47–62.
Numan, M., Stolzenberg, D.S., 2009. Medial preoptic area interactions with dopamine neural systems in the control of the onset and maintenance of maternal behavior in rats. Front. Neuroendocrinol. 30, 46–64.
Numan, M., Rosenblatt, J.S., Komisaruk, B.R., 1977. Medial preoptic area and onset of maternal behavior in rat. J. Comp. Neurol. 91, 146–164.
O'Connell, L.A., Hofmann, H.A., 2011. The vertebrate mesolimbic reward system and social behavior network: a comparative synthesis. J. Comp. Neurol. 519, 3599–3639.
O'Connell, L.A., Hofmann, H.A., 2012. Evolution of a vertebrate social decision-making network. Science 336, 1154–1157.
Oyegible, T.O., Marler, C.A., 2005. Winning fights elevates testosterone levels in California mice and enhance future ability to win fights. Horm. Behav. 48, 259–267.
Pak, T.R., Handa, R.J., 2007. Steroid hormone receptors and sex differences in behavior. In: Becker, J.B., Berkley, K.J., Geary, N., Hampson, E., Herman, J.P., Young, E.A. (Eds.), Sex Differences in the Brain: From Genes to Behavior. Oxford University Press, New York, pp. 109–138.
Panzica, G., Aste, N., Viglietti-Panzica, C., Ottinger, M., 1995. Structural sex differences in the brain: influence of gonadal steroids and behavioral correlates. J. Endocrinol. Invest. 18, 232–252.
Patisaul, H.B., Scordalakes, E., Young, L.J., Rissman, E.F., 2003. Oxytocin, but not oxytocin receptor, is regulated by oestrogen receptor β in the female mouse hypothalamus. J. Neuroendocrinol. 15, 787–793.
Pfaus, J.G., 1999. Revisiting the conceptor of sexual motivation. Annu. Rev. Sex Res. 10, 120–156.
Pfaus, J.G., Phillips, A.G., 1991. Role of dopamine in anticipatory and consummatory aspects of sexual behavior in the male rat. Behav. Neurosci. 105, 727–743.
Pfaus, J.G., Damsma, G., Nomikos, G.G., Wenkstern, D.G., Blaha, C.D., Phillips, A.G., Fibiger, H.C., 1990. Sexual behavior enhances central dopamine transmission in the male rat. Brain Res. 530, 345–348.
Pfaus, J.G., Kleopoulos, S.P., Mobbs, C.V., Gibbs, R.B., Pfaff, D.W., 1993. Sexual stimulation activates c-fos within estrogen-concentrating regions of the female rat forebrain. Brain Res. 624, 253–267.
Phelix, C.F., Liposits, Z., Paull, W.K., 1992. Serotonin-CRF interaction in the bed nucleus of the stria terminalis: a light microscopic double-label immunocytochemical analysis. Brain Res. Bull. 28, 943–948.
Phoenix, C.H., 1961. Hypothalamic regulation of sexual behavior in male guinea pigs. J. Comp. Physiol. Psychol. 54, 72.
Potegal, M., Ferris, C., Hebert, M., Meyerhoff, J., Skaredoff, L., 1996. Attack priming in female Syrian golden hamsters is associated with a *c-fos*-coupled process within the corticomedial amygdala. Neuroscience 75, 869–880.
Powers, B., Valenstein, E.S., 1972. Sexual receptivity: facilitation by medial preoptic lesions in female rats. Science 175, 1003–1005.
Powers, J.B., Newman, S.W., Bergondy, M.L., 1987. MPOA and BNST lesions in male Syrian hamsters: differential effects on copulatory and chemoinvestigatory behaviors. Behav. Brain Res. 23, 181–195.
Prendergast, B., Nelson, R., Zucker, I., 2009. 14 mammalian seasonal rhythms: behavior and neuroendocrine substrates. Hormones 1, 507–538.
Razzoli, M., Andreoli, M., Michielin, F., Quarta, D., Sokal, D.M., 2011. Increased phasic activity of VTA dopamine neurons in mice 3 weeks after repeated social defeat. Behav. Brain Res. 218, 253–257.

Resko, J.A., Pereyra-Martinez, A.C., Stadelman, H.L., Roselli, C.E., 2000. Region-specific regulation of cytochrome P450 aromatase messenger ribonucleic acid by androgen in brains of male rhesus monkeys. Biol. Reprod. 62, 1818–1822.

Rey, C.D., Lipps, J., Shansky, R.M., 2014. Dopamine D1 receptor activation rescues extinction impairments in low-estrogen female rats and induces cortical layer-specific activation changes in prefrontal-amygdala circuits. Neuropsychopharmacology 39, 1282–1289.

Ribble, D.O., Salvioni, M., 1990. Social organization and nest co-ocupancy in *Peromyscus californicus*, a monogamous rodent. Behav. Ecol. Sociobiol. 26, 9–15.

Rompré, P.P., Boye, S.M., 1989. Localization of reward-relevant neurons in the pontine tegmentum: a moveable electrode mapping study. Brain Res. 496, 295–302.

Roselli, C.E., Horton, L.E., Resko, J.A., 1985. Distribution and regulation of aromatase activity in the rat hypothalamus and limbic system. Endocrinology 117, 2471–2477.

Roselli, C.E., Klosterman, S.A., Fasasi, T.A., 1996. Sex differences in androgen responsiveness in the rat brain: regional differences in the induction of aromatase activity. Neuroendocrinology 64, 139–145.

Roselli, C.E., Abdelgadir, S.E., Resko, J.A., 1997. Regulation of aromatase gene expression in the adult rat brain. Brain Res. Bull. 44, 351–357.

Roselli, C.E., Larkin, K., Resko, J.A., Stellflug, J.N., Stormshak, F., 2004. The volume of a sexually dimorphic nucleus in the ovine medial preoptic area/anterior hypothalamus varies with sexual partner preference. Endocrinology 145, 478–483.

Rosvall, K.A., Bergeon Burns, C.M., Barske, J., Goodson, J.L., Schlinger, B.A., Sengelaub, D.R., Ketterson, E.D., 2012. Neural sensitivity to sex steroids predicts individual differences in aggression: implications for behavioural evolution. Proc. R. Soc. Lond. B Biol. Sci. 279 (1742), 3547–3555.

Sartor, G.C., Aston-Jones, G., 2012. Regulation of the ventral tegmetal area by the bed nucleus of the stria terminalis is require for expression of cocaine preference. Eur. J. Neurosci. 36, 3549–3558.

Schumacher, M., Balthazart, J., 1986. Testosterone-induced brain aromatase is sexually dimorphic. Brain Res. 370, 285–293.

Scordalakes, E.M., Imwalle, D.B., Rissman, E.F., 2002. Oestrogen's masculine side: mediation of mating in male mice. Reproduction 124, 331–338.

Scott, C.J., Tilbrook, A.J., Simmons, D.M., Rawson, J.A., Chu, S., Fuller, P.J., Ing, N.H., Clarke, I.J., 2000. The distribution of cells containing estrogen receptors α (ERα) and ERβ messenger ribonucleic acid in the preoptic area and hypothalamus of the sheep: comparison of males and females. Endocrinology 141, 2951–2962.

Shah, N.M., Pisapia, D.J., Maniatis, S., Mendelsohn, M.M., Nemes, A., Axel, R., 2004. Visualizing sexual dimorphism in the brain. Neuron 43, 313–319.

Shaham, Y., Shalev, U., Lu, L., De Wit, H., Stewart, C., 2003. The reinstatement model of drug relapse: history, methodology and major findings. Psychopharmacology 168, 3–20.

Sheridan, P.J., 1979. The nucleus interstitialis striae terminalis and the nucleus amygdaloideus medialis: prime targets for androgen in the rat forebrain. Endocrinology 104, 130–136.

Shibata, S.T., Yamamoto, T.Y., Ueki, S., 1982. Differential effects of medial, central, and basolateral amygdaloid lesions on four models of experimentally-induced aggression in rats. Physiol. Behav. 28, 289–294.

Shirayama, Y., Chen, A.C.-H., Nakagawa, S., Russell, D.S., Duman, R.S., 2002. Brain-derived neurotrophic factor produces antidepressant effects in behavioral models of depression. J. Neurosci. 22, 3251–3261.

Shughrue, P.J., Lane, M.V., Scrimo, P.J., Merchenthaler, I., 1998. Comparative distribution of estrogen receptor-α (ER-α) and β (ER-β) mRNA in the rat pituitary, gonad, and reproductive tract. Steroids 63, 498–504.

Shuster, S.M., Wade, M.J., 2003. Mating Systems and Strategies. Princeton University Press, Princeton, NJ.

Silva, A.L., Fry, W.H.D., Sweeney, C., Trainor, B.C., 2010. Effects of photoperiod and experience on aggressive behavior in female California mice. Behav. Brain Res. 208, 528–534.

Simerly, R., Swanson, L., 1987. Castration reversibly alters levels of cholecystokinin immunoreactivity within cells of three interconnected sexually dimorphic forebrain nuclei in the rat. Proc. Natl. Acad. Sci. 84, 2087–2091.

Simerly, R., Swanson, L., Chang, C., Muramatsu, M., 1990. Distribution of androgen and estrogen receptor mRNA-containing cells in the rat brain: an *in situ* hybridization study. J. Comp. Neurol. 294, 76–95.

Snyder, M.A., Smejkalvoa, T., Forlano, P.M., Woolley, C.S., 2010. Multiple ERbeta antisera label in ERbeta knockout and null mouse tissues. J. Neurosci. Methods 188, 226–234.

Spiteri, T., Musatov, S., Ogawa, S., Ribeiro, A., Pfaff, D.W., Agmo, A., 2010. The role of the estrogen receptor a in the medial amygdala and ventromedial nucleus of the hypothalamus in social recognition, anxiety, and aggression. Behav. Brain Res. 210, 211–220.

Steinman, M.Q., Laredo, S.A., Lopez, E.M., Manning, C.E., Hao, R.C., Doig, I.E., Campi, K.L., Flowers, A.E., Knight, J.K., Trainor, B.C., 2015. Hypothalamic vasopressin systems are more sensitive to social defeat in males versus females. Psychoneuroendocrinology 51, 122–134.

Stolzenberg, D.S., Numan, M., 2011. Hypothalamic interaction with the mesolimbic DA system in the control of the maternal and sexual behaviors in rats. Neurosci. Biobehav. Rev. 35, 826–847.

Stolzenberg, D.S., Stevens, J.S., Rissman, E.F., 2012. Experience-facilitated improvements in pup retrieval: evidence for an epigenetic effect. Horm. Behav. 62, 128–135.

Stolzenberg, D.S., Stevens, J.S., Rissman, E.F., 2014. Histone deacetylase inhibition induces long-lasting changes in maternal behavior and gene expression in female mice. Endocrinology 155, 3674–3683.

Swanson, L., 1982. The projections of the ventral tegmental area and adjacent regions: a combined fluorescent retrograde tracer and immunofluorescence study in the rat. Brain Res. Bull. 9, 321–353.

Takahashi, L.K., Gladstone, C.D., 1988. Medial amygdaloid lesions and the regulation of sociosexual behavioral patterns across the estrous cycle in female golden hamsters. Behav. Neurosci. 106, 162–171.

Takeo, T., Chiba, Y., Sakuma, Y., 1993. Suppression of the lordosis reflex of female rats by efferents of the medial preoptic area. Physiol. Behav. 53, 831–838.

Tetel, M.J., Getzinger, M.J., Blaustein, J.D., 1993. Fos expression in the rat brain following vaginal-cervical stimulation by mating and manual probing. J. Neuroendocrinology 5, 397–404.

Thompson, T.L., Moss, R.L., 1994. Estrogen regulation of dopamine release in the nucleus accumbens: genomic and nongenomic-mediated effects. J. Neurochem. 62, 1750–1756.

Trainor, B.C., 2011. Stress responses and the mesolimbic dopamine system: social contexts and sex differences. Horm. Behav. 60 (5), 457–469.

Trainor, B.C., Marler, C.A., 2002. Testosterone promotes paternal behaviour in a monogamous mammal via conversion to oestrogen. Proc. R. Soc. Lond. B Biol. Sci. 269, 823–829.

Trainor, B.C., Bird, I.M., Alday, N.A., Schlinger, B.A., Marler, C.A., 2003. Variation in aromatase activity in the medial preoptic area and plasma progesterone is associated with the onset of paternal behavior. Neuroendocrinology 78, 36–44.

Trainor, B.C., Greiwe, K.M., Nelson, R.J., 2006. Individual differences in estrogen receptor α in select brain nuclei are associated with individual differences in aggression. Horm. Behav. 50, 338–345.

Trainor, B.C., Lin, S., Finy, M.S., Rowland, M.R., Nelson, R.J., 2007a. Photoperiod reverses the effects of estrogens on male aggression via genomic and nongenomic pathways. Proc. Natl. Acad. Sci. 104, 9840–9845.

Trainor, B.C., Rowland, M.R., Nelson, R.J., 2007b. Photoperiod affects estrogen receptor α, estrogen receptor β and aggressive behavior. Eur. J. Neurosci. 26, 207–218.

Trainor, B.C., Finy, M.S., Nelson, R.J., 2008a. Paternal aggression in a biparental mouse: parallels with maternal aggression. Horm. Behav. 53, 200–207.

Trainor, B.C., Sima Finy, M., Nelson, R.J., 2008b. Rapid effects of estradiol on male aggression depend on photoperiod in reproductively non-responsive mice. Horm. Behav. 53, 192–199.

Trainor, B.C., Crean, K.K., Fry, W.H., Sweeney, C., 2010. Activation of extracellular signal-regulated kinases in social behavior circuits during resident-intruder aggression tests. Neuroscience 165, 325–336.

Trainor, B.C., Pride, M.C., Villalon Landeros, R., Knoblauch, N.W., Takahashi, E.Y., Silva, A.L., Crean, K.K., 2011. Sex differences in social interaction behavior following social defeat stress in the monogamous California mouse (*Peromyscus californicus*). PLOS One 6, e17405.

Trainor, B.C., Takahashi, E.Y., Campi, K.L., Florez, S.A., Greenberg, G.D., Laman-Maharg, A., Laredo, S.A., Orr, V.N., Silva, A.L., Steinman, M.Q., 2013. Sex differences in stress-induced social withdrawal: independence from adult gonadal hormones and inhibition of female phenotype by corncob bedding. Horm. Behav. 63, 543–550.

Tsuneoka, Y., Maruyama, T., Yoshida, S., Nishimori, K., Kato, T., Numan, M., Kuroda, K.O., 2013. Functional, anatomical, and neurochemical differentiation of medial preoptic area subregions in relation to maternal behavior in the mouse. J. Comp. Neurol. 521, 1633–1663.

Van Dis, H., Larsson, K., 1971. Induction of sexual arousal in the castrated male rat by intracranial stimulation. Physiol. Behav. 6, 85–86.

Viglietti-Panzica, C., Panzica, G.C., Fiori, M.G., Calcagni, M., Anselmetti, G.C., Balthazart, J., 1986. A sexually dimorphic nucleus in the quail preoptic area. Neurosci. Lett. 64, 129–134.
Vochteloo, J.D., Koolhaas, J.M., 1987. Medial amygdala lesions in male rats reduce aggressive behavior: interference with experience. Physiol. Behav. 41, 99–102.
Wade, J., Huang, J., Crews, D., 1993. Hormonal control of sex differences in the brain, behavior and accessory sex structures of whiptail lizards (*Cnemidophorus*) species. J. Neuroendocrinol. 5, 81–93.
Walker, Q.D., Rooney, M.B., Wightman, R.M., Kuhn, C.M., 1999. Dopamine release and uptake are greater in female than male rat striatum as measured by fast cyclic voltammetry. Neuroscience 95, 1061–1070.
Wang, Y., He, Z., Zhao, C., Lei, L., 2013. Medial amygdala lesions modify aggressive behavior and immediate early gene expression in oxytocin and vasopressin neurons during intermale exposure. Behav. Brain Res. 245, 42–49.
Whitney, J.F., 1986. Effect of medial preoptic lesions on sexual behavior of female rats is determined by test situation. Behav. Neurosci. 100, 230–235.
Wise, D.A., 1974. Aggression in the female golden hamster: effects of reproductive state and social isolation. Horm. Behav. 5, 235–250.
Wissman, A.M., May, R.M., Woolley, C.S., 2012. Ultrastructual analysis of sex differences in nucleus accumbens synaptic connectivity. Brain Struct. Funct. 217, 181–190.
Wood, R.I., 1996. Estradiol, but not dihydrotestosterone, in the medial amygdala facilitates male hamster sex behavior. Physiol. Behav. 59, 833–841.
Wood, R.I., Newman, S.W., 1993. Mating activates androgen receptor-containing neurons in chemosensory pathways of the male Syrian hamster brain. Brain Res. 614, 65–77.
Wood, R.I., Newman, S.W., 1995. The medial amygdaloid nucleus and medial preoptic area mediate steroidal control of sexual behavior in the male Syrian hamster. Horm. Behav. 29, 3380353.
Wu, M.V., Manoli, D.S., Fraser, E.J., Coats, J.K., Tollkuhn, J., Honda, S.-I., Harada, N., Shah, N.M., 2009. Estrogen masculinizes neural pathways and sex-specific behaviors. Cell 139, 61–72.
Xu, X., Coats, J.K., Yang, C.F., Wang, A., Ahmed, O.M., Alvarado, M., Izumi, T., Shah, N.M., 2012. Modular genetic control of sexually dimorphic behaviors. Cell 148, 596–607.

# CHAPTER 5

# Sexual Dimorphisms in Psychosis Risk: A Neurodevelopmental Perspective

**Deborah J. Walder, Beril Yaffe, Yosefa Ehrlich**
Brooklyn College of the City University of New York, Department of Psychology, Brooklyn, NY, USA

## 1 INTRODUCTION

Schizophrenia is a heterogeneous psychiatric disorder that is widely accepted as having early neurodevelopmental origins (Murray and Lewis, 1987; Weinberger, 1987). Consistent with a diathesis–stress model, latent vulnerability later manifests in illness expression via perturbations during normal neuromaturational processes (Walker and Bollini, 2002; Walker et al., 2004). Despite a relatively low prevalence rate of approximately 1% (Kendler et al., 1996), psychosis (e.g., schizophrenia and related psychotic spectrum disorders) has galvanized increased attention in the field of mental health for several reasons. First, in addition to prominent psychotic symptomatology (e.g., hallucinations, delusions, and behavioral deficits such as anhedonia and avolition, disorganized behavior) (Andreasen and Olsen, 1982), characteristic features such as impairments in neurocognition (Heinrichs, 2005; Nuechterlein et al., 2004; Walder et al., 2000), and social cognition (Pinkham et al., 2014) adversely impact "social role and global functioning" (Cornblatt et al., 2007). In turn, these deficits may translate into disabilities in employment, interpersonal relationships, and quality of life. Second, empirical evidence over the past few decades has brought into focus the need for accurate early identification of persons at high risk for psychosis, during incipient stages prior to illness onset (the "psychosis prodrome"). Preventive strategies hold potential to mitigate (if not thwart) the course and potentially devastating effects of illness. The corresponding body of scientific literature has emphasized the convergence of neurobiological and environmental precursors to psychosis risk, as well as measurable phenotypic biomarkers (which validate the high-risk state) that may aid early detection.

Building on earlier conceptualizations of psychosis etiology (Murray and Lewis, 1987; Walker and Diforio, 1997; Zubin and Spring, 1977) we proposed and provided support for the early components of a "polygenic neurodevelopmental diathesis–stress model" (Walder et al., 2014). Accordingly, psychosis vulnerability involves the temporally sensitive, independent, and synergistic confluence of numerous biological (e.g., genetic, perinatal) and environmental (e.g., stress exposure) factors, which impact brain maturation across sensitive developmental windows; namely, the prenatal period through adolescence (Walder et al., 2014). In turn, key neuromaturational processes (e.g., synaptic

*Sex Differences in the Central Nervous System.* http://dx.doi.org/10.1016/B978-0-12-802114-9.00005-6

pruning, neurohormonal surges) during these periods modify psychosis risk and features of illness.

As we have reported (Walder et al., 2013), newer on the scene is consideration of sexual dimorphisms and the modulating role of one's biological sex and neurohormones (e.g., by-products of the hypothalamic–pituitary–adrenal (HPA) and hypothalamic–pituitary–gonadal (HPG) axes) across the neurodevelopmental trajectory in conferring psychosis risk. This line of investigation has been prompted, in part, by well-documented sex differences in psychosis such as age at onset, premorbid history, clinical course, and prevalence (Walker et al., 2002; Goldstein and Walder, 2006; Salem and Kring, 1998; Seeman, 1982). Converging evidence points to sex differences among neurodevelopmental factors that give rise to, or reflect, compromises in central nervous system (CNS) integrity ("precursors" and "phenotypic traits," respectively), and play a role in psychosis risk. For example, the male (compared with female) brain is more vulnerable to disruptions during critical periods that have been established as contributing to psychosis risk. For example, obstetric complications have been more strongly linked with: (1) soft neurological signs (Hadders-Algra et al., 1988), greater in males than females; and (2) greater neuropsychological deficits (Casar et al., 1996) and neurological soft signs (Lane et al., 1996) in males (than females) with schizophrenia (Walker et al., 2002b). Exemplary biomarker candidates that appear to be associated with sexually differentiated effects in psychosis susceptibility include neurocognition, neurohormonal systems (e.g., HPA activity), perinatal complications, neuroanatomic structure and function, and genetic liability. These biomarkers are of interest given they exert effects (albeit not exclusively) during early development. Mechanisms of sex effects linked with such biomarkers, however, remain not fully explained.

In this chapter we provide a cursory overview of sex differences in clinical phenomenology, social cognition, and role functioning (and their neurobiological underpinnings), and select key neurodevelopmental factors implicated in psychosis, as well as theoretical conceptualizations, with an emphasis on risk. Of note, several methodological approaches (and terms) are used in the literature to investigate psychosis "risk." High-risk paradigms are advantageous by circumventing shortcomings in clinical research such as confounding effects of medication and hospitalization. One high-risk study design targets samples at-risk based on clinical criteria (e.g., prodromal, ultra high risk, clinical high risk); that is, individuals who demonstrate a recent functional decline or subthreshold psychotic-like symptoms. Alternative approaches include identification of individuals at familial/genetic high risk (those who have biological relatives with psychosis) or psychometric high risk (individuals from the general population who score high on self-report measures of psychotic-like traits) (Lenzenweger, 1994). In this chapter we use the term "high risk" to refer to all such at-risk samples and study designs.

Rationale for the current focus on high-risk literature is multifold. First, sex differences in psychosis are already well documented (Walker et al., 2002; Salem and Kring, 1998;

Seeman, 1982), whereas examination of sexual differentiation among individuals at high risk of psychosis is in its nascent stage. Preliminary findings suggest effects across the developmental trajectory. Second, there is a compelling need for research to better characterize the psychosis prodrome, with a focus on enhancing tools aimed at early detection and individually tailored preventive intervention. This is a priority given current psychosis prevention strategies emphasize late-emerging positive symptoms that occur after disability are relatively well established (Cornblatt et al., 2007, 2011). Third, sex and gonadal hormones have scarcely been considered as risk modifiers, despite evidence in preclinical animal (Farrell et al., 2015; Shansky et al., 2004, 2006; Shansky and Morrison, 2009) and nonclinical human studies. This is due to inherent challenges confronting sex differences research, such as funding limitations and methodological obstacles (see Becker et al., 2005 for a guideline of scientific strategies to investigate sex differences). Recent funding initiatives have been aimed at reducing impediments such as the imbalance of sex in scientific research (see McCullough et al., 2014). Among initiatives are National Institutes of Health (NIH) Administrative Supplements for Research on Sex/Gender Differences, and the Office of Research on Women's Health Specialized Centers of Research on Sex Differences (cofunded by several NIH institutes), who promote integration of basic, clinical, and translational research that incorporates sex and gender foci. Finally, research aimed at uncovering sexual dimorphisms during psychosis risk holds important implications for better understanding etiology, antecedents, and course of illness, building more robust predictive models of illness conversion and developing preventive interventions better tailored to meet individual (including sex-specific and sex-dependent) needs.

## 2 THEORETICAL CONCEPTUALIZATIONS

Early competing theories explaining sexually dimorphic patterns in schizophrenia (Castle et al., 1995) considered the protective role of gonadal hormones (e.g., estrogen hypothesis; Häfner et al., 1991; Riecher-Rössler and Häfner, 1993; Seeman and Lang, 1990) versus subtypes differentially vulnerable by sex. Castle and Murray (1991) proposed a greater susceptibility among males to a "neurodevelopmental" subtype distinguished by a more notable deficit syndrome. More contemporary integrative models incorporate a broader landscape of factors. We recently speculated that sexually dimorphic, genetically linked influences underlie sexually differentiated subtypes (or manifestations) of psychosis (Walder et al., 2013) which may be modulated by gonadal hormones (Walder et al., 2012b). The focus on gonadal hormones is primarily based on findings demonstrating a role of sex hormones in psychosis (Seeman and Lang, 1990) and psychosis risk (Walder et al., 2006). Moreover, the psychosis prodrome (which typically occurs during adolescence) is characterized by prominent hormonal fluctuations. Specifically, cortisol (a by-product of the HPA axis, the primary system in the mammalian biological stress response) is implicated in psychosis (Walder et al., 2000), sexually differentiated

(Kajantie and Phillips, 2006) and greater among high-risk youth who transition to psychosis (Walker et al., 2010). Additionally, in line with the neural diathesis–stress model (see Walker and Diforio, 1997), cortisol is believed to augment dopaminergic function (which potentiates psychotic symptoms).

## 3 SEX DIFFERENCES IN CLINICAL PHENOMENOLOGY AND FUNCTIONING

The few high-risk studies to date suggest a pattern of sexual dimorphisms in phenomenology that parallels that observed in acute illness. The general pattern of findings in psychosis is earlier age at onset, more severe negative and less severe affective symptoms, as well as poorer premorbid history, clinical course, and prognosis among males than females (Walker et al., 2002; Salem and Kring, 1998; Seeman, 1982). Sex distribution with respect to incidence and prevalence rates varies, depending on factors such as stringency of diagnostic criteria applied (for review see Abel et al., 2010). One recent meta-analysis revealed a male:female mean ratio incidence of 1.4:1, with a lower estimate when using less restrictive criteria (Aleman and Selten, 2003). However, this pattern appears to shift from the early twenties to older ages when there is a female predominance in incidence (Leung and Chue, 2000). In the high-risk literature, perhaps most compelling is evidence from the North American Prodrome Longitudinal Study (NAPLS), a consortium of eight research centers that ascertained and followed a cohort of individuals at clinical high risk for psychosis for up to 2 years (Addington et al., 2007). The NAPLS study represents one of the largest, longitudinal, high-risk databases worldwide. One NAPLS report demonstrated high-risk males experienced more severe negative and disorganized symptoms, with no sexual differentiation in positive and mood symptoms (Holtzman et al., in review). Another smaller, clinical, high-risk study showed more negative symptoms among males than females (Corcoran et al., 2011), which replicated and built upon an earlier retrospective report using smaller samples (Willhite et al., 2008). One more recent NAPLS report demonstrated baseline positive prodromal symptoms predicted conversion to psychosis (Walder et al., 2013). In sum, clinical phenomenology in psychosis risk appears analogous to that observed in psychosis.

## 4 SEX DIFFERENCES IN COGNITION

### 4.1 Neurocognition

Neurocognitive dysfunction represents one well-established characteristic feature of schizophrenia (Green, 2006; Heinrichs, 2005; Nuechterlein et al., 2004; Walder et al., 2000), notably in the domains of attention, processing speed, working memory and learning, and verbal and executive functions (Gold, 2004; Kuperberg and Heckers, 2000; Walder et al., 2000). Consistent with Meehl (1962, 1990), a similar pattern holds largely across the psychosis spectrum, including among first-episode patients (Mesholam-Gately et al., 2009), individuals at clinical (Brewer et al., 2006; Hawkins et al., 2004; Niendam et al., 2006; Seidman et al., 2010; Trotman et al., 2006; for a meta-analysis see Fusar-Poli et al., 2012;), genetic (Cannon et al., 1994), and psychometric (e.g., Cohen et al., 2006; Gooding et al., 2006) high risk, and during the premorbid period (for a meta-analytic review, see Aylward et al., 1984; Reichenberg et al., 2006). Increasingly, high-risk reports show cognitive deficits predict conversion to psychosis (Keefe et al., 2006; Lencz et al., 2006; Seidman et al., 2010). Neural abnormalities in prefrontal areas have been observed among individuals at psychometric risk (Nelson et al., 2012), clinical high risk (Fusar-Poli et al., 2012), ultra high risk (Fusar-Poli et al., 2011), as well as first-episode (Andreasen et al., 1997) and chronic schizophrenia patients (Kim et al., 2000), potentially accounting for the cognitive abnormalities observed in the schizophrenia spectrum (Kim et al., 2000).

Sex differences in cognition also are well-established in schizophrenia (Goldstein and Walder, 2006; Hoff and Kremen, 2002; Lewine et al., 1996; Walder et al., 2006; for review, see Leung and Chue, 2000). In particular, there is a female verbal advantage and evidence that normal sex differences (e.g., in language functions) are disrupted in schizophrenia (Walder et al., 2006) with neuroanatomic substrates (Walder et al., 2007). Surprisingly few studies have examined sex differences in neurocognition prior to illness onset during the premorbid and clinical high-risk periods (Aylward et al., 1984; Bilder et al., 2000; Erlenmeyer-Kimling and Cornblatt, 1984; Ittig et al., 2015; Walder et al., 2008; Jones and Done, 1997). Overall, studies point to better cognitive (particularly verbal) functions among females (compared to males) during the premorbid period (among those who subsequently develop schizophrenia) and clinical high-risk phase. Even fewer high-risk studies have examined sex differences in baseline cognition as a predictor of longitudinal conversion to psychosis. One study demonstrated an absence of significant differences in neurocognition between high-risk youth who converted to psychosis and those who did not (Walder et al., 2008). However, within-sex comparisons revealed a sex-specific effect among females only, whereby high-risk youth who converted to psychosis performed worse than high-risk youth who did not convert. Collectively, findings in this emerging literature bolster the use of neurocognitive profiles as sexually differentiated biomarkers of psychosis risk.

## 4.2 Social cognition

Psychosis research has increased its focus on social cognition; that is, how individuals detect, understand, perceive, interpret, and utilize social information (Penn et al., 1997; Pinkham et al., 2014). The realm of social cognition is wide and includes various skills (Green et al., 2005) such as emotion processing, understanding the thoughts and intentions of others (theory of mind), recognition of emotion (e.g., facial expression), social perception (interpretation of social cues), social knowledge, and attributional style. Social cognition is closely related to neurocognition, though is generally believed to constitute a distinct construct (Green et al., 2005; Penn et al., 1996). Importantly, the association of social cognitive impairments in schizophrenia (Horan et al., 2008) with functional outcome appears to supersede the contribution of basic cognitive deficits (Penn et al., 1996;

Pinkham and Penn, 2006). Related is empirical support (Addington et al., 2006; Brekke et al. 2005) of theoretical models that posit social cognition mediates neurocognition in relation to functional status in schizophrenia (Green and Nuechterlein, 1999). Interestingly, social cognition is among current competing candidate predictors of functional outcome in psychosis (Couture et al., 2006; Irani et al., 2012). Moreover, it has been suggested that neural networks subserving social cognition (medial prefrontal cortex, superior temporal sulcus, fusiform gyrus, amygdala, ventromedial prefrontal cortex) may underlie social dysfunction in patients (Pinkham et al., 2014). Additionally, these specific areas have been shown to be impaired in patients (Pinkham et al., 2014).

Dove-tailing the clinical literature, the majority of psychosis clinical high-risk research (Addington et al., 2008; Fusar-Poli et al., 2012; Shim et al., 2008; Yong et al., 2014) and high-risk research using other paradigms (familial risk, schizotypal personality disorder, psychometric high risk in the general population) (for brief overview see Walder et al., 2014) has likewise demonstrated impaired social cognition, though not consistently. This, together with evidence that social cognition (i.e., facial affect recognition; FAR) is linked with social functioning in psychosis (Couture et al., 2006) and among high-risk groups (for overview see Statucka and Walder, 2013). We recently argued that early social cognitive remediation (e.g., FAR remediation) is important, particularly among high-risk individuals, as it may diminish functional impairments and/or hinder psychosis onset (Statucka and Walder, 2013).

FAR is an aspect of social cognition demonstrated to be sexually differentiated; namely, healthy females outperform healthy males (Hall, 1978; McClure, 2000). Given recent evidence of sex differences in early social and role dysfunction among individuals at clinical high risk for psychosis (Walder et al., 2013), we emphasized the need for examination of sexual differentiation in social cognitive remediation strategies along the psychosis continuum (Walder et al., 2014). Toward better delineating types of sex-specific (or sex-dependent) interventions, we first looked more closely at FAR across the psychosis continuum. Interestingly, the healthy female FAR advantage does not appear to hold among schizophrenia spectrum disorders (Hofer et al., 2009; Kohler et al., 2003; for a meta-analysis, see Chan et al., 2010), healthy relatives (Kee et al., 2004), and high-risk individuals (Brown and Cohen, 2010; Jahshan and Sergi, 2007; Toomey and Schuldberg, 1995); though there are methodological limitations in this literature. Likewise, Kurtz and Richardson (2012) meta-analysis did not reveal sex as a moderator; however, sex was not equally distributed in many of the included studies, yielding a potential bias. We argued that use of broader samples that are more equally distributed by sex might yield inherent sex effects.

Taken together, as with the neurocognitive literature, research (albeit limited) suggests the normal pattern of sexual dimorphism of social cognition (specifically for FAR) across the psychosis spectrum is disrupted. This adds to mounting evidence suggesting that sex effects are an important consideration, especially with respect to future interventions given that social cognitive deficits may interfere with establishment of healthy, sustained social relationships. Of course, this line of investigation is in its early stages, warranting more rigorous future research.

## 5 SEX DIFFERENCES IN OTHER NEUROBIOLOGICAL PRECURSORS

As per our recent review (Walder et al., 2013), functional impairments not only accompany psychosis, but also manifest in years preceding clinical onset (Cornblatt et al., 2003, 2011), including social and role functioning deficits among individuals at clinical high risk (Corcoran et al., 2011; Cornblatt et al., 2011; Woods et al., 2009). Among the few longitudinal, prospective studies in this area, evidence shows worse baseline functioning predicts transition to psychosis (Corcoran et al., 2011; Dragt et al., 2011; Tarbox et al., 2013; Yung et al., 2003). In tandem, the NAPLS study demonstrated poorer childhood social functioning (Tarbox et al., 2013), and baseline social (and less robust role) functioning (Cornblatt et al., 2011) predicted conversion among individuals at clinical high risk. The even fewer studies directly examining sexual dimorphisms found social (Thorup et al., 2007) and premorbid (Walker et al., 2002) functional impairment in psychosis. Males demonstrate more premorbid academic and occupational deficits (McGlashan and Bardenstein, 1990). We are aware of two relevant clinical high-risk studies. One study demonstrated no sex differences in functioning, though males demonstrated a more insidious and prolonged deterioration than females (Lemos-Giráldez et al., 2009). The NAPLS study (Walder et al., 2013) revealed (as with patients) sex-dependent findings; namely, more impaired baseline social and role functioning in high-risk males than females (though early childhood academic, social, and total adjustments were comparable between sexes). Moreover, there was a sexually dimorphic pattern of associations among variables. Finally, social and role functioning predicted conversion, in a sex-dependent manner (e.g., poorer baseline social functioning and positive prodromal symptoms predicted greater conversion risk among males only; no predictors among females).

Overall, findings suggest sexually dimorphic patterns of baseline global functioning prior to psychosis onset and in relation to prodromal symptoms. Moreover, social (more than role) functioning appears to be a key predictor of conversion, in a manner consistent with the clinical literature (greater deficits among males) (McGlashan and Bardenstein, 1990).

### 5.1 Neurobiological stress

Another area experiencing an acceleration of research over the past few decades is examination of stress influences on physical and mental health, with an emphasis on neurobiological pathways and (increasingly) sexual differentiation. The HPA axis plays a key role in maintaining homeostasis. HPA dysregulation, beginning as early as the prenatal period (during fetal programming) through to adolescence and adulthood, can contribute to illness risk via homeostatic disruptions. Given known sex differences in HPA activity and sensitivity (Kudielka and Kirschbaum, 2005), it is not surprising that medical conditions (e.g., cardiovascular disease, multiple sclerosis, Alzheimer's disease) and various neuropsychiatric disorders (e.g., psychosis, depression, autism) that are marked by sex differences (in prevalence, clinical presentation, and/or treatment response) also involve HPA dysregulation. HPA integrity is particularly salient to understanding psychosis risk for several reasons. As reflected by the Walker and Diforio (1997) neural diathesis–stress model of schizophrenia, HPA dysreguation (e.g., heightened cortisol secretion) is evident in psychotic patients (Guest et al., 2011; Steen et al., 2011; Yıldırım et al., 2011) and believed to augment dopaminergic activity (Schatzberg et al., 1985; Wolkowitz, 1994), which in turn may trigger latent psychotic symptomatology. With respect to understanding psychosis risk, evidence largely supports a role for both behavioral and neurobiological stress sensitivity, particularly during the adolescent period. This makes sense given adolescence is characterized by marked neurohormonal surges (including by-products of the HPA and HPG axes; cortisol and gonadal hormones, respectively) and neuromaturational processes – as well as a period of heightened risk for transition to psychosis. Specifically, the prodromal phase has been demonstrated to involve elevated stress exposure and stress–sensitivity, and stressful life events tend to be experienced as more subjectively stressful (Trotman et al., 2014). In terms of neurobiology, there is mounting evidence of heightened cortisol secretion among clinical high-risk individuals, and nonspecific associations of cortisol with symptom severity and progression (Walker et al., 2013).

Mechanisms of action of the HPA system are multifactorial and complex, particularly given its dynamic interplay with a wide range of other neurobiological systems (e.g., regulation by the hippocampus via negative feedback). In this section, we circumscribe the focus to a brief overview of sex differences in HPA activity (sensitivity and responsivity) in nonclinical and clinical literature, given the absence of studies directly examining HPA sexual dimorphisms in psychosis risk. This work offers insight regarding the potential for sexually differentiated HPA activity in psychosis susceptibility. We also briefly address the importance of HPA synergistic activity with the HPG axis in conceptualizing biobehavioral mechanisms of psychosis risk.

Human research in the general population has evidenced sex differences in cortisol secretion (Kudielka and Kirschbaum, 2005) vary as a function of female menstrual cycle status (Kajantie and Phillips, 2006). Generally, males show a more robust stress response than females (Kajantie and Phillips, 2006), between-sex cortisol response is comparable when females are in the luteal phase (Kirschbaum et al., 1999), including post-stress (Kajantie and Phillips, 2006) and cortisol is greater among males than females who are in the follicular phase (Kirschbaum et al., 1999). Moreover, among females there is a greater evoked cortisol response during the luteal (as compared to the follicular) phase (Andreano et al., 2008; Kirschbaum et al., 1999) (summarized in Walder et al., 2012b).

Although sex differences in cortisol have not been examined in high-risk individuals, psychiatric symptoms are known to vary across the menstrual cycle phase (see Steiner et al., 2003). In one study using a nonclinical, adult sample (Walder et al., 2012b), cortisol levels were: (1) lowest during early-follicular, intermediary during late-follicular, and highest during mid-luteal phases (albeit not significantly); and (2) higher among males, particularly when females were in the early-follicular phase when estrogen levels are expected to be relatively low (compared with late-follicular and mid-luteal phases, when estrogen peaks then plateau). Moreover, cortisol was inversely related to nonclinical psychiatric symptoms during the early-follicular phase, and directly related to late-follicular/mid-luteal symptoms. Overall, results suggested differential effects of menstrual cycle phase on nonclinical psychiatric symptoms and the cortisol stress response. We speculated a dose–response cortisol effect on symptoms, modulated by endogenous gonadal hormones via gene expression. These findings together with sex differences in stress among individuals at clinical high risk render further exploration of sex differences in HPA activity compelling.

It is worth noting promising new directions regarding neuromodulatory actions of hormones on brain structure, function, and sex effects; particularly mechanisms by which hormonal effects on gene expression may influence psychosis risk. A comprehensive review of this topic can be found elsewhere (see Walker et al., 2011; Trotman et al., 2013). To highlight some key points, it is now known that estradiol exerts dual (synergistic) effects on sexual brain dimorphisms. First, through slow, enduring genomic effects via nuclear transcription; second, through more rapid, transient membrane-mediated effects that signal transduction cascades (Falkenstein et al., 2000; McEwen, 1991; see McCarthy et al., 2009). Moreover, evidence suggests differential messenger ribonucleic acid expression of glucocorticoid receptors across the menstrual cycle (Altemus et al., 1997). Finally, estrogen and progesterone regulate glucocorticoid and mineralocorticoid receptors in the brain; in turn, this regulates stress responsivity as a function of menstrual phase and brain region (Burgess and Handa, 1992; Patchev et al., 1995; see Kajantie and Phillips, 2006). As previously discussed (Walder et al., 2012b), this holds potential implications for brain regions involved in stress responsivity and psychosis (e.g., hippocampus, amygdala, hypothalamus; see Viau, 2002) that: (1) are normally sexually dimorphic; (2) show disrupted sexually dimorphic patterns in psychosis (Goldstein et al., 2002); and (3) contain high hormone receptor concentrations. One example is the hippocampus, which has a high concentration of estrogen receptors, plays an integral role in HPA regulation and psychosis risk, and is modulated by synaptic neurotransmission (via alterations in dendritic spine density) across the estrous cycle (e.g., Woolley and McEwen, 1992). Also of import is the notion that hormones yield both organizational and activational effects on the brain not only during fetal development, but also during adolescence (Arnold and Breedlove, 1985; Brown and Spencer, 2013; Charmandari et al., 2002). Taken together, this theoretical and empirical evidence is encouraging for future investigations of at-risk youth, toward understanding the underpinnings and effects of sexually dimorphic neurodevelopmental risk factors.

## 5.2 Perinatal complications

Various adverse perinatal events that can alter neurodevelopment have been linked with psychosis risk (Susser et al., 1996; Yolken and Torrey, 1995; for review see Cannon et al., 2002 and Walder et al., 2012a), including in a sex-dependent manner (Cannon et al., 2002; Hultman et al., 1999; for review see Goldstein and Walder, 2006). Perinatal complications vary in nature and are believed to bear differential behavioral and neuromaturational consequences contingent on severity, duration, and timing (e.g., occurrence during first, second, or third trimester, around delivery, and early postnatally), and the fetus's biological sex (males are believed to be more vulnerable to *in utero* insults). Examples of perinatal complications include prenatal maternal stress exposure, teratogen exposure, obstetric complications (preeclampsia, viral exposure, malnutrition), and maternal stress exposure. Also important are prenatal hormones, which exert both organizational and activational effects on neural development and impact sexual brain dimorphisms (see discussion of hormones in earlier and subsequent sections of this chapter). Disruptions in the prenatal gonadal hormone milieu have been implicated in psychosis risk (Walder et al., 2006).

As reviewed previously (Walder et al., 2012a), evidence also links obstetric complications with neurobiological (neuropsychological functioning; Casar et al., 1996), neurological soft signs (Lane et al., 1996), and behavioral (Hadders-Algra et al., 1988) deficits in psychosis, to a greater extent in males than females. Evidence of sexual dimorphisms in behavioral sequelae of prenatal insults is commensurate with greater male (than female) fetus vulnerability to prenatal gonadal hormone disruptions and adverse perinatal events (Hultman et al., 1999; see Goldstein and Walder, 2006). This may serve as the conduit to psychosis (Dalman et al., 2001; O'Callaghan et al., 1992; Rosso et al., 2000). There is compelling evidence of sex differences in rates of occurrence of obstetric complications in psychosis (males greater than females) (Foerster et al., 1991; Matsumoto et al., 2001) (albeit not consistently; see Goldstein and Walder, 2006), and greater risk of schizophrenia among boys exposed to obstetric complications (Hultman et al., 1999). In terms of sexually differentiated timing effects, Susser et al. (1996) found that prenatal exposure to severely malnourished mothers as a result of the 1944–1945 Dutch Hunger Winter were at a twofold increased risk for schizophrenia; and exposure during the second and third (not first) trimesters yielded sex effects in risk.

New to the field is emerging evidence that perinatal complications are linked with psychosis risk, in the context of familial high-risk studies (Gilbert et al., 2003; Walder et al., 2014). Yet, there are no known published studies to date examining sex effects in perinatal complications associated with clinical high risk for psychosis. This line of investigation holds promise in light of respective findings in the animal literature. Specifically, in a unique and elegant study, Piontkewitz et al. (2011) tracked neuroanatomic and behavioral abnormalities across the developmental trajectory (from adolescence to young adulthood) in rats born to mothers exposed to a viral mimic during pregnancy. Results indicated sex-specific effects; namely, onset of neuroanatomic abnormalities in frontocortical and ventricular regions, and behavioral abnormalities were delayed among females. However, hippocampal and striatal volumetric reductions preceded behavioral abnormalities in both sexes.

Further elaboration of this complex area falls beyond the scope of this chapter. Importantly, the cumulative research highlights the role of biological sex and prenatal hormones in modulating the impact of prenatal insults on brain development, behavior, and psychosis risk. Expansion of this line of investigation, with an even greater focus on the prodromal phase of psychosis risk (and interactions among variables) is warranted.

## 5.3 Genetic liability

Schizophrenia is known to have a high heritability estimate of approximately 0.65–0.70 (Gottesman and Shields, 1967), confirmed by national population-based and registry studies (Lichtenstein et al., 2009; Wray and Gottesman, 2012). Molecular genetics studies further indicate the genetic architecture of schizophrenia involves complex polygenic influences such as single nucleotide variants and copy number variants (Smoller et al., 2013; Tsuang et al., 2001). Currently, the search for genetic biomarkers focuses on polygenic influences (combined effects of several common alleles of small to moderate effects and possible rare events, e.g., copy number variants) (Gottesman and Shields, 1967; Wang et al., 2005) with increased attention on modifier (rather than susceptibility/candidate) genes. Despite high heritability, concordance among monozygotic twins approximating 41–65% (Cardno and Gottesman, 2000) implies environmental influences. Accordingly, contemporary polygenic models consider these genetic factors in concert with epigenetic mechanisms (reversible changes in gene expression imposed by environmental influences) (Tsuang et al., 2001).

In light of a well-established sex difference in the familial transmission of psychosis, it is not surprising that sexually dimorphic patterns have subsequently emerged in the context of molecular genetic studies (including genome-wide association studies; GWAS). Recent literature offers a review of sex differences in genetic risk for psychosis (Walder et al., 2012a; Goldstein et al., 2013), including among patterns of familial transmission, linkage and twin (behavioral genetic) studies, and GWAS (molecular genetic) studies. In this section, we briefly summarize cumulative evidence pointing to both sex-specific (apparent in only one sex) and sex-dependent (quantitative between-sex differences) risks for psychosis. Surprisingly, there is a notable absence of literature examining sexually differentiated genetic risk factors among individuals at clinical high risk for psychosis; this remains an important avenue for future research, particularly now that large-scale collaborative research efforts tend toward more robust sample sizes and, in turn, better statistically powered studies.

Findings among studies of familial transmission are mixed, and complicated by a wide range of methodological issues. Collectively, however, evidence suggests higher risk for psychotic illness among relatives of female probands, versus higher risk for forms of illness that do not involve acute psychosis (such as schizotypal personality disorder) among relatives of male probands; moreover, mothers are more likely to have affected male (18.8%) than female (9.5%) offspring, whereas fathers are more likely to have affected female (15.2%) than male (3.1%) offspring (Goldstein et al., 2013). The minimal investigation of sex-dependent effects in current twin studies (with one exception, in which sex-specific genetic effects were not found; Cannon et al., 1998) has not helped clarify sex-related genetic risk for psychosis (Goldstein et al., 2013). Another candidate genetic factor is the sex chromosomes, which were initially of interest given that the greatest sex difference in genetics resides with the X and Y chromosomes. This literature (including GWAS studies) is wide and marked by discrepancies across studies due to persistence of Type I error (e.g., small sample size yielding underpowered statistical analyses), Type II error, heterogeneous samples, and inadequate statistical methods to account for complexities in polygenic transmission (Bearden et al., 2004; Porteous et al., 2003) (reviewed in Goldstein et al., 2013). Future examination of the X chromosome using the Psychiatric Genetics Consortium database (which pools together multiple databases) may prove fruitful in addressing such limitations.

Candidate gene studies have indicated sex-specific and sex-dependent effects, though again Type I error renders results tenuous. As reviewed by Walder et al. (2012a) and Goldstein et al. (2013): (1) exemplary sex-dependent findings include an increased risk for schizophrenia associated with *SPEC2*, *PDZ-GEF2*, and *ACSL6* genes on chromosome 5q22-33 (Chen et al., 2006); (2) exemplary female-specific findings included *DISC-1* (Hennah et al., 2003), *COMT* (Shifman et al., 2002), *Nogo* gene (Tan et al., 2005), *GPR50* gene (Thomson et al., 2005), and *DRD3* gene (Godlewska et al., 2010); (3) exemplary male-specific findings include lipoprotein lipase gene (Xie et al., 2011); and (4) sex-specific findings using GWAS have primarily involved identification of female-specific genes such as Reeling gene (Shifman et al., 2008), *RYR2* gene (Wang et al., 2011), and *PBEF1* gene (Wang et al., 2011), which parallels familial transmission studies (whereby female probands come from the greatest multiply affected families).

In sum, there appears to be a role of sex-linked variation in genetic factors in psychosis risk, though findings remain inconclusive given inconsistent methodological approaches. The advent of GWAS and emerging genetic approaches hold promise for more effective, systematic future studies involving greater statistical power. Moreover, in light of the potential role of epigenetic influences (e.g., gonadal and adrenal hormone induced alterations in gene expression without alteration of DNA sequence; Sisk and

Zehr, 2005) on psychosis risk (see Holtzman et al., 2013), and observed sex differences in epigenetics (e.g., microRNA regulation of gene expression in the brain; Morgan and Bale, 2012), future research may also benefit from pursuit of the role of sex differences in epigenetic factors in psychosis risk.

## 6 CONCLUSIONS

Sex differences are consistently reported in schizophrenia across various neurobiological and behavioral factors that emerge as early as the prenatal period and yield direct and/or indirect effects on neurodevelopment. This includes programming of the CNS during the earliest stages of development in a manner that may yield long-term effects on neural and behavioral outcomes across the lifecycle. Consistent with a neural diathesis–stress perspective (Walker and Diforio, 1997), neurohormones critically impact CNS integrity – including via organizational and activational effects – during prenatal development. Gonadal hormones shape various morphological traits – including the brain – in a time- and dose-sensitive manner, yielding sexually dimorphic effects. Males are believed to be more vulnerable to adverse prenatal events (Hultman et al., 1999). Moreover, some prenatal disruptions (e.g., neurohormonal milieu and obstetric complications) linked with schizophrenia (Matsumoto et al., 2001) are sexually dimorphic and believed to impact schizophrenia manifestation (Walker and Walder, 2002; Goldstein and Walder, 2006; Walder et al., 2006). This fits with neuroanatomic findings of disrupted normal sexual dimorphisms (Goldstein et al., 2002), including regions implicated in schizophrenia (e.g., hippocampus). Structural and functional abnormalities in these regions partake in neural networks that either subserve – or bidirectionally influence – several factors believed to play a key role in the neurodevelopment of psychosis. Not surprisingly, many of these factors are sexually dimorphic in psychosis. Candidate sexually dimorphic (neurodevelopmental) biomarkers implicated in psychosis that are particularly noteworthy include neurocognition, the HPA axis, perinatal complications, and genetic liability. Moreover, consideration of sex differences in the neurobiological substrates of sex differences in clinical phenomenology, social cognition, and social and role functioning in psychosis are also of primary interest.

Though in its nascent stage, as researchers develop optimal strategies to recruit individuals at high risk for psychosis, the field is ripening with opportunity to better characterize sexually dimorphic patterns among individuals at high risk, with an eye toward maximizing our understanding of pathophysiology and identifying sexually dimorphic putative biomarkers of illness. Mounting evidence suggests that males and females may experience differential etiology and manifest illness differently. Moreover, the sexes may differentially benefit from a range of interventions (psychotropic medication, cognitive-behavioral therapy, social cognitive remediation, cognitive remediation), and/or variations in dosage, duration, or other treatment characteristics. Together, such sexually dimorphic effects fuel the need for research carefully designed to more aggressively, directly, and comprehensively investigate sexual variation (sex-dependent and sex-specific) in psychosis risk.

## REFERENCES


Abel, K.M., Drake, R., Goldstein, J., 2010. Sex differences in schizophrenia. Int. J. Soc. Psychiatr. 22 (5), 417–428.

Addington, J., Cadenhead, K.S., Cannon, T.D., Cornblatt, B., McGlashan, T.H., Perkins, D.O., et al., 2007. North American prodrome longitudinal study: a collaborative multisite approach to prodromal schizophrenia research. Schizophr. Bull. 33 (3), 665–672.

Addington, J., Penn, D., Woods, S.W., Addington, D., Perkins, D.O., 2008. Social functioning in individuals at clinical high risk for psychosis. Schizophr. Res. 99 (1–3), 119–124.

Addington, J., Saeedi, H., Addington, D., 2006. Influence of social perception and social knowledge on cognitive and social functioning in early psychosis. Br. J. Psychiatr. 189 (4), 373–378.

Aleman, A., Kahn, R.S., Selten, J., 2003. Sex differences in the risk of schizophrenia: evidence from meta-analysis. Arch. Gen. Psychiatr. 60 (6), 565–571.

Altemus, M., Redwine, L., Leong, Y.-M., Yoshikawa, T., Yehuda, R., Detera-Wadleigh, S., Murphy, D.L., 1997. Reduced sensitivity to glucocorticoid feedback and reduced glucocorticoid receptor mRNA expression in the luteal phase of the menstrual cycle. Neuropsychopharmacology 17 (2), 100–109.

Andreano, J.M., Arjomandi, H., Cahill, L., 2008. Menstrual cycle modulation of the relationship between cortisol and long-term memory. Psychoneuroendocrinology 33 (6), 874–882.

Andreasen, N.C., O'Leary, D.S., Flaum, M., Nopoulos, P., Watkins, G.L., Boles Ponto, L.L., Hichwa, R.D., 1997. Hypofrontality in schizophrenia: distributed dysfunctional circuits in neuroleptic-naive patients. Lancet 349 (9067), 1730–1734.

Andreasen, N.C., Olsen, S., 1982. Negative v positive schizophrenia: definition and validation. Arch. Gen. Psychiatr. 39 (7), 789–794.

Arnold, A.P., Breedlove, S.M., 1985. Organizational and activational effects of sex steroids on brain and behavior: a reanalysis. Horm. Behav. 19 (4), 469–498.

Aylward, E., Walker, E., Bettes, B., 1984. Intelligence in schizophrenia: meta-analysis of the research. Schizophr. Bull. 10 (3), 430–459.

Bearden, C.E., Reus, V.I., Freimer, N.B., 2004. Why genetic investigation of psychiatric disorders is so difficult. Curr. Opin. Genet. Dev. 14 (3), 280–286.

Becker, J.B., Arnold, A.P., Berkley, K.J., Blaustein, J.D., Eckel, L.A., Hampson, E., et al., 2005. Strategies and methods for research on sex differences in brain and behavior. Endocrinology 146 (4), 1650–1673.

Bilder, R.M., Goldman, R.S., Robinson, D., Reiter, G., Bell, L., Bates, J.A., et al.,2000. Neuropsychology of first-episode schizophrenia: initial characterization and clinical correlates. Am. J. Psychiatr. 157 (4), 549–559.

Brekke, J., Kay, D.D., Lee, K.S., Green, M.F., 2005. Biosocial pathways to functional outcome in schizophrenia. Schizophr. Res. 80 (2–3), 213–225.

Brewer, W.J., Wood, S.J., Phillips, L.J., Francey, S.M., Pantelis, C., Yung, A.R., Cornblatt, B., McGorry, P.D., 2006. Generalized and specific cognitive performance in clinical high-risk cohorts: a review highlighting potential vulnerability markers for psychosis. Schizophr. Bull. 32 (3), 538–555.

Brown, G.R., Spencer, K.A., 2013. Steroid hormones, stress and the adolescent brain: a comparative perspective. Neuroscience 249, 115–128.

Brown, L.A., Cohen, A.S., 2010. Facial emotion recognition in schizotypy: the role of accuracy and social cognitive bias. J. Int. Neuropsychol. Soc. 16 (3), 474–483.

Burgess, L.H., Handa, R.J., 1992. Chronic estrogen-induced alterations in adrenocorticotropin and corticosterone secretion, and glucocorticoid receptor-mediated functions in female rats. Endocrinology 131 (3), 1261–1269.

Cannon, M., Jones, P.B., Murray, R.M., 2002. Reviews and overviews obstetric complications and schizophrenia: historical and meta-analytic review. Am. J. Psychiatr. 159 (July), 1080–1092.

Cannon, T.D., Gur, R.C., Zorrilla, L.E., Shtasel, D., Raquel, E., Marco, J., et al., 1994. Neuropsychological functioning in siblings discordant for schizophrenia and healthy volunteers. Arch. Gen. Psychiatr. 51, 651–661.

Cannon, T.D., Kaprio, J., Lönnqvist, J., Huttunen, M., Koskenvuo, M., 1998. The genetic epidemiology of schizophrenia in a Finnish twin cohort: a population-based modeling study. Arch. Gen. Psychiatr. 55 (1), 67–74.

Cardno, A.G., Gottesman, I.I., 2000. Twin studies of schizophrenia: from bow-and-arrow concordances to star wars Mx and functional genomics. Am. J. Med. Genet. 97 (1), 12–17.

Casar, C., Artamendi, M., Gutiérrez, M., Gil, P., García, M.J., Cuesta, M.J., 1996. Neuropsychological deficits, obstetric complications and premorbid adjustment in patients with the first psychotic episode. Actas Luso Esp. Neurol. Psiquiatr. Cienc. Afines 25 (5), 303–307.

Castle, D.J., Abel, K., Takei, N., Murray, R.M., 1995. Gender differences in schizophrenia. Schizophr. Bull. 21 (1), 1–12.

Castle, D.J., Murray, R., 1991. The neurodevelopmental basis of sex differences in schizophrenia. Psychol. Med. 21, 565–575.

Chan, R.C.K., Li, H., Cheung, E.F.C., Gong, Q., 2010. Impaired facial emotion perception in schizophrenia: a meta-analysis. Psychiatr. Res. 178 (2), 381–390.

Charmandari, E., Kino, T., Souvatzoglou, E., Chrousos, G.P., 2002. Pediatric stress: hormonal mediators and human development. Horm. Res. 59 (4), 161–179.

Chen, X., Wang, X., Hossain, S., O'Neill, F.A., Walsh, D., Pless, L., et al., 2006. Haplotypes spanning SPEC2, PDZ-GEF2 and ACSL6 genes are associated with schizophrenia. Hum. Mol. Genet. 15 (22), 3329–3342.

Cohen, A.S., Leung, W.W., Saperstein, A.M., Blanchard, J.J., 2006. Neuropsychological functioning and social anhedonia: results from a community high-risk study. Schizophr. Res. 85 (1–3), 132–141.

Corcoran, C.M., Kimhy, D., Parrilla-Escobar, M.A., Cressman, V.L., Stanford, A.D., Thompson, J., et al., 2011. The relationship of social function to depressive and negative symptoms in individuals at clinical high risk for psychosis. Psychol. Med. 41 (2), 251–261.

Cornblatt, B.A., Auther, A.M., Niendam, T., Smith, C.W., Zinberg, J., Bearden, C.E., Cannon, T.D., 2007. Preliminary findings for two new measures of social and role functioning in the prodromal phase of schizophrenia. Schizophr. Bull. 33 (3), 688–702.

Cornblatt, B.A., Carrión, R.E., Addington, J., Seidman, L., Walker, E.F., Cannon, T.D., et al., 2011. Risk factors for psychosis: impaired social and role functioning. Schizophr. Bull. 38 (6), 1247–1257.

Cornblatt, B.A., Lencz, T., Smith, C.W., Correll, C.U., Auther, A.M., Nakayama, E., 2003. The schizophrenia prodrome revisited: a neurodevelopmental perspective. Schizophr. Bull. 29 (4), 633–651.

Couture, S.M., Penn, D.L., Roberts, D.L., 2006. The functional significance of social cognition in schizophrenia: a review. Schizophr. Bull. 32 (Suppl. 1), S44–S63.

Dalman, C., Thomas, H.V., David, A.S., Gentz, J., Lewis, G., Allebeck, P., 2001. Signs of asphyxia at birth and risk of schizophrenia: population-based case-control study. Br. J. Psychiatr. Suppl. 179, 403–408.

Dragt, S., Nieman, D.H., Veltman, D., Becker, H.E., van de Fliert, R., de Haan, L., Linszen, D.H., 2011. Environmental factors and social adjustment as predictors of a first psychosis in subjects at ultra high risk. Schizophr. Res. 125 (1), 69–76.

Erlenmeyer-Kimling, L., Cornblatt, B., 1984. Biobehavioral risk factors in children of schizophrenic parents. J. Autism Dev. Disord. 14 (4), 357–374.

Falkenstein, E., Tillmann, H.-C., Christ, M., Feuring, M., Wehling, M., 2000. Multiple actions of steroid hormones – a focus on rapid, nongenomic effects. Pharmacol. Rev. 52 (4), 513–556.

Farrell, M.R., Gruene, T.M., Shansky, R.M., 2015. The influence of stress and gonadal hormones on neuronal structure and function. Horm. Behav. [Epub ahead of print]

Foerster, A., Lewis, S.W., Owen, M.J., Murray, R.M., 1991. Low birth weight and a family history of schizophrenia predict poor premorbid functioning in psychosis. Schizophr. Res. 5 (1), 13–20.

Fusar-Poli, P., Bonoldi, I., Yung, A.R., Borgwardt, S., Kempton, M.J., Valmaggia, L., et al., 2012. Predicting psychosis. Arch. Gen. Psychiatr. 69 (3), 220–229.

Fusar-Poli, P., Broome, M.R., Woolley, J.B., Johns, L.C., Tabraham, P., Bramon, E., et al., 2011. Altered brain function directly related to structural abnormalities in people at ultra high risk of psychosis: longitudinal VBM-fMRI study. J. Psychiatr. Res. 45 (2), 190–198.

Gilbert, A.R., Montrose, D.M., Sahni, S.D., Diwadkar, V.A., Keshavan, M.S., 2003. Obstetric complications correlate with neurobehavioral and brain structural alterations in young relatives at risk for schizophrenia. Ann. NY Acad. Sci. 1008 (1), 269–275.

Godlewska, B.R., Olajossy-Hilkesberger, L., Limon, J., Landowski, J., 2010. Ser9Gly polymorphism of the DRD3 gene is associated with worse premorbid social functioning and an earlier age of onset in female but not male schizophrenic patients. Psychiatr. Res. 177 (1–2), 266–267.

Gold, J.M., 2004. Cognitive deficits as treatment targets in schizophrenia. Schizophr. Res. 72 (1), 21–28.

Goldstein, J.M., Cherkerzian, S., Tsuang, M.T., Petryshen, T.L., 2013. Sex differences in the genetic risk for schizophrenia: history of the evidence for sex-specific and sex-dependent effects. Am. J. Med. Genet. B Neuropsychiatr. Genet. 162 (7), 698–710.

Goldstein, J.M., Seidman, L.J., O'Brien, L.M., Horton, N.J., Kennedy, D.N., Makris, N., et al., 2002. Impact of normal sexual dimorphisms on sex differences in structural brain abnormalities in schizophrenia assessed by magnetic resonance imaging. Arch. Gen. Psychiat. 59 (2), 154–164.

Goldstein, J.M., Walder, D.J., 2006. Sex differences in schizophrenia: the case for developmental origins. In: Sharma, T., Harvey, P. (Eds.), The Early Course of Schizophrenia. Oxford University Press, New York, NY, pp. 143–173.

Gooding, D.C., Matts, C.W., Rollmann, E.A., 2006. Sustained attention deficits in relation to psychometrically identified schizotypy: evaluating a potential endophenotypic marker. Schizophr. Res. 82 (1), 27–37.

Gottesman, I.I., Shields, J., 1967. A polygenic theory of schizophrenia. Proc. Natl. Acad. Sci. USA 58 (1), 199–205.

Green, M.F., 2006. Cognitive impairment and functional outcome in schizophrenia and bipolar disorder. J. Clin. Psychiatr. 67 (10), e12–e112.

Green, M.F., Nuechterlein, K.H., 1999. Should schizophrenia be treated as a neurocognitive disorder? Schizophr. Bull. 25 (2), 309–319.

Green, M.F., Olivier, B., Crawley, J.N., Penn, D.L., Silverstein, S., 2005. Social cognition in schizophrenia: recommendations from the measurement and treatment research to improve cognition in schizophrenia new approaches conference. Schizophr. Bull. 31 (4), 882–887.

Guest, P.C., Schwarz, E., Krishnamurthy, D., Harris, L.W., Leweke, F.M., Rothermundt, M., et al., 2011. Altered levels of circulating insulin and other neuroendocrine hormones associated with the onset of schizophrenia. Psychoneuroendocrinology 36 (7), 1092–1096.

Hadders-Algra, M., Huisjes, H.J., Touwen, B.C.L., 1988. Perinatal correlates of major and minor neurological dysfunction at school age: a multivariate analysis. Dev. Med. Child Neurol. 30 (4), 472–481.

Häfner, H., Behrens, S., De Vry, J., Gattaz, W.F., 1991. An animal model for the effects of estradiol on dopamine-mediated behavior: implications for sex differences in schizophrenia. Psychiatr. Res. 38 (2), 125–134.

Hall, J.A., 1978. Gender effects in decoding nonverbal cues. Psychol. Bull. 85 (4), 845–857.

Hawkins, K.A., Addington, J., Keefe, R.S.E., Christensen, B., Perkins, D.O., Zipurksy, R., et al., 2004. Neuropsychological status of subjects at high risk for a first episode of psychosis. Schizophr. Res. 67 (2-3), 115–122.

Heinrichs, R.W., 2005. The primacy of cognition in schizophrenia. Am. Psychol. 60 (3), 229–242.

Hennah, W., Varilo, T., Kestilä, M., Paunio, T., Arajärvi, R., Haukka, J., Partonen, T., 2003. Haplotype transmission analysis provides evidence of association for DISC1 to schizophrenia and suggests sex-dependent effects. Hum. Mol. Genet. 12 (23), 3151–3159.

Hofer, A., Benecke, C., Edlinger, M., Huber, R., Kemmler, G., Rettenbacher, M.A., et al., 2009. Facial emotion recognition and its relationship to symptomatic, subjective, and functional outcomes in outpatients with chronic schizophrenia. Eur. Psychiatr. 24 (1), 27–32.

Hoff, A., Kremen, W., 2002. Sex differences in neurocognitive function in schizophrenia. In: Lewis-Hall, F., Williams, T.S., Panetta, J.A., Herrera, J.H. (Eds.), Psychiatric Illness in Women: Emerging Treatments and Research. American Psychiatric Publishing, Washington, DC, pp. 215–238.

Horan, W.P., Kern, R.S., Green, M.F., Penn, D.L., 2008. Social cognition training for individuals with schizophrenia: emerging evidence. Am. J. Psychiatr. Rehabil. 11 (3), 205–252.

Holtzman, C., Macdonald, A., Esterberg, M., et al., in review. Sex differences in symptom presentation in individuals at ultra high risk for psychosis.

Holtzman, C.W., Trotman, H.D., Goulding, S.M., Ryan, A.T., Macdonald, A.N., Shapiro, D.I., Brasfield, J.L., Walker, E.F., 2013. Stress and neurodevelopmental processes in the emergence of psychosis. Neuroscience 249, 172–191.

Hultman, C.M., Geddes, J., Sparén, P., Takei, N., Murray, R.M., Cnattingius, S., 1999. Prenatal and perinatal risk factors for schizophrenia, affective psychosis, and reactive psychosis of early onset: case-control study. BMJ 318 (7181), 421–426.

Irani, F., Seligman, S., Kamath, V., Kohler, C., Gur, R.C., 2012. A meta-analysis of emotion perception and functional outcomes in schizophrenia. Schizophr. Res. 137 (1–3), 203–211.

Ittig, S., Studerus, E., Papmeyer, M., Uttinger, M., Koranyi, S., Ramyead, A., Riecher-Rössler, A., 2015. Sex differences in cognitive functioning in at-risk mental state for psychosis, first episode psychosis and healthy control subjects. Eur. Psychiatr. 30 (2), 242–250.

Jahshan, C.S., Sergi, M.J., 2007. Theory of mind, neurocognition, and functional status in schizotypy. Schizophr. Res. 89 (1–3), 278–286.

Jones, P.B., Done, D.J., 1997. From birth to onset: a developmental perspective of schizophrenia in two national birth cohorts. In: Keshavan, M.S., Murray, R.M. (Eds.), Neurodevelopment and Adult Psychopathology. Cambridge University Press, Cambridge.

Kajantie, E., Phillips, D.I.W., 2006. The effects of sex and hormonal status on the physiological response to acute psychosocial stress. Psychoneuroendocrinology 31 (2), 151–178.

Kee, K.S., Horan, W.P., Mintz, J., Green, M.F., 2004. Do the siblings of schizophrenia patients demonstrate affect perception deficits? Schizophr. Res. 67 (1), 87–94.

Keefe, R.S.E., Perkins, D.O., Gu, H., Zipursky, R.B., Christensen, B.K., Lieberman, J.A., 2006. A longitudinal study of neurocognitive function in individuals at-risk for psychosis. Schizophr. Res. 88 (1–3), 26–35.

Kendler, K.S., Gallagher, T.J., Abelson, J.M., Kessler, R.C., 1996. Lifetime prevalence, demographic risk factors, and diagnostic validity of nonaffective psychosis as assessed in a US community sample: the National Comorbidity Survey. Arch. Gen. Psychiatr. 53 (11), 1022–1031.

Kim, J.J., Mohamed, S., Andreasen, N.C., O'Leary, D.S., Watkins, G.L., Boles Ponto, L.L., Hichwa, R.D., 2000. Regional neural dysfunctions in chronic schizophrenia studied with positron emission tomography. Am. J. Psychiatr. 157 (4), 542–548.

Kirschbaum, C., Kudielka, B.M., Gaab, J., Schommer, N.C., Hellhammer, D.H., 1999. Impact of gender, menstrual cycle phase, and oral contraceptives on the activity of the hypothalamus-pituitary-adrenal axis. Psychosom. Med. 61 (2), 154–162.

Kohler, C.G., Turner, T.H., Bilker, W.B., Brensinger, C.M., Siegel, S.J., Kanes, S.J., et al.,2003. Facial emotion recognition in schizophrenia: intensity effects and error pattern. Am. J. Psychiatr. 160 (10), 1768–1774.

Kudielka, B.M., Kirschbaum, C., 2005. Sex differences in HPA axis responses to stress: a review. Biol. Psychol. 69 (1), 113–132.

Kuperberg, G., Heckers, S., 2000. Schizophrenia and cognitive function. Curr. Opin. Neurobiol. 10 (2), 205–210.

Kurtz, M.M., Richardson, C.L., 2012. Social cognitive training for schizophrenia: a meta-analytic investigation of controlled research. Schizophr. Bull. 38 (5), 1092–1104.

Lane, A., Colgan, K., Moynihan, F., Burke, T., Waddington, J.L., Larkin, C., O'Callaghan, E., 1996. Schizophrenia and neurological soft signs: gender differences in clinical correlates and antecedent factors. Psychiatr. Res. 64 (2), 105–114.

Lemos-Giráldez, S., Vallina-Fernández, O., Fernández-Iglesias, P., Vallejo-Seco, G., Fonseca-Pedrero, E., Paíno-Piñeiro, M., et al., 2009. Symptomatic and functional outcome in youth at ultra-high risk for psychosis: a longitudinal study. Schizophr. Res. 115 (2-3), 121–129.

Lencz, T., Smith, C.W., McLaughlin, D., Auther, A., Nakayama, E., Hovey, L., Cornblatt, B.A., 2006. Generalized and specific neurocognitive deficits in prodromal schizophrenia. Biol. Psychiatr. 59 (9), 863–871.

Lenzenweger, M.F., 1994. Psychometric high-risk paradigm, perceptual aberrations, and schizotypy: an update. Schizophr. Bull. 20 (1), 121–135.

Leung, A., Chue, P., 2000. Sex differences in schizophrenia, a review of the literature. Acta Psychiatr. Scand. Suppl. 401, 3–38.

Lewine, R.R.J., Walker, E.F., Shurett, R., Caudle, J., Haden, C., 1996. Sex differences in neuropsychological functioning among schizophrenic patients. Am. J. Psychiatr. 153 (9), 1178–1184.

Lichtenstein, P., Yip, B.H., Björk, C., Pawitan, Y., Cannon, T.D., Sullivan, P.F., Hultman, C.M., 2009. Common genetic determinants of schizophrenia and bipolar disorder in Swedish families: a population-based study. Lancet 373 (9659), 234–239.

Matsumoto, H., Takei, N., Saito, F., Kachi, K., Mori, N., 2001. The association between obstetric complications and childhood-onset schizophrenia: a replication study. Psychol. Med. 31 (5), 907–914.

McCarthy, M.M., Auger, A.P., Bale, T.L., De Vries, G.J., Dunn, G.A., Forger, N.G., et al., 2009. The epigenetics of sex differences in the brain. J. Neurosci. 29 (41), 12815–12823.

McClure, E.B., 2000. A meta-analytic review of sex differences in facial expression processing and their development in infants, children, and adolescents. Psychol. Bull. 126 (3), 424–453.

McCullough, L.D., de Vries, G.J., Miller, V.M., Becker, J.B., Sandberg, K., McCarthy, M.M., 2014. NIH initiative to balance sex of animals in preclinical studies: generative questions to guide policy, implementation, and metrics. Biol. Sex Diff. 5 (1), 15.

McEwen, B.S., 1991. Non-genomic and genomic effects of steroids on neural activity. Trends Pharmacol. Sci. 12, 141–147.

McGlashan, T.H., Bardenstein, K.K., 1990. Gender differences in affective, schizoaffective, and schizophrenic disorders. Schizophr. Bull. 16 (2), 319–329.

Meehl, P.E., 1962. Schizotaxia, schizotypy, schizophrenia. Am. Psychol. 17 (12), 827–838.

Meehl, P.E., 1990. Toward an integrated theory of schizotaxia, schizotypy, and schizophrenia. J. Pers. Disord. 4 (1), 1–99.

Mesholam-Gately, R.I., Giuliano, A.J., Goff, K.P., Faraone, S.V., Seidman, L.J., 2009. Neurocognition in first-episode schizophrenia: a meta-analytic review. Neuropsychology 23 (3), 315–336.

Morgan, C.P., Bale, T.L., 2012. Sex differences in microRNA regulation of gene expression: no smoke, just miRs. Biol. Sex Diff. 3 (1), 22.

Murray, R.M., Lewis, S.W., 1987. Is schizophrenia a neurodevelopmental disorder? BMJ 295 (3), 681–682.

Nelson, B., Fusar-Poli, P., Yung, A.R., 2012. Can we detect psychotic-like experiences in the general population? Curr. Pharmaceut. Des. 18 (4), 376–385.

Niendam, T.A., Bearden, C.E., Johnson, J.K., McKinley, M., Loewy, R., O'Brien, M., et al., 2006. Neurocognitive performance and functional disability in the psychosis prodrome. Schizophr. Res. 84 (1), 100–111.

Nuechterlein, K.H., Barch, D.M., Gold, J.M., Goldberg, T.E., Green, M.F., Heaton, R.K., 2004. Identification of separable cognitive factors in schizophrenia. Schizophr. Res. 72 (1), 29–39.

O'Callaghan, E., Gibson, T., Colohan, H.A., Buckley, P., Walshe, D.G., Larkin, C., Waddington, J.L., 1992. Risk of schizophrenia in adults born after obstetric complications and their association with early onset of illness: a controlled study. BMJ 305 (6864), 1256–1259.

Patchev, V.K., Hayashi, S., Orikasa, C., Almeida, O.F., 1995. Implications of estrogen-dependent brain organization for gender differences in hypothalamo-pituitary-adrenal regulation. FASEB J. 9 (5), 419–423.

Penn, D.L., Corrigan, P.W., Bentall, R.P., Racenstein, J.M., Newman, L., 1997. Social cognition in schizophrenia. Psychol. Bull. 121 (1), 114–132.

Penn, D.L., Spaulding, W., Reed, D., Sullivan, M., 1996. The relationship of social cognition to ward behavior in chronic schizophrenia. Schizophr. Res. 20 (3), 327–335.

Pinkham, A.E., Penn, D.L., 2006. Neurocognitive and social cognitive predictors of interpersonal skill in schizophrenia. Psychiatr. Res. 143 (2–3), 167–178.

Pinkham, A.E., Penn, D.L., Green, M.F., Buck, B., Healey, K., Harvey, P.D., 2014. The social cognition psychometric evaluation study: results of the expert survey and RAND panel. Schizophr. Bull. 40 (4), 813–823.

Piontkewitz, Y., Arad, M., Weiner, I., 2011. Abnormal trajectories of neurodevelopment and behavior following *in utero* insult in the rat. Biol. Psychiatr. 70 (9), 842–851.

Porteous, D.J., Evans, K.L., Millar, J.K., Pickard, B.S., Thomson, P.A., James, R., et al., 2003. Genetics of schizophrenia and bipolar affective disorder: strategies to identify candidate genes. Cold Spring Harbor Symp. on Quant. Biol. 68, 383–394.

Reichenberg, A., Weiser, M., Caspi, A., Knobler, H.Y., Lubin, G., Harvey, P.D., et al., 2006. Premorbid intellectual functioning and risk of schizophrenia and spectrum disorders. J. Clin. Expt. Neuropsychol. 28 (2), 193–207.

Riecher-Rössler, A., Häfner, H., 1993. Schizophrenia and oestrogens – is there an association? Eur. Arch. Psychiatr. Clin. Neurosci. 242 (6), 323–328.

Rosso, I.M., Cannon, T.D., Huttunen, T., Huttunen, M.O., Jouko, L., Gasperoni, T.L., 2000. Obstetric risk factors for early-onset schizophrenia in a Finnish birth cohort. Am. J. Psychiatr. 157 (5), 801–807.
Salem, J.E., Kring, A.M., 1998. The role of gender differences in the reduction of etiologic heterogeneity in schizophrenia. Clin. Psychol. Rev. 18 (7), 795–819.
Schatzberg, A.F., Rothschild, A.J., Langlais, P.J., Bird, E.D., Cole, J.O., 1985. A corticosteroid/dopamine hypothesis for psychotic depression and related states. J. Psychiatr. Res. 19 (1), 57–64.
Seeman, M.V., 1982. Gender differences in schizophrenia. Can. Child Adolesc. Psychiatr. Rev. 27 (2), 107–112.
Seeman, M.V., Lang, M., 1990. The role of estrogens in schizophrenia gender differences. Schizophr. Bull. 16 (2), 185–194.
Seidman, L., Giuliano, A., Walker, E., 2010. Neuropsychology of the prodrome to psychosis in the NAPLS Consortium. Arch. Gen. Psychiatr. 67 (6), 578–588.
Sisk, C.L., Zehr, J.L., 2005. Pubertal hormones organize the adolescent brain and behavior. Front. Neuroendocrinol. 26 (3–4), 163–174.
Shansky, R.M., Glavis-Bloom, C., Lerman, D., McRae, P., Benson, C., Miller, K., et al., 2004. Estrogen mediates sex differences in stress-induced prefrontal cortex dysfunction. Mol. Psychiatr. 9 (5), 531–538.
Shansky, R.M., Morrison, J.H., 2009. Stress-induced dendritic remodeling in the medial prefrontal cortex: effects of circuit, hormones and rest. Brain Res. 1293, 108–113.
Shansky, R.M., Rubinow, K., Brennan, A., Arnsten, A.F.T., 2006. The effects of sex and hormonal status on restraint-stress-induced working memory impairment. Behav. Brain Funct. 2, 8.
Shifman, S., Bronstein, M., Sternfeld, M., Pisante-Shalom, A., Lev-Lehman, E., Weizman, A., et al., 2002. A highly significant association between a COMT haplotype and schizophrenia. Am. J. Hum. Genet. 71 (6), 1296–1302.
Shifman, S., Johannesson, M., Bronstein, M., Chen, S.X., Collier, D.A., Craddock, N.J., et al., 2008. Genome-wide association identifies a common variant in the reelin gene that increases the risk of schizophrenia only in women. PLoS Genet. 4 (2), e28.
Shim, G., Kang, D.-H., Sun Chung, Y., Young Yoo, S., Young Shin, N., Soo Kwon, J., 2008. Social functioning deficits in young people at risk for schizophrenia. Aust. NZJ. Psychiatr. 42 (8), 678–685.
Smoller, J.W., Craddock, N., Kendler, K., Lee, P.H., Neale, B.M., Nurnberger, J.I., et al., 2013. Cross-disorder group of the psychiatric genomics consortium. Identification of risk loci with shared effects on five major psychiatric disorders: a genome-wide analysis. Lancet 381 (9875), 1371–1379.
Statucka, M., Walder, D.J., 2013. Efficacy of social cognition remediation programs targeting facial affect recognition deficits in schizophrenia: a review and consideration of high-risk samples and sex differences. Psychiatr. Res. 206 (2–3), 125–139.
Steen, N.E., Lorentzen, S., Barrett, E.A., Lagerberg, T.V., Hope, S., Larsson, S., et al., 2011. Sex-specific cortisol levels in bipolar disorder and schizophrenia during mental challenge – relationship to clinical characteristics and medication. Prog. Neuropsychopharmacol. Biol. Psychiatr. 35 (4), 1100–1107.
Steiner, M., Dunn, E., Born, L., 2003. Hormones and mood: from menarche to menopause and beyond. J. Affect. Disord. 74 (1), 67–83.
Susser, E., Neugebauer, R., Hoek, H.W., Brown, A.S., Lin, S., Labovitz, D., Gorman, J.M., 1996. Schizophrenia after prenatal famine. Further evidence. Arch. Gen. Psychiatr. 53 (1), 25–31.
Tan, E.C., Chong, S.A., Wang, H., Chew-Ping Lim, E., Teo, Y.Y., 2005. Gender-specific association of insertion/deletion polymorphisms in the nogo gene and chronic schizophrenia. Mol. Brain Res. 139 (2), 212–216.
Tarbox, S.I., Addington, J., Cadenhead, K.S., Cannon, T.D., Cornblatt, B.A., Perkins, D.O., et al., 2013. Premorbid functional development and conversion to psychosis in clinical high-risk youths. Dev. Psychopathol. 25 (4 Pt. 1), 1171–1186.
Thomson, P.A., Wray, N.R., Thomson, A.M., Dunbar, D.R., Grassie, M.A., Condie, A., et al., 2005. Sex-specific association between bipolar affective disorder in women and GPR50, an X-linked orphan G protein-coupled receptor. Mol. Psychiatr. 10 (5), 470–478.
Thorup, A., Petersen, L., Jeppesen, P., Ohlenschlæger, J., Christensen, T., Krarup, G., et al., 2007. Gender differences in young adults with first-episode schizophrenia spectrum disorders at baseline in the danish OPUS study. J. Nerv. Ment. Dis. 195 (5). Available from: http://journals.lww.com/jonmd/Fulltext/2007/05000/Gender_Differences_in_Young_Adults_With.6.aspx.

Toomey, R., Schuldberg, D., 1995. Recognition and judgment of facial stimuli in schizotypal subjects. J. Commun. Disord. 28 (3), 193–203.
Trotman, H.D., Holtzman, C.W., Ryan, A.T., Shapiro, D.I., MacDonald, A.N., Goulding, S.M., et al., 2013. The development of psychotic disorders in adolescence: a potential role for hormones. Horm. Behav. 64 (2), 411–419.
Trotman, H.D., Holtzman, C.W., Walker, E.F., Addington, J.M., Bearden, C.E., Cadenhead, K.S., et al., 2014. Stress exposure and sensitivity in the clinical high-risk syndrome: initial findings from the North American Prodrome Longitudinal Study (NAPLS). Schizophr. Res. 160 (1-3), 104–109.
Trotman, H., McMillan, A., Walker, E., 2006. Cognitive function and symptoms in adolescents with schizotypal personality disorder. Schizophr. Bull. 32 (3), 489–497.
Tsuang, M.T., Stone, W.S., Faraone, S.V., 2001. Genes, environment and schizophrenia. Br. J. Psychiatr. Suppl. 178 (40), s18–s24.
Viau, V., 2002. Functional cross-talk between the hypothalamic-pituitary-gonadal and -adrenal axes. J. Neuroendocrinol. 14 (6), 506–513.
Walder, D.J., Andersson, T.L.C., McMillan, A.L., Breedlove, S.M., Walker, E.F., 2006. Sex differences in digit ratio (2D:4D) are disrupted in adolescents with schizotypal personality disorder: altered prenatal gonadal hormone levels as a risk factor. Schizophr. Res. 86 (1–3), 118–122.
Walder, D.J., Seidman, L.J., Makris, N., Tsuang, M.T., Kennedy, D.N., Goldstein, J.M., 2007. Neuroanatomic substrates of sex differences in language dysfunction in schizophrenia: a pilot study. Schizophr. Res. 90 (1-3), 295–301.
Walder, D.J., Mittal, V., Trotman, H.D., McMillan, A.L., Walker, E.F., 2008. Neurocognition and conversion to psychosis in adolescents at high-risk. Schizophr. Res. 101 (1–3), 161–168.
Walder, D.J., Faraone, S.V., Glatt, S.J., Tsuang, M.T., Seidman, L.J., 2014. Genetic liability, prenatal health, stress and family environment: risk factors in the harvard adolescent family high risk for schizophrenia study. Schizophr. Res. 157 (1–3), 142–148.
Walder, D.J., Holtzman, C.W., Addington, J., Cadenhead, K., Tsuang, M., Cornblatt, B., et al., 2013. Sexual dimorphisms and prediction of conversion in the NAPLS psychosis prodrome. Schizophr. Res. 144 (1-3), 43–50.
Walder, D.J., Ospina, L., Daly, M.P., Statucka, M., Raparia, E., 2012a. Early neurodevelopment and psychosis risk: role of neurohormones and biological sex in modulating genetic, prenatal and sensory processing factors in brain development. In: X. Anastassiou-Hadjicharalambous, (Ed.), Psychosis: Causes, Diagnosis and Treatment. Nova Science.
Walder, D.J., Statucka, M., Daly, M.P., Axen, K., Haber, M., 2012b. Biological sex and menstrual cycle phase modulation of cortisol levels and psychiatric symptoms in a non-clinical sample of young adults. Psychiatr. Res. 197 (3), 314–321.
Walder, D.J., Walker, E.F., Lewine, R.J., 2000. Cognitive functioning, cortisol release, and symptom severity in patients with schizophrenia. Biol. Psychiatr. 48 (12), 1121–1132.
Walker, E., Bollini, A.M., 2002. Pubertal neurodevelopment and the emergence of psychotic symptoms. Schizophr. Res. 54 (1–2), 17–23.
Walker, E.F., Brennan, P.a., Esterberg, M., Brasfield, J., Pearce, B., Compton, M.T., 2010. Longitudinal changes in cortisol secretion and conversion to psychosis in at-risk youth. J. Abnorm. Psychol. 119 (2), 401–408.
Walker, E.F., Diforio, D., 1997. Schizophrenia: a neural diathesis-stress model. Psychol. Rev. 104 (4), 667–685.
Walker, E.F., Trotman, H.D., Pearce, B.D., Addington, J., Cadenhead, K.S., Cornblatt, B.A., et al., 2013. Cortisol levels and risk for psychosis: initial findings from the North American Prodrome Longitudinal Study. Biol. Psychiatr. 74 (6), 410–417.
Walker, E., Walder, D., 2002. Neurohormonal aspects of the development of psychotic disorders. In: Cichetti, D., Walker, E. (Eds.), Neurodevelopmental Mechanisms in Psychopathology. Cambridge University Press, Boston.
Walker, E.F., Walder, D.J., Lewine, R., Loewy, R., 2002. Sex differences in the origins and premorbid development of schizophrenia. In: Lewis-Hall, F., Williams, T.S., Panetta, J.A., Herrera, J.M. (Eds.), Psychiatric Illness in Women: Emerging Treatments and Research. American Psychiatric Publishing, Inc, Arlington, VA, US.
Walker, E., Kestler, L., Bollini, A., Hochman, K.M., 2004. Schizophrenia: etiology and course. Ann. Rev. Psychol. 55, 401–430.

Walker, E., Trotman, H., Brasfield, J., Esterberg, M., Larsen, M., 2011. Stress hormones, genes, and neuronal signaling in adolescence. The perfect storm for vulnerability to psychosis. The Dynamic Genome and Mental Health: The Role of Genes and Environments in Youth Development. Oxford University Press, Inc., New York.

Wang, K., Liu, X., Zhang, Q., Aragam, N., Pan, Y., 2011. Genome-wide association analysis of age at onset in schizophrenia in a European-American sample. Am. J. Med. Genet. B Neuropsychiatr. Genet. 156 (6), 671–680.

Wang, W.Y., Barratt, B.J., Clayton, D.G., Todd, J.A., 2005. Genome-wide association studies: theoretical and practical concerns. Nat. Rev. Genet. 6 (2), 109–118.

Weinberger, D., 1987. Implications of normal brain development for the pathogenesis of schizophrenia. Arch. Gen. Psychiatr. 44 (7), 660–669.

Willhite, R.K., Niendam, T.A., Bearden, C.E., Zinberg, J., O'Brien, M.P., Cannon, T.D., 2008. Gender differences in symptoms, functioning and social support in patients at ultra-high risk for developing a psychotic disorder. Schizophr. Res. 104 (1–3), 237–245.

Wolkowitz, O.M., 1994. Prospective controlled studies of the behavioral and biological effects of exogenous corticosteroids. Psychoneuroendocrinology 19 (3), 233–255.

Woods, S.W., Addington, J., Cadenhead, K.S., Cannon, T.D., Cornblatt, B.A., Heinssen, R., et al., 2009. Validity of the prodromal risk syndrome for first psychosis: findings from the North American Prodrome Longitudinal Study. Schizophr. Bull. 35 (5), 894–908, Available from: http://schizophreniabulletin.oxfordjournals.org/content/35/5/894.abstract.

Woolley, C.S., McEwen, B.S., 1992. Estradiol mediates fluctuation in hippocampal synapse density during the estrous cycle in the adult rat. J. Neurosci. 12 (7), 2549–2554.

Wray, N.R., Gottesman, I.I., 2012. Using summary data from the danish national registers to estimate heritabilities for schizophrenia, bipolar disorder, and major depressive disorder. Front. Genet. 3, 118.

Xie, C., Wang, Z.C., Liu, X.F., Wang, L., Yang, M.S., 2011. Association between schizophrenia and single nucleotide polymorphisms in lipoprotein lipase gene in a Han Chinese population. Psychiatr. Genet. 21 (6), 307–314.

Yıldırım, O., Dogan, O., Semiz, M., Kilicli, F., 2011. Serum cortisol and dehydroepiandrosterone-sulfate levels in schizophrenic patients and their first-degree relatives. Psychiatr. Clin. Neurosci. 65 (6), 584–591.

Yolken, R.H., Torrey, E.F., 1995. Viruses, schizophrenia, and bipolar disorder. Clin. Microbiol. Rev. 8 (1), 131–145.

Yong, E., Barbato, M., Penn, D.L., Keefe, R.S., Woods, S.W., Perkins, D.O., Addington, J., 2014. Exploratory analysis of social cognition and neurocognition in individuals at clinical high risk for psychosis. Psychiat. Res. 218 (1-2), 39–43.

Yung, A.R., Phillips, L.J., Yuen, H.P., Francey, S.M., McFarlane, C.A., Hallgren, M., McGorry, P.D., 2003. Psychosis prediction: 12-month follow up of a high-risk ("prodromal") group. Schizophr. Res. 60 (1), 21–32.

Zubin, J., Spring, B., 1977. Vulnerability: a new view of schizophrenia. J. Abnorm. Psychol. 86 (2), 103–126.

# CHAPTER 6

# Sex Differences and Addiction

**Adam N. Perry*, Christel Westenbroek**, Jill B. Becker**,†,‡**
*University of Texas at El Paso, Department of Biology, El Paso, TX, USA
**University of Michigan, Molecular and Behavioral Neuroscience Institute, Ann Arbor, MI, USA
†University of Michigan, Department of Psychiatry, Ann Arbor, MI, USA
‡University of Michigan, Department of Psychology, Ann Arbor, MI, USA

## 1 INTRODUCTION

History is full of examples of people using alcohol, opiates, cocaine, and other mind-altering substances. It is difficult to discern from these documents the extent that men and women differ in their use of drugs. Qualitative data about the causes of drug use and addiction are colored by perceived gender roles as well as the positions of men and women in society at the time. Therefore, sex differences in problematic drug use can appear to vary over time and across cultures (Becker et al., 2012).

Quantitative data on the number of people addicted to any drug did not exist before late in the twentieth century. Even in these data when they are stratified by gender, however, it is still difficult to get an accurate index of the extent of sex differences in drug use and addiction, due to variation in sampling methods as well as the populations assessed. For example, surveys of the general population typically show relatively low rates of use and abuse and large sex differences for most substances. On the other hand, studies on marginalized populations (e.g., the incarcerated and homeless) or adolescents typically reveal equivalent rates of drug use and abuse in men and women.

Rather than focus on whether there are absolute differences in the number of men and women addicted, it is more important to recognize that addiction is a complex disorder with multiple causes and consequences, which can affect men and women differently. Understanding both shared and sex-specific aspects of addiction will be essential for identifying effective individualized treatments that are optimized for both men and women.

In this chapter, we will examine the evidence from epidemiological studies to assess whether patterns of drug use and its consequences are different in men and women and then highlight findings from clinical and preclinical studies that demonstrate how addiction manifests in the brains of each sex and how these neurobiological changes might relate to various aspects of the addiction cycle.

There are many aspects of drug addiction that can be examined for sex differences. At the most basic level, one can crunch the numbers to determine whether men and women have different rates of drug use, abuse, and dependence. This information is generally the easiest to extract from various sources; however, the results often depend

*Sex Differences in the Central Nervous System*. http://dx.doi.org/10.1016/B978-0-12-802114-9.00006-8

on the population being sampled. Beyond these numbers, one can ask whether men and women experience similar consequences from their drug use. The primary consequences of drug use result directly from the actions of drugs on mental and physical health, including the development and/or worsening of various psychopathologies (e.g., addiction, psychotic symptoms, bipolar disorder, anxiety, and depression) and physical problems (e.g., organ damage and excessive weight loss) (Chen et al., 2011; Degenhardt and Hall, 2012; Grella and Lovinger, 2012; Santolaria-Fernández et al., 1995).

The secondary consequences of drug use may also be different for men and women. These are indirect and largely reflect the actions of an individual, such as how they procure and use drugs and their behavior while intoxicated. For example, intravenous drug injection results in collapsed veins, skin infections, and abscesses, and practices such as sharing needles and other apparatus for drug preparation increase the risk of viral infections (e.g., hepatitis B, C, and human immunodeficiency virus) (Evans et al., 2003; Harris and Rhodes, 2013; Tracy et al., 2014). In addition to these physical consequences, drug use can also lead to the disruption of important social relationships, unemployment, incarceration, and homelessness, which can in turn exacerbate drug use and its consequences.

The following discussion of sex differences in addiction will be framed within the context of comparing males and females (and highlighting when and where they are different). The antecedents and consequences of drug use can be very similar for both sexes, even when the underlying biological processes mediating the effects of drug use may be different. Thus, addiction is a serious problem for both men and women, and neither sex "is better or worse off" when it comes to drug addiction. By examining drug use in both sexes, it should be possible to find more effective treatments for addiction, which are currently sorely lacking. Research needs to determine whether treatments should be "sex-specific" or whether sex/gender should be viewed as a continuum rather than a bimodal factor and treatment tailored to the specific needs of various subpopulations irrespective of their sex/gender.

## 2 SEX DIFFERENCES IN ADDICTION IN HUMANS

### 2.1 Epidemiological studies: sex differences in the rates of drug use, abuse, and dependence

The question "Are there sex differences in addiction?" is deceptively simple. In theory, one only needs to know how many men and women use a substance and what proportions meet abuse and dependence criteria. Crunching these numbers, one can easily determine if there are sex differences in the likelihood of using a given drug, and once correcting for the number of users, whether men and women are equally vulnerable to becoming an addict.

Surveys conducted on the general population demonstrate that across a variety of substances, men have higher rates of lifetime use relative to women: alcohol, 85 versus

80%; marijuana, 47 versus 38%; cocaine, 18 versus 11%; crack, 5 versus 2%; heroin, 3 versus 1% (Substance Abuse and Mental Health Services Administration, 2013). A large part of the sex difference in drug use relates to the fact that males generally have more and/or earlier opportunities to use most types of illicit drugs than females. Once you correct for opportunity, there is very little difference between males and females in whether or not they will use illicit drugs (Van Etten et al., 1999).

Table 6.1 displays the proportions of people in the general population (≥12 years of age) that have used various substances in the past year averaged over the past 10 years of available survey data (2003 to 2012). For all of these substances, men show higher rates of use than women, which is consistent with the lifetime rates of use mentioned earlier. Women often face greater social sanctions against alcohol and drug use, whereas traditional gender roles often promote drinking and use of other substances in men (Schulte et al., 2009). This could be seen in the early part of the twentieth century when the Harrison Tax Act was enacted in the United States. This act limited the nonmedical use of opiates and cocaine, frequently included as ingredients in patent medicines that were freely available. The proportion of women using opiates declined from 80% to less than 50% almost immediately, and then to 25–30% of the population by the beginning of

**Table 6.1** Past year substance use, abuse, and dependence

| | Use* | Abuse** | Dependence** | Abuse/dependence** |
|---|---|---|---|---|
| **Alcohol** | | | | |
| Men | 69.9 (0.2) | 8.0 (0.2) | 6.2 (0.1) | 14.2 (0.3) |
| Women | 62.5 (0.3)† | 4.0 (0.1)† | 3.8 (0.1)† | 7.8 (0.1)† |
| **Marijuana** | | | | |
| Men | 13.7 (0.3) | 6.8 (0.3) | 10.4 (0.2) | 17.2 (0.4) |
| Women | 8.4 (0.2)† | 4.4 (0.2)† | 8.6 (0.3)† | 13.1 (0.5)† |
| **Cocaine** | | | | |
| Men | 2.9 (0.1) | 8.4 (0.4) | 16.7 (1.1) | 25.1 (1.2) |
| Women | 1.4 (0.1)† | 6.0 (0.6)† | 19.3 (1.2) | 25.3 (1.2) |
| **Heroin** | | | | |
| Men | 0.3 (0.02) | 8.7 (0.9) | 53.6 (2.6) | 62.3 (2.9) |
| Women | 0.1 (0.01)† | 4.9 (0.9)† | 57.8 (2.1) | 62.7 (2.6) |

Mean (SEM). United States Department of Health and Human Services. Substance Abuse and Mental Health Services Administration. Center for Behavioral Health Statistics and Quality. National Survey on Drug Use and Health, 2003–2012. ICPSR34933-v2. Inter-university Consortium for Political and Social Research [distributor], Ann Arbor, MI, 2014-10-06. http://doi.org/10.3886/ICPSR34933.v2

* Percent of total.

** Percent of users.

† Men versus women ($t$-test, $p < 0.05$).

Substance Abuse and Mental Health Services Administration (2013).

World War II (Califano, 2006). Interestingly, the modest gender gap for alcohol use suggests a weakening of these restrictions in women and/or a restructuring of gender roles. In this regard, greater legalization and/or decriminalization of marijuana may also eventually lead to a closing of the gender gap in marijuana use.

When looking at what percentage of drug users actually meet the criteria for abuse and dependence, we see that men that have used alcohol, marijuana, cocaine, or heroin are more likely than women to meet abuse criteria. Men were also more likely to meet dependence criteria for alcohol and marijuana, whereas there is no apparent sex difference in the percentage of users that become dependent on cocaine or heroin. Interestingly, the percentage of women users that meet dependence criteria for cocaine and heroin tends to be higher than men, such that one can look back across the decade and find years in which women have much higher rates of cocaine dependence, which could explain some reports of greater dependence in women (Cotto et al., 2010).

It is also interesting that when the percentage of users meeting abuse or dependence criteria are combined, the differences between men and women are small for cocaine and heroin, whereas men continue to show greater abuse/dependence of alcohol and marijuana. It is unclear whether the apparently greater vulnerability of men for addiction to alcohol and marijuana also reflects sociocultural (i.e., greater sanctions, stigma, and social contraints against use in women) or biological factors. One study found that the age of onset of use and dependence was similar for men and women; however, women attained remission from dependence at an earlier age (Khan et al., 2013).

Given that the criteria for abuse and dependence have been merged into a single continuum of substance use disorders (SUD) in the Diagnostic and Statistical Manual of Mental Disorders, Fifth Edition (DSM-V) (Peer et al., 2013), it is likely that rates of SUD for cocaine and heroin will continue to be similar between the sexes, but with possibly greater severity in women.

Thus, even though there are some sex differences in drug use and dependence within the general population, the behavior and vulnerability of men and women appears to be similar across a wide variety of substances.

## 2.2 Incarceration, victimization, violence, and trauma

In the general population, approximately 13% of men and 6% of women exhibit drug dependence. Incarcerated individuals have much higher rates that are roughly similar in men and women (approximately 60–70% of the prison population) (James and Karberg, 2005; Lewis, 2006). The high prevalence of drug use and dependence in prison most likely reflects several interrelated factors, including the criminalization of drug possession and many activities related to the procurement of drugs (e.g., street crime, dealing, and prostitution) (Inciardi and Surratt, 2001).

Incarceration can also interact with other risk/environmental factors. For example, in a study of homeless youths in Canada, those that had been incarcerated reported more drug use and higher rates of psychopathologies (Saddichha et al., 2014). Homeless youths that had been incarcerated were also more likely to be male and to have grown up in foster care, which suggests that early life trauma may lead to cycles of substance abuse, homelessness, and incarceration that renders individuals vulnerable to mental illness and drug abuse.

Many experiences may be shared by men and women, although to varying degrees, and the shared experiences are relevant to an individual's drug addiction for both sexes. For example, there are clear relationships between substance use disorders and previous exposure to victimization, violence, and trauma in both men and women. Oftentimes (but not always) the exact nature of the adverse experiences may differ between men and women and its consequences can be influenced by the age at which it was experienced and the frequency of exposure (Afful et al., 2010; Blain et al., 2010; Lejuez et al., 2007); however, both sexes clearly have the potential for such adversity to be imprinted into their neurobiology. Importantly, exactly how such trauma affects the brains of men and women is also probably different (e.g., the sexual victimization occurs much more frequently in women than in men, and the number of women addicts that report a history of sexual abuse is well above the general population: 50–70% in many studies (Afful et al., 2010; Khoury et al., 2010; Roy et al., 2011; Shand et al., 2011; Torchalla et al., 2011).

## 2.3 Why do some people use drugs?

Are abused drugs so highly addictive that even a single use can lead to an individual getting hooked? Perhaps in some cases, but most epidemiological studies suggest that only 16% of cocaine users will develop an addiction (Wagner and Anthony, 2002). Given these numbers, it would seem more worthwhile to investigate the circumstances underlying the transition from regulated to dysregulated drug use.

There are several prevalent themes in the literature as to why addicts first start using drugs, which largely fall under the umbrellas of "sensation-seeking" and "self-medication for mental health issues." These categories are not mutually exclusive, as addicts generally endorse multiple reasons for initiating drug use. Irrespective or their reasons for first using, continued drug use is largely thought to be maintained by a combination of the reinforcing and conditioning effects of drugs combined with the negative consequences of not using the drug (i.e., withdrawal and lack of "high"). Epidemiological studies describe rates of drug use and addiction in men and women; however, they rarely delve into the reasons behind why people initiate and continue using drugs, which makes it difficult to discern whether men and women use drugs for similar reasons.

There are two fundamental questions that the preclinical literature rarely addresses: why are drugs of abuse used initially; and why are drugs of abuse used habitually? We suggest that the relative lack of information on these topics in the preclinical literature stems not from lack of interest, but from a failure of animal models to adequately capture the human condition. Not that all preclinical research dismisses the notion that individual experiences and social conditions can affect drug use. Studies examining the effects of social defeat or social support on drug self-administration have demonstrated important consequences of the social environment in animal models on self-administration behavior (Miczek et al., 2008; Westenbroek et al., 2013). With these concerns about the current animal models, we turn to a discussion of the preclinical literature.

## 3 SEX DIFFERENCES IN ADDICTION IN PRECLINICAL STUDIES

One of the most striking sex differences in addiction is that women progress through the landmark stages from initial use to dependence at a faster rate than men (Brady and Randall, 1999; Kosten et al., 1993). Females also have greater motivation to obtain and/or use drugs of abuse (Becker and Hu, 2008) and escalate drug use more rapidly after relapse (Paliwal et al., 2008). Importantly, rats exhibit these same sex differences in addiction. Female rats acquire drug taking more rapidly than males, progress to addictive-like behavior more rapidly than males, and are more highly motivated to take cocaine than are male rats (Becker and Hu, 2008). This demonstrates that even though there are cultural factors influencing addiction in humans, fundamental biological processes also contribute to sex differences in addiction.

Before we examine the results of sex differences in addiction in preclinical studies we want to examine what that means. Sex differences can exist where the behaviors exhibited by males and females are qualitatively different (e.g., in sexual behavior females exhibit lordosis and males exhibit mounting behavior). Alternatively, there can be population differences where the distribution of a trait is different for males and females. In other words, are all female rats (and by extension women) at higher risk for addiction than men, or are there just more addicted individuals in the female group driving the overall difference with males, but the neural processes of those who are more at risk are the same? Finally it could be that there are differences in the distribution of males and females that are at greater risk for addiction, and this could be due to basic underlying sex differences in the neural systems mediating addiction. Failure to include females in preclinical research means that we do not know whether there are fundamental sex differences in the neural mechanisms that mediate addiction in males and females, or a population difference in addiction liability that is mediated by the same neural changes.

## 4 SEX DIFFERENCES IN THE MECHANISMS OF ADDICTION

Individuals differ in the risk for transitioning from occasional drug use to drug abuse and dependence, with only a small percentage (16–17%) of people who have used cocaine developing dependence (Wagner and Anthony, 2002). Several behavioral traits, like impulsivity and risk-taking behavior, have been associated with the risk for transition from voluntary recreational use to compulsive drug taking (Kreek et al., 2005).

The dopaminergic reward system is one of the systems thought to mediate these "vulnerable" behavioral traits. Furthermore, differences in this system have been found to be involved in individual variability in sensitivity to drugs of abuse in humans, nonhuman primates, and rodents (Volkow et al., 1999; for review see Belin and Everitt, 2008).

The nucleus accumbens (NAc), dorsomedial striatum (DMS), and dorsolateral striatum (DLS) are thought to be crucial sites within the reward system that are involved in the transition from drug taking to avid drug seeking and addiction. The NAc and DMS are important for acquisition of drug taking, while the DLS is thought to be involved in escalated drug taking. When drug intake switches from casual use to avid drug intake there is enhanced dopamine (DA) transmission in the DLS (Willuhn et al., 2012). It is postulated that neural control, or dominance, shifts from the NAc and DMS, to the DLS as behaviors are learned and addiction develops (Belin and Everitt, 2008).

Estradiol treatment of ovariectomized female rats enhances the cocaine-induced DA increase in DLS (Cummings et al., 2014). Estradiol enhances psychostimulant-induced increases in DA from striatum only in females, not in males. Interestingly, we find a significantly lower cocaine-induced increase in DA from NAc of ovariectomized females compared to males with or without estradiol treatment (Cummings et al., 2014). A similar sex difference has been reported in the NAc core DA response to amphetamine (Gillies et al., 2014; Virdee et al., 2014), whereas the ventral striatal DA response to nicotine and alcohol is greater in men than in women (Cosgrove et al., 2014; Urban et al., 2010).

In the NAc and striatum, DA gates the activity of medium spiny projection neurons (MSNs) through the activation of D1 and D2 dopamine receptors (D1DR and D2DR). Elevation in DA leads to the preferential activation of D1DR-containing MSNs in the direct pathway and the inhibition of D2DR-containing MSNs in the indirect pathway, collectively resulting in appetitive/approach responses. Conversely, reduction in basal DA levels leads to less inhibition of D2DR-containing MSNs and less inhibition of competing behavioral activities. The direct pathway is important for initiation of appetitive behaviors and the indirect pathway inhibits competing repetitive behaviors. Within the striatum, plasticity in the MSN neurons in the direct versus indirect pathways occurs in opposite directions with greater activation of the direct pathway driving motivated behaviors and reduced activity in the indirect pathway, which results in greater addiction-like behaviors, especially when the DLS becomes involved. This "reciprocal plasticity" in D1/D2 has been demonstrated in DMS during acquisition of goal-directed behavior (Shan et al., 2014) and we hypothesize that it also occurs in the DLS with addiction-like behavior.

We propose that females acquire cocaine-taking more rapidly and take more cocaine because there is less of an increase in DA/infusion in NAc (compared with males). More cocaine is needed to achieve comparable increases in DA. We further hypothesize that the transition to compulsive drug taking in females is facilitated by enhanced DA transmission in DLS relative to NAc (see Figure 6.1).

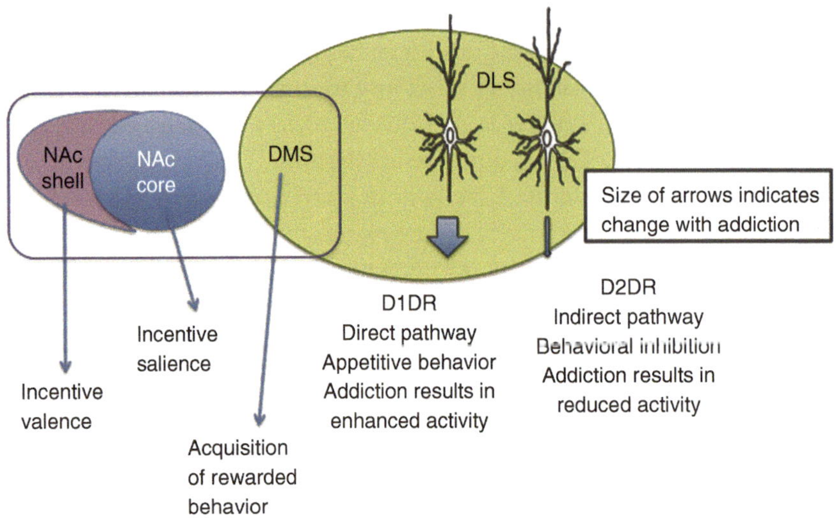


**Figure 6.1** ***In our model, the initial motivation for and acquisition of appetitive behaviors is mediated by the NAc and DMS.*** A lower DA increase with cocaine in these regions (box on left) relative to the DLS is hypothesized to mediate a more rapid transition to addiction because more cocaine is needed to produce the same DA response in NAc, but there is a greater cocaine-induced DA response in DLS. In DLS, the direct pathway is important for appetitive behaviors and the indirect pathway inhibits competing repetitive behaviors. Plasticity in the MSN neurons in the direct versus the indirect pathways occurs in opposite directions during addiction with greater activation of the direct pathway (D1DR) driving motivated behaviors and reduced indirect pathway activity (D2DR) resulting in greater addiction-like behaviors. This hypothesized D1/D2 reciprocal plasticity in DLS has been demonstrated in DMS during acquisition of goal-directed behavior (Shan et al., 2014).

Male rats with lower NAc DA are more impulsive and more likely to develop addiction-like behaviors (Belin and Everitt, 2008; Dalley et al., 2007; Deroche-Gamonet et al., 2004). Thus, even though there are more females with lower NAc DA, both males and females that are found to have lower NAc DA relative to DLS DA are predicted to be at greater risk to develop cocaine self-administration behavior and addiction-like behavior if the sex differences in addiction vulnerability are due to a population difference.

Previous studies have identified effects of estradiol in the DA system that may contribute to the increased rewarding effects of drugs in females. Estradiol enhances the behavioral response to amphetamine and cocaine as indicated by greater stereotypy, locomotion and rotational behavior after estradiol (Becker, 1999; Becker et al., 2001, 1982; Becker and Cha, 1989; Camp et al., 1986; Jackson et al., 2006; Robinson et al., 1982; Zhou et al., 2002). Estradiol also modulates the striatal DA response to amphetamine and cocaine (Becker, 1990a, b; Becker and Ramirez, 1981; Becker and Rudick, 1999; Cummings et al., 2014; Xiao and Becker, 1994). Intriguingly, the NAc DA response is not modulated by estradiol to the same extent as is the DLS in female rats and estradiol does not modulate DA release in NAc or DLS of males (Cummings et al., 2014).

Expression of striatal D1 receptors is 10% higher in males than in females, and although there are no sex differences in D2DR densities, there is a sexually dimorphic effect of estradiol on D2DR, where estradiol rapidly downregulates D2 binding in females but not males (Bazzett and Becker, 1994). In cell culture, pretreatment with estradiol reduces D2 receptor inhibition of adenylate cyclase activity, and enhances D1 receptor activation of adenylate cyclase (Maus et al., 1989). Additionally, estradiol treatment decreases expression of regulator of RGS9-2, a GTPase activating protein associated with D2 receptor signaling (Silverman and Koenig, 2007).

## 5 SEX DIFFERENCES IN PRECLINICAL RESEARCH

From the earliest laboratory studies in the 1960s and 1970s, abused drugs were linked to their ability to reinforce behavior in animal models. Several paradigms have been developed to examine the reinforcing effects of drugs in laboratory animals, including operant self-administration and conditioned place preferences. However, self-administration studies are generally considered the gold standard for preclinical addiction research, as the drug is voluntarily administered by the subject in a manner that resembles drug use by addicts, and as such these will be the focus of this discussion.

Self-administration studies are useful in that they provide a window into the animal's motivation to obtain the drug. Thus, by employing different schedules of reinforcement (e.g., fixed ratio and progressive ratio), it is possible to assess patterns of drug intake and how hard they will work to self-administer the drug (Richardson and Roberts, 1996), both are at the core of the diagnostic criteria for substance use disorders.

### 5.1 Acquisition of drug self-administration

When rats are placed in a self-administration chamber most animals readily acquire the behavioral response required to earn drug delivery (e.g., nose poking or lever pressing). Interestingly, several studies have shown that female rats acquire self-administration of cannabinoids, cocaine, and heroin more readily than males (Fattore et al., 2009; Lynch, 2008; Lynch and Carroll, 1999). More rapid acquisition by females has nearly always been interpreted as reflecting their greater vulnerability to drug abuse. Studies with different criteria and/or higher doses of drug have reported greater rates of acquisition in males, females, or neither sex (Caine et al., 2004). Additionally, rapid intake acquisition can be dissociated from escalation and dysregulated drug taking behavior, at least in males (Beckmann et al., 2012). Whether this is also true in females has not been established. Therefore, acquisition should not be the only criteria used to evaluate sex differences in addiction liability.

Numerous studies have demonstrated that rapid acquisition of drug self-administration in females is enhanced by estradiol in ovariectomized (OVX) females compared with OVX females treated with vehicle (Hu and Becker, 2008; Hu et al., 2003; Jackson et al., 2006; Perry et al., 2013a). Controlled studies such as these are useful for identifying the potential contributions of hormones to behavior. How these effects of estradiol are related to the escalation of drug taking in some females is an important next step that may help illuminate what leads to escalated drug taking in both females and males.

## 5.2 Patterns of drug intake and motivation

One of the earliest studies examining sex differences in motivation to self-administer cocaine reported that female rats were willing to work much harder than males, and that their motivation was highest during estrus (Roberts et al., 1989). The Becker laboratory also found female rats exhibit greater motivation at some doses of cocaine in rats that have been selectively bred to exhibit greater activity in a novel environment (Cummings et al., 2011). On the other hand, male and female rats have similar motivation before an extended discrete access condition, yet after 10 days of abstinence motivation increases in females but remains unchanged in males (Lynch and Taylor, 2004). In a different study, male and female rats had similar motivation following extended access to self-administration and 2 weeks of abstinence (Ramoa et al., 2013).

## 5.3 Modeling drug relapse: extinction and reinstatement

Drug addiction is associated with multiple bouts of intoxication, abstinence, and recidivism of drug use. Human addicts relapse for numerous reasons. Preclinical studies primarily rely on extinction and reinstatement methods for studying drug relapse in animal models (Epstein et al., 2006; Shaham et al., 2003). Extinction involves placing the animal in the self-administration chamber for several sessions in which operant responding no longer results in drug delivery or presentation of drug-associated cues. These conditions result in a relatively rapid decline in responding. Once the operant behavior has been extinguished, it can be reinstated by reactivation of the drug-associated cues (still in the absence of drug delivery) or a priming injection of the drug itself. Reinstatement of drug-seeking can also be induced by exposure to stressors, including foot shock and chemical triggers (e.g., yohimbine).

Female rats often display greater initial responding during extinction, particularly during estrus (Anker and Carroll, 2010; Fuchs et al., 2005; Kerstetter et al., 2008; Kippin et al., 2005). The slower extinction of females is often interpreted as greater addiction. The transition from drug self-administration to extinction may reflect the greater automation or habitualization of the response, which is hypothesized to mediate the sex difference in a more rapid transition to addicted behavior in females. These behaviors are more pronounced in females and influenced by estradiol (Colzato et al., 2010).

There are conflicting results in studies of sex differences in cue-induced reinstatement of drug seeking. Some studies report no sex differences or influence of ovarian hormones (Feltenstein et al., 2011). In contrast, the majority of studies indicate greater cocaine-induced reinstatement in females, again with the highest rates of responding occurring during estrus (Feltenstein et al., 2009; Kippin et al., 2005; Lynch and Carroll, 2000).

Only a few studies have examined sex differences in the neural responses to drug-associated cues in rodents. In general, male and female rats show similar patterns of cue-induced neural activation, even though females generally show elevated rates of responding (Zhou et al., 2013). The clinical literature suggests that cocaine-dependent men and women have similar cue-induced negative emotions, whereas stress imagery provokes greater negative emotions in women (Fox et al., 2013). Additionally, stress- and cue-induced craving are influenced by hormone fluctuations across the menstrual cycle (Sinha et al., 2007).

### 5.4 Distinguishing recreational drug use from addiction in preclinical models

The ease with which most animals learn to self-administer drugs that are abused is both a boon and a bane to the scientist. From a practical perspective having all of your subjects acquire the behavior is necessary for large sample sizes with which to examine how drug self-administration influences the brain. On the other hand, it is difficult to reconcile the high rates of drug self-administration in laboratory animals with the more modest rates at which human drug users develop addiction.

Different self-administration paradigms seek to distinguish rats self-administering drugs (i.e., recreational users) from those displaying compulsive (i.e., addicted rats). Most of these paradigms adopt a definition based on the diagnostic criteria for substance use disorders (e.g., use of a substance in larger amounts or over a longer period than was intended, which may also relate to the criteria for tolerance). Such definitions also fit nicely with current concepts of reward and reinforcement. Thus, regular/recreational drug use represents a normal goal-directed behavior that is largely reinforced by dopamine neurotransmission in the ventral striatum (i.e., nucleus accumbens). Conversely, compulsive drug use represents a neurobiological shift, such that drug use is viewed as a habit that is maintained by conditioned stimuli and dopamine neurotransmission in the dorsal striatum (Everitt et al., 2008; Willuhn et al., 2012).

## 6 NEW ANIMAL MODELS TO STUDY ADDICTION ARE NEEDED

In humans, approximately 10% of the population has a substance abuse problem and women who abuse drugs are thought to be more vulnerable to addiction. Of course cultural factors and availability of drugs also influence drug abuse in humans.

In order to better understand the mechanisms mediating individual differences in susceptibility for cocaine addiction and the extent that males and females differ in this regard, we have begun to employ novel self-administration paradigms to better model addiction. In the first paradigm, changes in the preference and motivation for cocaine compared with a palatable food reward (defined here as pellets that animals will work for even when not food deprived) are examined. Preferences were identified during self-administration on a fixed-ratio (FR) schedule with cocaine-only, pellet-only, and choice sessions, whereas motivation for the two rewards was examined concurrently on a progressive-ratio (PR) schedule (Perry et al., 2013b).

Cocaine-preferring rats increased their drug intake at the expense of pellets, displayed increased motivation for cocaine and attenuated motivation for pellets, and showed greater cue- and cocaine-induced reinstatement of drug seeking. Thus, the constellation of behavioral changes associated with the development of cocaine preferences encompasses several aspects of the clinical criteria for drug abuse/dependence (DSM-IV). Importantly, females were more likely to develop cocaine preferences than males (50–60% vs. 20–30%) (Perry et al., 2013b).

Using the choice paradigm, we find a reduction in cocaine-induced DA in dialysate from the NAc shell associated with the development of a cocaine preference (Perry et al., 2015). These findings are consistent with reports from other laboratories and other models where there is extended cocaine self-administration (Calipari et al., 2014; Gerrits et al., 2002; Willuhn et al., 2012, 2014). Importantly, individuals that self-administered cocaine, but that still retained a preference for the pellet, did not show a reduction of cocaine-induced DA in the NAc over the course of self-administration, even though many individuals took as much cocaine as some animals that formed a preference for cocaine and showed the reduction in DA overflow. Individuals that did not acquire cocaine self-administration even when they had the opportunity to do so and were tested for the same amount of time, also did not show a reduction in DA overflow in the NAc. Thus, the reduction in cocaine-induced DA overflow appears to be related specifically to the development of compulsive drug use, not just cocaine exposure or experience in the apparatus. Furthermore, the reduction in cocaine-induced DA overflow in the NAc shell was observed in both males and females, confirming that compulsive cocaine use is associated with attenuated DA signaling in both sexes (Perry et al., 2015).

In addition to serving as a method for identifying vulnerable and resilient individuals, the examination of cocaine and food self-administration simultaneously in the same individuals also permits more nuanced analyses of how drug use affects physiology and behavior, as well as sex differences in the self-administration of drug and nondrug rewards. For example, we reliably see effects of the estrous cycle on weight gain and pellet self-administration in female rats, including reduced intake on an FR5 schedule and reduced motivation on a PR schedule (Perry et al., 2013b). Thus, we have multiple pieces of evidence confirming that the estrous cycle is not disrupted by cocaine self-administration in these females, and yet we do not see changes in their intake or motivation to self-administer cocaine. The lack of estrous cycle effects on cocaine self-administration is consistent with several similar negative findings in rats, humans, and nonhuman primates (Cooper et al., 2013; Evans and Foltin, 2010; Kerstetter et al., 2012), but stands in contrast to the robust effects of estradiol treatment during acquisition in OVX females (Perry et al., 2013a; Ramoa et al., 2013; Caine et al., 2004).

There has not been a systematic evaluation of patterns of cocaine use in women addicts across the menstrual cycle, and the majority of menstrual cycle effects are found for the subjective effects of psychostimulants (Evans, 2007; Evans and Foltin, 2010; White et al., 2002), which may or may not translate into different patterns of actual cocaine intake. However, if the subjective effects of a given dose vary over the menstrual cycle, this may contribute to the finding that women are more likely to report "that was more cocaine than I intended to use" (Kennedy et al., 2013).

The choice paradigm we have employed is one of the few preclinical models that rely on changes in an individual's interest in cocaine and an alternative reward in order to identify vulnerable and resilient individuals. However, several studies have provided animals with access to alternative rewards and examined the effects on patterns of drug intake. Thus, similar to our results, access to sweet solutions reduces cocaine self-administration and/or drug seeking when offered as a direct alternative to drugs (Kerstetter and Kippin, 2011; Lenoir et al., 2007), or even when they are provided outside of the context of self-administration (Liu and Grigson, 2005).

Access to a running wheel inside the self-administration chamber or the home cage also reduces cocaine self-administration and reinstatement of drug seeking in both males and females (Cosgrove et al., 2002; Peterson et al., 2014; Smith and Pitts, 2011; Smith et al., 2012).

Interestingly, the protective effects of running wheel access appear to be more pronounced in females, particularly for those that are in estrous or treated with estradiol, which may be due to the competing effects of estradiol on general activity and locomotion. Thus, males and females may have unique profiles of protection and risk depending upon the type of alternative reward. In this regard, it would be interesting to examine whether the loss of interest in running wheels could also be used to identify vulnerable individuals, and whether males and females are differentially affected, as we have shown with the greater likelihood of losing interest in palatable food rewards displayed by females (Perry et al., 2013b). Furthermore, developing additional preclinical models with various alternative rewards would clarify whether the loss of interest in food rewards is associated with similar or different neurobiological adaptations as the loss of interest in other types of rewards, including running wheels and social partners.

The second paradigm we have developed examines manipulation of housing conditions to examine the role of environmental stress caused by isolation in drug taking behavior in males and females. Stress plays an important factor in the risk for drug abuse and comorbidity with stress-related psychopathologies is common (Briand and Blendy, 2010). Individual housing is the standard for drug self-administration studies primarily because of the complications associated with maintaining the indwelling catheter patent during the prolonged testing periods of the experiment. Most studies disregard the fact that isolation is a stressor, for female rats in particular (Brown and Grunberg, 1995; Westenbroek et al., 2003, 2005), that can affect drug-taking behavior (Covington and Miczek, 2005). Housing rats in pairs reduces the motivation to self-administer cocaine; however, only in female rats, and the motivation for taking cocaine is higher in females than in males, but only in isolated rats (Westenbroek et al., 2013). Therefore, housing conditions can affect drug-taking behavior in rats, and can also have a differential effect on females versus males and thus impact sex differences in drug taking.

The mechanisms of addiction are obviously complex. The behavioral testing models that are employed can enable investigators to distinguish individual animals based on drug taking behavior. Better animal models need to be developed to facilitate our understanding of sex differences in the neural mechanisms related to escalation of drug use, and development of the preference for cocaine. The new paradigms we have presented will potentially serve as a valuable model for testing the potential therapeutic efficacy of treatments for helping both males and females to reduce their consumption of cocaine while maintaining an interest in nondrug rewards is sustained.

## 7 CONCLUSIONS

Women exhibit more rapid escalation of drug taking than men. In females, estradiol enhances acquisition of drug taking and motivation for drugs of abuse, but how this is related to the sex difference in escalation in drug taking remains to be determined. Experiments from the Becker laboratory have demonstrated that estradiol enhances cocaine- or amphetamine-stimulated DA release in DLS but not the NAc. When behaviors, such as drug taking, become compulsive we hypothesize that this reflects enhanced DA release in the DLS and attenuation of NAc DA release. We propose that this pattern is seen more often in females (in a larger proportion of females) due to the effects of estradiol in DLS and this mediates the more rapid escalation of drug taking in some females. If our model is valid, this would predict that males who show rapid escalation would also exhibit the same pattern of low initial NAc DA response to cocaine and more rapid downregulation of the response with repeated cocaine exposure. If this is the case, then sex/gender may be a continuum rather than a bimodal factor and treatment will need to be tailored to the specific biological characteristics of this subpopulation, taking into consideration hormonal factors.

## ACKNOWLEDGMENTS

The research discussed has been supported by grants from the National Institutes of Health (R21-DA-032747 (JBB and ANP), R21-DA-032856 (JBB and CW), and R01-DA-012677 (JBB)) and the National Science Foundation (IOS1353263 to JBB).

## REFERENCES


Afful, S.E., Strickland, J.R., Cottler, L., Bierut, L.J., 2010. Exposure to trauma: a comparison of cocaine-dependent cases and a community-matched sample. Drug Alcohol Depend. 112 (1–2), 46–53.

Anker, J.J., Carroll, M.E., 2010. Sex differences in the effects of allopregnanolone on yohimbine-induced reinstatement of cocaine seeking in rats. Drug Alcohol Depend. 107 (2–3), 264–267.

Bazzett, T., Becker, J., 1994. Sex differences in the rapid and acute effects of estrogen on striatal D2 dopamine receptor binding. Brain Res. 637 (1–2), 163–172.

Becker, J.B., 1990a. Direct effect of 17 beta-estradiol on striatum: sex differences in dopamine release. Synapse 5 (2), 157–164.

Becker, J.B., 1990b. Estrogen rapidly potentiates amphetamine-induced striatal dopamine release and rotational behavior during microdialysis. Neurosci. Lett. 118 (2), 169–171.

Becker, J.B., 1999. Gender differences in dopaminergic function in striatum and nucleus accumbens. Pharmacol. Biochem. Behav. 64 (4), 803–812.

Becker, J., Cha, J., 1989. Estrous cycle-dependent variation in amphetamine-induced behaviors and striatal dopamine release assessed with microdialysis. Behav. Brain Res. 35 (2), 117–125.

Becker, J.B., Hu, M., 2008. Sex differences in drug abuse. Front. Neuroendocrinol. 29 (1), 36–47.

Becker, J.B., Ramirez, V.D., 1981. Sex differences in the amphetamine stimulated release of catecholamines from rat striatal tissue *in vitro*. Brain Res. 204 (2), 361–372.

Becker, J.B., Rudick, C.N., 1999. Rapid effects of estrogen or progesterone on the amphetamine-induced increase in striatal dopamine are enhanced by estrogen priming: a microdialysis study. Pharmacol. Biochem. Behav. 64 (1), 53–57.

Becker, J.B., Robinson, T.E., Lorenz, K.A., 1982. Sex differences and estrous cycle variations in amphetamine-elicited rotational behavior. Eur. J. Pharmacol. 80 (1), 65–72.

Becker, J.B., Molenda, H., Hummer, D.L., 2001. Gender differences in the behavioral responses to cocaine and amphetamine. Implications for mechanisms mediating gender differences in drug abuse. Ann. NY Acad. Sci. 937, 172–187.

Becker, J.N., Perry, A.N., Westenbroek, C., 2012. Sex differences in the neural mechanisms mediating addiction: a new synthesis and hypothesis. Biol. Sex Differ. 3 (1), 14–49.

Beckmann, J.S., Gipson, C.D., Marusich, J.A., Bardo, M.T., 2012. Escalation of cocaine intake with extended access in rats: dysregulated addiction or regulated acquisition? Psychopharmacology 222 (2), 257–267.

Belin, D., Everitt, B.J., 2008. Cocaine seeking habits depend upon dopamine-dependent serial connectivity linking the ventral with the dorsal striatum. Neuron 57 (3), 432–441.

Blain, L.M., Galovski, T.E., Robinson, T., 2010. Gender differences in recovery from posttraumatic stress disorder: a critical review. Aggress. Violent Behav. 15 (6), 463–474.

Brady, K.T., Randall, C.L., 1999. Gender differences in substance use disorders. Psychiatr. Clin. North Am. 22 (2), 241–252.

Briand, L.A., Blendy, J.A., 2010. Molecular and genetic substrates linking stress and addiction. Brain Res. 1314, 219–234.

Brown, K.J., Grunberg, N.E., 1995. Effects of housing on male and female rats: crowding stresses male but calm females. Physiol. Behav. 58 (6), 1085–1089.

Caine, S.B., Bowen, C.A., Yu, G., Zuzga, D., Negus, S.S., Mello, N.K., 2004. Effect of gonadectomy and gonadal hormone replacement on cocaine self-administration in female and male rats. Neuropsychopharmacology 29 (5), 929–942.

Califano, J.A., 2006. Women under the Influence. The Johns Hopkins University Press, Baltimore, MD.

Calipari, E.S., Ferris, M.J., Jones, S.R., 2014. Extended access of cocaine self-administration results in tolerance to the dopamine-elevating and locomotor-stimulating effects of cocaine. J. Neurochem. 128 (2), 224–232.

Camp, D.M., Becker, J.B., Robinson, T.E., 1986. Sex differences in the effects of gonadectomy on amphetamine-induced rotational behavior in rats. Behav. Neural Biol. 46 (3), 491–495.

Chen, K.W., Banducci, A.N., Guller, L., Macatee, R.J., Lavelle, A., Daughters, S.B., Lejuez, C.W., 2011. An examination of psychiatric comorbidities as a function of gender and substance type within an inpatient substance use treatment program. Drug Alcohol Depend. 118 (2–3), 92–99.

Colzato, L.S., Hertsig, G., van den Wildenberg, W.P.M., Hommel, B., 2010. Estrogen modulates inhibitory control in healthy human females: evidence from the stop-signal paradigm. Neuroscience 167 (3), 709–715.

Cooper, Z.D., Foltin, R.W., Evans, S.M., 2013. Effects of menstrual cycle phase on cocaine self-administration in rhesus macaques. Horm. Behav. 63 (1), 105–113.
Cosgrove, K.P., Hunter, R.G., Carroll, M.E., 2002. Wheel-running attenuates intravenous cocaine self-administration in rats: sex differences. Pharmacol. Biochem. Behav. 73 (3), 663–671.
Cosgrove, K.P., Wang, S., Kim, S.J., McGovern, E., Nabulsi, N., Gao, H., et al., 2014. Sex differences in the brain's dopamine signature of cigarette smoking. J. Neurosci. 34 (50), 16851–16855.
Cotto, J.H., Davis, E., Dowling, G.J., Elcano, J.C., Staton, A.B., Weiss, S.R., 2010. Gender effects on drug use, abuse, and dependence: a special analysis of results from the national survey on drug use and health. Gend. Med. 7 (5), 402–413.
Covington, H.E., Miczek, K.A., 2005. Intense cocaine self-administration after episodic social defeat stress, but not after aggressive behavior: dissociation from corticosterone activation. Psychopharmacology 183 (3), 331–340.
Cummings, J.A., Gowl, B.A., Westenbroek, C., Clinton, S.M., Akil, H., Becker, J.B., 2011. Effects of a selectively bred novelty-seeking phenotype on the motivation to take cocaine in male and female rats. Biol. Sex Differ. 2 (1), 3.
Cummings, J.A., Jagannathan, L., Jackson, L.R., Becker, J.B., 2014. Sex differences in the effects of estradiol in the nucleus accumbens and striatum on the response to cocaine: neurochemistry and behavior. Drug Alcohol Depend. 135, 22–28.
Dalley, J.W., Fryer, T.D., Brichard, L., Robinson, E.S.J., Theobald, D.E.H., Lääne, K., et al., 2007. Nucleus accumbens D2/3 receptors predict trait impulsivity and cocaine reinforcement. Science 315 (5816), 1267–1270.
Degenhardt, L., Hall, W., 2012. Extent of illicit drug use and dependence, and their contribution to the global burden of disease. Lancet 379 (9810), 55–70.
Deroche-Gamonet, V., Belin, D., Piazza, P., 2004. Evidence for addiction-like behavior in the rat. Science 305 (5686), 1014.
Epstein, D.H., Preston, K.L., Stewart, J., Shaham, Y., 2006. Toward a model of drug relapse: an assessment of the validity of the reinstatement procedure. Psychopharmacology 189 (1), 1–16.
Evans, S.M., 2007. The role of estradiol and progesterone in modulating the subjective effects of stimulants in humans. Exp. Clin. Psychopharmacol. 15 (5), 418–426.
Evans, S.M., Foltin, R.W., 2010. Does the response to cocaine differ as a function of sex or hormonal status in human and non-human primates? Horm. Behav., 13–21.
Evans, J.L., Hahn, J.A., Page-Shafer, K., Lum, P.J., Stein, E.S., Davidson, P.J., Moss, A.R., 2003. Gender differences in sexual and injection risk behavior among active young injection drug users in San Francisco (the UFO Study). J. Urban Health 80 (1), 137–146.
Everitt, B.J., Belin, D., Economidou, D., Pelloux, Y., Dalley, J.W., Robbins, T.W., 2008. Review. Neural mechanisms underlying the vulnerability to develop compulsive drug-seeking habits and addiction. Philos. Trans. R. Soc. Lond. B Biol. Sci. 363 (1507), 3125–3135.
Fattore, L., Spano, M.S., Altea, S., Angius, F., Fadda, P., Fratta, W., 2009. Cannabinoid self-administration in rats: sex differences and the influence of ovarian function. Br. J. Pharmacol. 152 (5), 795–804.
Feltenstein, M.W., Byrd, E.A., Henderson, A.R., See, R.E., 2009. Attenuation of cocaine-seeking by progesterone treatment in female rats. Psychoneuroendocrinology 34 (3), 343–352.
Feltenstein, M.W., Henderson, A.R., See, R.E., 2011. Enhancement of cue-induced reinstatement of cocaine-seeking in rats by yohimbine: sex differences and the role of the estrous cycle. Psychopharmacology 216 (1), 53–62.
Fox, H.C., Sofuoglu, M., Morgan, P.T., Tuit, K.L., Sinha, R., 2013. The effects of exogenous progesterone on drug craving and stress arousal in cocaine dependence: impact of gender and cue type. Psychoneuroendocrinology, 1–13.
Fuchs, R.A., Evans, K.A., Mehta, R.H., Case, J.M., See, R.E., 2005. Influence of sex and estrous cyclicity on conditioned cue-induced reinstatement of cocaine-seeking behavior in rats. Psychopharmacology 179 (3), 662–672.
Gerrits, M.A.F.M., Petromilli, P., Westenberg, H.G.M., Di Chiara, G., van Ree, J.M., 2002. Decrease in basal dopamine levels in the nucleus accumbens shell during daily drug-seeking behaviour in rats. Brain Res. 924 (2), 141–150.

Gillies, G.E., Virdee, K., Mcarthur, S., Dalley, J.W., 2014. Sex-dependent diversity in ventral tegmental dopaminergic neurons and developmental programing: a molecular, cellular and behavioral analysis. Neuroscience 282 (C), 69–85.

Grella, C.E., Lovinger, K., 2012. Gender differences in physical and mental health outcomes among an aging cohort of individuals with a history of heroin dependence. Addict. Behav. 37 (3), 306–312.

Harris, M., Rhodes, T., 2013. Injecting practices in sexual partnerships: hepatitis C transmission potentials in a "risk equivalence" framework. Drug Alcohol Depend. 132 (3), 617–623.

Hu, M., Becker, J.B., 2008. Acquisition of cocaine self-administration in ovariectomized female rats: effect of estradiol dose or chronic estradiol administration. Drug Alcohol Depend. 94 (1–3), 56–62.

Hu, M., Crombag, H.S., Robinson, T.E., Becker, J.B., 2003. Biological basis of sex differences in the propensity to self-administer cocaine. Neuropsychopharmacology 29 (1), 81–85.

Inciardi, J.A., Surratt, H.L., 2001. Drug use, street crime, and sex-trading among cocaine-dependent women: implications for public health and criminal justice policy. J. Psychoactive Drugs 33 (4), 379–389.

Jackson, L.R., Robinson, T.E., Becker, J.B., 2006. Sex differences and hormonal influences on acquisition of cocaine self-administration in rats. Neuropsychopharmacology 31 (1), 129–138.

James, D. J., Karberg. J.C., 2005. Substance dependence, abuse, and treatment of jail inmates, 2002 (No. NCJ 209588) (pp. 1–12). The U.S. Department of Justice, Office of Justice Programs, Bureau of Justice Statistics. Retrieved from http://www.bjs.gov/index.cfm?ty=pbdetail&iid=1128.

Kennedy, A.P., Epstein, D.H., Phillips, K.A., Preston, K.L., 2013. Sex differences in cocaine/heroin users: drug-use triggers and craving in daily life. Drug Alcohol Depend. 132 (0), 29–37.

Kerstetter, K.A., Kippin, T.E., 2011. Impact of sex and gonadal hormones on cocaine and food reinforcement paradigms. J. Addict. Res. Ther. Suppl. 4 (2).

Kerstetter, K.A., Aguilar, V.R., Parrish, A.B., Kippin, T.E., 2008. Protracted time-dependent increases in cocaine-seeking behavior during cocaine withdrawal in female relative to male rats. Psychopharmacology 198 (1), 63–75.

Kerstetter, K.A., Ballis, M.A., Duffin-Lutgen, S., Carr, A.E., Behrens, A.M., Kippin, T.E., 2012. Sex differences in selecting between food and cocaine reinforcement are mediated by estrogen. Neuropsychopharmacology 37 (12), 2605–2614.

Khan, S.S., Secades-Villa, R., Okuda, M., Wang, S., Pérez-Fuentes, G., Kerridge, B.T., Blanco, C., 2013. Gender differences in cannabis use disorders: results from the National Epidemiologic Survey of Alcohol and Related Conditions. Drug Alcohol Depend. 130 (1–3), 101–108.

Khoury, L., Tang, Y.L., Bradley, B., Cubells, J.F., Ressler, K.J., 2010. Substance use, childhood traumatic experience, and posttraumatic stress disorder in an urban civilian population. Depress. Anxiety 27 (12), 1077–1086.

Kippin, T.E., Fuchs, R.A., Mehta, R.H., Case, J.M., Parker, M.P., Bimonte-Nelson, H.A., See, R.E., 2005. Potentiation of cocaine-primed reinstatement of drug seeking in female rats during estrus. Psychopharmacology 182 (2), 245–252.

Kosten, T.A., Gawin, F.H., Kosten, T.R., Rounsaville, B.J., 1993. Gender differences in cocaine use and treatment response. J. Subst. Abuse Treat. 10 (1), 63–66.

Kreek, M.J., Nielsen, D.A., Butelman, E.R., LaForge, K.S., 2005. Genetic influences on impulsivity, risk taking, stress responsivity and vulnerability to drug abuse and addiction. Nat. Neurosci. 8 (11), 1450–1457.

Lejuez, C.W., Bornovalova, M.A., Reynolds, E.K., Daughters, S.B., Curtin, J.J., 2007. Risk factors in the relationship between gender and crack/cocaine. Exp. Clin. Psychopharmacol. 15 (2), 165–175.

Lenoir, M., Serre, F., Cantin, L., Ahmed, S., 2007. Intense sweetness surpasses cocaine reward. PLoS ONE 2 (8).

Lewis, C., 2006. Treating incarcerated women: gender matters. Psychiatr. Clin. North Am. 29 (3), 773–789.

Liu, C., Grigson, P.S., 2005. Brief access to sweets protect against relapse to cocaine-seeking. Brain Res. 1049 (1), 128–131.

Lynch, W.J., 2008. Acquisition and maintenance of cocaine self-administration in adolescent rats: effects of sex and gonadal hormones. Psychopharmacology 197 (2), 237–246.

Lynch, W.J., Carroll, M.E., 1999. Sex differences in the acquisition of intravenously self-administered cocaine and heroin in rats. Psychopharmacology 144 (1), 77–82.

Lynch, W.J., Carroll, M.E., 2000. Reinstatement of cocaine self-administration in rats: sex differences. Psychopharmacology 148 (2), 196–200.

Lynch, W.J., Taylor, J.R., 2004. Sex differences in the behavioral effects of 24-h/day access to cocaine under a discrete trial procedure. Neuropsychopharmacology 29 (5), 943–951.
Maus, M., Cordier, J., Glowinski, J., Premont, J., 1989. 17-beta oestradiol pretreatment of mouse striatal neurons in culture enhances the responses of adenylate cyclase sensitive to biogenic amines. Eur. J. Neurosci. 1 (2), 154–161.
Miczek, K.A., Yap, J.J., Covington, III, H.E., 2008. Social stress, therapeutics and drug abuse: preclinical models of escalated and depressed intake. Pharmacol. Ther. 120 (2), 102–128.
Paliwal, P., Hyman, S.M., Sinha, R., 2008. Craving predicts time to cocaine relapse: further validation of the now and brief versions of the cocaine craving questionnaire. Drug Alcohol Depend. 93 (3), 252–259.
Peer, K., Rennert, L., Lynch, K.G., Farrer, L., Gelernter, J., Kranzler, H.R., 2013. Prevalence of DSM-IV and DSM-5 alcohol, cocaine, opioid, and cannabis use disorders in a largely substance dependent sample. Drug Alcohol Depend. 127 (1–3), 215–219.
Perry, A.N., Westenbroek, C., Becker, J.B., 2013a. Impact of pubertal and adult estradiol treatments on cocaine self-administration. Horm. Behav. 64 (4), 573–578.
Perry, A.N., Westenbroek, C., Becker, J.B., 2013b. The development of a preference for cocaine over food identifies individual rats with addiction-like behaviors. PLoS ONE 8 (11), e79465.
Perry, A.N., Westenbroek, C., Jagannathan, L., Becker, J.B., 2015. The roles of dopamine and α1-adrenergic receptors in cocaine preferences in female and male rats. Neuropsychopharmacology. Advance online publication. doi: 10.1038/npp.2015.116.
Peterson, A.B., Hivick, D.P., Lynch, W.J., 2014. Dose-dependent effectiveness of wheel running to attenuate cocaine-seeking: impact of sex and estrous cycle in rats. Psychopharmacology 231 (13), 2661–2670.
Ramoa, C.P., Doyle, S.E., Naim, D.W., Lynch, W.J., 2013. Estradiol as a mechanism for sex differences in the development of an addicted phenotype following extended access cocaine self-administration. Neuropsychopharmacology 38 (9), 1698–1705.
Richardson, N.R., Roberts, D.C., 1996. Progressive ratio schedules in drug self-administration studies in rats: a method to evaluate reinforcing efficacy. J. Neurosci. Methods 66 (1), 1–11.
Roberts, D., Bennett, S., Vickers, G., 1989. The estrous cycle affects cocaine self-administration on a progressive ratio schedule in rats. Psychopharmacology 98 (3), 408–411.
Robinson, T.E., Camp, D.M., Jacknow, D.S., Becker, J.B., 1982. Sex differences and estrous cycle dependent variation in rotational behavior elicited by electrical stimulation of the mesostriatal dopamine system. Behav. Brain Res. 6 (3), 273–287.
Roy, Ã.L., Boivin, J.-F., Leclerc, P., 2011. Initiation to drug injection among street youth: a gender-based analysis. Drug Alcohol Depend. 114 (1), 49–54.
Saddichha, S., Fliers, J.M., Frankish, J., Somers, J., Schuetz, C.G., Krausz, M.R., 2014. Homeless and incarcerated: an epidemiological study from Canada. Int. J. Soc. Psychiatry 60 (8), 795–800.
Santolaria-Fernández, F.J., Gómez-Sirvent, J.L., González-Reimers, C.E., Batista-López, J.N., Jorge-Hernández, J.A., Rodríguez-Moreno, F., et al., 1995. Nutritional assessment of drug addicts. Drug Alcohol Depend. 38 (1), 11–18.
Schulte, M.T., Ramo, D., Brown, S.A., 2009. Gender differences in factors influencing alcohol use and drinking progression among adolescents. Clin. Psychol. Rev. 29 (6), 535–547.
Shaham, Y., Shalev, U., Lu, L., de Wit, H., Stewart, J., 2003. The reinstatement model of drug relapse: history, methodology and major findings. Psychopharmacology 168 (1–2), 3–20.
Shan, Q., Ge, M., Christie, M.J., Balleine, B.W., 2014. The acquisition of goal-directed actions generates opposing plasticity in direct and indirect pathways in dorsomedial striatum. J. Neurosci. 34 (28), 9196–9201.
Shand, F.L., Degenhardt, L., Slade, T., Nelson, E.C., 2011. Sex differences amongst dependent heroin users: histories, clinical characteristics and predictors of other substance dependence. Addict. Behav. 36 (1–2), 27–36.
Silverman, J.L., Koenig, J.I., 2007. Evidence for the involvement of ERβ and RGS9-2 in 17-β estradiol enhancement of amphetamine-induced place preference behavior. Horm. Behav. 52 (2), 146–155.
Sinha, R., Fox, H., Hong, K.-I., Sofuoglu, M., Morgan, P.T., Bergquist, K.T., 2007. Sex steroid hormones, stress response, and drug craving in cocaine-dependent women: implications for relapse susceptibility. Exp. Clin. Psychopharmacol. 15 (5), 445–452.
Smith, M.A., Pitts, E.G., 2011. Access to a running wheel inhibits the acquisition of cocaine self-administration. Pharmacol. Biochem. Behav. 100 (2), 237–243.

Smith, M.A., Pennock, M.M., Walker, K.L., Lang, K.C., 2012. Access to a running wheel decreases cocaine-primed and cue-induced reinstatement in male and female rats. Drug Alcohol Depend. 121 (1–2), 54–61.

Substance Abuse and Mental Health Services Administration, 2013. Results from the 2012 National Survey on Drug Use and Health: Summary of National Findings, NSDUH Series H-46, HHS Publication No. (SMA) 13-4795. Substance Abuse and Mental Health Services Administration, Rockville, MD.

Torchalla, I., Strehlau, V., Li, K., Krausz, M., 2011. Substance use and predictors of substance dependence in homeless women. Drug Alcohol Depend. 118 (2-3), 173–179.

Tracy, D., Hahn, J.A., Fuller Lewis, C., Evans, J., Briceño, A., Morris, M.D., et al., 2014. Higher risk of incident hepatitis C virus among young women who inject drugs compared with young men in association with sexual relationships: a prospective analysis from the UFO Study cohort. BMJ Open 4 (5), e004988.

Urban, N.B.L., Kegeles, L.S., Slifstein, M., Xu, X., Martinez, D., Sakr, E., et al., 2010. Sex differences in striatal dopamine release in young adults after oral alcohol challenge: a positron emission tomography imaging study with [$^{11}$C]raclopride. Biol. Psychiatry 68 (8), 689–696.

Van Etten, M.L., Neumark, Y.D., Anthony, J.C., 1999. Male-female differences in the earliest stages of drug involvement. Addiction 94 (9), 1413–1419.

Virdee, K., McArthur, S., Brischoux, F., Caprioli, D., Ungless, M.A., Robbins, T.W., et al., 2014. Antenatal glucocorticoid treatment induces adaptations in adult midbrain dopamine neurons, which underpin sexually dimorphic behavioral resilience. Neuropsychopharmacology 39 (2), 339–350.

Volkow, N.D., Wang, G.J., Fowler, J.S., Logan, J., Gatley, S.J., Gifford, A., et al., 1999. Prediction of reinforcing responses to psychostimulants in humans by brain dopamine D2 receptor levels. Am. J. Psychiat. 156 (9), 1440–1443.

Wagner, F.A., Anthony, J.C., 2002. From first drug use to drug dependence: developmental periods of risk for dependence upon marijuana, cocaine, and alcohol. Neuropsychopharmacology 26 (4), 479–488.

Westenbroek, C., Horst, G.J.T., Roos, M.H., Kuipers, S.D., Trentani, A., den Boer, J.A., 2003. Gender-specific effects of social housing in rats after chronic mild stress exposure. Prog. Neuropsychopharmacol. Biol. Psychiat. 27 (1), 21–30.

Westenbroek, C., Snijders, T.A.B., den Boer, J.A., Gerrits, M., Fokkema, D.S., Ter Horst, G.J., 2005. Pair-housing of male and female rats during chronic stress exposure results in gender-specific behavioral responses. Horm. Behav. 47 (5), 620–628.

Westenbroek, C., Perry, A.N., Becker, J.B., 2013. Pair housing differentially affects motivation to self-administer cocaine in male and female rats. Behav. Brain Res. 252, 68–71.

White, T.L., Justice, A.J.H., de Wit, H., 2002. Differential subjective effects of D-amphetamine by gender, hormone levels and menstrual cycle phase. Pharmacol. Biochem. Behav. 73 (4), 729–741.

Willuhn, I., Burgeno, L.M., Everitt, B.J., Phillips, P.E.M., 2012. Hierarchical recruitment of phasic dopamine signaling in the striatum during the progression of cocaine use. Proc. Natl. Acad. Sci. 109 (50), 20703–20708.

Willuhn, I., Burgeno, L.M., Groblewski, P.A., Phillips, P.E.M., 2014. Excessive cocaine use results from decreased phasic dopamine signaling in the striatum. Nat. Neurosci. 17 (5), 704–709.

Xiao, L., Becker, J.B., 1994. Quantitative microdialysis determination of extracellular striatal dopamine concentration in male and female rats: effects of estrous cycle and gonadectomy. Neurosci. Lett. 180 (2), 155–158.

Zhou, W., Cunningham, K.A., Thomas, M.L., 2002. Estrogen regulation of gene expression in the brain: a possible mechanism altering the response to psychostimulants in female rats. Brain Res. Mol. Brain Res. 100 (1–2), 75–83.

Zhou, L., Pruitt, C., Shin, C.B., Garcia, A.D., Zavala, A.R., See, R.E., 2013. Fos expression induced by cocaine-conditioned cues in male and female rats. Brain Struct. Funct. 219 (5), 1831–1840.

CHAPTER 7

# Stress and Emotional Learning in Humans: Evidence for Sex Differences

**Christian J. Merz, Oliver T. Wolf**
Ruhr-University Bochum, Institute of Cognitive Neuroscience, Department of Cognitive Psychology, Bochum, Germany

## 1 INTRODUCTION

Stress has been conceptually defined as a response to a threat to homeostasis. In humans, psychosocial stressors are especially powerful. A threat to the social self (social evaluative threat) in combination with uncontrollability of the situation is especially potent in prompting stress (Dickerson and Kemeny, 2004). In the laboratory, stress can be induced by means of public speaking paradigms such as the Trier Social Stress Test (TSST; Kirschbaum et al., 1993), or by painful manipulations (immersion of the hand into ice water) such as the cold pressor test (CPT) or the later-developed socially evaluated CPT (Schwabe et al., 2008).

It has been suggested that women and men differ in how they respond to stressors, based upon endocrinological and behavioral responses (Taylor et al., 2000). These differences might translate into vulnerabilities for distinct stress-associated psychiatric disorders. Compared with men, women have, for example, a higher risk for major depression, posttraumatic stress disorder (PTSD), and several anxiety disorders, but a lower prevalence in schizophrenia, substance abuse, and autism (Cover et al., 2014).

When discussing possible sex differences in how stressors affect learning and memory, two possible scenarios should be considered. On the one hand, sex differences might occur because the two sexes differ in their endocrinological response to a stressor. Alternatively, or additionally, sex differences might reflect a different responsivity of the brain to the same neuroendocrine stress signal (e.g., glucocorticoids).

Endocrinologically, both sexes generally respond to stressors with an activation of the two major stress systems, the sympathetic nervous system (SNS) and the hypothalamic–pituitary–adrenocortical (HPA) axis (see Figure 7.1). The SNS initiates a rapid first stress response wave, which is dominated by the effects of (nor)adrenaline released from the adrenal medulla. Increases in heart rate and breathing frequency prepare the body for action. This initial response has been conceptualized as the fight or flight response (Cannon, 1932).

Slightly delayed, at least with respect to its adrenal end product, the HPA axis also responds. Corticotropin-releasing hormone (CRH), released from the hypothalamus, works to stimulate the secretion of adrenocorticotropin (ACTH) in conjunction with

*Sex Differences in the Central Nervous System.* http://dx.doi.org/10.1016/B978-0-12-802114-9.00007-X

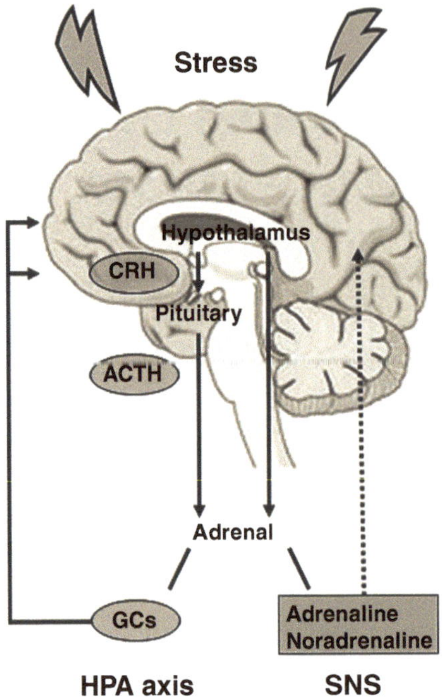


**Figure 7.1** ***Stress activates two lines of defense mechanisms: the rapidly acting sympathetic nervous system (SNS) and the slower hypothalamic–pituitary–adrenocortical (HPA) axis.*** Activation of the hypothalamus stimulates the SNS to secrete (nor)adrenaline from the adrenal medulla. The hypothalamus also releases CRH, which stimulates the secretion of adrenocorticotropin (ACTH) from the anterior pituitary gland into the blood stream. ACTH stimulates the adrenal cortex to release glucocorticoids (GCs), which can easily pass the blood–brain barrier and modulate brain functions involved in learning and memory. GCs exert negative feedback effects on the hypothalamus and the pituitary gland, leading to reduced activity of the HPA axis.

arginine vasopressin (Joëls and Baram, 2009). ACTH in turn stimulates the synthesis and release of glucocorticoids (GCs; mainly cortisol in humans) from the adrenal cortex. The neuroendocrine stress response is regulated by circulating GCs via negative feedback mechanisms targeting the pituitary, the hypothalamus, the hippocampus, and prefrontal areas (Ulrich-Lai and Herman, 2009). This negative feedback loop is essential for the regulation of both the HPA axis and the stress response (De Kloet et al., 2005). GCs mediate their effects by binding to two subtypes of intracellular receptors, the mineralocorticoid receptor (MR) and the glucocorticoid receptor (GR). These two receptors differ in their affinity and distribution within the brain: while MRs are mainly located in the hippocampus, GRs are expressed throughout the brain, for example, in the prefrontal cortex. In addition, membrane-bound GRs and MRs have also been identified (Joëls et al., 2008; Roozendaal et al., 2010). Due to their prominence throughout the brain, corticoid receptors modulate several cognitive processes, including memory. While most

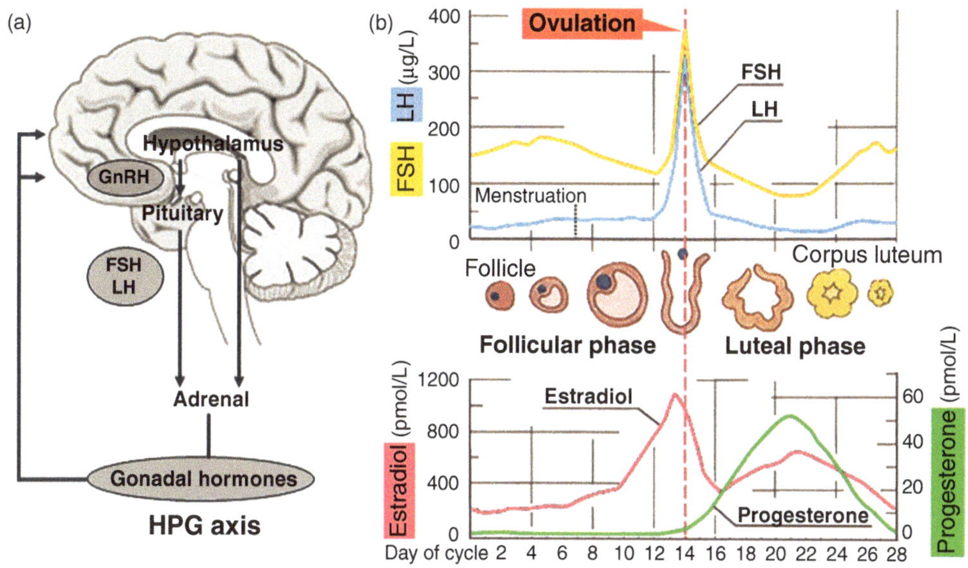


**Figure 7.2** (a) The hypothalamus controls the release of gonadal hormones via the hypothalamic–pituitary–gonadal (HPG) axis. The hypothalamus secretes gonadotropin-releasing hormone (GnRH), which stimulates the secretion of follicle stimulating hormone (FSH) and luteinizing hormone (LH) from the anterior pituitary gland into the blood stream. The gonads secrete gonadal hormones, which exert negative feedback effects on the hypothalamus and the pituitary gland. (b) In women, secretion of hormones from the HPG axis is subject to changes over the menstrual cycle. In the follicular phase, low levels of the gonadal hormone estradiol are present, which rise midcycle to initiate ovulation. After that, the following luteal phase is characterized by higher estradiol as well as progesterone concentrations. The pituitary hormones LH and FSH show peak concentrations during ovulation.

of the effects associated with GCs – especially when related to stress – have been attributed to GR, the importance of MRs has also been emphasized (Joëls et al., 2008). The newly discovered membrane bound MR in particular might be responsible for some of the rapid effects of acute stress on memory discussed in the present chapter.

The hypothalamic–pituitary–gonadal (HPG) axis controls the release of the gonadal hormones estradiol, progesterone, and testosterone from the ovaries and testes. Regarding the course of the menstrual cycle, estradiol concentrations peak in the middle of the cycle around ovulation, while progesterone concentrations increase in the second half of the cycle (in the luteal phase; see Figure 7.2). The HPG axis is influenced by acute and, even more pronounced, chronic stress. In turn, it also influences the HPA axis.

It is well established that the HPA axis response is modulated by gonadal steroids (Kajantie and Phillips, 2006; Kudielka and Kirschbaum, 2005; Taylor et al., 2000). Experimental studies in humans using psychosocial laboratory stressors such as the TSST have often observed that men show a stronger HPA axis response to the stressor than women (Kajantie and Phillips, 2006; Kudielka and Kirschbaum, 2005). However, this might depend on the specific paradigm used (Stroud et al., 2002). No strong overall influence of sex on the cortisol response to laboratory stressors could be detected in a large meta-analysis, which, however, was not able to investigate a possible influence of menstrual cycle and hormonal contraceptives (Dickerson and Kemeny, 2004).

In women, fluctuations of gonadal hormones during the menstrual cycle seem to modulate the HPA axis response. A more pronounced HPA axis response to stressors is observed during the luteal phase (Kajantie and Phillips, 2006; Kudielka and Kirschbaum, 2005), characterized by elevated progesterone and estradiol levels. The situation in humans is further complicated by the fact that hormonal contraceptives appear to dampen the free (unbound, i.e., biologically active) cortisol stress response by increasing cortisol-binding globulin (Kirschbaum et al., 1999; Rohleder et al., 2003). The typical, blunted salivary free cortisol response of women using hormonal contraceptives is displayed in Figure 7.3.

In conclusion, gonadal steroids impact HPA axis reactivity in humans as a result of the complex interaction between the HPA and HPG axes. Furthermore, hormones released by both axes can influence the brain and cognition. However, many experimental studies in humans have been conducted exclusively with men. Moreover, in studies with

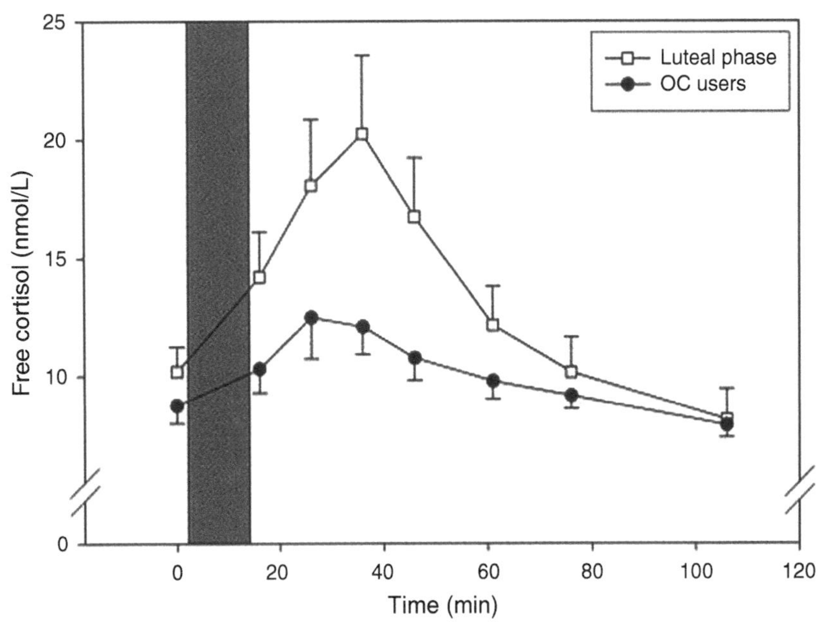


**Figure 7.3** ***Exposure to a standardized psychosocial stress protocol (Trier Social Stress Test, TSST) leads to increases in free cortisol concentrations over time.*** Women in the luteal phase of the menstrual cycle exhibit higher cortisol increases compared with women taking oral contraceptives (OCs) who typically show blunted cortisol responses in saliva. *(Reprinted from Rohleder et al. (2003). With permission from Elsevier.)*

women, information about menstrual cycle phase and/or hormonal contraception is often not taken into account in the experimental design (Beckner et al., 2006; Smeets et al., 2006). Similarly, most studies in rodents focus exclusively on males when examining stress effects on memory.

In the present chapter, we will review the literature concerning the impact of stress hormones on learning and memory processes in humans. Particularly, we will focus on episodic memory and fear conditioning, two forms of long-term memory, for which some data are available on the influence of gonadal hormones and the intake of hormonal contraceptives.

## 2 EPISODIC MEMORY

Episodic memory relies heavily on intact medial temporal lobes, including the most recognized structure for memory: the hippocampus (Nadel and Moscovitch, 1997). Personal individual events of the past, such as the party on your 18th birthday or the day of your driving test, are stored in your episodic memory. In general, three stages of memory processing can be identified (Tulving, 1983): (1) encoding: transformation of incoming information into a memory representation, (2) consolidation: modification and stabilization of this representation, and (3) retrieval: reactivation of the stored memory trace.

Stress and stress hormones exert numerous effects on these memory stages in humans. Typically, stress impairs memory retrieval, but enhances consolidation processes. The effects on encoding seem to depend on various factors. Emotionally arousing stimuli in particular are affected by stress, which is explained by an interaction between GCs and an activation of the SNS affecting the amygdala and the hippocampus (Roozendaal et al., 2009). In the following sections, we will delineate these different effects. After brief introductions to the overall picture in each memory stage, we will concentrate on experimental studies focusing on sex differences in episodic memory after exposure to stress or administration of cortisol.

### 2.1 Encoding

Generally, findings on pre-encoding stress have revealed enhancing, impairing, and absent effects of stress. Several ideas have been proposed to explain these results. The first refers to the timing of the stressor relative to the learning session. Exposure to stress a relatively long time before encoding exerts detrimental effects on memory, whereas stress taking place directly before or as part of the learning session facilitates memory encoding (for reviews see Joëls, 2006; Schwabe and Wolf, 2013). Stress within the learning context situation helps the individual to better retrieve this memory at a later point in time. The second idea is time of day: a meta-analysis found impaired memory when acute cortisol administration was set before encoding in the morning, but the reversed effect was found when cortisol administration took place in the afternoon (Het et al., 2005). Since cortisol concentrations are subject to a circadian rhythm, with high levels in the morning declining over the course of the day, adding cortisol to these already relatively high endogenous concentrations in the morning seems to have negative effects on memory. The third idea concerns emotionality of the stimulus material: pre-encoding stress or cortisol application leads to a superior encoding and/or consolidation of emotionally arousing material (Buchanan and Lovallo, 2001; Kuhlmann and Wolf, 2006; Payne et al., 2007; Rimmele et al., 2003; Wolf, 2012), whereas nonarousing information is stored less efficiently (Kuhlmann and Wolf, 2006; Payne et al., 2007; Rimmele et al., 2003; Smeets et al., 2006). Thus, a stressor can potentiate the impact of stimulus emotionality on long-term memory.

Now let us turn to those studies reporting sex differences and the impact of circulating gonadal hormones. Cornelisse et al. (2011) investigated men and women encoding neutral and emotional pictures after stress induction using the TSST. One week later, men in the stress condition showed better memory for emotional pictures compared with men in the control condition. This effect was not found for neutral pictures or in the female group. The latter fact is interesting – why should women not display a stress effect? One explanation could be that more than two-thirds of the female participants used hormonal contraceptives, which are associated with a substantially blunted free cortisol stress response (see "Introduction"). In both sexes, the cortisol stress response was correlated with the enhanced emotional memory recognition suggesting that indeed differences in cortisol concentrations underlie the observed sex differences in this study rather than different effects of cortisol on the brain.

In general, pooling groups of women with and without hormonal contraceptive usage into one group might cancel out potential effects of circulating (endogenous and exogenous) gonadal hormones. Indeed, a separation of women into different groups of free-cycling women and women taking hormonal contraceptives is needed to allow for a more detailed understanding of the influence of circulating gonadal hormones. Unfortunately, only few studies have included a closer description of their female sample up to now, so that the picture remains inconclusive, at least at the moment.

One experiment in men and women taking oral contraceptives was able to show that cold pressor stress before the learning session increased memory in men and women taking oral contraceptives (Schwabe et al., 2008), suggesting that, regarding cognitive performance, both groups might respond to stress in a similar fashion. However, this pattern was not confirmed in a study clearly differentiating women according to the stage of their menstrual cycle (follicular and luteal phase) and the intake of oral contraceptives (Espin et al., 2013). Under control conditions, all of these female groups remembered more neutral words compared with men. Exposure to stress before encoding and directly following retrieval disrupted this sex difference, which could be traced back to an enhanced performance after stress in men, but no change after stress in the three groups of women. However, stress induction only increased cortisol concentrations in women in the luteal phase (limiting the conclusions to be drawn from this study). In the same group of women, a correlation was found between the cortisol increase after stress and poorer performance. This finding contrasts a previous study showing that stress-induced cortisol increases were negatively correlated with memory performance in men, but not in women tested in the luteal phase (Wolf et al., 2001).

In conclusion, many studies investigating stress effects on encoding have investigated men only (Nater et al., 2007; Quaedflieg et al., 2013; Tops et al., 2003) or men and women but without reporting details regarding the female sample and also without explicitly testing for sex differences (Rimmele et al., 2003; Zoladz et al., 2011). Only a small number of experiments have outlined details on menstrual cycle or usage of hormonal contraceptives in women and at the same time looked at potential group differences, but these studies do not yield a consistent picture. A tentative conclusion based on the available data is that the effects of stress on encoding are more robust in men compared with women.

## 2.2 Consolidation

Experimental elevation of stress hormones after a learning episode typically leads to enhanced memory for this information, especially for emotionally arousing material (Wolf, 2009). This effect can be observed when female and male participants are exposed to stress (Beckner et al., 2006; Cahill et al., 2003; Preuss and Wolf, 2009; Smeets et al., 2006). From an evolutionary point of view, the facilitation of memory consolidation after a stressful event clearly serves adaptation: better memory for a dangerous situation helps individuals to initiate adequate survival mechanisms when later encountering a comparable threatening situation (De Kloet et al., 2005; McEwen, 1998).

Looking closer at potential underlying sex differences, the positive effect of postencoding stress on later familiarity-based recognition has been reported by one research group to occur in men, but not in women (McCullough and Yonelinas, 2013; Yonelinas et al., 2011). It is worth mentioning that men and women exhibited similar cortisol increases in response to stress in both of the studies mentioned, so an explanation for the above-mentioned discrepancy cannot be derived from differing cortisol responses. Thus, differences in the sensitivity to stress hormones appear to be the most likely explanation.

In addition, another study found an inverted U relationship between a stress-induced cortisol increase and memory for a neutral story in men (Andreano and Cahill, 2006), while no relationship was evident in women. Felmingham et al. (2012b) found a similar enhancing effect of postlearning stress for nonarousing images in men. In free-cycling women, on the other hand, the increase in stress hormones after encoding was positively related to later memory recall of emotionally arousing images. Again, cortisol responses to the stressor were comparable in men and women. A similar study revealed that stress enhances memory consolidation for emotionally arousing images when women are tested in the midluteal phase, in which estradiol and progesterone concentrations are high, but not in women tested in other phases of the menstrual cycle, when progesterone is low (Felmingham et al., 2012a). In line with these results, a positive relationship between cortisol concentrations after encoding and an enhanced memory performance was only observed in women tested in the midluteal phase (Andreano et al., 2008; see Figure 7.4), but not in the early or late follicular phase. Thus, progesterone might represent an important player in modulating stress effects on memory consolidation in women. Of course, it cannot be ruled out that the effects observed in the midluteal phase are not solely due to progesterone increases in this phase. Relatively high estradiol levels might exert their effects only in the presence of high progesterone concentrations; both are present in the midluteal phase. Pharmacological interventions directly manipulating progesterone and/or

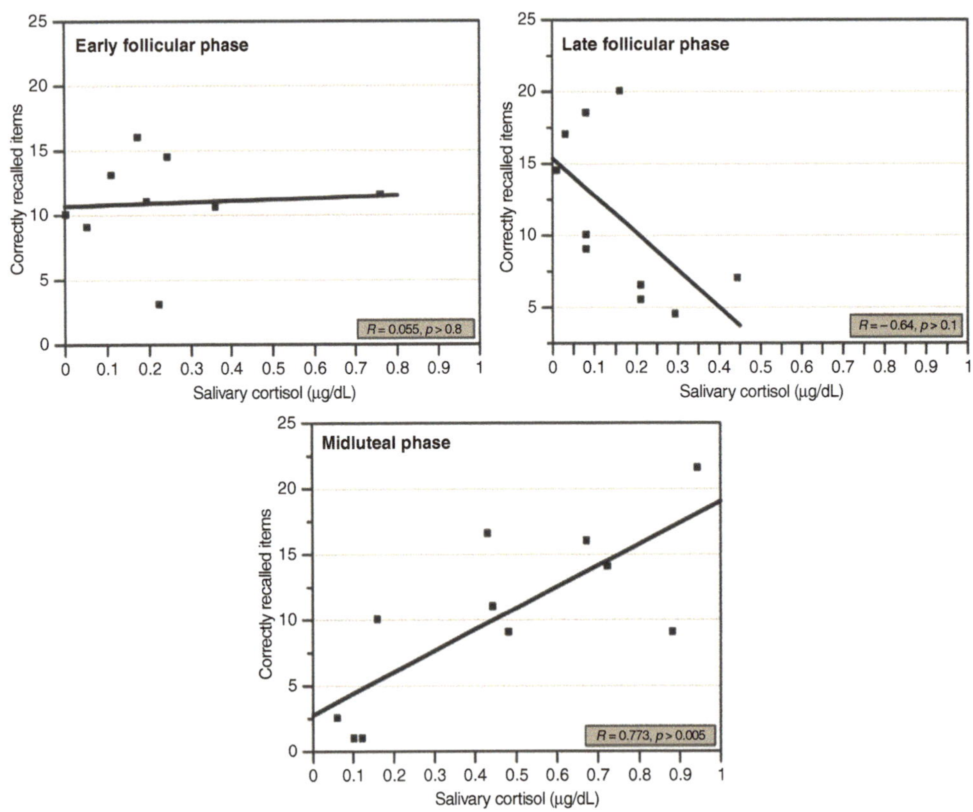


**Figure 7.4** ***Association between salivary cortisol concentrations after stress induction (postlearning stress) and memory retrieval 1 week later in women in the early follicular, late follicular, and midluteal phases.*** A significant positive relationship occurred in the midluteal phase, whereas a trend toward the negative direction was observed in the early follicular phase. *(Slightly modified from Andreano et al. (2008). With permission from Elsevier.)*

estradiol levels would be desirable in order to understand the exact underlying gonadal hormone-related mechanisms.

Apart from differences due to gonadal hormones, the impact of oral contraceptives in stress effects on consolidation processes has also been investigated recently. In a study by Nielsen et al. (2013), free-cycling women and women taking hormonal contraceptives viewed neutral and emotionally arousing images before being exposed to cold pressor stress or a control procedure. The stressor was more effective in increasing cortisol concentrations in free-cycling women than in women taking hormonal contraceptives, who typically display relatively blunted free cortisol responses (Kirschbaum et al., 1999; Rohleder et al., 2003). Stress hormones did not facilitate memory consolidation of emotional pictures in any free-cycling women, but recall of nonarousing pictures 1 week later was heightened in those free-cycling women responding to the stressor with increased SNS and HPA axis activation. Women taking hormonal contraceptives tended to retrieve more emotionally arousing images when the SNS, but not the HPA axis, was activated. This rather complex pattern at least suggests that, in addition to the particular menstrual cycle phase, the interplay between the SNS and the HPA axis needs to be taken into account when trying to understand stress effects on memory consolidation in women. The second study on this issue once again confirmed that stress leads to enhanced memory consolidation for emotional material in free-cycling women (Nielsen et al., 2014). However, women taking hormonal contraceptives were not influenced in their memory performance by stress exposure at all. It seems that consolidation processes are subject to modulatory influences of stress and gonadal hormones in complex ways. This issue certainly needs to be investigated in future studies to gain greater insight into how memory is differentially consolidated in men and women under stressful conditions, which will also contribute to the understanding of factors leading to psychopathology after confrontation with a traumatic situation.

Taken together, stress hormones enhance memory consolidation, especially for emotionally arousing material. With regard to the potential influence of gonadal hormones, the limited literature on this issue assumes an enhancing effect of stress on consolidation in women with higher progesterone concentrations as present in the midluteal phase. In contrast, postencoding stress might cause a detrimental influence on memory performance in women in the late follicular phase. These presumably opposing effects could explain why sometimes no impact of stress on memory consolidation has been reported in women: the negative and positive effects of postlearning stress across the cycle may have cancelled each other out. These opposing influences occurring in different stages of the menstrual cycle must be further examined since they rely on studies with rather small sample sizes. In addition, the impact of hormonal contraceptives needs to be investigated in future studies; first findings indicate a differential effect of stress on memory consolidation in free-cycling women and in women taking hormonal contraceptives.

## 2.3 Retrieval

Generally, memory retrieval is reduced when persons are stressed (Kuhlmann et al., 2005b; Merz et al., 2010b; Smeets et al., 2006; Tollenaar et al., 2008) or when cortisol is administered prior to retrieval (De Quervain et al., 2000; see Het et al., 2005 for a meta-analysis). As in consolidation, emotionally arousing material is particularly affected (Wolf, 2009). From an evolutionary point of view, this stress-induced retrieval deficit might serve an adaptive purpose (Roozendaal, 2002): retrieval processes for prior information interfere with the encoding of new information (Allan and Allen, 2005); thus, a temporary reduction in retrieval allows for appropriate and efficient encoding and consolidation of the stressful and potentially threatening new event itself.

What do we know about the potential impact of sex differences and gonadal hormones on stress effects on memory retrieval? When men and free-cycling women across the whole menstrual cycle are studied together, there does not seem to be much of a difference in the overall picture (Rohleder et al., 2009; Schönfeld et al., 2014; Schwabe and Wolf, 2009; Smeets, 2011; Young et al., 2011). When women are tested in the follicular phase, the finding of cortisol-induced memory reduction of emotional material stays the same (Kuhlmann et al., 2005a). One study demonstrated that cortisol reduces the retrieval of emotional and neutral information in women investigated in the luteal phase and during menses but not in women taking oral contraceptives (Kuhlmann and Wolf, 2005; see Figure 7.5). These data have been interpreted to support the notion that

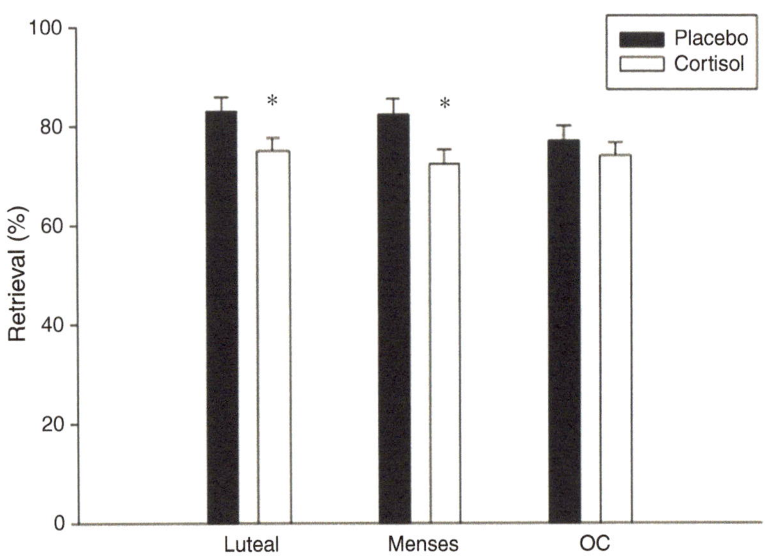


**Figure 7.5** ***Effects of an oral administration of 30 mg cortisol on delayed free retrieval of words in women tested in the luteal phase, during menses, and in women taking oral contraceptives (OC).*** Cortisol significantly reduced memory retrieval in women in the luteal phase and during menses, but not in OC women. *$p < 0.05$. *(Reprinted from Kuhlmann and Wolf (2005). With kind permission from Springer Science and Business Media.)*

women using oral contraceptives exhibit a slightly reduced central GC sensitivity to cortisol. Oral contraceptives contain exogenous gonadal hormones, which could either directly or indirectly affect the sensitivity of the brain to GCs. This issue needs to be tackled in future work.

Apart from cortisol administration, exposure to stress also led to an interesting result in one study challenging some of the previously made conclusions. This study failed to observe the adverse effect of stress on memory retrieval in a group of women tested in the luteal phase (Schoofs and Wolf, 2009). In the luteal phase, stress-induced increases in free cortisol concentrations are heightened (Kirschbaum et al., 1999), while HPA feedback and peripheral GC sensitivity are decreased (Altemus et al., 1997; Rohleder et al., 2001). In light of results obtained by Schoofs and Wolf (2009), it could be speculated whether these reductions in peripheral GC sensitivity also transfer to central GC sensitivity (Rohleder et al., 2009). The absent effects of stress on memory retrieval in women investigated in the luteal phase can be interpreted as a reduced central sensitivity to stress-induced cortisol effects (Schoofs and Wolf, 2009). Cortisol administration inhibited memory retrieval in the luteal phase (Kuhlmann and Wolf, 2005), whereas stress did not (Schoofs and Wolf, 2009). Thus, GC sensitivity in women in the luteal phase might be slightly reduced thereby abolishing effects on memory retrieval induced by stress-related cortisol increases while not affecting effects induced by supraphysiological cortisol elevations after pharmacological administration.

All in all, it has been found that stress is typically associated with poorer memory retrieval, especially for emotionally arousing information. A small number of studies support the idea that the menstrual cycle and the intake of contraceptives might modulate the observed findings. In one of these studies, cortisol administration did not influence retrieval performance in women taking oral contraceptives, while in the other women in the luteal phase were not affected in memory retrieval after stress induction. Of course, these findings need to be replicated, but they emphasize the importance of paying attention to the modulatory impact of gonadal steroids on retrieval processes.

## 2.4 Interim conclusion on episodic memory

Taken together, we can conclude that stress differentially affects the three stages of episodic memory processing. Stress enhances the encoding of emotionally arousing material, whereas, at the same time, it might impair the encoding of nonarousing information. The timing of the stressor relative to the encoding is also critical for stress effects to occur. On the one hand, stress facilitates memory when applied shortly before or during learning, and on the other hand, it deteriorates memory when applied before encoding (e.g., at the peak of cortisol concentrations). Few studies have addressed the issue of possible sex differences of pre-encoding stress. The usage of hormonal contraceptives and its associated blunted free cortisol stress response appears to cause weaker effects of stress on memory.

Stress enhances consolidation processes but typically reduces retrieval, particularly retrieval of emotional material. Regarding consolidation, it has been suggested that progesterone plays a critical role in the enhancing effects of stress; thus, women seem to benefit from postencoding stress especially when they are in the midluteal phase. At the same time, a different study reported that women in the luteal phase are not susceptible to the impairing effects of psychosocial stress (but not cortisol administration) on memory retrieval. Thus, women in the luteal phase benefit from the positive effect of stress on consolidation, but also from not being affected by the stress-inhibiting effect on memory retrieval. In addition, the usage of hormonal contraceptives appears to weaken the effects of cortisol on memory retrieval.

All in all, the existing literature indicates that stress and cortisol administration lead to more variable and smaller effects on long-term memory in women compared with men. The use of hormonal contraceptives, as well as elevated progesterone levels during the luteal phase, are important mediators. Interestingly, there is little overall evidence for an enhanced sensitivity to stress or stress hormones in women when it comes to hippocampal-based episodic memory. In Chapter 8, we will proceed to fear conditioning, another form of long-term memory relying on amygdala activity.

## 3 FEAR CONDITIONING

During classical conditioning, individuals learn that a typically neutral stimulus is paired with an aversive or pleasant event (unconditioned stimulus, UCS), which leads to an unconditioned response (e.g., fear; Pavlov, 1927). As a result of a couple of pairings of both stimuli, the originally neutral stimulus alone can trigger parts of the unconditioned response and is now termed the conditioned stimulus (CS). In this section, only fear conditioning studies using aversive events as the UCS (such as an electrical stimulation) will be presented. Fear conditioning processes are assumed to play a major role in the development of anxiety disorders as well as PTSD (Graham and Milad, 2011; Mineka and Oehlberg, 2008).

After initial fear memory formation with repeated couplings of the CS with the UCS, recurrent presentations of the CS without the UCS result in fear extinction leading to the suppression of conditioned fear. The most effective psychotherapeutic strategy to treat anxiety disorders, namely exposure therapy, in which patients are confronted with their phobic stimuli or situations, relies on fear extinction processes (Vervliet et al., 2013). But extinction training does not erase fear forever – fear is just temporarily not expressed. Individuals often experience a relapse of (pathological) fear, for example, after a change in context (renewal) (Bouton, 2004).

The amygdala is assumed to be the interspecies, core structure responsible for fear conditioning (LeDoux, 2000). In addition, the anterior cingulate cortex, the insula, the ventromedial prefrontal cortex, and the hippocampus constitute integral parts of the fear and extinction network (Mechias et al., 2010; Sehlmeyer et al., 2009). Intriguingly, patients with anxiety disorders or PTSD exhibit abnormalities in this fear circuitry

(Etkin and Wager, 2007; Graham and Milad, 2011). Because the prevalence of anxiety disorders and PTSD is substantially different in men and women (Cover et al., 2014), it is important to understand the neurobiological mechanisms of these sex differences in the underlying fear conditioning processes. Moreover, the impact of stress and stress hormones on the fear and extinction network as a function of sex needs to be elucidated. Generally, sex hormones seem to influence cortisol effects on fear learning. Additionally, high cortisol concentrations also inhibit fear retrieval. First, we will present results on the sex-dependent impact of stress hormones on fear memory formation and consolidation, and second, on extinction learning and the return of fear.

## 3.1 Fear memory formation and consolidation

Controversial findings exist regarding the question of whether and how stress and stress hormones affect fear learning; some studies have reported increased, others have reported reduced, conditioned fear memory after exposure to stress. Again, the timing of the stress exposure appears to be important. For example, exposure to stress rapidly leads to SNS activity and the accompanying release of (nor)adrenaline, which seems to be associated with increased conditioned fear (Antov et al., 2013). In line with these results, enhanced noradrenergic stimulation by the α2-adrenoreceptor antagonist, yohimbine, strengthens subsequent fear memory as shown by slower subsequent extinction learning, heightened fear retrieval after reinstatement, and augmented fear reacquisition (Soeter and Kindt, 2011).

In contrast to the influence of the rapidly developing SNS response, the slower cortisol response to a stressor appears to be negatively correlated with fear acquisition (Antov et al., 2013). This is consistent with findings showing that the cortisol response 30 min after stress exposure is negatively associated with amygdala activation (Oei et al., 2012). However, the above-mentioned results were obtained in men only, and there is, in line with animal data on eye-blink and fear conditioning (Dalla and Shors, 2009), accumulating evidence for significant sex differences in stress effects on fear conditioning. Exposure to stress 1 h before fear conditioning increased fear conditioned responses in men, but seemed to inhibit fear learning in women (Jackson et al., 2006). Correspondingly, stress-induced cortisol elevations observed after fear conditioning strengthened fear memory consolidation in men, but not in women (Zorawski et al., 2005, 2006). Critically, information on the usage of oral contraceptives was not given in these studies. Nevertheless, the same effect was also seen when cortisol was administered after fear acquisition in men (Merz et al., 2014b). When fear learning takes place at a time when stress-induced cortisol concentrations are back to baseline, no differences between groups with differing gonadal hormone status occurred in fear acquisition and subsequent extinction (Antov and Stockhorst, 2014). However, on the next day, stress led to higher fear recovery in women tested in the follicular phase compared with men. Extinction memory was also reduced in women in the follicular phase compared with women investigated midcycle.

In addition, pharmacological studies propose that GCs may further modulate fear learning and memory processes differently in men and women. Pharmacological elevations of cortisol diminish neuronal fear responses in men and free-cycling women (tested in the follicular and luteal phase), but heighten fear-related brain activity in women taking oral contraceptives (Merz et al., 2010a, 2012; Stark et al., 2006; Tabbert et al., 2010; see Figure 7.6). Importantly, psychosocial stress induction led to the same pattern of results: whereas stress attenuated conditioned brain activation in men, it increased fear responses in women taking oral contraceptives, for example, in the amygdala (Merz et al., 2013). These studies using functional neuroimaging discovered that GC administration and exposure to stress influenced the fear network sex hormone dependently, affecting, among others, the amygdala and the hippocampus, critical for emotional memory formation. In the case of the fear conditioning study using psychosocial stress induction (Merz et al., 2013), it needs to be mentioned that women taking oral contraceptives showed significantly lower increases in cortisol concentrations than men (Kirschbaum et al., 1999; Rohleder et al., 2003). Because the pattern of results after stress

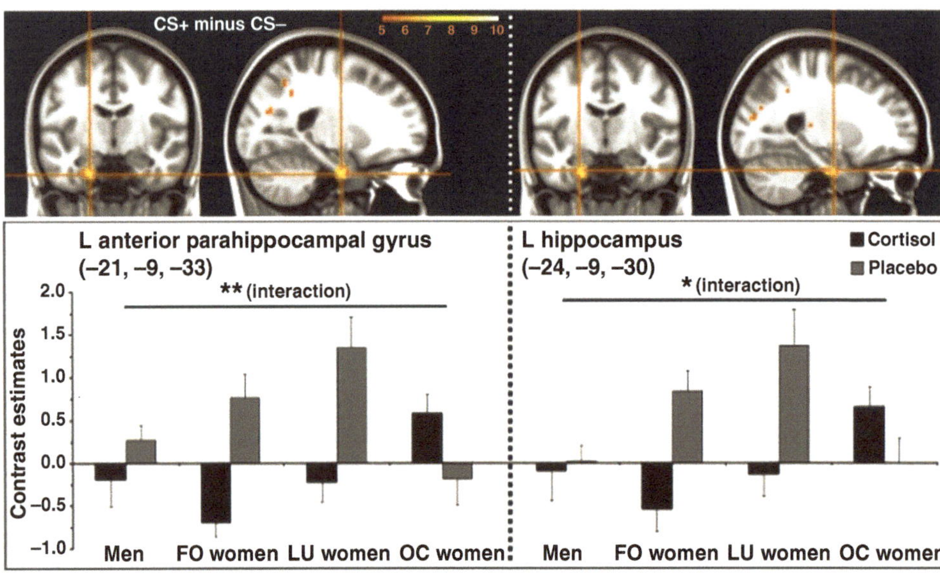


**Figure 7.6** ***Sex hormone-dependent effect of the administration of 30 mg cortisol on fear memory formation (contrast CS+ minus CS−) in the left anterior parahippocampal gyrus and the left hippocampus.*** For illustration reasons, a threshold of F≥5.0 was applied to data (see color bar for exact F values), which were then displayed on the standard MNI brain template. In the bar graphs, mean contrast estimates to CS+ minus CS− are given for men, women in the follicular (FO) phase, in the luteal (LU) phase, and women taking oral contraceptives (OC) separately for the cortisol and the placebo group in the respective peak voxel. In both brain regions, cortisol reduced the CS+/CS− differentiation in men, FO, and LU women, but enhanced it in OC women. $*p < 0.05$; $**p < 0.005$ for the treatment × sex interaction. *(Reprinted from Merz et al. (2012). With permission from Elsevier.)*

induction is similar to the pattern obtained after GC administration (leading to comparable supraphysiological cortisol concentrations in all sex hormone status groups), an underlying difference in GC sensitivity in women taking oral contraceptives compared with men does not seem very likely. Oral contraceptives contain ethinylestradiol, which binds to estrogen receptors, and a gestagenic component, which binds to progestin receptors. A higher estradiol and/or progesterone binding after continuous oral contraceptive intake might lead to a subsequent downregulation and/or to a desensitization of these receptors in various brain structures such as the hippocampus. Stress or cortisol administration may abolish this reduced excitability enabling more pronounced learning processes compared with normal (stress-free) conditions. In men and free-cycling women, these receptors are not continuously affected by oral contraceptives, thus enabling learning and memory processes to function properly.

Furthermore, cortisol weakened fear contextualization and amplified fear generalization in women taking oral contraceptives, whereas the opposite pattern was observed in men (Van Ast et al., 2012). More precisely, in women taking oral contraceptives, cortisol increased fear toward cues signaling danger in both threatening and safe contexts, and toward safety signals in a threatening context. Such deficits in fear contextualization might be interpreted as a vulnerability factor in the development of anxiety disorders and PTSD.

In conclusion, activation of the SNS seems to enhance fear memory formation. When cortisol concentrations peak shortly before or during fear acquisition, they seem to inhibit the fear circuit surrounding the amygdala and the hippocampus. In contrast, fear-related brain activation in women taking oral contraceptives is enhanced by stress or cortisol administration. It remains to be tested how exactly gonadal and stress hormones interact in different brain areas, and, in particular, how fear memory consolidation is affected.

## 3.2 Fear extinction and retrieval

Statements on the effects of stress hormones on extinction and retrieval processes can only be made cautiously. Nevertheless, interesting preliminary findings have emerged in the past few years, pointing to potential influences of gonadal hormones on extinction learning, too. In a study investigating only women taking oral contraceptives, administration of cortisol before fear acquisition enhanced neuronal activation during fear learning, but reduced activation of the amygdala and the hippocampus during subsequent extinction learning (Tabbert et al., 2010). Similarly, cortisol administration directly after fear learning, 45 min before the fear extinction session (Merz et al., 2014a,b), reduced activation of the amygdala, the medial prefrontal cortex, and the nucleus accumbens in men, which was paralleled by an enhanced electrodermal fear response during extinction. Presumably, cortisol disrupted the interchange between these structures, which may have delayed fear extinction learning. Another experiment in healthy humans examined how stress affects the extinction of a conditioned fear memory acquired the day before (Bentz et al., 2013). Exposure to stress attenuated fear retrieval (expressed as UCS expectancy) in men but not in women taking oral contraceptives.

Extinction learning is the experimental analogue to exposure therapy; we will now focus on clinical studies applying cortisol administration prior to exposure in men and women with spider phobia, height phobia, or social phobia (De Quervain et al., 2011; Soravia et al., 2006, 2014). Patients receiving cortisol reported less fear after these sessions as well as at a later point in time without medication. The explanation for these important findings was derived from the literature on episodic memory (see previous section; De Quervain and Margraf, 2008): while cortisol attenuated (fear) retrieval when patients encountered their feared stimulus, they could better consolidate the corrected information (less fear) at the same time. In addition to cortisol, pharmacologically increased noradrenergic activity before exposure therapy also reduced fear at a 1-week follow-up in male and female acrophobic patients (Powers et al., 2009). Unfortunately, all of these clinical studies did not differentiate between women in different phases of the menstrual cycle or according to the usage of oral contraceptives. It seems that these findings apply to both men and women; however, opposing or null effects in a distinct phase of the menstrual cycle or under the usage of contraceptives cannot be excluded.

A study extended these findings in female spider phobics taking oral contraceptives (Lass-Hennemann and Michael, 2014). The authors utilized the circadian cortisol rhythm, with high levels in the morning and low levels in the afternoon, when conducting the exposure session. Indeed, phobic fear was more reduced in the group in which the exposure session took place under high compared with low endogenous cortisol concentrations. The therapist thus might benefit from conducting exposure sessions in the morning (and not in the afternoon) in order to make them more effective.

A minor drawback of these studies is that the effects of stress or cortisol administration influence the retrieval of fear during the first encounter with the feared stimulus and also enhance consolidation at the same time; a clear distinction between these different memory phases cannot be made. This issue was addressed in the first study with humans to test the impact of stress on the consolidation of extinction in a renewal design, with fear learning toward discrete cues in context A, extinction in context B, and retrieval in both contexts realized on 3 consecutive days respectively (Hamacher-Dang et al., 2015). Stress was applied after the extinction session. In the retrieval test phase, the authors observed increased conditioned responses in the stress group in the acquisition, but not in the extinction context, leading to the assumption that exposure to stress augmented the integration of contextual cues into long-term memory. With the same design, the effects of stress on fear retrieval were tested using a slightly different timing (Merz et al., 2014a). Changing the timing of stress from postextinction to preretrieval led to a change in the results: stress attenuated fear retrieval in the acquisition context and also generally lowered skin conductance responses in the extinction context, in line with the aforementioned reports from clinical samples undergoing exposure sessions (De Quervain et al., 2011; Lass-Hennemann and Michael, 2014; Soravia et al., 2006, 2014). However, male participants only were tested in both studies, therefore conclusions regarding sex differences or the impact of oral contraceptives cannot be drawn.

As the study of stress effects on fear conditioning is still in its early stages, modulating variables will need to be identified to deepen our understanding of the results obtained so far. At least, there exist clinical studies suggesting an anxiolytic effect of cortisol administration and of the utilization of the circadian rhythm on phobic fear. One promising avenue to follow in this context will undoubtedly be the investigation of sex differences and of circulating gonadal hormones in interaction with stress hormones.

### 3.3 Interim conclusion on fear conditioning

In summary, preliminary evidence exists for sex differences in fear learning after stress or cortisol administration. Neuroimaging studies have revealed that increases in cortisol attenuate the fear circuitry in men and free-cycling women, but increase conditioned fear in women taking oral contraceptives. Thus, these basic learning mechanisms seem to vary substantially in women depending on the individual status of gonadal hormones at the time of testing. From a clinical perspective, it would be desirable to investigate whether these altered fear-learning processes translate into vulnerability factors for the acquisition of an anxiety disorder or PTSD.

Regarding extinction and its application in exposure therapy, high cortisol levels, as observed after pharmacological administration or during the circadian rhythm, seem to exert beneficial effects: patients report less fear, and this effect transfers to follow-up sessions without medication. Moreover, stress hormones appear to inhibit fear retrieval and enhance the consolidation of fear extinction, similar to the impact of stress on episodic memory. Both effects are in play when the confrontation of phobic patients with their feared stimulus leads to reduced fear. However, the impact of gonadal hormones on fear extinction and retrieval processes needs to be addressed in future studies to better understand the underlying mechanisms and whether they work in all patients.

## 4 GENERAL CONCLUSIONS AND SUGGESTIONS FOR FUTURE WORK

Stress exerts manifold effects on learning and memory processes, both in the domain of episodic memory as well as in the domain of fear conditioning. Cortisol appears to reduce retrieval and enhance consolidation in episodic memory, particularly for emotionally arousing information. Similar effects have been reported for fear conditioning, which induces emotional arousal as well. In both domains, preliminary evidence suggests that the intake of hormonal contraceptives can either weaken the observed effects (such as in episodic memory) or even reverse them (such as in fear conditioning). However, the influence of sex and gonadal hormones has been largely ignored in this area, even though initial studies have found important interactions between the HPA and the HPG axes. These experiments underline the importance of investigating women, and especially of examining the impact of circulating gonadal hormones on different learning and memory processes more closely. Sometimes, opposing effects in different groups of women have been observed, which need to be considered when trying to understand the basic mechanisms of stress-related clinical disorders as well as daily phenomena such as studying for an exam at school or university.

In some of the studies mentioned, it remains unclear whether the observed memory effects are due to differing cortisol responses in men and women or due to an altered central GC sensitivity. In the future, this issue clearly needs to be addressed. Furthermore, future studies examining the impact of stress hormones on episodic memory and conditioning processes should try to focus more on the investigation of women in different stages of the menstrual cycle and of women taking hormonal contraceptives. Crucially, free-cycling women should be tested several times to prove that both episodic memory consolidation and retrieval after stress or cortisol application change during different stages of the menstrual cycle. Accordingly, within-subject designs are highly desirable to explore women in their stress-related memory performance when they do not take oral contraceptives, and at a later point in time, when they have begun to take them (and later still, when they no longer take them). These suggested study designs are work-intense but undoubtedly important for a deeper insight into the modulation of episodic memory and fear conditioning processes by stress and gonadal hormones.

Fear conditioning processes in particular should be further explored to better understand critical factors facilitating extinction learning, such as fluctuating gonadal hormone concentrations over the course of the menstrual cycle. Future studies might derive neurobiological explanations for the different prevalence in men and women to acquire stress-associated disorders such as PTSD. These basic insights into fear learning and memory processes are also important for the amelioration of exposure therapy in the context of individualized treatment of patients with anxiety disorders or PTSD.

## ACKNOWLEDGMENTS

We thank Eve-Mariek Hessas for language editing. The work of the authors was supported by grants from the German Research Foundation.

## REFERENCES

Allan, K., Allen, R., 2005. Retrieval attempts transiently interfere with concurrent encoding of episodic memories but not vice versa. J. Neurosci. 25, 8122–8130.

Altemus, M., Redwine, L., Leong, Y.M., Yoshikawa, T., Yehuda, R., Detera-Wadleigh, S., Murphy, D.L., 1997. Reduced sensitivity to glucocorticoid feedback and reduced glucocorticoid receptor mRNA expression in the luteal phase of the menstrual cycle. Neuropsychopharmacology 17, 100–109.

Andreano, J.M., Cahill, L.F., 2006. Glucocorticoid release and memory consolidation in men and women. Psychol. Sci. 17, 466–470.

Andreano, J.M., Arjomandi, H., Cahill, L.F., 2008. Menstrual cycle modulation of the relationship between cortisol and long-term memory. Psychoneuroendocrinology 33, 874–882.
Antov, M.I., Stockhorst, U., 2014. Stress exposure prior to fear acquisition interacts with estradiol status to alter recall of fear extinction in humans. Psychoneuroendocrinology 49, 106–118.
Antov, M.I., Wölk, C., Stockhorst, U., 2013. Differential impact of the first and second wave of a stress response on subsequent fear conditioning in healthy men. Biol. Psychol. 94, 456–468.
Beckner, V.E., Tucker, D.M., Delville, Y., Mohr, D.C., 2006. Stress facilitates consolidation of verbal memory for a film but does not affect retrieval. Behav. Neurosci. 120, 518–527.
Bentz, D., Michael, T., Wilhelm, F.H., Hartmann, F.R., Kunz, S., von Rohr, I.R.R., de Quervain, D.J.-F., 2013. Influence of stress on fear memory processes in an aversive differential conditioning paradigm in humans. Psychoneuroendocrinology 38, 1186–1197.
Bouton, M.E., 2004. Context and behavioral processes in extinction. Learn. Mem. 11, 485–494.
Buchanan, T.W., Lovallo, W.R., 2001. Enhanced memory for emotional material following stress-level cortisol treatment in humans. Psychoneuroendocrinology 26, 307–317.
Cahill, L.F., Gorski, L., Le, K., 2003. Enhanced human memory consolidation with post-learning stress: interaction with the degree of arousal at encoding. Learn. Mem. 10, 270–274.
Cannon, W.B., 1932. Wisdom of the Body. W.W. Norton & Company, United States.
Cornelisse, S., van Stegeren, Anda, H., Joëls, M., 2011. Implications of psychosocial stress on memory formation in a typical male versus female student sample. Psychoneuroendocrinology 36, 569–578.
Cover, K.K., Maeng, L.Y., Lebron-Milad, K., Milad, M.R., 2014. Mechanisms of estradiol in fear circuitry: implications for sex differences in psychopathology. Transl. Psychiat. 4, e422.
Dalla, C., Shors, T.J., 2009. Sex differences in learning processes of classical and operant conditioning. Physiol. Behav. 97, 229–238.
De Kloet, E.R., Joëls, M., Holsboer, F., 2005. Stress and the brain: from adaptation to disease. Nat. Rev. Neurosci. 6, 463–475.
De Quervain, D.J.-F., Margraf, J., 2008. Glucocorticoids for the treatment of post-traumatic stress disorder and phobias: a novel therapeutic approach. Eur. J. Pharmacol. 583, 365–371.
De Quervain, D.J.-F., Roozendaal, B., Nitsch, R.M., McGaugh, J.L., Hock, C., 2000. Acute cortisone administration impairs retrieval of long-term declarative memory in humans. Nat. Neurosci. 3, 313–314.
De Quervain, D.J.-F., Bentz, D., Michael, T., Bolt, O.C., Wiederhold, B.K., Margraf, J., Wilhelm, F.H., 2011. Glucocorticoids enhance extinction-based psychotherapy. Proc. Natl. Acad. Sci. USA 108, 6621–6625.
Dickerson, S.S., Kemeny, M.E., 2004. Acute stressors and cortisol responses: a theoretical integration and synthesis of laboratory research. Psychol. Bull. 130, 355–391.
Espin, L., Almela, M., Hidalgo, V., Villada, C., Salvador, A., Gomez-Amor, J., 2013. Acute pre-learning stress and declarative memory: impact of sex, cortisol response and menstrual cycle phase. Horm. Behav. 63, 759–765.
Etkin, A., Wager, T.D., 2007. Functional neuroimaging of anxiety: a meta-analysis of emotional processing in PTSD, social anxiety disorder, and specific phobia. Am. J. Psychiat. 164, 1476–1488.
Felmingham, K.L., Fong, W.C., Bryant, R.A., 2012a. The impact of progesterone on memory consolidation of threatening images in women. Psychoneuroendocrinology 37, 1896–1900.
Felmingham, K.L., Tran, T.P., Fong, W.C., Bryant, R.A., 2012b. Sex differences in emotional memory consolidation: the effect of stress-induced salivary alpha-amylase and cortisol. Biol. Psychol. 89, 539–544.
Graham, B.M., Milad, M.R., 2011. The study of fear extinction: implications for anxiety disorders. Am. J. Psychiat. 168, 1255–1265.
Hamacher-Dang, T.C., Merz, C.J., Wolf, O.T., 2015. Stress following extinction learning leads to a context-dependent return of fear. Psychophysiology 52, 489–498.
Het, S., Ramlow, G., Wolf, O.T., 2005. A meta-analytic review of the effects of acute cortisol administration on human memory. Psychoneuroendocrinology 30, 771–784.
Jackson, E.D., Payne, J.D., Nadel, L., Jacobs, W.J., 2006. Stress differentially modulates fear conditioning in healthy men and women. Biol. Psychiat. 59, 516–522.
Joëls, M., 2006. Corticosteroid effects in the brain: U-shape it. Trends Pharmacol. Sci. 27, 244–250.
Joëls, M., Baram, T.Z., 2009. The neuro-symphony of stress. Nat. Rev. Neurosci. 10, 459–466.
Joëls, M., Karst, H., de Rijk, R., de Kloet, E.R., 2008. The coming out of the brain mineralocorticoid receptor. Trends Neurosci. 31, 1–7.
Kajantie, E., Phillips, D.I.W., 2006. The effects of sex and hormonal status on the physiological response to acute psychosocial stress. Psychoneuroendocrinology 31, 151–178.

Kirschbaum, C., Pirke, K.-M., Hellhammer, D.H., 1993. The "Trier Social Stress Test" – a tool for investigating psychobiological stress responses in a laboratory setting. Neuropsychobiology 28, 76–81.
Kirschbaum, C., Kudielka, B.M., Gaab, J., Schommer, N.C., Hellhammer, D.H., 1999. Impact of gender, menstrual cycle phase, and oral contraceptives on the activity of the hypothalamus–pituitary–adrenal axis. Psychosom. Med. 61, 154–162.
Kudielka, B.M., Kirschbaum, C., 2005. Sex differences in HPA axis responses to stress: a review. Biol. Psychol. 69, 113–132.
Kuhlmann, S., Wolf, O.T., 2005. Cortisol and memory retrieval in women: influence of menstrual cycle and oral contraceptives. Psychopharmacology 183, 65–71.
Kuhlmann, S., Wolf, O.T., 2006. Arousal and cortisol interact in modulating memory consolidation in healthy young men. Behav. Neurosci. 120, 217–223.
Kuhlmann, S., Kirschbaum, C., Wolf, O.T., 2005a. Effects of oral cortisol treatment in healthy young women on memory retrieval of negative and neutral words. Neurobiol. Learn. Mem. 83, 158–162.
Kuhlmann, S., Piel, M., Wolf, O.T., 2005b. Impaired memory retrieval after psychosocial stress in healthy young men. J. Neurosci. 25, 2977–2982.
Lass-Hennemann, J., Michael, T., 2014. Endogenous cortisol levels influence exposure therapy in spider phobia. Behav. Res. Ther. 60, 39–45.
LeDoux, J.E., 2000. Emotion circuits in the brain. Ann. Rev. Neurosci. 23, 155–184.
McCullough, A.M., Yonelinas, A.P., 2013. Cold-pressor stress after learning enhances familiarity-based recognition memory in men. Neurobiol. Learn. Mem. 106, 11–17.
McEwen, B.S., 1998. Stress, adaptation, and disease – allostasis and allostatic load. Ann. NY Acad. Sci. 840, 33–44.
Mechias, M.-L., Etkin, A., Kalisch, R., 2010. A meta-analysis of instructed fear studies: implications for conscious appraisal of threat. Neuroimage 49, 1760–1768.
Merz, C.J., Tabbert, K., Schweckendiek, J., Klucken, T., Vaitl, D., Stark, R., Wolf, O.T., 2010a. Investigating the impact of sex and cortisol on implicit fear conditioning with fMRI. Psychoneuroendocrinology 35, 33–46.
Merz, C.J., Wolf, O.T., Hennig, J., 2010b. Stress impairs retrieval of socially relevant information. Behav. Neurosci. 124, 288–293.
Merz, C.J., Tabbert, K., Schweckendiek, J., Klucken, T., Vaitl, D., Stark, R., Wolf, O.T., 2012. Oral contraceptive usage alters the effects of cortisol on implicit fear learning. Horm. Behav. 62, 531–538.
Merz, C.J., Wolf, O.T., Schweckendiek, J., Klucken, T., Vaitl, D., Stark, R., 2013. Stress differentially affects fear conditioning in men and women. Psychoneuroendocrinology 11, 2529–2541.
Merz, C.J., Hamacher-Dang, T.C., Wolf, O.T., 2014a. Exposure to stress attenuates fear retrieval in healthy men. Psychoneuroendocrinology 41, 89–96.
Merz, C.J., Hermann, A., Stark, R., Wolf, O.T., 2014b. Cortisol modifies extinction learning of recently acquired fear in men. Soc. Cogn. Affect. Neurosci. 9, 1426–1434.
Mineka, S., Oehlberg, K., 2008. The relevance of recent developments in classical conditioning to understanding the etiology and maintenance of anxiety disorders. Acta Psychol. 127, 567–580.
Nadel, L., Moscovitch, M., 1997. Memory consolidation, retrograde amnesia and the hippocampal complex. Curr. Opin. Neurobiol. 7, 217–227.
Nater, U.M., Moor, C., Okere, U., Stallkamp, R., Martin, M., Ehlert, U., Kliegel, M., 2007. Performance on a declarative memory task is better in high than low cortisol responders to psychosocial stress. Psychoneuroendocrinology 32, 758–763.
Nielsen, S.E., Segal, S.K., Worden, I.V., Yim, I.S., Cahill, L., 2013. Hormonal contraception use alters stress responses and emotional memory. Biol. Psychol. 92, 257–266.
Nielsen, S.E., Ahmed, I., Cahill, L., 2014. Postlearning stress differentially affects memory for emotional gist and detail in naturally cycling women and women on hormonal contraceptives. Behav. Neurosci. 128, 482–493.
Oei, N.Y.L., Veer, I.M., Wolf, O.T., Spinhoven, P., Rombouts, S.A.R.B., Elzinga, B.M., 2012. Stress shifts brain activation towards ventral "affective" areas during emotional distraction. Soc. Cogn. Affect. Neurosci. 7, 403–412.
Pavlov, I.P., 1927. Conditioned Reflexes. Oxford University Press, London.
Payne, J.D., Jackson, E.D., Hoscheidt, S., Ryan, L., Jacobs, W.J., Nadel, L., 2007. Stress administered prior to encoding impairs neutral but enhances emotional long-term episodic memories. Learn. Mem. 14, 861–868.

Powers, M.B., Smits, J.A.J., Otto, M.W., Sanders, C., Emmelkamp, Paul, M.G., 2009. Facilitation of fear extinction in phobic participants with a novel cognitive enhancer: a randomized placebo controlled trial of yohimbine augmentation. J. Anxiety Disord. 23, 350–356.

Preuss, D., Wolf, O.T., 2009. Post-learning psychosocial stress enhances consolidation of neutral stimuli. Neurobiol. Learn. Mem. 92, 318–326.

Quaedflieg, C.W.E.M., Schwabe, L., Meyer, T., Smeets, T., 2013. Time dependent effects of stress prior to encoding on event-related potentials and 24 h delayed retrieval. Psychoneuroendocrinology 38, 3057–3069.

Rimmele, U., Domes, G., Mathiak, K., Hautzinger, M., 2003. Cortisol has different effects on human memory for emotional and neutral stimuli. NeuroReport 14, 2485–2488.

Rohleder, N., Schommer, N.C., Hellhammer, D.H., Engel, R., Kirschbaum, C., 2001. Sex differences in glucocorticoid sensitivity of proinflammatory cytokine production after psychosocial stress. Psychosom. Med. 63, 966–972.

Rohleder, N., Wolf, J.M., Piel, M., Kirschbaum, C., 2003. Impact of oral contraceptive use on glucocorticoid sensitivity of pro-inflammatory cytokine production after psychosocial stress. Psychoneuroendocrinology 28, 261–273.

Rohleder, N., Wolf, J.M., Kirschbaum, C., Wolf, O.T., 2009. Effects of cortisol on emotional but not on neutral memory are correlated with peripheral glucocorticoid sensitivity of inflammatory cytokine production. Int. J. Psychophysiol. 72, 74–80.

Roozendaal, B., 2002. Stress and memory: opposing effects of glucocorticoids on memory consolidation and memory retrieval. Neurobiol. Learn. Mem. 78, 578–595.

Roozendaal, B., McEwen, B.S., Chattarji, S., 2009. Stress, memory and the amygdala. Nat. Rev. Neurosci. 10, 423–433.

Roozendaal, B., Hernandez, A., Cabrera, S.M., Hagewoud, R., Malvaez, M., Stefanko, D.P., Haettig, J., Wood, M.A., 2010. Membrane-associated glucocorticoid activity is necessary for modulation of long-term memory via chromatin modification. J. Neurosci. 30, 5037–5046.

Schönfeld, P., Ackermann, K., Schwabe, L., 2014. Remembering under stress: different roles of autonomic arousal and glucocorticoids in memory retrieval. Psychoneuroendocrinology 39, 249–256.

Schoofs, D., Wolf, O.T., 2009. Stress and memory retrieval in women: no strong impairing effect during the luteal phase. Behav. Neurosci. 123, 547–554.

Schwabe, L., Wolf, O.T., 2009. The context counts: congruent learning and testing environments prevent memory retrieval impairment following stress. Cogn. Affect. Behav. Neurosci. 9, 229–236.

Schwabe, L., Wolf, O.T., 2013. Stress and multiple memory systems: from "thinking" to "doing". Trends Cogn. Sci. 17, 60–68.

Schwabe, L., Böhringer, A., Chatterjee, M., Schächinger, H., 2008. Effects of pre-learning stress on memory for neutral, positive and negative words: different roles of cortisol and autonomic arousal. Neurobiol. Learn. Mem. 90, 44–53.

Schwabe, L., Haddad, L., Schächinger, H., 2008. HPA axis activation by a socially evaluated cold-pressor test. Psychoneuroendocrinology 33, 890–895.

Sehlmeyer, C., Schöning, S., Zwitserlood, P., Pfleiderer, B., Kircher, T., Arolt, V., Konrad, C., 2009. Human fear conditioning and extinction in neuroimaging: a systematic review. PLoS ONE 4, e5865.

Smeets, T., 2011. Acute stress impairs memory retrieval independent of time of day. Psychoneuroendocrinology 36, 495–501.

Smeets, T., Jelicic, M., Merckelbach, H., 2006. The effect of acute stress on memory depends on word valence. Int. J. Psychophysiol. 62, 30–37.

Soeter, M., Kindt, M., 2011. Noradrenergic enhancement of associative fear memory in humans. Neurobiol. Learn. Mem. 96, 263–271.

Soravia, L.M., Heinrichs, M., Aerni, A., Maroni, C., Schelling, G., Ehlert, U., Roozendaal, B., de Quervain, D.J.-F., 2006. Glucocorticoids reduce phobic fear in humans. Proc. Natl. Acad. Sci. USA 103, 5585–5590.

Soravia, L.M., Heinrichs, M., Winzeler, L., Fisler, M., Schmitt, W., Horn, H., Dierks, T., Strik, W., Hofmann, S.G., de Quervain, D.J.-F., 2014. Glucocorticoids enhance *in vivo* exposure-based therapy of spider phobia. Depress. Anxiety 31, 429–435.

Stark, R., Wolf, O.T., Tabbert, K., Kagerer, S., Zimmermann, M., Kirsch, P., Schienle, A., Vaitl, D., 2006. Influence of the stress hormone cortisol on fear conditioning in humans: evidence for sex differences in the response of the prefrontal cortex. Neuroimage 32, 1290–1298.

Stroud, L.R., Salovey, P., Epel, E.S., 2002. Sex differences in stress responses: social rejection versus achievement stress. Biol. Psychiat. 52, 318–327.

Tabbert, K., Merz, C.J., Klucken, T., Schweckendiek, J., Vaitl, D., Wolf, O.T., Stark, R., 2010. Cortisol enhances neural differentiation during fear acquisition and extinction in contingency aware young women. Neurobiol. Learn. Mem. 94, 392–401.

Taylor, S.E., Klein, L.C., Lewis, B.P., Gruenewald, T.L., Gurung, R.A.R., Updegraff, J.A., 2000. Biobehavioral responses to stress in females: tend-and-befriend, not fight-or-flight. Psychol. Rev. 107, 411–429.

Tollenaar, M.S., Elzinga, B.M., Spinhoven, P., Everaerd, W.T.A.M., 2008. The effects of cortisol increase on long-term memory retrieval during and after acute psychosocial stress. Acta Psychol. 127, 542–552.

Tops, M., van der Pompe, G., Baas, D., Mulder, L.J.M., den Boer, J.A., Meijman, T.F., Korf, J., 2003. Acute cortisol effects on immediate free recall and recognition of nouns depend on stimulus valence. Psychophysiology 40, 167–173.

Tulving, E., 1983. Elements of Episodic Memory. Oxford University Press, New York.

Ulrich-Lai, Y.M., Herman, J.P., 2009. Neural regulation of endocrine and autonomic stress responses. Nat. Rev. Neurosci. 10, 397–409.

Van Ast, V.A., Vervliet, B., Kindt, M., 2012. Contextual control over expression of fear is affected by cortisol. Front. Behav. Neurosci. 6, 67.

Vervliet, B., Craske, M.G., Hermans, D., 2013. Fear extinction and relapse: state of the art. Ann. Rev. Clin. Psychol. 9, 215–248.

Wolf, O.T., 2009. Stress and memory in humans: twelve years of progress. Brain Res. 1293, 142–154.

Wolf, O.T., 2012. Immediate recall influences the effects of pre-encoding stress on emotional episodic long-term memory consolidation in healthy young men. Stress 15, 272–280.

Wolf, O.T., Schommer, N.C., Hellhammer, D.H., McEwen, B.S., Kirschbaum, C., 2001. The relationship between stress induced cortisol levels and memory differs between men and women. Psychoneuroendocrinology 26, 711–720.

Yonelinas, A.P., Parks, C.M., Koen, J.D., Jorgenson, J., Mendoza, S.P., 2011. The effects of post-encoding stress on recognition memory: examining the impact of skydiving in young men and women. Stress 14, 136–144.

Young, K., Drevets, W.C., Schulkin, J., Erickson, K., 2011. Dose-dependent effects of hydrocortisone infusion on autobiographical memory recall. Behav. Neurosci. 125, 735–741.

Zoladz, P.R., Clark, B., Warnecke, A., Smith, L., Tabar, J., Talbot, J.N., 2011. Pre-learning stress differentially affects long-term memory for emotional words, depending on temporal proximity to the learning experience. Physiol. Behav. 103, 467–476.

Zorawski, M., Cook, C.A., Kuhn, C.M., LaBar, K.S., 2005. Sex, stress, and fear: individual differences in conditioned learning. Cogn. Affect. Behav. Neurosci. 5, 191–201.

Zorawski, M., Blanding, N.Q., Kuhn, C.M., LaBar, K.S., 2006. Effects of stress and sex on acquisition and consolidation of human fear conditioning. Learn. Mem. 13, 441–450.

# CHAPTER 8

# Biological Underpinnings of Sex Differences in Eating Disorders

**Kristen M. Culbert*, Sarah E. Racine**, Kelly L. Klump†**
*University of Nevada Las Vegas, Department of Psychology, Las Vegas, NV, USA
**Ohio University, Department of Psychology, Athens, OH, USA
†Michigan State University, Department of Psychology, East Lansing, MI, USA

## 1 INTRODUCTION

Sex differences in eating disorders and their associated disordered eating symptoms are pronounced. Females outnumber males approximately 2:1 to 10:1 in eating disorder diagnoses (American Psychiatric Association, 2013; Hudson et al., 2007), and sex differences in disordered eating symptoms tend to be medium-to-large in magnitude (Culbert et al., 2013a). Traditional theories of sex-differentiated risk have largely focused on psychosocial explanations (e.g., sociocultural pressures for thinness in women; Striegel-Moore and Bulik, 2007). Indeed, in Western cultures, females are inundated with messages that beauty and attractiveness are closely tied to achieving the "thin ideal" – a slender physique with little body fat (Keel and Forney, 2013; Thompson and Stice, 2001). Pressure to attain the thin ideal, coupled with cognitive acceptance and adherence to this socially defined ideal of attractiveness, is presumed to increase risk for eating pathology in females relative to males (Thompson and Stice, 2001). Extant data demonstrate that these psychosocial factors are important (for reviews, see Hausenblas et al., 2013; Keel and Forney, 2013); however, animal and human data converge to suggest significant biological contributions to sex-differentiated eating disorder risk as well (Baker et al., 2012; Culbert et al., 2011; Klump et al., 2013c).

This chapter aims to review studies that have investigated sex differences in etiology of eating pathology, with a particular focus on biological factors. Nonetheless, it is recognized that the etiology of eating disorders is complex, in that it involves a number of biological (e.g., neurotransmitter and hormone disturbances), psychological (e.g., personality traits and mood symptoms), and sociocultural (e.g., pressure to attain thinness) factors (Culbert et al., 2015b). Biological vulnerabilities likely co-occur and interact with a number of psychological and social/environmental variables to contribute to within-sex and between-sex differences in risk for eating pathology. That said, current data on sex-differentiated risk effects are rather limited; the majority of prior research has exclusively focused on females, and few studies have directly explored sex differences in etiologic factors. For example, leading psychological candidates (e.g., negative

*Sex Differences in the Central Nervous System*. http://dx.doi.org/10.1016/B978-0-12-802114-9.00008-1

emotionality, perfectionism) have primarily been explored as prospective risk variables in studies examining adolescent and young adult females (Culbert et al., 2015b). Although abuse history (physical, sexual, or emotional abuse), a nonspecific environmental risk factor for eating disorders and other psychiatric disorders (e.g., depression, anxiety, personality disorders, and substance abuse), has shown stronger associations with eating disorder onset in women than men (Jonas et al., 2011), sex-differentiated effects have rarely been examined for psychosocial and environmental factors. Further, studies that have aimed to explore unique risk factors for eating pathology in men have often focused on sexual orientation. Although mean levels of disordered eating tend to be higher in gay men than heterosexual men, a number of heterosexual men also develop disordered eating and eating disorders (Russell and Keel, 2002; Jones and Morgan, 2010). In contrast, links between sexual orientation and eating pathology in women have been largely mixed (Bankoff and Pantalone, 2014).

Overall, the general lack of research on males and lack of direct comparisons of the sexes has resulted in significant gaps in our current understanding of mechanisms underlying sex-differentiated risk for eating pathology. We have focused this review on the relatively few biologically informative studies (behavior and genetic twin studies) that have directly examined sex differences by comparing females and males and on studies that have examined biological influences (gonadal hormones: estradiol, progesterone, testosterone; and gastrointestinal hormones: ghrelin and cholecystokinin) known to contribute to sex-differentiated eating behavior.

## 1.1 Definitions

Eating disorders can be broadly conceptualized as substantial disturbances in disordered eating cognitions (e.g., concerns about one's body weight/shape) and behaviors (e.g., over- and/or under-control of eating). The Diagnostic and Statistical Manual of Mental Disorders, Fifth Edition recognizes three primary eating disorder diagnoses: anorexia nervosa, bulimia nervosa, and binge eating disorder, in addition to residual categories describing symptom presentations that do not fit within these diagnoses. The core features of each disorder are presented in Table 8.1. Notably, despite the division of eating disorders into distinct diagnostic categories, it is apparent that several core eating disorder symptoms cut across diagnostic boundaries (see Table 8.1).

Focusing on key component symptoms using dimensional models (e.g., across a spectrum of severity) rather than categorical diagnoses (i.e., disorder as present vs. absent) has important implications for etiologic research – specific symptoms may be more closely linked to etiologic mechanisms than broad diagnoses (Cuthbert, 2005; Krueger and Markon, 2011). Moreover, the identification of etiologic factors underlying risk for eating disorders is complicated by: (1) low prevalence rates (0.3–3.5%) across diagnoses (Hudson et al., 2007; Smink et al., 2014; Swanson et al., 2011), especially in boys/men; and (2) difficulty delineating disease risk from disease consequence, as several

**Table 8.1** Descriptions of disordered eating symptoms and core features of eating disorder diagnoses

| | |
|---|---|
| **Disordered eating symptoms** | |
| ***Cognitive features*** | |
| Body dissatisfaction | Dissatisfaction with the size and/or shape of one's body or body parts (e.g., thighs, stomach). |
| Weight concerns | Preoccupation with weight, desire to lose weight, pursuit of thinness. |
| Overevaluation of body shape and weight | Undue influence of body shape and weight on one's scheme for self-evaluation. |
| ***Behavioral features*** | |
| Binge eating | Consumption of an amount of food in a short period of time (e.g., 2 h) that is definitely larger than most people would eat and a sense of a lack of control over eating during the binge episode. |
| Compensatory behaviors | Behaviors to compensate for food consumed and to prevent weight gain, such as self-induced vomiting, misuse of laxatives, diuretics, or other medications, fasting, or excessive exercise. |
| Dietary restriction | Behavioral attempts to restrict food intake (e.g., skipping meals, rules about what one should or should not eat in the pursuit of weight loss). |
| Low weight status | Low body weight considering one's age, sex, developmental trajectory, and physical health status. |
| Emotional eating | Increased eating in response to emotions, typically negative emotions (e.g., sadness). |
| **Core features of eating disorder diagnoses** | |
| Anorexia nervosa | Core features: low weight status and dietary restriction.<br>Typically present: overevaluation of body shape and weight, body dissatisfaction, weight concerns.<br>Sometimes present: binge eating and compensatory behaviors (i.e., binge eating/purging subtype). |
| Bulimia nervosa | Core features: binge eating, compensatory behaviors, and overevaluation of body shape and weight.<br>Typically present: body dissatisfaction, weight concerns, and dietary restriction.<br>Sometimes present: emotional eating. |
| Binge eating disorder | Core feature: binge eating (in absence of compensatory behaviors).<br>Sometimes present: emotional eating, overevaluation of body shape and weight, body dissatisfaction, weight concerns, dietary restriction. |

biological and cognitive factors change secondary to the illness (Klump et al., 2009). Disordered eating symptoms are strong precursors to the later onset of eating disorders (Jacobi et al., 2004) and are much more prevalent (Culbert et al., 2009; Jones et al., 2001) than full-threshold diagnoses. Thus, dimensional measures of these symptoms likely have etiologic relevance, and in comparison to categorical diagnoses, may increase statistical power to detect etiologic effects. Our review therefore includes studies that utilized dimensional symptom-based approaches as well as studies that focused on full-threshold syndromes.

## 2 BEHAVIOR GENETICS: SEX-DIFFERENTIATED GENETIC RISK FOR EATING PATHOLOGY

### 2.1 Etiologic effects on eating pathology in late adolescence and adulthood

Twin studies provide a powerful design for investigating genetic and environmental influences on eating disorders, and importantly, whether the magnitude and/or types of genetic and environmental influences on disordered eating symptoms differ between the sexes (Plomin et al., 2008). Moreover, genetic and environmental influences are estimated by comparing the degree of similarity on a trait (e.g., disordered eating) between twins that differ in their degree of genetic relatedness, i.e., identical twins (monozygotic) who share 100% of their segregating genes are compared with fraternal twins (dizygotic) who share approximately 50% of their segregating genes (Plomin et al., 2008). Twin models partition the variance on a phenotype (e.g., disordered eating) into additive genetic influences ($a^2$; the influence of many genes on a phenotype), shared environmental influences ($c^2$; factors that act to make twins in the same family more similar to one another on a phenotype), and nonshared environmental influences ($e^2$; factors that act to make twins in the same family more dissimilar on a phenotype).

Extant data demonstrate that eating disorders and their component symptoms are heritable in adolescent and adult females, with approximately 50% of variance accounted for by genetic factors (for review, see Trace et al., 2013). Far fewer twin studies have included males, and most studies have reported estimates for genetic and environmental influences on eating pathology separately for each sex rather than statistically testing for sex differences (Culbert et al., 2011). However, in the few studies that have statistically tested for sex differences in the magnitude of genetic and environmental effects on eating pathology, no significant differences were observed (see Table 8.2). Despite the consistency of these findings, it is important to acknowledge that larger sample sizes are required for the statistical detection of sex differences in genetic/environmental influences than for the detection of sex similarities (Plomin et al., 2008). The lack of sex differences in the magnitude of genetic and environmental

**Table 8.2** Genetic and environmental influences on disordered eating symptoms in males and females

| Disordered eating symptoms | | Females | | | Males | | | Sex difference? |
|---|---|---|---|---|---|---|---|---|
| Study | Measure | $a^2$ | $c^2$ | $e^2$ | $a^2$ | $c^2$ | $e^2$ | Yes/No |
| ***Binge eating or bulimic symptoms*** | | | | | | | | |
| Baker et al. (2009a) | EDI: bulimia | 0.16 (0.00–0.42) | 0.16 (0.00–0.36) | 0.69 (0.57–0.81) | 0.33 (0.00–0.44) | 0.00 (0.00–0.30) | 0.67 (0.56–0.80) | Not tested |
| Reichborn-Kjennerud et al. (2003) | 1 item: binge eating | 0.51 (0.43–0.58) | – | 0.49 (0.42–0.57) | 0.51 (0.43–0.58) | – | 0.49 (0.42–0.57) | No |
| Reichborn-Kjennerud et al. (2004a) | 1 item: binge eating | 0.41 (0.31–0.50) | – | 0.59 (0.50–0.69) | 0.41 (0.31–0.50) | – | 0.59 (0.50–0.69) | No |
| Root et al. (2010) | 2 item: binge eating | 0.70 (0.60–0.77) | – | 0.31 (0.23–0.40) | 0.74 (0.36–0.93) | – | 0.26 (0.07–0.64) | Not tested |
| ***Body shape concerns*** | | | | | | | | |
| Keski-Rahkonen et al. (2005a) | EDI: body dissatisfaction | 0.59 (0.53–0.65) | – | 0.41 (0.35–0.47) | – | 0.85 (0.83–0.87) | 0.15 (0.13–0.17) | Not tested |
| Baker et al. (2009a) | EDI: body dissatisfaction | 0.57 (0.30–0.70) | 0.07 (0.00–0.31) | 0.36 (0.30–0.44) | 0.40 (0.06–0.57) | 0.07 (0.00–0.35) | 0.53 (0.44–0.64) | Not tested |
| ***Body weight concerns*** | | | | | | | | |
| Baker et al. (2009a) | EDI: drive for thinness | 0.61 (0.33–0.68) | 0.01 (0.00–0.25) | 0.38 (0.31–0.46) | 0.20 (0.33–0.68) | 0.11 (0.00–0.35) | 0.69 (0.57–0.82) | Not tested |
| Keski-Rahkonen et al. (2005a) | EDI: drive for thinness | 0.51 (0.44–0.58) | – | 0.49 (0.43–0.56) | – | 0.86 (0.84–0.88) | 0.14 (0.12–0.16) | Not tested |
| Keski-Rahkonen et al. (2005b) | 1 item: intentional weight loss | 0.66 (0.55–0.75) | – | 0.34 (0.25–0.45) | 0.38 (0.19–0.55) | – | 0.62 (0.45–0.81) | Not tested |
| Reichborn-Kjennerud et al. (2004b) | 1 item: undue influence of body weight | – | 0.31 (0.24–0.38) | 0.69 (0.68–0.76) | – | 0.31 (0.24–0.38) | 0.69 (0.68–0.76) | No |

*(Continued)*

**Table 8.2** Genetic and environmental influences on disordered eating symptoms in males and females *(cont.)*

| Disordered eating symptoms | | Females | | | Males | | | Sex difference? |
|---|---|---|---|---|---|---|---|---|
| **Study** | **Measure** | $a^2$ | $c^2$ | $e^2$ | $a^2$ | $c^2$ | $e^2$ | **Yes/No** |
| ***Night eating*** | | | | | | | | |
| Root et al. (2010) | 2 items: night eating | 0.35 (0.17–0.52) | – | 0.65 (0.48–0.83) | 0.44 (0.24–0.61) | – | 0.56 (0.39–0.76) | Not tested |
| ***Overall disordered eating*** | | | | | | | | |
| Klump et al. (2012) | MEBS: total score (binge eating, compensatory behavior, body dissatisfaction, weight preoccupation) | 0.55 (0.37–0.63) | 0.00 (0.00–0.15) | 0.45 (0.37–0.54) | 0.55 (0.37–0.63) | 0.00 (0.00–0.15) | 0.45 (0.37–0.54) | No |
| Slof-Op't Landt et al. (2008) | 4 items: total score (dieting for weight loss, fear of weight gain, undue importance of body weight/shape, binge eating) | 0.65 (0.58–0.71) | – | 0.35 (0.29–0.42) | 0.39 (0.28–0.49) | – | 0.35 (0.29–0.42) | Not tested |

$a^2$, Additive or nonadditive genetic effects; $c^2$, shared environmental effects; $e^2$, nonshared environmental effects; EDI, eating disorder inventory (Garner, 1991); MEBS, Minnesota Eating Behaviors Survey (von Ranson et al., 2005). Ninety-five percent confidence intervals are included in parentheses. Parameter estimates are from the best-fitting model, and dashes indicate that the parameter was dropped in the best-fitting model. Results for Klump et al. (2012) reflect estimates from postpubertal and young adult male and female twins only. Parameter estimates reported for Slof-Op't Landt et al. (2008) are from a bivariate Cholesky analysis of body mass index and disordered eating symptoms, and notably, heritability estimates unique to disordered eating symptoms were 0.53 for women and 0.29 for men.

Adapted from Culbert et al. (2011). With kind permission from Springer Science and Business Media.

estimates in prior twin studies that tested for such effects could therefore be due to a lack of statistical power (Culbert et al., 2011). Interpretation of the pattern of estimates across all prior studies tentatively suggests that there may be sex differences in genetic/environmental influences for at least some disordered eating symptoms. For example, there appears to be a lack of sex differences in the magnitude of genetic and nonshared environmental influences on binge eating (heritability ~40–70% in both sexes), yet potential sex differences in the magnitude of genetic/environmental influences on body weight and shape concerns may be present (see Table 8.2). Specifically, estimates of genetic influences on body weight/shape concerns tend to be higher in females than in males (see Table 8.2).

In addition to sex differences in the magnitude of genetic and environmental influences on disordered eating, sex differences can exist in the types of genetic factors contributing to disordered eating in males versus females. Such effects are determined via the genetic correlation (i.e., an index of the extent to which genes that influence liability to disordered eating overlap in the two sexes ($r_g$)). Only three studies have examined such effects, but thus far, findings consistently suggest moderate overlap ($r_g$ = .49–.67) in the genetic influences contributing to disordered eating cognitions (Baker et al., 2009a) and behaviors (i.e., binge eating; Reichborn-Kjennerud et al., 2003, 2004a) in males and females. These findings are important in highlighting that at least some of the genetic risk for disordered eating is shared between the sexes; however, the lack of complete overlap highlights that a substantial proportion of genetic risk is qualitatively different in males versus females (Culbert et al., 2011).

Taken together, studies have not found significant sex differences in the magnitude of genetic/environmental influences on disordered eating symptoms in late adolescence or adulthood; however, data tentatively point to the possibility of sex differences in genetic/environmental effects on some key symptoms, such as body weight/shape concerns. Sex similarities and differences have also been observed for the types of genetic factors contributing to heritability indices within each sex. Nonetheless, it is important to reiterate that relatively few studies have examined sex difference effects, and sample sizes have generally been small (and thus potentially underpowered). Further, several prior studies (see Table 8.2) have assessed disordered eating constructs using only one or two items, which could increase measurement error (Culbert et al., 2011). This is an important issue since measurement error loads onto the nonshared environmental parameter, and inflation of nonshared environmental effects decreases estimates of genetic and shared environmental effects (Plomin et al., 2008). Finally, all of the aforementioned studies investigated sex differences in genetic and environmental influences on disordered eating in samples spanning late adolescence and/or adulthood, despite the fact that the phenotypic emergence of sex-differentiated risk (e.g., females > males on levels of eating pathology) occurs during early adolescence/puberty (Culbert et al., 2013a; Ferreiro et al., 2011).

## 2.2 Developmental considerations

To date, only one study has considered developmental moderation of sex differences in genetic and environmental influences on disordered eating symptoms. Findings indicated significant sex differences in the magnitude of genetic influences on disordered eating in prepuberty (heritability: ~0% in females, ~55% in males), but not midpuberty or young adulthood (heritability ~55% in females and males; Klump et al., 2012). Thus, sex-specific developmental shifts in genetic influences on disordered eating symptoms were observed in girls during puberty, whereas genetic effects accounted for approximately 55% of the variance in disordered eating in males, regardless of developmental stage (Klump et al., 2012). Sex differences in pubertal moderation of genetic influences on disordered eating were not accounted for by differences in age or body mass index, as these factors were controlled for in analyses (Klump et al., 2012).

The observed pubertal moderation of genetic influences on eating pathology in girls has been replicated in several studies (Culbert et al., 2009; Klump et al., 2003, 2007). For example, genetic influences on disordered eating are prominent in mid-to-late puberty but minimal in pre-to-early puberty (Culbert et al., 2009; Klump et al., 2003) and have been shown to increase linearly with advancing pubertal development (Klump et al., 2007). Findings have also demonstrated that pubertal shifts in genetic effects on disordered eating will be "missed" if a late indicator (onset of menarche) is used as the sole marker of pubertal development (Culbert et al., 2009; Rowe et al., 2002), suggesting that early-to-mid puberty is the developmental window that corresponds to changes in genetic risk (Culbert et al., 2009).

Researchers have begun to focus on identifying the factors that drive pubertal shifts in genetic effects on disordered eating. Estradiol has gained considerable attention given that it is a biological factor that dramatically increases in girls during early-to-mid puberty and is known to regulate gene transcription in key neurobiological systems relevant to eating disorders (e.g., serotonin; Ostlund et al., 2003). Thus, estradiol may serve to "activate" genetic vulnerability to disordered eating in girls during early-to-mid puberty by "turning on" risk genes within these neurobiological systems (e.g., Klump et al., 2012, 2013c), although specific genes have yet to be identified. Consistent with this hypothesis, Klump et al. (2010) found that levels of circulating estradiol moderate genetic influences on disordered eating symptoms in adolescent girls; genetic influences on disordered eating were minimal in girls with low levels of estradiol, whereas substantial genetic influences on disordered eating were found at high levels of estradiol. These effects were independent of several potentially confounding factors related to puberty (e.g., age, body mass index, and physical changes, such as breast development), pointing to a particular role of estradiol (Klump et al., 2010). Data collection is currently under way to replicate these findings in a much larger sample of female twins.

Overall, results from developmental twin studies are important in demonstrating sex-differentiated pubertal moderation of etiologic effects on disordered eating, with dramatic pubertal shifts in genetic influences on disordered eating in girls, but not boys. Initial data point to estradiol as a novel biological mechanism that might underlie these pubertal changes in genetic effects on disordered eating in girls. Together, these data enhance etiologic models of pubertal risk for eating pathology by demonstrating that traditional psychosocial theories (e.g., increases in body dissatisfaction and dieting due to physical changes of puberty that shift girls further away from the "thin ideal") cannot completely account for increased rates of disordered eating in girls during puberty; the "activation" of individual differences in genetic risk contributes as well (Klump, 2013c). Moving forward, it will be important for studies to replicate these effects using larger samples and to use longitudinal studies to confirm that developmental shifts in etiologic influences on disordered eating reflect within-person changes in these variables (e.g., disordered eating, pubertal maturation, estradiol). Identifying whether shifts in genetic influences on disordered eating in boys occur during other developmental periods also is a critical question for future research. For example, shifts in genetic influences on disordered eating in boys may emerge earlier in development, such as during adrenarche (e.g., maturational stage ~age 6–8 years that is marked by changes in the adrenal glands and rise in adrenal androgens, like dehydroepiandrosterone (DHEA) and DHEA sulfate; Grumbach and Styne, 1992) as opposed to during gonadarche (e.g., maturational stage ~ages 11–12 years that is marked by changes in the gonads and increases in the secretion of sex hormones, like testosterone; Grumbach and Styne, 1992).

## 3 GONADAL HORMONES AND SEX-DIFFERENTIATED RISK FOR EATING PATHOLOGY

Data from twin studies suggest that the exploration of biological factors relevant to sex differences and puberty may be of etiologic importance for disordered eating. As evidenced by estradiol's moderation of genetic effects on disordered eating (see previous section; Klump et al., 2010), gonadal hormones (e.g., estradiol, progesterone, and testosterone) have emerged as prime candidates (Baker et al., 2012). Indeed, there are substantial sex differences in gonadal hormone concentrations throughout development (e.g., testosterone: males > females; estradiol and progesterone: females > males; Collaer and Hines, 1995), and gonadal hormones exert effects on the central nervous system that facilitate the expression of sex-typical characteristics, including eating behavior (Arnold and Breedlove, 1985; Asarian and Geary, 2013). A growing line of research has aimed to investigate the general hypothesis that ovarian hormones (e.g., estradiol and progesterone) may act on the central nervous system to increase risk for disordered eating in females, whereas testosterone may exert protective effects that decrease risk for eating pathology in males.

## 3.1 Risk effects of ovarian hormones on eating pathology

Extant animal data have demonstrated substantial sex differences in eating behaviors – adult male rodents eat more, show greater meal frequency, and exhibit fewer day-to-day fluctuations in eating than adult females (Bell and Zucker, 1971; Laviano et al., 1996; Wade, 1972, 1975), and ovarian hormones are central to the regulation of female-typical patterns of food intake (for review, see Asarian and Geary, 2006, 2013). For example, following the experimental elimination of ovarian hormones (ovariectomy), adult female rodents show immediate and sustained increases in food consumption; the administration of estradiol reverses this effect by decreasing food intake (for review, see Asarian and Geary, 2006). Progesterone has stimulatory effects on food intake, but this appears to largely occur via its antagonizing effects on estradiol since progesterone has minimal effects on food intake in the absence of estradiol (Asarian and Geary, 2006; Wade, 1972). Interestingly, results for general eating behavior extend to animal models of binge eating. Sex differences in binge eating are prominent in adulthood, such that rates of binge eating proneness (i.e., the behavioral tendency to consistently consume a large amount of palatable food within a short period of time) are higher in female than male rats (Klump et al., 2013c). Ovarian hormones have been linked to the consumption of palatable foods typically consumed during binge behavior. Female rats show increases in palatable food consumption following ovariectomy (Klump et al., 2011), and subcutaneous injections of ovarian hormones (high estradiol/low progesterone) have been shown to reduce the consumption of high fat food in ovariectomized rats (Yu et al., 2008). Taken together, experimental animal data clearly demonstrate that ovarian hormones are key players in female-typical patterns of feeding behavior and binge eating proneness.

Ovarian hormones have also been implicated in patterns of food intake and disordered eating in women. The menstrual cycle provides a powerful quasiexperimental design to model within-person relationships between hormones and behavior since natural changes in levels of estradiol and progesterone occur in all regularly cycling women, and changes in ovarian hormones that precede changes in disordered eating (e.g., binge eating) likely reflect the influence of hormones on behavior, rather than the reverse (Klump et al., 2008; Racine et al., 2012). Studies have explored differences in general food intake, food preferences (e.g., high carbohydrate/sweet and fat), and dysregulated eating (e.g., emotional eating and binge eating) across menstrual cycle phases. A consistent pattern has emerged, such that general food intake and/or preferences for high carbohydrate and fat foods (Buffenstein et al., 1995; Martini et al., 1994; Pliner and Fleming, 1983), emotional eating (Klump et al., 2008, 2013a), and binge eating (Edler et al., 2007; Klump et al., 2014) have all been shown to be elevated in women during the midluteal phase (marked by high levels of estradiol and progesterone) compared with the follicular/ovulatory phases (marked by high levels of estradiol only). Cycle phase data indirectly point to a role of ovarian hormones in general feeding behavior and disordered eating symptomatology; additional studies that directly explored ovarian hormone-disordered eating associations complement these findings.

Natural fluctuations in ovarian hormones predict changes in binge eating and emotional eating across the menstrual cycle. Initial small-scale pilot studies ($N = 9$) found that decreases in levels of estradiol and increases in levels of progesterone were linked to increases in emotional eating in a community sample (Klump et al., 2008) and binge eating in a clinical sample (Edler et al., 2007) across the menstrual cycle, independent of body weight and changes in negative affect. These data highlight that within-person fluctuations in ovarian hormones predict within-person changes in emotional eating and binge eating across the menstrual cycle and have served as a foundation for further explorations of hormone–behavior relationships.

Indeed, larger scale studies have started to examine more complex relationships between ovarian hormones and dysregulated eating symptoms (e.g., emotional eating). For example, in a large community sample of women ($N = 196$), Klump et al. (2013a) examined the interactive effects between estradiol and progesterone on changes in emotional eating across the menstrual cycle. Emotional eating was highest during the midluteal phase of the cycle (Klump et al., 2013a) – similar to prior findings (Edler et al., 2007; Klump et al., 2008). Interestingly, significant interactions between estradiol and progesterone (i.e., high estradiol × high progesterone) explained this midluteal increase (Klump et al., 2013a), such that progesterone is likely to antagonize the inhibitory effects of estradiol on emotional eating during the midluteal phase. Body mass index or changes in negative affect did not account for ovarian hormone–emotional eating associations. These findings highlight that estradiol and progesterone likely act in concert, rather than alone, in predicting within-person changes in emotional eating across the menstrual cycle.

Further, data have also demonstrated individual differences in the magnitude of ovarian hormone influences on changes in dysregulated eating across the menstrual cycle. Ovarian hormone–emotional eating associations were stronger in women who exhibited more severe levels of emotional eating (e.g., top 33% and top 10% of sample) (Klump et al., 2013a). Thus, the strength of hormone–emotional eating associations may vary by symptom severity, and estradiol × progesterone interactions on emotional eating may be particularly relevant for women at the highest risk for binge eating (Klump et al., 2013a). Further research extended these findings by examining whether the strength or nature of associations between ovarian hormones and dysregulated eating (e.g., binge eating, emotional eating) vary between women with versus without clinically diagnosed episodes of binge eating (Klump et al., 2014). Stronger hormone–emotional eating associations were detected in women with clinical binge episodes compared with women without binge episodes; however, the nature of the associations varied between the groups (Klump et al., 2014). More complex effects were observed in women with clinically diagnosed episodes of binge eating. Low levels of estradiol exerted main effects on increased emotional eating and binge eating frequency in women with clinically diagnosed episodes of binge eating only, and notably, the interactive effects between progesterone and estradiol corresponded to a different hormone "risk" profile (i.e., low estradiol × low progesterone) in women with clinically diagnosed binge eating than the profile (i.e., high estradiol × high progesterone) observed in nonbinge eating women (Klump et al., 2014).

Taken together, the menstrual cycle has been used as a powerful methodological approach for the exploration of natural within-person fluctuations in ovarian hormones predicting changes in eating behaviors. Data have consistently demonstrated significant associations between ovarian hormones and dysregulated eating across the menstrual cycle, with elevations in dysregulated eating (e.g., emotional eating, binge frequency) during the midluteal phase. However, other data also highlight unique variations in the types of hormonal milieus contributing to increased risk for disordered eating in women with (i.e., low estradiol × low progesterone) versus those without (i.e., high estradiol × high progesterone) clinically significant binge episodes. Identifying the mechanisms underlying these unique relationships will be an important direction for future research.

Notably, it will also be important for future research to continue to explore individual difference variables that may serve to enhance or attenuate ovarian hormone-disordered eating associations and to establish whether the effects observed for binge eating and emotional eating extend to other core disordered eating symptoms as well. Only two studies have explored potential moderators of within-person associations between ovarian hormones and emotional eating, including facets of personality (e.g., negative urgency and negative emotionality; Racine et al., 2013), body mass index (Klump et al., 2013b), and dietary restraint (Klump et al., 2013b). No significant moderation effects emerged for these variables (Klump et al., 2013b; Racine et al., 2013), but trend-level estradiol × dietary restraint effects indicated a possible enhancement of estradiol's effects on emotional eating in individuals with high levels of dietary restraint (Klump et al., 2013b). Further, studies have examined associations between ovarian hormones and other eating disorder phenotypes (e.g., body dissatisfaction, weight concerns, dietary restraint). Using two small independent samples ($N < 10$) of women from the community, Racine et al. (2012) found that levels of body dissatisfaction and drive for thinness were highest during the midluteal phase of the cycle and were negatively associated with estradiol and positively associated with progesterone – effects consistent with those observed for dysregulated eating. Nonetheless, a study conducted in a much larger sample of women ($N = 352$) indicated that fluctuations in weight preoccupation across the menstrual cycle were primarily due to changes in emotional eating as opposed to ovarian hormones (Hildebrandt et al., 2014). Current data therefore suggest that ovarian hormone-disordered eating associations may be strongest and most relevant for facets of dysregulated eating (e.g., emotional eating, binge eating), but additional research is needed to examine the potential interplay between ovarian hormones, dysregulated eating, and other core symptoms such as body dissatisfaction.

## 3.2 Protective effects of testosterone on eating pathology

In addition to the effects of ovarian hormones on disordered eating in females, data from animal and human studies suggest that elevated perinatal (i.e., encompassing the prenatal period and period immediately following birth) exposure to testosterone may contribute to between-sex and within-sex variation in risk for eating pathology. Testosterone (and its metabolites, e.g., dihydrotestosterone) exerts organizational effects (i.e., long-lasting changes to the central nervous system that persist beyond initial hormone exposure) during perinatal development that drive sex differences in the central nervous system and behavior (Arnold and Breedlove, 1985). The absence of elevated testosterone during perinatal development averts neural and phenotypic masculinization and is typical of normal female development (Arnold and Breedlove, 1985). Consequently, female rodents that are exposed to elevated concentrations of testosterone during perinatal development, via exogenous administration of testosterone or intrauterine position effects (i.e., *in utero* exposure due to developing adjacent to males), show more masculinized characteristics, such as increased aggression, later in life (Ryan and Vandenbergh, 2002).

Importantly, perinatal exposure to testosterone has been shown to alter sex-differentiated patterns of food intake. Female rats treated with testosterone during perinatal development exhibit more male-like (i.e., increased) patterns of food consumption in adulthood (Donohoe and Stevens, 1983; Madrid et al., 1993), and male rats castrated on the day of birth show permanent decreases in food intake (Wade, 1972). Preliminary findings suggest that the effects of perinatal testosterone may extend to animal models of eating disorder phenotypes, as an indirect measure of elevated prenatal testosterone exposure (i.e., longer anogenital distance) predicted decreased susceptibility to activity-based anorexia in female mice (Culbert et al., 2013b). Future experiments are needed to directly confirm that differences in prenatal/perinatal testosterone exposure account for such effects, and data collection is currently under way to determine whether findings generalize to other animal models of eating pathology (e.g., binge eating proneness; research supported by Hilda and Preston Davis Foundation, PI: Culbert, K.M.; Mentors: Klump, K.L. and Sisk, C.L.).

Notably, experimental animal paradigms provide the strongest test for examining prenatal/perinatal testosterone exposure effects on eating disorder risk. Identifying whether differential perinatal testosterone exposure is a mechanism involved in disordered eating in humans is challenging given that: (1) hormone exposure cannot be directly manipulated; (2) methods of attaining prenatal testosterone levels (e.g., amniocentesis) are invasive and risky; and (3) natural conditions that are known to be linked to elevated prenatal/perinatal testosterone exposure (e.g., congenital adrenal hyperplasia) are rare in the population (Berenbaum and Beltz, 2011). To date, the examination of prenatal testosterone effects on disordered eating in human research has relied on two proxy measures of differential prenatal testosterone exposure: (1) finger-length ratios (index finger (2D)/ring finger (4D)), and (2) opposite-sex twin pairs.

2D:4D finger-length ratios are thought to be a biomarker of prenatal testosterone (or the balance between prenatal testosterone and estrogen exposure) because they have been shown to correlate with levels of prenatal testosterone as well as the relative ratio between testosterone and estradiol acquired from amniocentesis (Lutchmaya et al., 2004) and are sexually dimorphic as early as fetal development (Malas et al., 2006). In addition, experimental animal data have directly demonstrated that 2D:4D ratios are altered by prenatal testosterone and estrogen signaling (Zheng and Cohn, 2011). Interestingly, data from studies that used 2D:4D ratios have provided support for the hypothesis that prenatal testosterone has masculinizing or protective effects on disordered eating symptoms. Lower (more masculinized) 2D:4D finger-length ratios have been linked to lower levels of disordered eating in females (Klump et al., 2006; Oinonen and Bird, 2012) and males (Smith et al., 2010). Nonetheless, because these studies only include one of the sexes, findings can only speak to prenatal testosterone contributions to within-sex variation in risk for disordered eating and not whether prenatal testosterone decreases risk for eating pathology in men versus women.

Studies that utilize opposite-sex twin pairs have aimed to overcome this limitation by investigating whether prenatal testosterone exposure may underlie sex differences in disordered eating symptoms. Females from opposite-sex twin pairs develop *in utero* with a male co-twin, and thus, similar to intrauterine position effects observed in rodents, are thought to be exposed to elevated levels of testosterone (Miller, 1994). Testosterone transfer from the male co-twin to the female co-twin is thought to occur, based on animal data, via the mother's bloodstream (Meulenberg and Hofman, 1991) or diffusion across fetal membranes (Even et al., 1992). Evidence to support elevated exposure to prenatal testosterone in females from opposite-sex twin pairs comes from data showing that they tend to exhibit more masculinized characteristics (e.g., aggression; Cohen-Bendahan et al., 2005), including traits unlikely to be influenced by social mechanisms (e.g., auditory systems; McFadden, 1993). In addition, consistent with hypotheses, young adult females from opposite-sex twin pairs have been found to have more masculinized (decreased) levels of disordered eating symptoms (Culbert et al., 2008, 2013a). Specifically, levels of disordered eating were lowest in male twins, intermediate in females from opposite-sex twin pairs, and highest in female controls (females from same-sex twin pairs and nontwin females reared with brothers). A statistical trend toward lower rates of broadly defined anorexia nervosa (Raevuori et al., 2008) in females from opposite-sex twin pairs, compared with female controls (females from same-sex twin pairs), has also been found.

Despite these initial findings, it is important to acknowledge that prior research has not provided consistent support for prenatal testosterone's protective effects on disordered eating. For example, Baker et al. (2009b) found no evidence for prenatal testosterone's protective effects on disordered eating symptoms. Mixed findings have also emerged, as some studies have found protective effects of prenatal testosterone exposure (measured via 2D:4D or the opposite-sex twin paradigm) on only some eating disorder symptoms (e.g., intentional weight loss, but not bulimic symptoms; Raevuori et al., 2008) or lifetime diagnoses (e.g., for anorexia nervosa but not bulimia nervosa; Lydecker et al., 2012; Quinton et al., 2011; Raevuori et al., 2008). Several possibilities have been proposed to explain inconsistent findings, including potential specificity of prenatal testosterone's masculinizing effects on the type of disordered eating symptoms (e.g., cognitive vs. behavioral symptoms) or eating disorders assessed, the assessment of slightly different disordered eating constructs given the use of different questionnaires (Baker et al., 2009b), and/or lack of statistical power to examine eating disorder diagnoses given low prevalence rates (Raevuori et al., 2014).

Developmental differences in prenatal testosterone's protective effects on disordered eating have been proposed as an additional explanation for discrepant findings (Culbert et al., 2015a). Moreover, studies have varied in the age of participants assessed and have tended to detect prenatal testosterone effects on disordered eating in early adolescent (Culbert et al., 2013a) and young adult samples (Culbert et al., 2008; Klump et al., 2006; Oinonen and Bird, 2012; Raevuori et al., 2008; Smith et al., 2010), but not in late adolescence (Baker et al., 2009b). Culbert et al. (2015a) aimed to directly test the possibility that prenatal testosterone's protective effects on disordered eating vary by age using two different proxy markers of prenatal testosterone exposure (i.e., 2D:4D and females from opposite-sex twin pairs). Findings were consistent across methods, and indeed, higher prenatal testosterone predicted lower disordered eating in early adolescence and young adulthood only (Culbert et al., 2015a). Associations between prenatal testosterone and disordered eating symptoms were not observed during late adolescence (Culbert et al., 2015a). These data suggest that there may be developmental windows of expression for prenatal testosterone's protective effects on disordered eating and provide some indication that prior discrepant findings might be due to age differences between samples (Culbert et al., 2015a). Moving forward, it will be important to replicate these effects using longitudinal data to confirm that age-related differences reflect within-person changes in the expression of disordered eating across development (Culbert et al., 2015a). The factors contributing to an attenuation of prenatal testosterone's protective effects on disordered eating during late adolescence will also need to be further elucidated, but possibilities include environmental factors (e.g., increased pressures to develop autonomy, importance and pursuit of the thin ideal, and time spent with peers or dating partners; Field et al., 2001; Stice et al., 2010) and/or factors that are not purely environmental (e.g., changes in neural systems, heightened social sensitivity, and emotional processing; Jones et al., 2014; Somerville, 2013) that change during late adolescence (Culbert et al., 2015a).

Interestingly, developmental differences in prenatal testosterone's masculinizing effects on disordered eating have also been found as a function of pubertal status. Using the opposite-sex twin paradigm, no-sex or twin-type differences in levels of disordered eating were found in prepuberty (Culbert et al., 2013a). In contrast, during mid-to-late puberty, females from opposite-sex twin pairs exhibited more masculinized (lower) levels of disordered eating symptoms than female controls (Culbert et al., 2013a). Effects were not explained by several potential confounding factors, such as age, body mass index, anxiety, depression, autonomy difficulties, or being reared with a brother (Culbert et al., 2013a). Notably, the observed pubertal moderation of prenatal testosterone's masculinizing effects on disordered eating is consistent with animal data on food intake. In rodents, the masculinizing effects of early testosterone exposure become prominent during puberty when circulating gonadal hormones rise and facilitate the expression of sex-typical eating behavior (Asarian and Geary, 2006; Bell and Zucker, 1971; Wade, 1972). Female rats that are exposed to elevated testosterone during perinatal development exhibit masculinized eating behavior after pubertal onset (Bell and Zucker, 1971), and the exogenous administration of ovarian hormones in adulthood do not reverse the masculinizing effects of perinatal testosterone exposure on eating behavior (i.e., treatment with ovarian hormones does not facilitate the expression of female-typical eating; Donohoe and Stevens, 1983; Gentry and Wade, 1976; Zucker, 1969). Thus, perinatal testosterone exposure alters eating behavior, in part, by decreasing sensitivity to ovarian hormone effects on female-typical eating (Gentry and Wade, 1976; Zucker, 1969). Given these animal data and findings reviewed previously, one speculative possibility is that females from opposite-sex twin pairs may show decreased risk for disordered eating via decreased sensitivity to the genetic and phenotypic risk effects of ovarian hormones on disordered eating after pubertal onset (Culbert et al., 2013a). Future studies are needed to explore this possibility.

In addition to the aforementioned effects of prenatal testosterone, one recent study investigated whether pubertal increases in circulating testosterone contribute to lower disordered eating risk in boys. Similar to effects observed for eating behavior in male rodents (e.g., circulating testosterone facilitating male-typical patterns of eating after puberty; Asarian and Geary, 2006), higher levels of circulating testosterone predicted lower levels of disordered eating and effects became prominent with advancing pubertal maturation (Culbert et al., 2014). Results were robust, as they were not accounted for by several potential confounds (e.g., age, adiposity, depression, anxiety; Culbert et al., 2014) and were present across the full spectrum of severity. Specifically, the emergence of inverse associations between testosterone and disordered eating with advancing puberty were: (1) present in boys at the highest risk for an eating disorder (those who scored above clinical cut-offs), and (2) detected even after eliminating boys who scored above clinical cut-offs, demonstrating that effects were not merely accounted for by potential disorder sequelae (e.g., low testosterone production as a result of eating pathology; Culbert et al., 2014). Overall these findings are important in suggesting that elevated levels of circulating testosterone may decrease risk for eating pathology in boys during/after puberty; conversely, low levels of circulating testosterone may be a biological mechanism that helps to explain why some, although relatively few, males may be at increased risk for developing an eating disorder (Culbert et al., 2014). Nonetheless, these findings await replication, particularly using a longitudinal design to establish that pubertal changes in circulating levels of testosterone predict changes in disordered eating in boys rather than the reverse.

Taken together, animal data have demonstrated that elevated exposure to testosterone during perinatal development alters sex-differentiated eating behaviors and that circulating testosterone after pubertal onset further contributes to the facilitation of male-like patterns of eating. Nonetheless, whether these effects on general food intake extend to animal models of eating disorder phenotypes is largely unknown. Using experimental designs to directly test the effects of early testosterone exposure on eating disorder behaviors (e.g., binge eating) in animal models is an important direction for future research – especially given the challenges of establishing prenatal/perinatal effects of testosterone on eating pathology using human models. Moreover, human studies have had to rely on proxy measures of differential exposure to testosterone early in life, and findings across initial published reports were largely mixed. Subsequent analyses have, however, started to point to developmental factors such as age and puberty as key moderators that alter the magnitude of prenatal testosterone's masculinizing effects on disordered eating symptoms and appear to at least partially explain prior discrepant findings. Further confirmation of these effects (e.g., via animal models) and identification of the factors underlying developmental shifts will be important to more fully understand the role of early testosterone exposure on the expression of eating pathology across development. Altered sensitivity to circulating gonadal hormones (e.g., enhanced sensitivity to circulating testosterone and/or decreased sensitivity to ovarian hormones) during and after puberty is one biological mechanism that may underlie prenatal testosterone's masculinizing effects on disordered eating both between-the-sexes and within-sex, at least for moderation effects observed during puberty. The only study that has examined the effects of circulating testosterone on disordered eating in boys suggests that individual differences in levels of circulating testosterone predict differential risk for disordered eating during/after pubertal onset; however, whether variation in prenatal testosterone exposure alters the strength of these associations has yet to be explored.

## 4 APPETITE-REGULATING HORMONES: IMPLICATIONS FOR EATING PATHOLOGY?

In addition to gonadal hormones, animal data have highlighted sex differences in peripheral controls of eating (e.g., gastrointestinal hormones) that may have implications for sex differences in risk for eating disorders. Leading candidates include ghrelin and cholecystokinin (CCK), both of which interact with gonadal hormone systems (Asarian and Geary, 2013). Ghrelin has stimulatory effects on appetite and promotes gastric emptying, whereas as CCK plays an important role in satiation (Asarian and Geary, 2013; Monteleone and Maj, 2013). Notably, animal data suggest that estradiol inhibits the appetite stimulating effects of ghrelin and enhances the satiation effects of CCK (for review, see Asarian and Geary, 2013). For example, ghrelin administration resulted in significantly greater food intake in intact male rats and ovariectomized female rats compared with intact female rats or ovariectomized female rats treated with estradiol (Clegg et al., 2007). However, following estradiol administration, ghrelin's stimulatory effects on food intake in males and ovariectomized females were attenuated (Clegg et al., 2007). Interactions between estradiol and CCK have also been observed. Ovariectomized female rats treated with estradiol showed increased satiation to CCK injections (for review, see Asarian and Geary, 2013; Butera et al., 1993). Further, data indicate that ghrelin's stimulatory effects on food intake and CCK's satiation effects vary across the estrus cycle in female rats (ghrelin's effects enhanced at low levels of estradiol; CCK effects enhanced at high levels of estradiol; Asarian and Geary, 2013; Clegg et al., 2007; Eckel and Geary, 1999). Together, animal data highlight a role for ghrelin and CCK in key aspects of feeding behavior, such as consumption and satiation, and notably, the effects of ghrelin and CCK on eating behavior are at least partially dependent on gonadal hormones such as estradiol. Nonetheless, the effects of ghrelin and CCK were observed for general food intake and whether these findings extend to pathological eating has, to our knowledge, yet to be explored in animal models.

Human data are also limited. It is currently unknown if ghrelin and/or CCK disturbances contribute to sex-differentiated risk for eating disorders – prior studies have only examined women with eating disorders, and disturbances in ghrelin and CCK have primarily been examined in patients during the ill state. Meta-analytic data showed no group differences between women ill with an eating disorder versus healthy controls for ghrelin or CCK release to a test meal, yet women ill with an eating disorder showed substantially higher baseline ghrelin (large effect size: $d = 1.05$) and CCK (medium effect size: $d = .54$) concentrations than controls (Prince et al., 2009). These data indicate that disturbances in ghrelin and CCK are associated with eating disorders, but these pathophysiological effects may be secondary to illness (e.g., resulting from malnutrition, aberrant eating behavior, etc.). Indeed, studies tend to find no differences in ghrelin or CCK levels between patients recovered from an eating disorder and healthy controls (for review, see Monteleone and Maj, 2013). Future studies that directly examine whether ghrelin and CCK disturbances predict eating disorder onset are needed to determine whether these gastrointestinal hormones have etiologic relevance, as are studies that explore possible sex differences and gonadal hormone moderation of ghrelin/CCK–eating disorder associations. The examination of other appetite-regulating hormones, such as leptin, in the etiology of eating pathology may also be important. Nonetheless, there is currently limited evidence to support sex-differentiated leptin effects on eating behavior (for review, see Asarian and Geary, 2013), and data suggest that leptin disruptions are limited to the ill state of eating disorders (e.g., remit with weight gain and eating regularly; for review, see Monteleone and Maj, 2013).

## 5 BRAIN IMAGING: SEX-DIFFERENTIATED RISK FOR DISORDERED EATING

Despite the detection of sex-differentiated genetic and hormonal influences on disordered eating, very few studies have aimed to explore sex differences in other biological factors that could underlie the female preponderance of eating disorder risk. However, neuroimaging studies have begun to provide support for biological underpinnings of sex differences in neural responses to food-related processes and body image stimuli. Only a handful of studies have examined sex differences in food-related processes in nonclinical/community samples. Findings indicated sex-differentiated neural activation in several brain regions (e.g., dorsolateral prefrontal cortex, inferior occipitotemporal cortex) in response to food intake and visual food-related stimuli during a hunger (Del Parigi et al., 2002; Uher et al., 2006) and a satiated state (Del Parigi et al., 2002; Smeets et al., 2006; Uher et al., 2006). Although these studies did not explore whether observed effects were predictive of sex differences in rates of disordered eating symptomatology, additional data demonstrate that vulnerability to eating pathology may be related to similar neural regions, at least in females. For example, increased activity in regions of the interior occipitotemporal cortex, following the presentation of visual food-related stimuli, has been observed in females with eating disorders as compared to control females (Uher et al., 2004).

Factors often considered to be purely psychosocial (e.g., body image stimuli) have also been shown to have sex-differentiated biological underpinnings that may contribute to sex differences in eating disorder risk. For example, Shirao et al. (2005) examined whether sex differences in susceptibility to body weight and shape concerns could be due to sex differences in neural processing of body image stimuli (i.e., words concerning body image). Sex differences were observed in the activation of brain regions (e.g., only in males: left medial prefrontal cortex and hippocampus; only in females: left amygdala) and in the degree of activation, in that males showed significantly greater activation in the apical prefrontal region than females (Shirao et al., 2005). These data point to the prefrontal region as a key player in sex differences in neural responsiveness to body image stimuli, which may underlie sex-differentiated vulnerability to the development of eating pathology. Additional studies examining sex differences in brain function are needed, but initial results are promising in suggesting that males and females differ in neural activation to food-related and body image processes that are important for eating disorders.

## 6 CONCLUSIONS

Research into the effects of sex on etiologic risk for eating disorders and their associated symptoms is a relatively new yet promising area of research. Data indicate both sex differences and similarities in the magnitude and types of genetic influences contributing to disordered eating. The most dramatic sex-differentiated effects on genetic risk for disordered eating are observed during puberty. Genetic influences account for ~50% of the variability in disordered eating in boys, regardless of pubertal status. In contrast, genetic effects on disordered eating are negligible in girls during prepuberty, but increase linearly with advancing pubertal development and are particularly prominent during/after midpuberty (heritability ~50%). Gonadal hormones may contribute to sex differences in the magnitude and types of genetic effects on disordered eating. Estradiol activation at puberty appears to account for pubertal shifts in genetic influences on disordered eating in girls, and several studies demonstrate that both estradiol and progesterone contribute to the phenotypic expression of eating pathology in women. Exposure to testosterone during prenatal/perinatal development and puberty may explain the relatively stable and low risk for disordered eating in males across development, although this does not preclude the fact that variation in levels of testosterone exposure may contribute to individual differences in risk within males. Initial data also point to sex differences in neural responses to eating and body-related stimuli, which may serve as additional biological mechanisms that underlie sex differences in the development of disordered eating. Despite this growing literature, much more research is needed to further elucidate how biological mechanisms, such as gonadal hormones, genes, neural activation, etc., contribute to sex-differentiated risk for eating pathology across development.

## REFERENCES

American Psychiatric Association, 2013. The Diagnostic and Statistical Manual of Mental Disorders, fifth ed.: DSM 5. Arlington, VA.

Arnold, A.P., Breedlove, S.M., 1985. Organizational and activational effects of sex steroids on brain and behavior: a reanalysis. Horm. Behav. 19 (4), 469–498.

Asarian, L., Geary, N., 2006. Modulation of appetite by gonadal steroid hormones. Philos. Trans. R. Soc. Lond. B Biol. Sci. 361 (1471), 1251–1263.

Asarian, L., Geary, N., 2013. Sex differences in the physiology of eating. Am. J. Physiol. Regul. Integr. Comp. Physiol. 305 (11), R1215–R1267.

Baker, J.H., Maes, H.H., Lissner, L., Aggen, S.H., Lichtenstein, P., Kendler, K.S., 2009a. Genetic risk factors for disordered eating in adolescent males and females. J. Abnorm. Psychol. 118 (3), 576–586.

Baker, J.H., Lichtenstein, P., Kendler, K.S., 2009b. Intrauterine testosterone exposure and risk for disordered eating. Br. J. Psychiat. 194 (4), 375–376.
Baker, J.H., Girdler, S.S., Bulik, C.M., 2012. The role of reproductive hormones in the development and maintenance of eating disorders. Expert Rev. Obstet. Gynecol. 7 (6), 573–583.
Bankoff, S.M., Pantalone, D.W., 2014. Patterns of disordered eating behavior in women by sexual orientation: a review of the literature. Eat. Disord. 22 (3), 261–274.
Bell, D.D., Zucker, I., 1971. Sex differences in body weight and eating: organization and activation by gonadal hormones in the rat. Physiol. Behav. 7 (1), 27–34.
Berenbaum, S.A., Beltz, A.M., 2011. Sexual differentiation of human behavior: effects of prenatal and pubertal organizational hormones. Front. Neuroendocrinol. 32 (2), 183–200.
Buffenstein, R., Poppitt, S.D., McDevitt, R.M., Prentice, A.M., 1995. Food intake and the menstrual cycle: a retrospective analysis, with implications for appetite research. Physiol. Behav. 58 (6), 1067–1077.
Butera, P.C., Bradway, D.M., Cataldo, N.J., 1993. Modulation of the satiety effect of cholecystokinin by estradiol. Physiol. Behav. 53 (6), 1235–1238.
Clegg, D.J., Brown, L.M., Zigman, J.M., Kemp, C.J., Strader, A.D., Benoit, S., Woods, S.C., Mangiaracina, M., Geary, N., 2007. Estradiol-dependent decrease in the orexigenic potency of ghrelin in female rats. Diabetes 56 (4), 1051–1058.
Cohen-Bendahan, C.C.C., Buitelaar, J.K., van Goozen, S.H.M., Orlebeke, J.F., Cohen-Kettenis, P.T., 2005. Is there an effect of prenatal testosterone on aggression and other behavioral traits? A study comparing same-sex and opposite-sex twin girls. Horm. Behav. 47 (2), 230–237.
Collaer, M.L., Hines, M., 1995. Human behavioral sex differences: a role for gonadal hormones during early development? Psychol. Bull. 118 (1), 55–107.
Culbert, K.M., Breedlove, S.M., Burt, S.A., Klump, K.L., 2008. Prenatal hormone exposure and risk for eating disorders: a comparison of opposite-sex and same-sex twins. Arch. Gen. Psychiat. 65 (3), 329–336.
Culbert, K.M., Burt, S.A., McGue, M., Iacono, W.G., Klump, K.L., 2009. Puberty and the genetic diathesis of disordered eating attitudes and behaviors. J. Abnorm. Psychol. 118 (4), 788–796.
Culbert, K.M., Racine, S.E., Klump, K.L., 2011. The influence of gender and puberty on the heritability of disordered eating symptoms. In: Adan, R.A.H., Kaye, H. (Eds.), Behavioral Neurobiology of Eating Disorders. Springer, New York, pp. 177–185.
Culbert, K.M., Breedlove, S.M., Sisk, C.L., Burt, S.A., Klump, K.L., 2013a. The emergence of sex differences in risk for disordered eating attitudes during puberty: a role for prenatal testosterone exposure. J. Abnorm. Psychol. 122 (2), 420–432.
Culbert, K.M., Klenotich, S., Dulawa, S., Le Grange, D., 2013b, September. From benchside to bedside: prenatal testosterone exposure decreases severity of anorexic-like symptoms in adolescent female mice and girls. Poster presented at the Eating Disorders Research Society meeting. Bethesda, Maryland.
Culbert, K.M., Burt, S.A., Sisk, C.L., Nigg, J.T., Klump, K.L., 2014. The effects of circulating testosterone and pubertal maturation on risk for disordered eating symptoms in adolescent males. Psychol. Med. 44 (11), 2271–2286.
Culbert, K.M., Breedlove, S.M., Sisk, C.L., Keel, P.K., Neale, M.C., Boker, S.M., Burt, S.A., Klump, K.L., 2015a. Age differences in prenatal testosterone's protective effects on disordered eating symptoms: developmental windows of expression? Behav. Neurosci. 129 (1), 18–36.
Culbert, K.M., Racine, S.E., Klump, K.L., 2015b. What we have learned about the causes of eating disorders: a synthesis of sociocultural, psychological, and biological research. J. Child Psychol. Psychiatr., in press.
Cuthbert, B.N., 2005. Dimensional models of psychopathology: research agenda and clinical utility. J. Abnorm. Psychol. 114 (4), 565–569.
Del Parigi, A., Chen, K., Gautier, J.F., Salbe, A.D., Pratley, R.E., Ravussin, E., Reiman, E.M., Tataranni, P.A., 2002. Sex differences in the human brain's response to hunger and satiation. Am. J. Clin. Nutr. 75 (6), 1017–1022.
Donohoe, T.P., Stevens, R., 1983. Effects of ovariectomy, estrogen treatment and CI-628 on food intake and body weight in female rats treated neonatally with gonadal hormones. Physiol. Behav. 31 (3), 325–329.
Eckel, L.A., Geary, N., 1999. Endogenous cholecystokinin's satiating action increases during estrus in female rats. Peptides 20 (4), 451–456.
Edler, C., Lipson, S.F., Keel, P.K., 2007. Ovarian hormones and binge eating in bulimia nervosa. Psychol. Med. 37 (1), 131–141.

Even, M.D., Dhar, M.G., vom Saal, F.S., 1992. Transport of steroids between fetuses via amniotic fluid in relation to the intrauterine position phenomenon in rats. J. Reprod. Fertil. 96 (2), 709–716.
Ferreiro, F., Seoane, G., Senra, C., 2011. A prospective study of risk factors for the development of depression and disordered eating in adolescents. J. Clin. Child Adolesc. Psychol. 40 (3), 500–505.
Field, A.E., Camargo, C.A., Taylor, C.B., Berkey, C.S., Roberts, S.B., Colditz, G.A., 2001. Peer, parent, and media influences on the development of weight concerns and frequent dieting among preadolescent and adolescent girls and boys. Pediatrics 107 (1), 54–60.
Garner, D.M., 1991. Eating Disorder Inventory-2 manual. Psychological Assessment Resources, Odessa, FL.
Gentry, T., Wade, G.N., 1976. Sex differences in sensitivity of food intake, body weight, and running-wheel activity to ovarian steroids in rats. J. Comp. Physiol. Psychol. 90 (8), 747–754.
Grumbach, M., Styne, D., 1992. Puberty: ontogeny, neuroendocrinology, physiology, and disorders. In: Wilson, J., Foster, P. (Eds.), Williams Textbook of Endocrinology, eighth ed. W.B. Saunders, Philadelphia, pp. 1139–1231.
Hausenblas, H.A., Campbell, A., Menzel, J.E., Doughty, J., Levine, M., Thompson, J.K., 2013. Media effects of experimental presentation of the ideal physique on eating disorder symptoms: a meta-analysis of laboratory studies. Clin. Psych. Rev. 33 (1), 168–181.
Hildebrandt, B.A., Racine, S.E., Keel, P.K., Burt, S.A., Neale, M., Boker, S., Sisk, C.L., Klump, K.L., 2014. The effects of ovarian hormones and emotional eating on changes in weight preoccupation across the menstrual cycle. Int. J. Eat. Disord. Advance online publication. doi: 10.1002/eat.22326
Hudson, J.I., Hiripi, E., Pope, Jr., H.G., Kessler, R.C., 2007. The prevalence and correlates of eating disorders in the National Comorbidity Survey Replication. Biol. Psychiat. 61 (3), 348–358.
Jacobi, C., Hayward, C., de Zwaan, M., Kraemer, H.C., Stewart, W., 2004. Coming to terms with risk factors for eating disorders: application of risk terminology and suggestions for a general taxonomy. Psychol. Bull. 130 (1), 19–65.
Jonas, S., Bebbington, P., McManus, S., Meltzer, H., Jenkins, R., Kuiper, E., Cooper, C., King, M., Brugha, T., 2011. Sexual abuse and psychiatric disorder in England: results from the 2007 adult psychiatric morbidity survey. Psychol. Med. 41 (4), 709–719.
Jones, W., Morgan, J., 2010. Eating disorders in men: a review of the literature. J. Pub. Ment. Health 9 (2), 23–31.
Jones, J.M., Bennett, S., Olmsted, M.P., Lawson, M.L., Rodin, G., 2001. Disordered eating attitudes and behaviours in teenaged girls: a school-based study. Can. Med. Assoc. J. 165 (5), 547–552.
Jones, R.M., Somerville, L.H., Li, J., Ruberry, E.J., Powers, A., Mehta, N., Dyke, J., Casey, B.J., 2014. Adolescent-specific patterns of behavior and neural activity during social reinforcement learning. Cogn. Affect. Behav. Neurosci. 14 (2), 683–697.
Keel, P.K., Forney, K.J., 2013. Psychosocial risk factors for eating disorders. Int. J. Eat. Disord. 46 (5), 433–439.
Keski-Rahkonen, A., Bulik, C.M., Neale, B.M., Rose, R.J., Rissanen, A., Kaprio, J., 2005a. Body dissatisfaction and drive for thinness in young adult twins. Int. J. Eat. Disord. 37 (4), 188–199.
Keski-Rahkonen, A., Neale, B.M., Bulik, C.M., Pietiläinen, K.H., Rose, R.J., Kaprio, J., Rissanen, A., 2005b. Intentional weight loss in young adults: sex-specific genetic and environmental effects. Obes. Res. 13 (4), 745–753.
Klump, K.L., 2013c. Puberty as a critical risk period for eating disorders: a review of human and animal studies. Horm. Behav. 64 (2), 399–410.
Klump, K.L., McGue, M., Iacono, W.G., 2003. Differential heritability of eating attitudes and behaviors in prepubertal versus pubertal twins. Int. J. Eat. Disord. 33 (3), 287–292.
Klump, K.L., Gobrogge, K.L., Perkins, P.S., Thorne, D., Sisk, C.L., Breedlove, S.M., 2006. Preliminary evidence that gonadal hormones organize and activate disordered eating. Psychol. Med. 36 (4), 539–546.
Klump, K.L., Perkins, P.S., Burt, S.A., McGue, M., Iacono, W.G., 2007. Puberty moderates genetic influences on disordered eating. Psychol. Med. 37 (5), 627–634.
Klump, K.L., Keel, P.K., Culbert, K.M., Edler, C., 2008. Ovarian hormones and binge eating: exploring associations in community samples. Psychol. Med. 38 (12), 1749–1757.
Klump, K.L., Bulik, C.M., Kaye, W.H., Treasure, J., Tyson, E., 2009. Academy for eating disorders position paper: eating disorders are serious mental illnesses. Int. J. Eat. Disord. 42 (2), 97–103.
Klump, K.L., Keel, P.K., Sisk, C., Burt, S.A., 2010. Preliminary evidence that estradiol moderates genetic influences on disordered eating attitudes and behaviors during puberty. Psychol. Med. 40 (10), 1745–1753.

Klump, K.L., Suisman, J.L., Culbert, K.M., Kashy, D.A., Keel, P.K., Sisk, C.L., 2011. The effects of ovariectomy on binge eating proneness in adult female rats. Horm. Behav. 59 (4), 585–593.
Klump, K.L., Culbert, K.M., Slane, J.D., Burt, S.A., Sisk, C.L., Nigg, J.T., 2012. The effects of puberty on genetic risk for disordered eating: evidence for a sex difference. Psychol. Med. 42 (3), 627–637.
Klump, K.L., Keel, P.K., Racine, S.E., Burt, S.A., Neale, M., Sisk, C.L., Boker, S., Hu, J.Y., 2013a. The interactive effects of estrogen and progesterone on changes in emotional eating across the menstrual cycle. J. Abnorm. Psychol. 122 (1), 131–137.
Klump, K.L., Keel, P.K., Burt, S.A., Racine, S.E., Neale, M.C., Sisk, C.L., Boker, S., 2013b. Ovarian hormones and emotional eating associations across the menstrual cycle: an examination of the potential moderating effects of body mass index and dietary restraint. Int. J. Eat. Disord. 46 (3), 256–263.
Klump, K.L., Racine, S., Hildebrandt, B., Sisk, C.L., 2013c. Sex differences in binge eating patterns in male and female adult rats. Int. J. Eat. Disord. 46 (7), 729–736.
Klump, K.L., Racine, S.E., Hildebrandt, B., Burt, S.A., Neale, M., Sisk, C.L., Boker, S., Keel, P.K., 2014. Influences of ovarian hormones on dysregulated eating: a comparison of associations in women with versus women without binge episodes. Clin. Psychol. Sci. 2 (5), 545–559.
Krueger, R.F., Markon, K.E., 2011. A dimensional-spectrum model of psychopathology: progress and opportunities. Arch. Gen. Psychiat. 68 (1), 10–11.
Laviano, A., Meguid, M.M., Gleason, J.R., Yang, Z.J., Renvyle, T., 1996. Comparison of long-term feeding pattern between male and female Fischer 344 rats: influence of estrous cycle. Am. J. Physiol. 270 (2), R413–R419.
Lutchmaya, S., Baron-Cohen, S., Raggatt, P., Knickmeyer, R., Manning, J.T., 2004. 2nd to 4th digit ratios, fetal testosterone and estradiol. Early Hum. Dev. 77 (1–2), 23–28.
Lydecker, J.A., Pisetsky, E.M., Mitchell, K.S., Thornton, L.M., Kendler, K.S., Reichborn-Kjennerud, T., Lichenstein, P., Bulik, C.M., Mazzeo, S.E., 2012. Association between co-twin sex and eating disorders in opposite sex twin pairs: evaluations in North American, Norwegian, and Swedish samples. J. Psychosom. Res. 72 (1), 73–77.
Madrid, J.A., Lopez-Bote, C., Martín, E., 1993. Effect of neonatal androgenization on the circadian rhythm of feeding behavior in rats. Physiol. Behav. 53 (2), 329–335.
Malas, M.A., Dogan, S., Hilal Evcil, E., Desdicioglu, K., 2006. Fetal development of the hand, digits and digit ratio (2D:4D). Early Hum. Dev. 82 (7), 469–475.
Martini, M.C., Lampe, J.W., Slavin, J.L., Kurzer, M.S., 1994. Effect of the menstrual cycle on energy and nutrient intake. Am. J. Clin. Nutr. 60 (6), 895–899.
McFadden, D., 1993. A masculinizing effect on the auditory systems of human females having male co-twins. Proc. Natl. Acad. Sci. USA 90 (24), 11900–11904.
Meulenberg, P.M.M., Hofman, J.A., 1991. Maternal testosterone and fetal sex. J. Steroid Biochem. Mol. Biol. 39 (1), 51–54.
Miller, E.M., 1994. Prenatal sex hormone transfer: a reason to study opposite-sex twins. Pers. Individ. Dif. 17 (4), 511–529.
Monteleone, P., Maj, M., 2013. Dysfunctions of leptin, ghrelin, BDNF, and endocannabinoids in eating disorders: beyond the homeostatic control of food intake. Psychoneuroendocrinology 38 (3), 312–330.
Oinonen, K.A., Bird, J.L., 2012. Age at menarche and digit ratio (2D:4D): relationships with body dissatisfaction, drive for thinness, and bulimia symptoms in women. Body Image 9 (2), 302–306.
Ostlund, H., Keller, E., Hurd, Y., 2003. Estrogen receptor gene expression in relation to neuropsychiatric disorders. Ann. NY Acad. Sci. 1007, 54–63.
Pliner, P., Fleming, A.S., 1983. Food intake, body weight, and sweetness preferences over the menstrual cycle in humans. Physiol. Behav. 30 (4), 663–666.
Plomin, R., DeFries, J., McClearn, G., McGuffin, P., 2008. Behavioral Genetics, fifth ed. Worth, New York.
Prince, A.C., Brooks, S.J., Stahl, D., Treasure, J., 2009. Systematic review and meta-analysis of the baseline concentrations and physiologic responses of gut hormones to food in eating disorders. Am. J. Clin. Nutr. 89, 1–11.
Quinton, S.J., Smith, A.R., Joiner, T., 2011. The 2nd to 4th digit ratio (2D:4D) and eating disorder diagnosis in women. Pers. Individ. Dif. 51 (4), 402–405.
Racine, S.E., Culbert, K.M., Keel, P.K., Sisk, C.L., Burt, S.A., Klump, K.L., 2012. Differential associations between ovarian hormones and disordered eating symptoms across the menstrual cycle in women. Int. J. Eat. Disord. 45 (3), 333–344.

Racine, S.E., Keel, P.K., Burt, S.A., Sisk, C.L., Neale, M., Boker, S., Klump, K.L., 2013. Individual differences in the relationship between ovarian hormones and emotional eating across the menstrual cycle: a role for personality? Eat. Behav. 14 (2), 161–166.

Raevuori, A., Kaprio, J., Hoek, H.W., Sihvola, E., Rissanen, A., Keski-Rahkonen, A., 2008. Anorexia and bulimia nervosa in same-sex and opposite-sex twins: lack of association with twin type in a nationwide study of Finnish twins. Am. J. Psychiatry 165 (12), 1604–1610.

Raevuori, A., Linna, M.S., Keski-Rahkonen, A., 2014. Prenatal and perinatal factors in eating disorders: a descriptive review. Int. J. Eat. Disord. 47 (7), 676–685.

Reichborn-Kjennerud, T., Bulik, C.M., Kendler, K.S., Røysamb, E., Maes, H., Tambs, K., Harris, J.R., 2003. Gender differences in binge-eating: a population-based twin study. Acta Psychiatr. Scand. 108 (3), 196–202.

Reichborn-Kjennerud, T., Bulik, C.M., Tambs, K., Harris, J.R., 2004a. Genetic and environmental influences on binge eating in the absence of compensatory behaviors: a population-based twin study. Int. J. Eat. Disord. 36 (3), 307–314.

Reichborn-Kjennerud, T., Bulik, C.M., Kendler, K.S., Roysamb, E., Tambs, K., Torgersen, S., Harris, J.R., 2004b. Undue influence of weight on self-evaluation: a population-based twin study of gender differences. Int. J. Eat. Disord. 35 (2), 123–132.

Root, T.L., Thornton, L.M., Lindroos, A.K., Stunkard, A.J., Lichtenstein, P., Pedersen, N.L., Rasmussen, F., Bulik, C.M., 2010. Shared and unique genetic and environmental influences on binge eating and night eating: a Swedish twin study. Eat. Behav. 11 (2), 92–98.

Rowe, R., Pickles, A., Simonoff, E., Bulik, C.M., Silberg, J.L., 2002. Bulimic symptoms in the Virginia Twin Study of Adolescent Behavioral Development: correlates, comorbidity, and genetics. Biol. Psychiatry 51 (2), 172–182.

Russell, C.J., Keel, P.K., 2002. Homosexuality as a specific risk factor for eating disorders in men. Int. J. Eat. Disord. 31 (3), 300–306.

Ryan, B.C., Vandenbergh, J.G., 2002. Intrauterine position effects. Neurosci. Biobehav. Rev. 26 (6), 665–678.

Shirao, N., Okamoto, Y., Mantani, T., Okamoto, Y., Yamawaki, S., 2005. Gender differences in brain activity generated by unpleasant word stimuli concerning body image: an fMRI study. Br. J. Psychiat. 186, 48–53.

Slof-Op't Landt, M.C.T., Bartels, M., Van Furth, E.F., Van Beijsterveldt, C.E.M., Meulenbelt, I., Slagboom, P.E., Boomsma, D.I., 2008. Genetic influences on disordered eating behaviour are largely independent of body mass index. Acta Psychiatr. Scand. 117 (5), 348–356.

Smeets, P.A., de Graaf, C., Stafleu, A., van Osch, M.J., Nievelstein, R.A., van der Grond, J., 2006. Effect of satiety on brain activation during chocolate tasting in men and women. Am. J. Clin. Nutr. 83 (6), 1297–1305.

Smink, F.R.E., van Hoeken, D., Oldehinkel, A.J., Hoek, H.W., 2014. Prevalence and severity of DSM-5 eating disorders in a community cohort of adolescents. Int. J. Eat. Disord. 47 (6), 610–619.

Smith, A.R., Hawkeswood, S.E., Joiner, T.E., 2010. The measure of a man: associations between digit ratio and disordered eating in males. Int. J. Eat. Disord. 43 (6), 543–548.

Somerville, L.H., 2013. Special issue on the teenage brain: sensitivity to social evaluation. Curr. Dir. Psychol. Sci. 22 (2), 121–127.

Stice, E., Ng, J., Shaw, H., 2010. Risk factors and prodromal eating pathology. J. Child Psychol. Psychiat. 51 (4), 518–525.

Striegel-Moore, R.H., Bulik, C.M., 2007. Risk factors for eating disorders. Am. Psychol. 62 (3), 181–198.

Swanson, S., Crow, S.J., Le Grange, D., Swendsen, J., Merikangas, K., 2011. Prevalence and correlates of eating disorders in adolescents: results from the national comorbidity survey replication adolescent supplement. Arch. Gen. Psychiat. 68 (7), 714–723.

Thompson, J.K., Stice, E., 2001. Thin-ideal internalization: mounting evidence for a new risk factor for body-image disturbance and eating pathology. Curr. Dir. Psychol. Sci. 10 (5), 181–183.

Trace, S.E., Baker, J.H., Peñas-Lledó, E., Bulik, C.M., 2013. The genetics of eating disorders. Ann. Rev. Clin. Psychol. 9, 589–620.

Uher, R., Murphy, T., Brammer, M.J., Dalgleish, T., Phillips, M.L., Ng, V.W., Andrew, C.M., Williams, S.C.R., Campbell, I.C., Treasure, J., 2004. Medial prefrontal cortex activity associated with symptom provocation in eating disorders. Am. J. Psychiat. 161 (7), 1238–1246.

Uher, R., Treasure, J., Heining, M., Brammer, M.J., Campbell, I.C., 2006. Cerebral processing of food-related stimuli: effects of fasting and gender. Behav Brain Res. 169 (1), 111–119.
von Ranson, K.M., Klump, K.L., Iacono, W.G., McGue, M., 2005. The Minnesota eating behavior survey: a brief measure of disordered eating attitudes and behaviors. Eat. Behav. 6 (4), 373–392.
Wade, G.N., 1972. Gonadal hormones and behavioral regulation of body weight. Physiol. Behav. 8 (3), 523–534.
Wade, G.N., 1975. Some effects of ovarian hormones on food intake and body weight in female rats. J. Comp. Physiol. Psychol. 88 (1), 183–193.
Yu, Z., Geary, N., Corwin, R.L., 2008. Ovarian hormones inhibit fat intake under binge-type conditions in ovariectomized rats. Physiol. Behav. 95 (3), 501–507.
Zheng, Z., Cohn, M.J., 2011. Developmental basis of sexually dimorphic digit ratios. Proc. Natl. Acad. Sci. USA 108 (39), 16289–16294.
Zucker, I., 1969. Hormonal determinants of sex differences in saccharin preference, food intake and body weight. Physiol. Behav. 4 (4), 595–602.

CHAPTER 9

# The Maternal Brain: Short- and Long-Term Effects of Reproductive Experience on Hippocampus Structure and Function in Adulthood

**Meighen Roes*, Liisa A.M. Galea*,****
*University of British Columbia, Department of Psychology, Vancouver, BC, Canada
**University of British Columbia, Centre for Brain Health, Vancouver, BC, Canada

## 1 INTRODUCTION

Pregnancy and the postpartum period are accompanied by prolonged changes in steroid and peptide hormones (Casey et al., 1993). Estradiol, progesterone, and cortisol levels increase across pregnancy in women largely due to secretions from the placenta (Casey et al., 1993). During gestation, estradiol levels are approximately 200 times preconception levels by week 20 and prior to parturition can be 1000 times preconception levels (Casey et al., 1993). Progesterone levels are approximately 20 times preconception levels while cortisol levels are 2–3 times higher than preconception levels during gestation (Casey et al., 1993). Many other hormones are also elevated during pregnancy and parturition including estriol, oxytocin, and prolactin (Brett and Baxendale, 2001). At parturition, with the expulsion of the placenta, there are rapid decreases in steroid and peptide hormone levels and women are hypogonadal for up to 180 days postparturition (Brett and Baxendale, 2001). These hormonal changes correspond to increased behavioral and lactational demands to the mother in order to ensure the survival of her offspring. The hormones altered by pregnancy and parturition stimulate maternal behaviors in rodents (Rosenblatt, 1988), and these maternal hormones along with pup exposure induce permanent changes to maternal responsiveness termed maternal memory (Fleming et al., 1999; Numan, 2006; Scanlan et al., 2006; Stolzenberg et al., 2012). Given these dramatic changes in behavior and brain it is perhaps little wonder that pregnancy and early motherhood are times of remarkable plasticity for the maternal brain. In women, brain size decreases by as much as 7% over the course of pregnancy, reaching its nadir at delivery and returning to preconception size by 6 months after birth (Oatridge et al., 2002). This review focuses on changes in the maternal hippocampus, as the hippocampus is important for memory (Burgess, 2008; Maren, 2001) and the volume of the hippocampus has been used as an indicator of brain health (Ashburner et al., 2003).

*Sex Differences in the Central Nervous System.* http://dx.doi.org/10.1016/B978-0-12-802114-9.00009-3

The hippocampus maintains a high degree of plasticity throughout life and contains a large quantity of steroid hormone receptors, including glucocorticoid and estrogen receptors (McEwen, 1994; Weiland et al., 1997), thus suggesting the hippocampus can be modified by steroid hormones during pregnancy and postpartum. This review explores how reproductive experience alters hippocampal morphology and function during pregnancy, early postpartum period, and older age.

## 2 PREGNANCY AND PARTURITION: SHORT-TERM CHANGES IN COGNITION AND HIPPOCAMPUS

### 2.1 Cognitive changes in pregnancy and early postpartum

A majority of pregnant women experience subjective cognitive disturbance during pregnancy, with memory difficulties being the most common complaint (Brett and Baxendale, 2001). Other subjective impairments include forgetfulness, confusion, and poor concentration (Poser et al., 1986). While the colloquial term "baby brain" has been coined to refer to this common phenomenon, objective deficits in memory are not typically detected in women in the laboratory environment (Cuttler et al., 2011). Field tasks may be more sensitive to deficits, as they take place in the naturalistic context of multiple competing cognitive demands (Cuttler et al., 2011). Importantly, however, not all domains of memory are impaired during pregnancy and early postpartum. A meta-analysis (Henry and Rendell, 2007) found that working memory and free recall are impaired during pregnancy, but that alterations in short-term memory, recognition memory, and implicit memory are not reliably detected. Verbal memory impairment is a commonly detected deficit associated with pregnancy and early postpartum (Buckwalter, 1999; Crawley et al., 2003; de Groot et al., 2003, 2006; Eidelman et al., 1993; Glynn, 2010, 2012; Keenan et al., 1998; Sharp et al., 1993). Deficits in attention (de Groot et al., 2006), processing speed (Christensen et al., 2010), visual memory (Eidelman et al., 1993), and spatial ability (Adams et al., 2002; Woodfield, 1984) have also been reported.

Impairments in cognition during pregnancy are dependent on a number of factors, including trimester, fetal sex, and parity. The third trimester of pregnancy and the early postpartum period in particular are associated with verbal and spatial memory impairments relative to the first two trimesters of pregnancy (Darnaudery et al., 2007; Galea et al., 2000; Glynn, 2012). For example, in rodents, spatial ability is impaired during the third week of pregnancy (Galea et al., 2000; Keenan et al., 1998), while in women, verbal memory and processing speed are impaired in the third but not first two trimesters (Keenan et al., 1998; Christensen et al., 2010). Fetal sex alters cognition during pregnancy as women pregnant with girls perform more poorly than women pregnant with boys on working memory and spatial (mental rotation) tasks throughout gestation and into the postpartum period (Vanston and Watson, 2005). Finally, it is important to note that the adverse effects of pregnancy on cognition are compounded by successive pregnancies, at least in certain domains. Glynn (2012) found that verbal memory impairments in pregnancy and postpartum were more severe with increasing parity, such that multigravid mothers performed worse than bigravid (two pregnancies) mothers, who in turn performed worse than primigravid (single pregnancy) mothers tested at gestational week 34 and 3 months postpartum. This latter finding provides one indication that the effect of reproduction on brain structure and function is cumulative and enduring. The mechanisms for these effects are not known but hormone levels likely play a role as hormone levels are highest during the third trimester (Brett and Baxendale, 2001) and differ with fetal sex (Vanston and Watson, 2005) and parity (Bridges et al., 1993; Paris and Frye, 2008). For example, pregnant biparous rats (two reproductive experiences) have higher levels of estradiol, and lower levels of prolactin across gestation compared with primiparous (one reproductive experience) females (Bridges et al., 1993; Paris and Frye, 2008). Taken together, these studies indicate that cognitive impairments during pregnancy are observed on some tasks that depend on trimester, task, fetal sex, and parity.

Animal studies mirror findings in women, showing that the third week of gestation in rodents is associated with impairments in spatial ability (Darnaudery et al., 2007; Galea et al., 2000). However, it is important to note that there are differences in pregnancy hormone profiles in rat versus human (Brett and Baxendale, 2001; Galea et al., 2000; Pawluski et al., 2009). Galea et al. (2000) found that during the third week of pregnancy (approximately days 20 and 21), primigravid rats traveled longer distances to find a hidden platform in a working memory version of the Morris water maze compared to nulliparous (no reproductive experience) rats. However, the first and second weeks of pregnancy were associated with improved spatial working memory compared to nulliparous rats. Similarly, primigravid and bigravid rats outperform nulliparous rats in the first and second week (days 6–18) of gestation using an object placement task. Together, the animal studies suggest a beneficial effect of early pregnancy followed by a detrimental effect in the final days of pregnancy on spatial ability. As stated earlier, this may correspond to the different hormone profile seen in pregnant rats compared to humans. The first 18 days of rat pregnancy are characterized by low estradiol levels that correspond to the levels seen in diestrus (Anderson et al., 1996; Rosenblatt et al., 1988). Estradiol levels are increased in the last 3–4 days of rat pregnancy and correspond to levels seen in the proestrous phase of the rat estrous cycle (Anderson et al., 1996; Rosenblatt et al., 1988). This is a much different hormone profile than that of human pregnancy, in which estradiol levels steadily increase across the course of pregnancy and are elevated for prolonged periods relative to the increase seen in rats (Pawluski et al., 2009; Brett and Baxendale, 2001). Previous studies have shown that lower levels of estradiol are associated with better spatial ability, while higher levels of estradiol are associated with poorer spatial ability (Holmes et al., 2002), and as discussed in the following section, fluctuations in estradiol across pregnancy may account for pregnancy effects on spatial ability.

## 2.2 Possible mechanisms of cognitive changes during pregnancy: steroid hormones

Fluctuating hormone levels are implicated in cognitive changes throughout late pregnancy and early postpartum. For example, high levels of cortisol are associated with a number of cognitive deficits during pregnancy and the postpartum period, including low verbal recall scores, attention, and poor spatial abilities (Buckwalter, 1999; Glynn, 2010; Henry and Sherwin, 2012). Estradiol and prolactin levels have also been associated with paragraph recall and executive function, and attention, respectively, during pregnancy and postpartum (Henry and Sherwin, 2012). However, not all studies have found a significant relationship between hormone levels and cognition (e.g., Silber et al., 1990). Modulation of cognition by pregnancy hormone fluctuations is consistent with a considerable literature implicating steroid hormones in cognition (Barha and Galea, 2010; Buckwalter, 1999). For example, changes in cognitive performance vary across the menstrual cycle, with women performing better on spatial tasks during phases when estradiol levels are low and worse when estradiol levels are high (Hampson, 1990). Similarly, performance in the Morris water maze varies across the estrous cycle, with enhanced performance during estrus, when estradiol levels are low, and impaired performance displayed during proestrus when estradiol levels are high (Warren and Juraska, 1997). Thus, the differences in pregnancy-related cognitive changes in rats and humans mirror the species differences in the time course of estradiol elevations. However, it is important to consider that many other hormones are likely involved in cognitive changes during pregnancy, as progesterone (Freeman et al., 1992, 1993), estradiol/progesterone ratio (Carroll et al., 2007; Harburger et al., 2007), corticosterone/cortisol (Lupien et al., 2009), and oxytocin (Kirsch et al., 2005) have all been implicated in affecting cognition.

## 2.3 Possible mechanisms of cognitive changes during pregnancy and early postpartum: the hippocampus

Pregnancy and the postpartum period are characterized by changes in the structure of the hippocampus. In CA1 pyramidal cells, apical spine density increases during pregnancy and lactation in primiparous rats compared with nulliparous Sprague Dawley rats (Kinsley et al., 2006). The observed increase in spine density is likely induced by pregnancy-related hormone exposure, as simulating pregnancy in nulliparous rats with hormone injections causes the same increase in spine density (Kinsley et al., 2006). By the late postpartum period, however, the effect of parity on apical CA1 spine density may disappear as Pawluski and Galea (2005) found no effect of either primiparity or multiparity on this measure. Instead, in the late postpartum period, primiparous rats show reduced apical and basal branch points and dendritic length in CA3 and CA1 pyramidal neurons comparison with nulliparous and multiparous rats (Pawluski and Galea, 2005). Multiparous rats had greater spine density in the basal region of CA1 pyramidal neurons compared with nulliparous rats postweaning. However, this effect was likely due to differences in litter sex ratio, as spine density of the basal region of CA1 pyramidal neurons correlated positively with the number of males in a litter, and multiparous rats gave birth to more male pups than primiparous rats (Pawluski and Galea, 2005). Together, these findings indicate time- and parity-specific effects of reproductive experience on dendritic remodeling in the CA1 and CA3 regions of the hippocampus.

Coupled with these changes in morphology are changes in the electrophysiological properties of the CA1 region. Long-term potentiation (LTP) is viewed as a cellular model of learning and memory (Bliss and Collingridge, 1993; Kandel et al., 2014) and LTP is augmented along Schaffer collaterals in lactating biparous C57BL6 mice that have mothered for 3–6 days (Tomizawa et al., 2003). Tomizawa et al. propose that oxytocin mediates not only the observed increase in LTP, but also the early postpartum (days 3–7) enhancement in spatial reference memory in multiparous compared to nulliparous rats. They demonstrated that intracerebroventricular oxytocin improved spatial reference, but not spatial working, memory in a radial arm task in nulliparous mice. Conversely, an injection of an oxytocin antagonist in biparous mice decreased spatial reference memory compared with biparous controls. Incubation of hippocampal slices in oxytocin prolonged LTP in the hippocampus of nulliparous mice (Tomizawa et al., 2003). The researchers proposed that oxytocin-induced phosphorylation of cyclic adenosine monophosphate response element binding protein ultimately underlie the prolongation of LTP and enhancement in spatial reference memory seen during the early postpartum period.

Reproductive experience also alters neurogenesis levels in the dentate gyrus of the hippocampus (Barha and Galea, 2011; Darnaudery et al., 2007; Leuner et al., 2007; Pawluski and Galea, 2007). Neurogenesis is composed of four stages: cell proliferation, cell migration, cell differentiation, and cell survival (Ming and Song, 2005). Cell proliferation is the production of new cells, whereas cell survival refers to how many of these new neurons survive. Cell differentiation is the process by which a new cell differentiates into most typically a new neuron or glial cell, and cell migration is the process by which the new cells move to their new location (Ming and Song, 2005). A number of factors, including steroid hormones (Barha et al., 2009a), may affect each of these stages independently. Alterations in any of these four stages can alter the number of new neurons. For example, testosterone increases neurogenesis via alterations in cell survival independently of alterations in cell proliferation (Hamson et al., 2013; Spritzer and Galea, 2007), and prenatal ethanol exposure decreases neurogenesis in female offspring via alterations in cell differentiation (Uban et al., 2010). In young adult rats, approximately 9000 new cells are generated each day in the dentate gyrus, or 250,000 new cells per month (Cameron et al., 1993). Although the exact functions of this population of new neurons are not precisely understood, there is evidence that these new neurons are involved in hippocampus-dependent learning and memory (Deng et al., 2010). For example, hippocampal neurogenesis is required for pattern separation (Clelland et al., 2009).

New neurons are preferentially activated in response to spatial memory retrieval (Epp et al., 2011; Kee et al., 2007), further indicating a functional role of adult neurogenesis in hippocampus-dependent learning and memory. Importantly, adult hippocampal neurogenesis is modulated by estradiol (Galea et al., 2013) and corticosterone (Mirescu and Gould, 2006), and one study found that hippocampus neurogenesis is required for stress buffering (Snyder et al., 2011). For example, administration of estradiol in females increases cell proliferation and cell survival in a dose- and duration-specific manner (Galea, 2008), whereas exogenous and endogenous increases in corticosterone suppress cell proliferation and survival (Brummelte and Galea, 2010).

Adult neurogenesis is impacted by reproductive experience in a region-, time-, and parity-dependent manner. The subventricular zone (SVZ) undergoes an increase in cell proliferation during both early (Shingo et al., 2003) and late (Furuta and Bridges, 2005) pregnancy in the dam. Experimentally reducing prolactin during early pregnancy suppresses neurogenesis in the SVZ and subsequently impairs maternal behavior (Larsen and Grattan, 2010). In contrast, alterations in cell proliferation and cell survival are not detected in the dentate gyrus during pregnancy in the primigravid or bigravid dam (Banasr et al., 2001; Pawluski et al., 2010). However, the survival of new hippocampal neurons born during mid-pregnancy may be reduced, as Rolls et al. (2008) found that primigravid mice have fewer numbers of new neurons surviving for 2 weeks after gestational days 11–12 compared with nulliparous wildtype mice. The survival of new neurons returned to control levels by postpartum day 21 (Rolls et al., 2008). Intriguingly, this parity-related transient decrease in new neuron survival is reduced in T cell deficient mice and is partially restored by reconstitution of T cell populations (Rolls et al., 2008). The authors speculate that T cell populations may contribute to transient decreases in new neuron survival during pregnancy (Rolls et al., 2008). In contrast, Banasr et al. (2001) found enhanced expression of polysialylated neural cell adhesion molecule (PSA–NCAM) in the dentate gyrus of pregnant rats at gestation day 18. PSA–NCAM is involved in cell migration and synaptic reorganization (Bonfanti, 2006); therefore, plasticity in the dentate gyrus may be altered at specific time points during pregnancy independent of alterations in cell proliferation or survival. Additionally, Pawluski et al. (2010) found that primigravid and bigravid females have significantly fewer pyknotic cells in the granule cell layer than nulliparous females. Collectively these findings suggest that there are perhaps more dramatic changes in neurogenesis during pregnancy in the SVZ than in the hippocampus. The cells produced in the SVZ migrate to the olfactory bulbs and possibly assist in maternal identification of offspring (Lévy et al., 2011; Shingo et al., 2003).

In contrast to findings during pregnancy, neurogenesis levels, via cell proliferation, are dramatically reduced during the postpartum period. Both primiparous and biparous rats show decreased cell proliferation 24 h after giving birth compared to nulliparous rats (Darnaudery et al., 2007; Pawluski and Galea, 2007). Cell proliferation in the dentate gyrus remains suppressed during lactation at least until postpartum day 8 and rebound to normal levels 28 days after parturition (Hillerer et al., 2014; Leuner et al., 2007). The decreased cell proliferation levels in the early postpartum are related to elevated levels of corticosterone (Leuner et al., 2007) as removing pups after birth reduces basal corticosterone and prevents the decrease in cell proliferation (Leuner et al., 2007). However, when reductions in cell proliferation are accounted for, Pawluski and Galea (2007) found that biparous rats had greater new neuron survival compared to nulliparous and primiparous rats. Furthermore, only primiparous rats show reduced survival of adult-generated hippocampal neurons over the 21-day postpartum period compared with both biparous and nulliparous rats (Pawluski and Galea, 2007). This reduction in new neuron survival in primiparous rats is attributable to pregnancy alone, as primiparous rats have similar levels of new neuron survival regardless of whether or not pups are removed 2 days after birth (Pawluski and Galea 2007). Interestingly, exposure of nulliparous rats to rat pups increased cell proliferation and new neuron survival across the 23 days of exposure to pups, indicating that "mothering" alone produces increased hippocampal neurogenesis unlike the reductions in neurogenesis seen with primiparity (Pawluski and Galea, 2007). Thus, the postpartum period is characterized by reduced neurogenesis in the early postpartum that appears to return to preconception levels by postpartum day 28. This may mirror the transient reduction in brain size seen with pregnancy in women (Oatridge et al., 2002).

## 2.4 Effects of parity on cognition beyond the postpartum

The cognitive impairments seen in women in late pregnancy and the early postpartum period are transient, and some findings suggest they may give way to enhanced cognition within 2 years of parturition. Verbal memory and attention skills in postpartum women are impaired immediately postpartum (Buckwalter, 1999; Crawley et al., 2003; de Groot et al., 2006; Eidelman et al., 1993; Glynn, 2010, 2012) but are once again equivalent to nonmothers 10–13 months after delivery (Crawley et al., 2003) and may even be enhanced by 2 years postpartum (Buckwalter et al., 2001). Consistent with a cognitive benefit emerging after the postpartum period, a number of animal studies find that reproductive experience enhances hippocampus-dependent learning and memory after the pups have been weaned. For example, primiparous rats outperform nulliparous rats on an object placement task 42 days postpartum (Cost et al., 2014). The effects of parity on cognition after weaning depend on both pregnancy and mothering experience. Kinsley et al. (1999) found that primiparous dams with 16 days of mothering experience had shorter latencies to escape a dry land maze compared with nulliparous females. Continuous exposure to foster pups for 16 days caused nulliparas to perform like primiparous rats (Kinsley et al., 1999). Similarly, Lambert et al. (2005) found that primiparous rats that mothered their offspring for 21 days had an advantage (measured using latency) in a spatial version of the dry land maze over nulliparous rats. Further, primiparous dams outperformed primiparous rats that did not mother their offspring and nulliparas that had been sensitized to show maternal behavior through exposure to pups (Lambert et al., 2005). However, it is important to note that the latency measure used by Kinsley et al. (1999) and Lambert et al. (2005) may be conflated with differences in motivation or motor activity. Pawluski et al. (2006a) found that either immediately or 60 days after weaning primiparous rats made fewer spatial working and reference memory errors in a radial arm maze task compared with biparous or nulliparous rats. Furthermore the advantage for primiparous rats was attributable to both pregnancy and mothering as they outperformed both pregnant only and sensitized nulliparous rats 60 days after weaning (Pawluski et al., 2006b). Intriguingly, primiparous rats whose offspring were removed at birth failed to complete the task, had longer latencies to reach the goal arm, and in a separate experiment showed increased depressive-like behavior, implicating noncognitive impairments in performance of these "pregnant only" females (Pawluski et al., 2009). While primiparous rats demonstrated enhanced spatial working and reference memory, biparous rats tended toward improved spatial memory regardless of error type compared to nulliparous rats (Pawluski et al., 2006b). Interestingly, aspects of maternal behavior (duration of licking and nursing) correlated positively with number of reference memory errors but did not significantly associate with working or working/memory performance. These findings further suggest that pup interactions may impact reference memory, perhaps through oxytocin activity (Pawluski et al., 2006b). Together, these studies suggest that the improved spatial performance long after weaning associated with parity requires the combined effects of pregnancy and mothering.

## 3 LONG-TERM CHANGES IN COGNITION WITH AGING ARE ALTERED BY REPRODUCTIVE FACTORS AND PARITY

Reproductive experience alters cognitive function into middle age in both humans and rodents. McLay et al. (2003) longitudinally studied 361 postmenopausal women for a median of 12.8 years and found that nulliparity was associated with a small but significant reduction in global cognitive decline (approximately 1 point on the mini-mental state examination, MMSE). As detailed cognitive assessments or diagnoses were not made, this study was unable to differentiate cognitive decline due to "normal" age-related changes from that due to dementia or other pathological processes. As a result, the question of whether the "normal" aging trajectory is altered by motherhood is not fully addressed by this design. Nevertheless, these data indicate that the relationship between reproductive experience and cognition might change as a woman enters middle age and older adulthood, with negative effects of parity as a woman ages. Nulliparity is assumed to be related to increased cumulative lifetime estrogen exposure (Heys et al., 2011; Ryan et al., 2008, 2009). The assumption that parity reduces lifetime estrogen exposure appears to stem from the observations that: (1) nulliparous women are at increased risk for breast cancer, and breast cancer is associated with greater exposure to ovarian hormones (Kalache et al., 1993), (2) lactation suppresses ovarian hormone levels (McNeilly, 2001), and (3) there are subtle and specific age- and menstrual/estrous cycle- dependent alterations in ovarian hormone fluctuations associated with parity (Bridges and Byrnes, 2006; Dorgan et al., 1995). However, it must also be noted that parity involves periods of sustained elevated hormones during pregnancy and is associated with later onset of menopause in women (Gold et al., 2001). Despite these complexities, it is fairly common to read of parity-related differences being attributed to reduced lifetime exposure to estrogens.

Smith et al. (1999) created an index of estrogen exposure (IEE) by combining z-scores for age at menarche and menopause, parity, duration of estrogen hormone therapy, postmenopausal weight, and number of years since menopause. The IEE correlated with a measure of global cognitive function (MMSE) and correlated with verbal attention (Boston Naming Test) after controlling for age and depression. These findings are consistent with effects of reproductive factors on cognitive outcomes during aging, but parity did not significantly associate with these cognitive outcomes on its own. Other investigations of the effects of endogenous estrogen exposure have found that breastfeeding (which suppresses ovulation and reduces estradiol levels) impairs cognition (Hesson, 2012; Heys et al., 2011). An adapted IEE that included duration of breastfeeding (Hesson, 2012) found a significant relationship with this index and a measure of prospective but not retrospective memory. Longer reproductive period (menarche to menopause) is associated with better visual memory, working memory, and verbal memory among older nondemented postmenopausal women when controlling for reproductive factors including parity (Tierney et al., 2013). Furthermore, longer reproductive period is associated with greater telomere length and lower telomerase activity, two biomarkers of cellular aging (Lin et al., 2011). Together, these studies suggest that reproductive factors that alter endogenous estrogen exposure may impact cognitive aging in women. However, specific investigations of the effects of parity are warranted, as parity and/or amount of parity may alter outcomes independent of or in addition to altered cumulative exposure.

Ryan et al. (2009) used detailed cognitive assessments and a design with both cross-sectional and longitudinal (4-year follow-up) features and importantly excluded cases of probable pre-existing dementia. Whereas length of reproductive period and age at first menses, and estradiol-based hormone therapy were associated with performance on a number of cognitive tasks, parity level was not. None of the reproductive variables were associated with cognitive performance or dementia incidence during the short follow-up period. Interestingly, among mothers, the age of first motherhood was related with global cognitive function (MMSE), verbal fluency, visual memory, and psychomotor speed. Women who gave birth between the ages of 21 and 29 years had better cognitive performance compared to women who gave birth at a younger age (younger than 21). Older age (>30) at first childbirth did not differ significantly from young first-time childbearers. Heys et al. (2011) studied a large cohort of Chinese women from the Guangzhou Biobank Cohort Study. They included only parous women in their design and found that lower parity and shorter average duration of breastfeeding per child were associated with better MMSE scores and delayed recall. In contrast to the findings of Ryan et al. (2009), older age at first pregnancy was associated with poorer cognitive function. It is unclear why these studies diverge regarding the effects of pregnancy age and parity on cognition. However, Heys et al. (2011) used structural equation modeling that may not have been sensitive to a U-shaped relationship between age at first pregnancy and cognitive outcome. Furthermore, their investigation focused on a population with different sociocultural and reproductive histories to those usually studied. The women participating in their study (Heys et al., 2011) had relatively high fertility and breastfeeding rates, and may have been subject to unique cohort effects. Furthermore, endogenous estrogen levels are lower in Chinese compared with Caucasian women, especially postmenopausally (Key et al., 1990). Rasgon et al. (2005) analyzed data from a Swedish population-based cohort study of cognitive impairment and found that reproductive period, hormone therapy, and age at menopause were inversely associated with risk of cognitive impairment. In this study, having five or more children increased the odds ratio of cognitive impairment. McLay et al. (2003) also investigated birth control use, surgical menopause, and age at natural menopause, and found that late menopause similarly reduced cognitive decline. In summary, human data concerning the effects of parity on normal cognitive decline have largely stemmed from an interest in cumulative exposure to estrogens. While some studies (Heys et al., 2011; McLay et al., 2003) have found that increased parity is associated with cognitive impairment, others have found no association (Ryan et al., 2009), or associations only after a relatively high number of pregnancies (Rasgon et al., 2005). Age at first parity may influence cognition independent of number of children born, but the effects of age at first pregnancy have differed across studies, perhaps due to sample characteristics (Heys et al., 2011; Rasgon et al., 2005).

## 4 GREATER ENDOGENOUS OVARIAN HORMONE EXPOSURE MAY IMPROVE COGNITIVE AGING

### 4.1 Reproductive experience, aging, and cognition: animal studies

A number of animal studies indicate that lower parity can positively affect hippocampus-dependent cognition long after weaning. Parous (one or more litters) wildtype mice have better spatial working and reference memory in the Barnes maze at 12 months, and have enhanced spontaneous alternation in the Y maze compared to nulliparous controls at 12 months of age (Li et al., 2013). Lemaire et al. (2006) found that primiparous Wistar rats that were retested at 22 months of age had enhanced spatial memory compared to nulliparous rats. Gatewood et al. (2005) used repeated testing in the dry land maze and found that both primiparous and biparous Sprague Dawley rats had lower latencies to find a baited food well when repeatedly tested 12, 18, and 24 months of age compared with nulliparas, and also had lower latencies to find a food reward in a reversal task in which the location of the baited well was changed (Gatewood et al., 2005). In contrast to the findings of Gatewood et al. (2005), Love et al. (2005) found no overall effect of parity on reversal learning or on latency to find baited wells during reacquisition, though biparous females outperformed nulliparous rats in the latter task at age 17 months. Probe trials conducted at each time point revealed age-specific effects of parity on spatial memory retrieval. Whereas primiparous and biparous rats outperformed nulliparous animals at a younger age (5 months), they did not consistently do so at older ages. The parity advantage was absent by age 17 months. By age 21 months, primiparous rats showed a trend toward enhanced performance relative to nulliparous and multiparous rats (Love et al., 2005). These results serve as a caution that different aspects of spatial cognition (acquisition/reacquisition vs. memory retrieval) may be altered differently by parity. It should be noted that the repeated testing protocols used by Gatewood et al. (2005) and Love et al. (2005) place different demands on animals than the learning of new tasks. Additionally, these studies used latency as a measure of cognition, which is influenced by other nonmnemonic factors such as locomotor activity and motivation. These findings point to the intriguing possibility that some aspects of cognition may be altered in an age-dependent way, with early benefits to spatial memory receding or even reversing in older age. Therefore, age at assessment may be an important factor in detecting cognitive changes related to reproductive experience. Primiparity did not alter working or reference memory performance at 18 months, nor was spatial memory retrieval affected (Zimberknopf et al., 2011). Older (11-month) nulliparous Fischer-344 rats show impaired performance on the object placement task compared to multiparous rats (Macbeth et al., 2008), although there is controversy regarding the extent of hippocampal involvement in this task (Barker and Warburton, 2011). In summary, whereas a number of studies have found long-term improvements in hippocampus-dependent learning and memory long after weaning, others have not. Parity, age at testing, strain differences, and task demands may contribute to the varied findings in rodents. It is important to remember that cognition is not uniformly altered with parity, as studies demonstrate dissociations in how parity alters spatial working or reference memory, and in addition whether the task chosen requires integrity of the hippocampus (Love et al., 2005; Zimberknopf et al., 2011). Importantly, the impact of reproductive experience on the trajectory of cognitive aging may be complex, with diminished or perhaps even reversed benefits as the female ages.

## 4.2 Neural correlates of reproductive experience in later life

Reproductive experience is associated with long-term changes in plasticity, synaptic proteins, and neurotrophic factors in the hippocampus. The enhancement in LTP demonstrated in young biparous mice (Tomizawa et al., 2003) is also seen in 22-month-old primiparous Wistar rats 16 months after parturition, suggesting that the electrophysiological changes associated with early mothering persist into later life (Lemaire et al., 2006). Mice that have had one or more litters have increased expression of the presynaptic marker synaptophysin compared with nulliparas when tested at 12 months of age (Li et al., 2013). Brain-derived neurotrophic factor (BDNF) is involved in neuronal plasticity and hippocampus-dependent spatial memory (Allen and Dawbarn, 2006), and many of estradiol's effects in the hippocampus are mediated by BDNF (Scharfman and MacLusky, 2006). BDNF levels are elevated in the septum and in the CA1 region of the hippocampus (but not the dentate gyrus (DG) or CA3 region) in multiparous (5–6 litters) rats compared with nulliparous 12-month-old rats (Macbeth et al., 2008). Additionally, amyloid precursor protein immunoreactivity is significantly reduced in the CA1 and DG regions of biparous 24-month-old rats compared to nulliparous rats (Gatewood et al., 2005). Lastly, multiparous rats have more immature neurons in the DG than nulliparous rats in middle age (Barha et al., in press). These findings suggest that increased parity is associated with increased neuroplasticity markers dependent on hippocampal region.

## 4.3 Sensitivity to estrogens changes with parity and aging

Changes in sensitivity to estrogens may also contribute to the hippocampal changes observed in pregnancy, postpartum, and beyond. In young nulliparous rats, cell proliferation in the hippocampus is upregulated by acute estradiol (Barha et al., 2009b; Chiba et al., 2007; Ormerod and Galea, 2001; Ormerod et al., 2003) but this sensitivity declines with age as middle-aged nulliparous rats fail to show increased cell proliferation in response to estradiol (Barha and Galea, 2011; Chiba et al., 2007) or estrone (Barha and Galea, 2011). The ability of estradiol to influence spatial cognition in nulliparas shows a similar decline with aging. In young female rats, estradiol replacement reverses ovariectomy-induced impairments in spatial learning and memory in the Morris water maze (Talboom et al., 2008). In contrast, middle-aged nulliparous rats are not impaired by ovariectomy, and aged nulliparous rats are not responsive to either ovariectomy or estradiol (Talboom et al., 2008). Multiparous rats, in contrast, demonstrate sustained sensitivity to estrogens in middle age, with increases in cell proliferation in the dentate gyrus after acute exposure to estradiol or estrone, much like young adult female rats (Barha and Galea, 2011). Together, these findings suggest that reproductive experience maintains the response to estrogens seen in younger females. However, to date little is known about how estrogens affect learning or other components of neurogenesis (differentiation, migration, survival) during aging, and how parity might alter this trajectory.

Although studies in aged animals are lacking, investigations in young rats indicate that parity increases sensitivity to estradiol. Young multiparous rats demonstrate greater prolactin secretion in response to high doses of estradiol (Bridges and Byrnes, 2006) 2 weeks after weaning. Young adult primiparous rats are more sensitive than nulliparous rats to the anxiolytic-like effects of a repeatedly administered estrogen receptor alpha (ERα) antagonist propyl-pyrazole triol (PPT) (Byrnes et al., 2013). Reproductive experience alters behavioral and neuroendocrine reactivity to acute ERα antagonism as well in young adult females, with PPT-treated primiparous rats tested on the elevated plus maze showing fewer anxiety-like behaviors and less corticotrophin-releasing hormone messenger RNA expression in the paraventricular nucleus compared with nulliparous counterparts (Byrnes et al., 2013). During late pregnancy, primigravid and bigravid females have fewer ERα-expressing cells in the CA3 region of the dorsal hippocampus than nulliparous females (Pawluski et al., 2010). It is not known whether reproductive experience has enduring effects on ER expression in the hippocampus later in life, but long-term upregulation of ER expression has been detected in the medial preoptic area and lateral amygdala in middle-aged primiparous compared with nulliparous rats (Byrnes et al., 2009). Overall, these data indicate that motherhood alters the cellular, neuroendocrine, and behavioral response to estradiol activity early and later in life.

## 5 THE HIPPOCAMPUS AND ALZHEIMER'S DISEASE

Alzheimer's disease (AD) is the most common type of progressive dementia in the elderly, and is characterized neuropathologically by β-amyloid deposits and neurofibrillary tangles (Serrano-Pozo et al., 2011). The integrity of the hippocampus is profoundly altered during the progression of AD. For example, postmortem analysis indicates that AD is associated with a 32% reduction in volume of the hippocampus and a 23% reduction in neuron number in the CA1 and DG regions of the hippocampus (Simic et al., 1997). The exact reductions depend on progression of the disease; there are significant aberrations in hippocampal neurogenesis in AD, with both increases (Jin et al., 2004) and decreases (Chen et al., 2008) in neurogenesis detected depending on progression of AD (Bobinski et al., 1998) or Braak stage (Ekonomou et al., 2014; Perry et al., 2012). For example, neurogenesis is increased in the early stages of disease, but decreased in later stages of the disease (Ekonomou et al., 2014; Perry et al., 2012). Animal models of AD also suggest that early increases in cell proliferation can also give way to decreased neurogenesis as the disease progresses (Lopez-Toledano et al., 2010). The increased neurogenesis in early stages of the disease may be a compensatory mechanism, and indeed mirrors findings that early in the disease reductions in volume and cell number can be detected in the CA1 region but not in the DG region (Simic et al., 1997). In patients with mild AD, the volume of the CA1 region, but not the DG or CA3 regions, correlated with delayed memory recall (Kerchner et al. 2012). These findings are consistent with other reports of correlations between overall CA1 size and memory performance (Apostolova et al., 2010; Atienza et al., 2011; Scheff et al., 2007). Together, these studies implicate hippocampal deficits, and CA1 atrophy in particular, in the early stages of AD.

# 6 REPRODUCTIVE EXPERIENCE AND SUSCEPTIBILITY TO ALZHEIMER'S DISEASE

## 6.1 Introduction

Women are at greater risk for Alzheimer's disease (AD) even after controlling for age and level of education (Andersen et al., 1999; Anderson et al., 1996; Hy and Keller, 2000; Letenneur et al., 1994), pointing to a possible role of hormonal factors in risk for AD incidence. A number of observational studies indicate that reproductive experience may contribute to the progression of the disease in women. Colucci et al. (2006) found that women diagnosed with probable AD had on average a higher number of pregnancies and lower frequency of nulliparity compared with control women. Within the AD group, a higher number of pregnancies correlated with an earlier onset of disease, and patients that had more than three pregnancies had a 3.9-year earlier average onset of disease compared with other AD patients (Colucci et al., 2006). These findings are perhaps consistent with McLay's findings of better cognitive outcomes in nulliparous women (McLay et al., 2003). Similarly, Ptok et al. (2002) found that in women, but not men, having had children was found to be associated with a diagnosis of AD. However, in that study, unlike the findings of Colucci et al. (2006), the number of children was not significantly related to AD diagnosis. Sobow and Kloszewska (2004) conducted a small case series of 45 women meeting criteria for AD with age at onset above 65. Using retrospective reports from patients and collateral informants, they found that women who had been pregnant at least once had younger age at onset of AD, and each successive pregnancy was associated with a decrease of almost 3 years in age of onset. Seemingly in contrast to these findings, Fox et al. (2013a) found that women with more than the cohort median number of months pregnant had lower AD risk compared to women who had below median number of months pregnant. Parity may be associated differently with AD progression/onset versus risk. A number of studies of find associations of parity with age of onset but not incidence of dementia depending on genetic risk, suggesting that reproductive experience may have a particular impact on the progression of already-occurring pathological brain aging (discussed further). Fox et al. (2013a) estimated the relative risk for AD across different ages by back-extrapolating age of onset from Dementia Rating Scale scores at the time of interview. However, if parity alters progression of AD, there may be systematic errors in extrapolated age of onset and thus calculation of risk.

Parity is also associated with increased Alzheimer's-like neuropathology with aging (Beeri et al., 2009; Chan et al., 2012). Beeri et al. (2009) found that increasing parity was significantly associated with increased Alzheimer's-like neuropathology using a composite measure of neurofibrillary tangle and neuritic plaque severity. Importantly, the correlations held for women but not men, and were significant after partialling out age at death and controlling for level of antemortem dementia as measured by a Clinical Dementia Rating (CDR) scale. This study oversampled individuals with a diagnosis of AD (84% of the sample consisted of individuals diagnosed with definite or probable AD, whereas 8% were categorized as having a normal brain by the Consortium to Establish a Registry for Alzheimer's Disease neuropathological battery). Consequently, the average CDR score was in the moderate to severely impaired range. While these studies identify a relationship of reproductive experience with AD incidence, progression, and brain pathology, their correlational and retrospective nature prohibits inferences of causation. For example, it is likely that the AD groups in these studies had greater pre-existing risk (e.g., genetic) for AD compared with healthy controls. The observed relationship between AD onset and parity may then, to a degree, represent the effects of reproductive experience within the context of pre-existing risk for pathological aging.

A new development in this field concerns fetal microchimerism, the transplantation of fetal cells to the mother during pregnancy (Chan et al., 2012). Male fetal microchimerism has been detected in peripheral tissues decades after pregnancy (Bianchi et al., 1996; Khosrotehrani et al., 2004) and is widely distributed in the human female brain including the hippocampus (Chan et al., 2012; Zeng et al., 2010). Interestingly, AD brains have a lower concentration of male microchimerism than healthy brains (Chan et al., 2012). It is currently unknown how or whether these cells of fetal origin might relate to maternal dementia or cognition (Chan and Nelson, 2013). However, fetal microchimerisms in mice have been found to have the capacity to differentiate and mature into neurons in much the same way adult neurogenesis occurs (Zeng et al., 2010). Furthermore the possibility that carrying a male fetus may somehow protect against AD is consistent with Vanston and Watson (2005) who found that being pregnant with a boy improves working memory compared with being pregnant with a girl. All in all, these findings suggest that not only parity, the amount of parity, and age of first pregnancy, but also the sex of children may play a role in AD and healthy brain aging in women.

## 6.2 Reproductive experience interacts with genetic susceptibility for AD

Possessing the *APOE4* allele is a well-established genetic risk factor for susceptibility to nonfamilial AD (Ashford, 2004), a lower age of onset and higher risk of AD (Blacker et al., 1997; Kamboh, 2004; Meyer et al., 1998). Corbo et al. (2001) found that parity is associated with lower age of onset of AD in women who have the alleles *APOE2* and *3,* but not the *APOE4* allele; however, this latter finding may be due to a floor effect. The majority of women possess the *APOE2* and *3* alleles that Corbo et al. (2001) found interact with previous parity (Corbo and Scacchi, 1999), suggesting that for many women, reproductive experience alters subsequent risk for AD to develop at an earlier age. Corbo et al. (2009) subsequently analyzed single-nucleotide polymorphisms in another suspected genetic risk factor for AD, *CYP19*. This gene encodes aromatase, an enzyme involved in the conversion of testosterone into estradiol, and thus may moderate estradiol levels. They found that age of onset (but not risk) for AD was significantly associated with *CYP19* variations in parous women. The polymorphism that associated with an earlier age of onset, rs4646, has been associated with variation in estradiol levels, with the TT genotype related to lower estradiol levels (Haiman et al., 2003). Interestingly, among nulliparous women, mean differences favored an older age at onset with this genotype (Corbo et al. 2009). Corbo et al. (2009) hypothesize that nulliparity buffers against effects of the TT rs4646 genotype due to higher endogenous estradiol. Alternatively, parity-related alterations in sensitivity to estrogens (and possibly therefore to deficits in estrogens) might explain these findings.

It is worth noting that the studies mentioned previously found that the relationship of parity to AD diagnosis occurred independently of more direct measures of ovarian hormone exposure. For example, Colucci et al. (2006) found that the AD group did not differ from the control group in age at menarche and menopause or reproductive period. In their regression model predicting AD risk that included age, education, and age at menopause, Sobow and Kloszewska (2004) found that the number of pregnancies independently predicted AD risk more strongly than age at menopause alone. Indeed, reproductive factors over the lifespan may themselves have different effects on AD risk depending on genetic background: Geerlings et al. (2003) found in a prospective study that longer reproductive periods increased the risk of developing dementia particularly for women with the *APOE4* genotype. Breastfeeding is associated with decreased risk of AD in women without a parent or sibling with dementia (Fox et al., 2013b) but does not confer significant benefit to women with a parent or sibling who had dementia. This is curious as breastfeeding duration was found to be negatively associated with cognition outside the context of AD (Hesson, 2012; Heys et al., 2011). These studies suggest that reproductive factors interact with pre-existing genetic risk for AD. Therefore, although cumulative hormone exposure remains a plausible contributor to AD, it is not a logical conclusion of the above-discussed studies that parity affects brain aging solely through this mechanism.

These studies suggest that reproductive experience may alter the trajectory of cognitive decline differently depending on genetic background. Normal cognitive aging may differ qualitatively from the declines seen in neurodegenerative disorders including Alzheimer's disease. In the context of genetic risk for pathological brain aging, motherhood may have a negative effect on cognition during aging. In contrast, reproductive experience may positively affect cognition in the absence of neurodegenerative processes. Animal studies are required to further investigate this possibility. Though human studies of the genetic interaction with parity help to refine understanding of the nature of the relationship between parity and AD progression, they do not address the directionality of this association. For example, *ApoE4* allele status is related to reduced fertility (Corbo et al., 2008). Several animal models have been used to experimentally study the neuropathology and cognitive impairments of AD. Cui et al. (2013) found that 12-month-old wildtype breeder mice that had had one or more litters outperformed age-matched nonbreeders in multiple cognitive tests. However, this effect reversed direction in *APP23* mice, as *APP23* breeders were impaired compared to nonbreeders on tests of spatial reference and working memory and spontaneous alternation. By 12 months of age, *APP23* breeders also developed larger and more numerous amyloid plaques compared to nonbreeders in the hippocampus and cortex. These findings are consistent with an interaction of genetic risk with parity. They also align with the findings of Beeri et al. (2009) that within women with AD, motherhood exerts negative effects on brain pathology.

## 7 CONCLUSIONS

Pregnancy and the postpartum period are times of maximal plasticity in both behavior and in the maternal brain. Pregnancy and mothering are associated with time- and parity-dependent changes in cognition and hippocampal neurogenesis, morphology, and electrophysiology both in the short and long term. The endocrine correlates of these changes include alterations in cumulative estrogen exposure over the lifespan and increased sensitivity to estradiol. The cognitive deficits of late pregnancy and the early postpartum period give way to modest improvements in verbal and spatial learning and memory in the late postpartum and during lactation in the mother. The cognitive effects of reproductive experience appear to be cumulative and enduring, as reproductive experience alters learning and memory in rats long after weaning in a parity-dependent fashion. Animal models have found domain-specific cognitive benefits of reproductive experience even into reproductive senescence, but there are indications that these benefits may decline or even reverse in old age. Possibly consistent with this observation, investigations in humans have found negative or no effects of reproductive experience on cognition in older age. Age at first pregnancy may moderate cognitive effects of reproductive experience on cognitive outcomes in later life, as well as the sex ratio of offspring, but these factors are not adequately studied. Reproductive experience is associated with higher prevalence of AD, earlier age at onset, and increased AD neuropathology in observational studies. Further research is needed to determine whether reproductive experience alters risk for AD or instead modifies the progression of the disorder. Importantly, a number of studies indicate that reproductive factors interact with genetic risk for AD, as parity interacts with genetic risk factors and with factors that alter estrogen exposure. Therefore, the trajectory of cognitive decline may be altered differently by reproductive experience depending on whether pathological or normal aging is occurring simultaneously. In conclusion, reproductive experience has long-lasting effects on brain health in both rodents and humans, and further studies are needed to fully characterize the impact of reproductive experience in aging women.

## REFERENCES


Adams, M.M., Fink, S.E., Shah, R.A., Janssen, W.G., Hayashi, S., Milner, T.A., McEwen, B.S., Morrison, J.H., 2002. Estrogen and aging affect the subcellular distribution of estrogen receptor-alpha in the hippocampus of female rats. J. Neurosci. 22, 3608–3614.

Allen, S.J., Dawbarn, D., 2006. Clinical relevance of the neurotrophins and their receptors. Clin. Sci. (Lond.) 110, 175–195.

Andersen, K., Launer, L.J., Dewey, M.E., Letenneur, L., Ott, A., Copeland, J.R.M., Dartigues, J-F., Kragh-Sorensen, P., Baldereschi, M., Brayne, C., Lobo, A., Martinez-Lage, J.M., Stijnen, T., Hofman, A., 1999. Gender differences in the incidence of AD and vascular dementia: the EURODEM Studies. EURODEM Incidence Research Group. Neurology 53, 1992–1997.

Anderson, S.M., Saviolakis, G.A., Bauman, R.A., Chu, K.Y., Ghosh, S., Kant, G.J., 1996. Effects of chronic stress on food acquisition, plasma hormones, and the estrous cycle of female rats. Physiol. Behav. 60, 325–329.

Apostolova, L.G., Mosconi, L., Thompson, P.M., Green, A.E., Hwang, K.S., Ramirez, A., Mistur, R., Tsui, W.H., de Leon, M.J., 2010. Subregional hippocampal atrophy predicts Alzheimer's dementia in the cognitively normal. Neurobiol. Aging 31, 1077–1088.

Ashburner, J., Csernansky, J.G., Davatzikos, C., Fox, N.C., Frisoni, G.B., Thompson, P.M., 2003. Computer-assisted imaging to assess brain structure in healthy and diseased brains. Lancet Neurol. 2, 79–88.

Ashford, J.W., 2004. APOE genotype effects on Alzheimer's disease onset and epidemiology. J. Mol. Neurosci. 23, 157–165.

Atienza, M., Atalaia-Silva, K.C., Gonzalez-Escamilla, G., Gil-Neciga, E., Suarez-Gonzalez, A., Cantero, J.L., 2011. Associative memory deficits in mild cognitive impairment: the role of hippocampal formation. Neuroimage 57, 1331–1342.

Banasr, M., Hery, M., Brezun, J.M., Daszuta, A., 2001. Serotonin mediates oestrogen stimulation of cell proliferation in the adult dentate gyrus. Eur. J. Neurosci. 14, 1417–1424.

Barha, C.K., Galea, L.A., 2010. Influence of different estrogens on neuroplasticity and cognition in the hippocampus. Biochim. Biophys. Acta 1800, 1056–1067.

Barha, C.K., Galea, L.A.M., 2011. Motherhood alters the cellular response to estrogens in the hippocampus later in life. Neurobiol. Aging 32, 2091–2095.

Barha, C.K., Barker, J.M., Brummelte, S., Epp, J.R., Galea, L.A.M., 2009a. Hormone regulation of adult hippocampal neurogenesis in the mammalian brain. In: Pfaff, D.W., Arnold, A.P., Etgen, A.M., Fahrbach, S.E., Rubin, T.R. (Eds.), Hormones Brain and Behaviour. Academic Press, San Diego, pp. 2165–2197.

Barha, C.K., Lieblich, S.E., Galea, L.A., 2009b. Different forms of oestrogen rapidly upregulate cell proliferation in the dentate gyrus of adult female rats. J. Neuroendocrinol. 21, 155–166.

Barha, C.K., Lieblich, S.E., Chow, C.; Galea, L.A.M. (in press). Multiparity induced enhancement of hippocampal neurogenesis and spatial memory depends on ovarian hormone status in middle age. Neurobiol. Aging doi:10.1016/j.neurobiolaging.2015.04.007.

Barker, G.R., Warburton, E.C., 2011. When is the hippocampus involved in recognition memory? J. Neurosci. 31, 10721–10731.

Beeri, M.S., Rapp, M., Schmeidler, J., Reichenberg, A., Purohit, D.P., Perl, D.P., Grossman, H.T., Prohovnik, I., Haroutunian, V., Silverman, J.M., 2009. Number of children is associated with neuropathology of Alzheimer's disease in women. Neurobiol. Aging 30, 1184–1191.

Bianchi, D.W., Zickwolf, G.K., Weil, G.J., Sylvester, S., DeMaria, M.A., 1996. Male fetal progenitor cells persist in maternal blood for as long as 27 years postpartum. Proc. Natl. Acad. Sci. USA 93, 705–708.

Blacker, D., Haines, J.L., Rodes, L., Terwedow, H., Go, R.C.P., Harrell, L.E., Perry, R.T., Bassett, S.S., Chase, G., Meyers, D., Albert, M.S., Tanzi, R., 1997. ApoE-4 and age at onset of Alzheimer's disease. Neurology 48, 139–147.

Bliss, T.V., Collingridge, G.L., 1993. A synaptic model of memory: long-term potentiation in the hippocampus. Nature 361, 31–39.

Bobinski, M., de Leon, M.J., Tarnawski, M., Wegiel, J., Reisberg, B., Miller, D.C., Wisniewski, H.M., 1998. Neuronal and volume loss in CA1 of the hippocampal formation uniquely predicts duration and severity of Alzheimer disease. Brain Res. 805, 267–269.

Bonfanti, L., 2006. PSA–NCAM in mammalian structural plasticity and neurogenesis. Prog. Neurobiol. 80, 129–164.
Brett, M., Baxendale, S., 2001. Motherhood and memory: a review. Psychoneuroendocrinology 26, 339–362.
Bridges, R.S., Byrnes, E.M., 2006. Reproductive experience reduces circulating 17 beta-estradiol and prolactin levels during proestrus and alters estrogen sensitivity in female rats. Endocrinology 147, 2575–2582.
Bridges, R.S., Felicio, L.F., Pellerin, L.J., Stuer, A.M., Mann, P.E., 1993. Prior parity reduces post-coital diurnal and nocturnal prolactin surges in rats. Life Sci. 53, 439–445.
Brummelte, S., Galea, L.A., 2010. Chronic high corticosterone reduces neurogenesis in the dentate gyrus of adult male and female rats. Neuroscience 168, 680–690.
Buckwalter, G., 1999. Pregnancy, the postpartum, and steroid hormones: effects on cognition and mood. Psychoneuroendocrinology 24, 69–84.
Buckwalter, J.G., Buckwalter, D.K., Bluestein, B.W., Stanczyk, F.Z., 2001. Pregnancy and post partum: changes in cognition and mood. Prog. Brain Res. 133, 303–319.
Burgess, N., 2008. Spatial cognition and the brain. Ann. NY Acad. Sci. 1124, 77–97.
Byrnes, E.M., Babb, J.A., Bridges, R.S., 2009. Differential expression of oestrogen receptor alpha following reproductive experience in young and middle-aged female rats. J. Neuroendocrinol. 21, 550–557.
Byrnes, E.M., Casey, K., Carini, L.M., Bridges, R.S., 2013. Reproductive experience alters neural and behavioural responses to acute oestrogen receptor alpha activation. J. Neuroendocrinol. 25, 1280–1289.
Cameron, H.A., Woolley, C.S., McEwen, B.S., Gould, E., 1993. Differentiation of newly born neurons and glia in the dentate gyrus of the adult rat. Neuroscience 56, 337–344.
Carroll, J.C., Rosario, E.R., Chang, L., Stanczyk, F.Z., Oddo, S., LaFerla, F.M., Pike, C.J., 2007. Progesterone and estrogen regulate Alzheimer-like neuropathology in female 3xTg-AD mice. J. Neurosci. 27, 13357–13365.
Casey, M.L., MacDonald, P.C., Sargent, I.L., Starkey, P.M., 1993. Placental Endocrinology. Blackwell Scientific Publications, Oxford.
Chan, W.F., Nelson, J.L., 2013. Microchimerism in the human brain: more questions than answers. Chimerism 4, 32–33.
Chan, W.F., Gurnot, C., Montine, T.J., Sonnen, J.A., Guthrie, K.A., Nelson, J.L., 2012. Male microchimerism in the human female brain. PLoS One 7, e45592.
Chen, Q., Nakajima, A., Choi, S.H., Xiong, X., Sisodia, S.S., Tang, Y.P., 2008. Adult neurogenesis is functionally associated with AD-like neurodegeneration. Neurobiol. Dis. 29, 316–326.
Chiba, S., Suzuki, M., Yamanouchi, K., Nishihara, M., 2007. Involvement of granulin in estrogen-induced neurogenesis in the adult rat hippocampus. J. Reprod. Dev. 53, 297–307.
Christensen, H., Leach, L.S., Mackinnon, A., 2010. Cognition in pregnancy and motherhood: prospective cohort study. Br. J. Psychiatry 196, 126–132.
Clelland, C.D., Choi, M., Romberg, C., Clemenson, G.D., Fragniere, A., Tyers, P., Jessberger, S., Saksida, L.M., Barker, R.A., Gage, F.H., Bussey, T.J., 2009. A functional role for adult hippocampal neurogenesis in spatial pattern separation. Science 325, 210–213.
Colucci, M., Cammarata, S., Assini, A., Croce, R., Clerici, F., Novello, C., Mazzella, L., Dagnino, N., Mariani, C., Tanganelli, P., 2006. The number of pregnancies is a risk factor for Alzheimer's disease. Eur. J. Neurol. 13 (12), 1374–1377.
Corbo, R.M., Gambina, G., Ulizzi, L., Monini, P., Broggio, E., Rosano, A., Scacchi, R., 2001. Combined effect of apolipoprotein e genotype and past fertility on age at onset of Alzheimer's disease in women. Dement. Geriatr. Cogn. Disord. 2007, 24, 32–35.
Corbo, R.M., Scacchi, R., 1999. Apolipoprotein E (APOE) allele distribution in the world. Is APOE*4 a "thrifty" allele? Ann. Hum. Genet. 63, 201–310.
Corbo, M., Ulizzi, L., Piombo, L., Sacchi, R., 2008. Study on a possible effect of four longevity candidate genes (ACE, PON1, PPAR-gamma, and APOE) on human fertility. Biogerontology 9, 317–323.
Corbo, R.M., Gambina, G., Ulizzi, L., Moretto, G., Scacchi, R., 2009. Genetic variation of CYP19 (aromatase) gene influences age at onset of Alzheimer's disease in women. Dement. Geriatr. Cogn. Disord. 2009, 27, 513–518.

Cost, K.T., Lobell, T.D., Williams-Yee, Z.N., Henderson, S., Dohanich, G., 2014. The effects of pregnancy, lactation, and primiparity on object-in-place memory of female rats. Horm. Behav. 65, 32–39.
Crawley, R.A., Dennison, K., Carter, C., 2003. Cognition in pregnancy and the first year post-partum. Psychol. Psychother. 76, 69–84.
Cui, J., Jothishankar, B., He, P., Staufenbiel, M., Shen, Y., Li, R., 2014. Amyloid precursor protein mutation disrupts reproductive experience-enhanced normal cognitive development in a mouse model of Alzheimer's disease. Mol. Neurobiol. 49 (1), 103–112.
Cuttler, C., Graf, P., Pawluski, J.L., Galea, L.A.M., 2011. Everyday life memory deficits in pregnant women. Can. J. Exp. Psychol. 65, 27.
Darnaudery, M., Perez-Martin, M., Del Favero, F., Gomez-Roldan, C., Garcia-Segura, L.M., Maccari, S., 2007. Early motherhood in rats is associated with a modification of hippocampal function. Psychoneuroendocrinology 32, 803–812.
de Groot, R.H., Hornstra, G., Roozendaal, N., Jolles, J., 2003. Memory performance, but not information processing speed, may be reduced during early pregnancy. J. Clin. Exp. Neuropsychol. 25, 482–488.
de Groot, R.H., Vuurman, E.F., Hornstra, G., Jolles, J., 2006. Differences in cognitive performance during pregnancy and early motherhood. Psychol. Med. 36, 1023–1032.
Deng, W., Aimone, J.B., Gage, F.H., 2010. New neurons and new memories: how does adult hippocampal neurogenesis affect learning and memory? Nat. Rev. Neurosci. 11, 339–350.
Dorgan, J.F., Reichman, M.E., Judd, J.T., Brown, C., Longcope, C., Schatzkin, A., Campbell, W.S., Franz, C., Kahle, L., Taylor, P.R., 1995. Relationships of age and reproductive characteristics with plasma estrogens and androgens in premenopausal women. Cancer Epidemiol. Biomarkers Prev. 4, 381–386.
Eidelman, A.I., Hoffman, N.W., Kaitz, M., 1993. Cognitive deficits in women after childbirth. Obstet. Gynecol. 81, 764–767.
Ekonomou, A., Savva, G.M., Brayne, C., Forster, G.F., P.T., Johnson M.P., E. K., Attems, J., Somani, A., Minger, SLB., C.G., 2014. Stage-specific changes in neurogenic and glial markers in Alzheimer's disease. Biol. Psychiatr. 8 (77), 711–719.
Epp, J.R., Haack, A.K., Galea, L.A., 2011. Activation and survival of immature neurons in the dentate gyrus with spatial memory is dependent on time of exposure to spatial learning and age of cells at examination. Neurobiol. Learn. Mem. 95, 316–325.
Fleming, A.S., O'Day, D.H., Kraemer, G.W., 1999. Neurobiology of mother-infant interactions: experience and central nervous system plasticity across development and generations. Neurosci. Biobehav. Rev. 23, 673–685.
Fox, M., Berzuini, C., Knapp, L.A., 2013a. Cumulative estrogen exposure, number of menstrual cycles, and Alzheimer's risk in a cohort of British women. Psychoneuroendocrinology 38, 2973–2982.
Fox, M., Berzuini, C., Knapp, L.A., 2013b. Maternal breastfeeding history and Alzheimer's disease risk. J. Alzheimers Dis. 37, 809–821.
Freeman, E.W., Weinstock, L., Rickels, K., Sondheimer, S.J., Coutifaris, C., 1992. A placebo-controlled study of effects of oral progesterone on performance and mood. Br. J. Clin. Pharmacol. 33, 293–298.
Freeman, E.W., Purdy, R.H., Coutifaris, C., Rickels, K., Paul, S.M., 1993. Anxiolytic metabolites of progesterone: correlation with mood and performance measures following oral progesterone administration to healthy female volunteers. Neuroendocrinology 58, 478–484.
Furuta, M., Bridges, R.S., 2005. Gestation-induced cell proliferation in the rat brain. Devel. Brain Res. 156, 61–66.
Galea, L.A., 2008. Gonadal hormone modulation of neurogenesis in the dentate gyrus of adult male and female rodents. Brain Res. Rev. 57, 332–341.
Galea, L.A., Ormerod, B.K., Sampath, S., Kostaras, X., Wilkie, D.M., Phelps, M.T., 2000. Spatial working memory and hippocampal size across pregnancy in rats. Horm. Behav. 37, 86–95.
Galea, L.A., Wainwright, S.R., Roes, M.M., Duarte-Guterman, P., Chow, C., Hamson, D.K., 2013. Sex, hormones and neurogenesis in the hippocampus: hormonal modulation of neurogenesis and potential functional implications. J. Neuroendocrinol. 25, 1039–1061.
Gatewood, J.D., Morgan, M.D., Eaton, M., McNamara, I.M., Stevens, L.F., Macbeth, A.H., Meyer, E.A.A., Lomas, L.M., Kozub, F.J., Lambert, K.G., Kinsley, C.H., 2005. Motherhood mitigates aging-related decrements in learning and memory an positively affects brain aging in the rat. Brain Res. Bull. 66, 91–98.

Geerlings, M.I., Launer, L.J., de Jong, F.H., Ruitenberg, A., Stijnen, T., van Swieten, J.C., Hofman, A., Witteman, J.C., Pols, H.A., Breteler, M.M., 2003. Endogenous estradiol and risk of dementia in women and men: the Rotterdam Study. Ann. Neurol. 53, 607–615.

Glynn, L.M., 2010. Giving birth to a new brain: hormone exposures of pregnancy influence human memory. Psychoneuroendocrinology 35, 1148–1155.

Glynn, L.M., 2012. Increasing parity is associated with cumulative effects on memory. J. Womens Health 21, 1038–1045.

Gold, E.B., Bromberger, J., Crawford, S., Samuels, S., Greendale, G.A., Harlow, S.D., Skurnick, J., 2001. Factors associated with age at natural menopause in a multiethnic sample of midlife women. Am. J. Epidemiol. 153, 865–874.

Haiman, C.A., Stram, D.O., Pike, M.C., Kolonel, L.N., Burtt, N.P., et al., 2003. A comprehensive haplotype analysis of CYP19 and breast cancer risk: the Multiethnic Cohort. Hum. Mol. Genet. 12 (20), 2679–2692.

Hampson, E., 1990. Variations in sex-related cognitive abilities across the menstrual cycle. Brain Cogn. 14, 26–43.

Hamson, D.K., Wainwright, S.R., Taylor, J.R., Jones, B.A., Watson, N.V., Galea, L.A., 2013. Androgens increase survival of adult-born neurons in the dentate gyrus by an androgen receptor-dependent mechanism in male rats. Endocrinology 154, 3294–3304.

Harburger, L.L., Bennett, J.C., Frick, K.M., 2007. Effects of estrogen and progesterone on spatial memory consolidation in aged females. Neurobiol. Aging 28, 602–610.

Henry, J.D., Rendell, P.G., 2007. A review of the impact of pregnancy on memory function. J. Clin. Exp. Neuropsychol. 29, 793–803.

Henry, J.F., Sherwin, B.B., 2012. Hormones and cognitive functioning during late pregnancy and postpartum: a longitudinal study. Behav. Neurosci. 126, 73.

Hesson, J., 2012. Cumulative estrogen exposure and prospective memory in older women. Brain Cogn. 80, 89–95.

Heys, M., Jiang, C., Cheng, K.K., Zhang, W., Au Yeung, S.L., Lam, T.H., Leung, G.M., Schooling, C.M., 2011. Life long endogenous estrogen exposure and later adulthood cognitive function in a population of naturally postmenopausal women from Southern China: the Guangzhou Biobank Cohort Study. Psychoneuroendocrinology 36, 864–873.

Hillerer, K.M., Neumann, I.D., Couillard-Despres, S., Aigner, L., Slattery, D.A., 2014. Lactation-induced reduction in hippocampal neurogenesis is reversed by repeated stress exposure. Hippocampus 24, 673–683.

Holmes, M.M., Wide, J.K., Galea, L.A., 2002. Low levels of estradiol facilitate, whereas high levels of estradiol impair, working memory performance on the radial arm maze. Behav. Neurosci. 116, 928–934.

Hy, L.X., Keller, D.M., 2000. Prevalence of AD among whites: a summary by level of severity. Neurology 52 (2), 198–204.

Jin, K., Peel, A.L., Mao, X.O., Xie, L., Cottrell, B.A., Henshall, D.C., Greenberg, D.A., 2004. Increased hippocampal neurogenesis in Alzheimer's disease. Proc. Natl. Acad. Sci. USA 101, 343–347.

Kalache, A., Maguire, A., Thompson, S.G., 1993. Age at last full-term pregnancy and risk of breast cancer. Lancet 341, 33–36.

Kamboh, M.I., 2004. Molecular genetics of late-onset Alzheimer's disease. Ann. Hum. Genet. 68, 381–404.

Kandel, E.R., Dudai, Y., Mayford, M.R., 2014. The molecular and systems biology of memory. Cell 157, 163–186.

Kee, N., Teixeira, C.M., Wang, A.H., Frankland, P.W., 2007. Preferential incorporation of adult-generated granule cells into spatial memory networks in the dentate gyrus. Nat. Neurosci. 10, 355–362.

Keenan, P.A., Yaldoo, D.T., Stress, M.E., Fuerst, D.R., Ginsburg, K.A., 1998. Explicit memory in pregnant women. Am. J. Obstet. Gynecol. 179, 731–737.

Key, T.J., Chen, J., Wang, D.Y., Pike, M.C., Boreham, J., 1990. Sex hormones in women in rural China and in Britain. Br. J. Cancer 62, 631–666.

Khosrotehrani, K., Johnson, K.L., Cha, D.H., Salomon, R.N., Bianchi, D.W., 2004. Transfer of fetal cells with multilineage potential to maternal tissue. JAMA 292, 75–80.

Kinsley, C.H., Madonia, L., Gifford, G.W., Tureski, K., Griffin, G.R., Lowry, C., Williams, J., Collins, J., McLearie, H., Lambert, K.G., 1999. Motherhood improves learning and memory. Nature 402, 137–138.

Kinsley, C.H., Trainer, R., Stafisso-Sandoz, G., Quadros, P., Marcus, L.K., Hearon, C., Meyer, E.A.A., Hester, N., Morgan, M., Kozub, F.J., Lambert, K.G., 2006. Motherhood and the hormones of pregnancy modify concentrations of hippocampal neuronal dendritic spines. Horm. Behav. 49, 131–142.

Kirsch, P., Esslinger, C., Chen, Q., Mier, D., Lis, S., Siddhanti, S., Gruppe, H., Mattay, V.S., Gallhofer, B., Meyer-Lindenberg, A., 2005. Oxytocin modulates neural circuitry for social cognition and fear in humans. J. Neurosci. 25, 11489–11493.

Lambert, K.G., Berry, A.E., Griffins, G., Amory-Meyers, E., Madonia-Lomas, L., Love, G., Kinsley, C.H., 2005. Pup exposure differentially enhances foraging ability in primiparous and nulliparous rats. Physiol. Behav. 84, 799–806.

Larsen, C.M., Grattan, D.R., 2010. Prolactin-induced mitogenesis in the subventricular zone of the maternal brain during early pregnancy is essential for normal postpartum behavioral responses in the mother. Endocrinology 151, 3805–3814.

Lemaire, V., Billard, J.M., Dutar, P., George, O., Piazza, P.V., Epelbaum, J., Le Moal, M., Mayo, W., 2006. Motherhood-induced memory improvement persists across lifespan in rats but is abolished by a gestational stress. Eur. J. Neurosci. 23, 3368–3374.

Letenneur, L., Commenges, D., Dartigues, J.F., Barberger-Gateau, P., 1994. Incidence of dementia and Alzheimer's disease in elderly community residents of South-Western France. Int. J. Epidemiol. 23, 1256–1261.

Leuner, B., Mirescu, C., Noiman, L., Gould, E., 2007. Maternal experience inhibits the production of immature neurons in the hippocampus during the postpartum period through elevations in adrenal steroids. Hippocampus 17, 434–442.

Lévy, F., Gheusi, G., Keller, M., 2011. Plasticity of the parental brain: a case for neurogenesis. J. Neuroendocrinol. 23, 984–993.

Li, R., Cui, J., Jothishankar, B., Shen, J., He, P., Shen, Y., 2013. Early reproductive experiences in females make differences in cognitive function later in life. J. Alzheimers Dis. 34, 589–594.

Lin, J., Kroenke, C.H., Epel, E., Kenna, H.A., Wolkowitz, O.M., Blackburn, E., Rasgon, N.L., 2011. Greater endogenous estrogen exposure is associated with longer telomeres in postmenopausal women at risk for cognitive decline. Brain Res. 1379, 224–231.

Lopez-Toledano, M.A., Ali Faghihi, M., Patel, N.S., Wahlestedt, C., 2010. Adult neurogenesis: a potential tool for early diagnosis in Alzheimer's disease? J. Alzheimers Dis. 20, 395–408.

Love, G., Torrey, N., McNamara, I., Morgan, M., Banks, M., Hester, N.W., Glasper, E.R., DeVries, A.C., Kinsley, C.H., Lambert, K.G., 2005. Maternal experience produces long-lasting behavioral modifications in the rat. Behav. Neurosci. 119, 1084–1096.

Lupien, S.J., McEwen, B.S., Gunnar, M.R., Heim, C., 2009. Effects of stress throughout the lifespan on the brain, behaviour and cognition. Nat. Rev. Neurosci. 10, 434–445.

Macbeth, A.H., Scharfman, H.E., MacLusk, N.J., Gautreaux, C., Luine, V.N., 2008. Effects of multiparity on recognition memory, monoaminergic neurotransmitters, and brain-derived neurotrophic factor (BDNF). Horm. Behav. 54, 7–17.

Maren, S., 2001. Neurobiology of Pavlovian fear conditioning. Annu. Rev. Neurosci. 24, 897–931.

McEwen, B.S., 1994. Corticosteroids and hippocampal plasticity. Ann. NY Acad. Sci. 746, 134–142, discussion 142–144, 178–179.

McLay, R.N., Maki, P.M., Lyketsos, C.G., 2003. Nulliparity and late menopause are associated with decreased cognitive decline. J. Neuropsychiat. Clin. Neurosci. 15, 161–167.

McNeilly, A.S., 2001. Neuroendocrine changes and fertility in breast-feeding women. Prog. Brain Res. 133, 207–214.

Meyer, M.R., Tschanz, J.T., Norton, M.C., Welsh-Bohmer, K.A., Steffens, D.C., Wyse, B.W., Breitner, J.C.S., 1998. APOE genotype predicts when – not whether – one is predisposed to develop Alzheimer disease. Nat. Genet. 19, 321–322.

Ming, G.I., Song, H., 2005. Adult neurogenesis in the mammalian central nervous system. Ann. Rev. Neurosci. 28, 223–250.

Mirescu, C., Gould, E., 2006. Stress and adult neurogenesis. Hippocampus 16, 233–238.

Numan, M., 2006. Hypothalamic neural circuits regulating maternal responsiveness toward infants. Behav. Cogn. Neurosci. Rev. 5, 163–190.

Oatridge, A., Holdcroft, A., Saeed, N., Hajnal, J.V., Puri, B.K., Fusi, L., Bydder, G.M., 2002. Change in brain size during and after pregnancy: study in healthy women and women with preeclampsia. Am. J. Neuroradiol. 23, 19–26.

Ormerod, B.K., Galea, L.A., 2001. Reproductive status influences cell proliferation and cell survival in the dentate gyrus of adult female meadow voles: a possible regulatory role for estradiol. Neuroscience 102, 369–379.

Ormerod, B.K., Lee, T.T., Galea, L.A., 2003. Estradiol initially enhances but subsequently suppresses (via adrenal steroids) granule cell proliferation in the dentate gyrus of adult female rats. J. Neurobiol. 55, 247–260.

Paris, J.J., Frye, C.A., 2008. Estrous cycle, pregnancy, and parity enhance performance of rats in object recognition or object placement tasks. Reproduction 136, 105–115.

Pawluski, J.L., Galea, L.A.M., 2005. Hippocampal morphology is differentially affected by reproductive experience in the mother. J. Neurobiol. 66, 71–81.

Pawluski, J.L., Galea, L.A.M., 2007. Reproductive experience alters hippocampal neurogenesis during the postpartum period in the dam. Neuroscience 149, 53–67.

Pawluski, J.L., Vanderbyl, B.L., Ragan, K., Galea, L.A.M., 2006a. First reproductive experience persistently affects spatial reference and working memory in the mother and these effects are not due to pregnancy or "mothering" alone. Behav. Brain Res. 175, 157–165.

Pawluski, J.L., Walker, S.K., Galea, L.A., 2006b. Reproductive experience differentially affects spatial reference and working memory performance in the mother. Horm. Behav. 49, 143–149.

Pawluski, J.L., Brummelte, S., Barha, C.K., Crozier, T.M., Galea, L.A., 2009. Effects of steroid hormones on neurogenesis in the hippocampus of the adult female rodent during the estrous cycle, pregnancy, lactation and aging. Front. Neuroendocrinol. 30, 343–357.

Pawluski, J.L., Barakauskas, V.E., Galea, L.A., 2010. Pregnancy decreases oestrogen receptor alpha expression and pyknosis, but not cell proliferation or survival, in the hippocampus. J. Neuroendocrinol. 22, 248–257.

Perry, E.K., Johnson, M., Ekonomou, A., Perry, R.H., Ballard, C., Attems, J., 2012. Neurogenic abnormalities in Alzheimer's disease differ between stages of neurogenesis and are partly related to cholinergic pathology. Neurobiol. Dis. 47, 155–162.

Poser, C.M., Kassirer, M.R., Peyser, J.M., 1986. Benign encephalopathy of pregnancy. Preliminary clinical observations. Acta Neurol. Scand. 73, 39–43.

Ptok, U., Barkow, K., Heun, R., 2002. Fertility and number of children in patients with Alzheimer's disease. Arch. Womens Ment. Health 5, 83–86.

Rasgon, N.L., Magnusson, C., Johansson, A.L.V., Pedersen, N.L., Elman, S., Gatz, M., 2005. Endogenous and exogenous hormone exposure and risk of cognitive impairment in Swedish twins: a preliminary study. Psychoneuroendocrinology 30, 558–567.

Rolls, A., Schori, H., London, A., Schwartz, M., 2008. Decrease in hippocampal neurogenesis during pregnancy: a link to immunity. Mol Psychiatr. England 13, 468–469.

Rosenblatt, J.S., 1988. Psychoneuroendocrine aspects of maternal behavior. Psychoneuroendocrinology 13, 3–212.

Rosenblatt, J.S., Mayer, A.D., Giordano, A.L., 1988. Hormonal basis during pregnancy for the onset of maternal behavior in the rat. Psychoneuroendocrinology 13, 29–46.

Ryan, J., Scali, J., Carriere, I., Ritchie, K., Ancelin, M.L., 2008. Hormonal treatment, mild cognitive impairment and Alzheimer's disease. Int. Psychogeriatr. 20, 47–56.

Ryan, J., Carriere, I., Scali, J., Ritchie, K., Ancelin, M.L., 2009. Life-time estrogen exposure and cognitive functioning in later life. Psychoneuroendocrinology 34, 287–298.

Scanlan, V.F., Byrnes, E.M., Bridges, R.S., 2006. Reproductive experience and activation of maternal memory. Behav. Neurosci. 120, 676–686.

Scharfman, H.E., MacLusky, N.J., 2006. Estrogen and brain-derived neurotrophic factor (BDNF) in hippocampus: complexity of steroid hormone-growth factor interactions in the adult CNS. Front. Neuroendocrinol. 27, 415–435.

Scheff, S.W., Price, D.A., Schmitt, F.A., DeKosky, S.T., Mufson, E.J., 2007. Synaptic alterations in CA1 in mild Alzheimer disease and mild cognitive impairment. Neurology 68, 1501–1508.

Serrano-Pozo, A., Frosch, M.P., Masliah, E., Hyman, B.T., 2011. Neuropathological alterations in Alzheimer disease. Cold Spring Harb. Perspect. Med. 1, a006189.

Sharp, K., Brindle, P.M., Brown, M.W., Turner, G.M., 1993. Memory loss during pregnancy. Br. J. Obstet. Gynaecol. 100, 209–215.

Shingo, T., Gregg, C., Enwere, E., Fujikawa, H., Hassam, R., Geary, C., Cross, J.C., Weiss, S., 2003. Pregnancy-stimulated neurogenesis in the adult female forebrain mediated by prolactin. Science 299, 117–120.

Silber, M., Almkvist, O., Larsson, B., Uvnäs-Moberg, K., 1990. Temporary peripartal impairment in memory and attention and its possible relation to oxytocin concentration. Life Sci. 47, 57–65.

Simic, G., Kostovic, I., Winblad, B., Bogdanovic, N., 1997. Volume and number of neurons of the human hippocampal formation in normal aging and Alzheimer's disease. J. Comp. Neurol. 379, 482–494.

Smith, C.A., McCleary, C.A., Murdock, G.A., Wilshire, T.W., Buckwalter, D.K., Bretsky, P., Marmol, L., Gorsuch, R.L., Buckwalter, J.G., 1999. Lifelong estrogen exposure and cognitive performance in elderly women. Brain Cogn. 39, 203–218.

Snyder, J.S., Soumier, A., Brewer, M., Pickel, J., Cameron, H.A., 2011. Adult hippocampal neurogenesis buffers stress responses and depressive behaviour. Nature 476, 458–461.

Sobow, T., Kloszewska, I., 2004. Parity, number of pregnancies, and the age of onset of Alzheimer's disease. J. Neuropsychiat. Clin. Neurosci. 16, 120–121.

Spritzer, M.D., Galea, L.A., 2007. Testosterone and dihydrotestosterone, but not estradiol, enhance survival of new hippocampal neurons in adult male rats. Dev. Neurobiol. 67, 1321–1333.

Stolzenberg, D.S., Stevens, J.S., Rissman, E.F., 2012. Experience-facilitated improvements in pup retrieval; evidence for an epigenetic effect. Horm. Behav. 62, 128–135.

Talboom, J.S., Williams, B.J., Baxley, E.R., West, S.G., Bimonte-Nelson, H.A., 2008. Higher levels of estradiol replacement correlate with better spatial memory in surgically menopausal young and middle-aged rats. Neurobiol. Learn. Mem. 90, 155–163.

Tierney, M.C., Ryan, J., Ancelin, M.L., Moineddin, R., Rankin, S., Yao, C., MacLusky, N.J., 2013. Lifelong estrogen exposure and memory in older postmenopausal women. J. Alzheimers Dis. 34, 601–608.

Tomizawa, K., Iga, N., Lu, Y.F., Moriwaki, A., Matsushita, M., Li, S.T., Miyamoto, O., Itano, T., Matsui, H., 2003. Oxytocin improves long-lasting spatial memory during motherhood through MAP kinase cascade. Nat. Neurosci. 6, 384–390.

Uban, K.A., Sliwowska, J.H., Lieblich, S., Ellis, L.A., Yu, W.K., Weinberg, J., Galea, L.A., 2010. Prenatal alcohol exposure reduces the proportion of newly produced neurons and glia in the dentate gyrus of the hippocampus in female rats. Horm. Behav. 58, 835–843.

Vanston, C.M., Watson, N.V., 2005. Selective and persistent effect of foetal sex on cognition in pregnant women. Neuroreport 16, 779–782.

Warren, S.G., Juraska, J.M., 1997. Spatial and nonspatial learning across the rat estrous cycle. Behav. Neurosci. 111, 259–266.

Weiland, N.G., Orikasa, C., Hayashi, S., McEwen, B.S., 1997. Distribution and hormone regulation of estrogen receptor immunoreactive cells in the hippocampus of male and female rats. J. Comp. Neurol. 388, 603–612.

Woodfield, R.L., 1984. Embedded Figures Test performance before and after childbirth. Br. J. Psychol. 75 (Pt 1), 81–88.

Zeng, X.X., Tan, K.H., Yeo, A., Sasajala, P., Tan, X., Xiao, Z.C., Dawe, G., Udolph, G., 2010. Pregnancy-associated progenitor cells differentiate and mature into neurons in the maternal brain. Stem Cells Dev. 19, 1819–1830.

Zimberknopf, E., Xavier, G.F., Kinsley, C.H., Felicio, L.F., 2011. Prior parity positively regulates learning and memory in young and middle-aged rats. Comp. Med. 61, 366–377.

CHAPTER 10

# Sex and the Developing Brain

**Jaclyn M. Schwarz**
University of Delaware, Department of Psychological and Brain Sciences, Newark, DE, USA

## 1 INTRODUCTION

Our sex, whether male or female, is one of the most important defining factors of ourselves because it influences our lives in so many different ways. Our sex is determined at the earliest moment of conception via a process of random chance that almost belies the importance it has on our physiology and behavior. From the moment a mother finds out that she is expecting a new baby, the first question every family member and friend asks is, "Is it a boy or a girl?" The answer to that simple question dictates the name of the baby, the clothes he/she wears, the toys he/she receives, and the expectations that family and friends have for many aspects of the baby's future personality from childhood into adolescence. And yet early in development, it is just a fetus, and its sex is not defined by the external social factors that many of us associate with being either a "boy" or a "girl." Instead, the sex of the developing fetus is determined by the individual sex chromosomes found in each and every one of its cells, and the hormones that the developing gonads produce. From the moment of conception, those sex chromosomes and hormones differentiate the developing body, brain, and future behavior of the developing baby. The sex of an individual also influences many aspects of our health and disease from the earliest moments of fetal development, and many of these "programming" effects on health and disease are maintained throughout life. Thus, it is critical that we explore the underlying processes of fetal sex determination and sexual differentiation of the developing brain with the goal of understanding how these processes influence our health and impact our risk of disease.

## 2 SEX DETERMINATION AND SEXUAL DIFFERENTIATION OF THE BRAIN

Sex determination is the process at fertilization when the sex chromosomes of an individual are determined. We all have 23 pairs of chromosomes in our DNA. Twenty-two of these chromosome pairs are called autosomes and one of these pairs is the sex chromosomes – only the sex chromosomes are distinct between males and females. In humans, males have one X and one Y chromosome, while females have two X chromosomes. The process of sex determination occurs immediately at fertilization, when the mother's ovum (or egg) contributes one X chromosome and the father's sperm contributes either

*Sex Differences in the Central Nervous System.* http://dx.doi.org/10.1016/B978-0-12-802114-9.00010-X

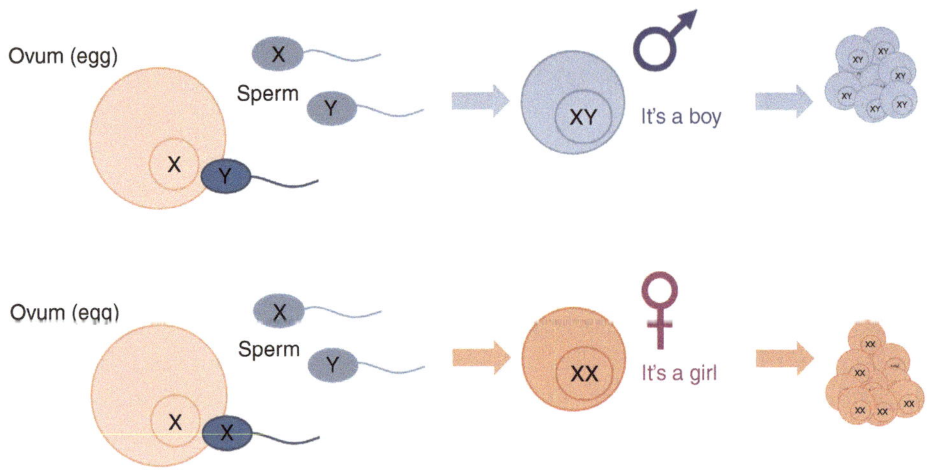


**Figure 10.1** ***The process of sex determination.*** At the time of fertilization, the ovum (egg) contributes one X chromosome to the developing fetus. The sperm that fertilizes the egg provides either an X or a Y chromosome. This process results in a male zygote that contains an X and a Y chromosome or a female zygote that contains two X chromosomes. Each cell in the developing embryo and brain maintains this chromosomal sex.

an X or a Y chromosome, thereby establishing the chromosomal sex of the embryo as either male (XY) or female (XX) (Figure 10.1). As that single fertilized egg divides into a growing embryo, each of these cells maintains that chromosomal sex determination, meaning that each cell of the developing embryo has a chromosomal sex, either two X chromosomes (female cells) or an X and a Y chromosome (male cells) in addition to the 22 autosomes of the DNA.

After sex determination, sexual differentiation is the process by which the developing fetus becomes either male or female. This process is dependent upon the expression of sex-specific genes produced by the sex chromosomes and then the hormones produced by the developing gonads. The gonads begin as bipotential organs, meaning they can be differentiated as either male or female based on cues from the developing fetus during the process of sexual differentiation. The gonads are differentiated first by the presence or absence of the *SRY* (sex determining region of the Y chromosome) gene found on the Y chromosome of males. The *SRY* gene encodes for a protein called the testes determining factor, which differentiates the bipotential gonads into the testes (Figure 10.2). In the absence of an *SRY* gene, in females, the bipotential gonads are formed into ovaries (Koopman et al., 1990; Sinclair et al., 1990). This process constitutes the sexual differentiation of the gonads.

After the sexual differentiation of the gonads, the developing testes of the male begin to produce testosterone embryonically (Jost, 1947; Rhoda et al., 1984; Weisz and Ward, 1980). In humans, this occurs early in embryonic development around the

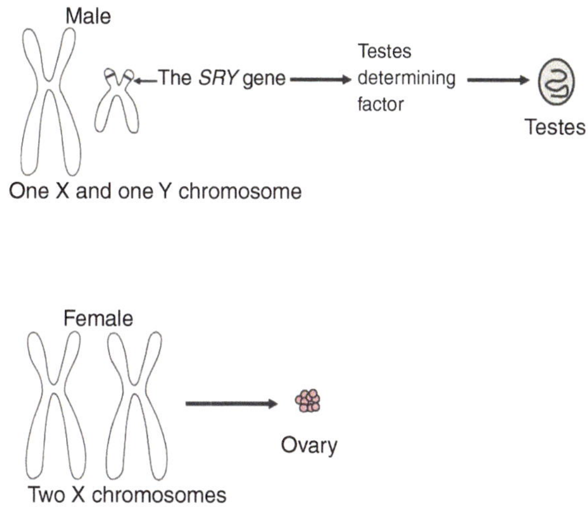


**Figure 10.2** ***Sexual differentiation of the gonads.*** In the developing male fetus, the Y chromosome contains the sex-determining region of the Y chromosome (*SRY* gene), which transcribes the testes determining factor. The testes determining factor protein differentiates a primordial gonad into the testes. In the absence of the Y chromosome, the primordial gonad differentiates into an ovary.

beginning of the second trimester (13 weeks' gestation; Hollier et al., 2014), in rodents this occurs at embryonic day 18 (Jost, 1947; Rhoda et al., 1984; Weisz and Ward, 1980). Testosterone produced in the male fetus further differentiates the bipotential sex organs into the sex organs, including the penis, prostate, seminal vesicles, and vas deferens. In contrast to the male testes, the female ovaries produce no hormones during development, and thus in the absence of testosterone production, the bipotential sex organs differentiate into the sex organs of the female, including the fallopian tubes, uterus, cervix, and vagina. As a result, during a mother's first ultrasound, at around 18 weeks, she can already find out whether her growing baby will be a boy or a girl because this process of gonadal differentiation has already occurred.

## 3 THE ORGANIZATION AND ACTIVATIONAL HYPOTHESIS OF SEX DIFFERENCES IN THE BRAIN AND BEHAVIOR

Importantly, this differential exposure to testosterone that occurs between the sexes during early fetal development occurs at a time when the brain is also particularly sensitive to hormone exposure, and this period of time is defined as the critical period for sexual differentiation of the brain. A critical period is defined as a window wherein a system maintains a heightened sensitivity to particular stimuli in order to develop in a functional manner. The concept of a critical developmental period applies to multiple neurological endpoints, including the visual system (Desai et al., 2002; Wong, 1999; Zhang and Poo, 2001), development of motor neurons (Hanson and Landmesser, 2004; Haverkamp and Oppenheim, 1986), the control of whisker movement in the rat (Schlaggar and O'Leary, 1991, 1993), and even some aspects of language development in humans (Ruben, 1997). Sexual differentiation of the bipotential brain, just as the gonads, occurs during a critical period in all species examined to date, including primates, although the timing of various aspects can vary substantially within and between species (Arnold and Gorski, 1984; MacLusky and Naftolin, 1981; Nevison et al., 1997; Swaab, 2004).

The laboratory rat is arguably one of the best characterized species for studying the sexual differentiation of the developing brain that occurs during the critical period for two primary reasons. First, the critical period of brain sexual differentiation is very well defined in the rat. Secondly, the behavioral endpoints associated with sexual differentiation of the rat brain are very well characterized. The male rat testes produce testosterone as early as embryonic day 18, and this exposure begins the process of sexual differentiation of the male brain (Figure 10.3). In the absence of hormone exposure in the female, the bipotential brain differentiates into a female brain. The organization of the brain that occurs during the critical period, followed by the activational effects of hormones on sex-specific sexual behaviors in adulthood, constitute what has been named the organizational/activational hypothesis of the brain and behavior (Arnold and Breedlove, 1985) (Figure 10.4). The rat is an exceptional model to study sexual differentiation of the brain because hormonal treatment during the organizational critical period can induce robust and reliable sex differences in the brain and the functional output can be tested by the administration of hormones in adulthood and the examination of sexual behaviors in adulthood. For example, from embryonic day 18 until around postnatal

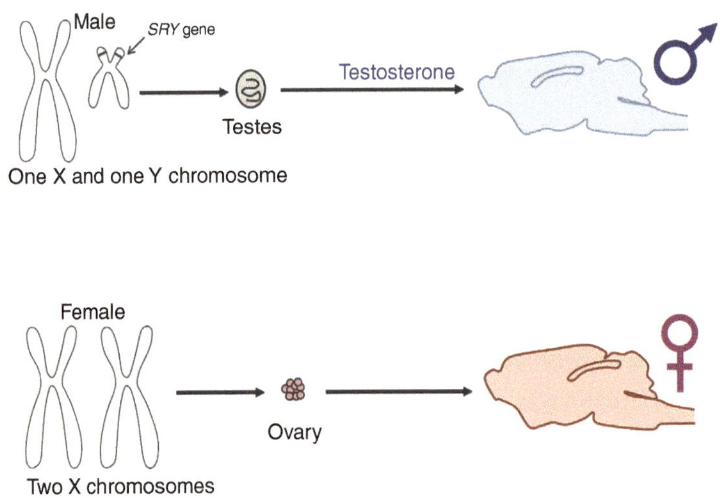


**Figure 10.3** ***Sexual differentiation of the brain.*** Following the differentiation of the gonads, the developing testes begin to produce testosterone, which is necessary for masculinizing the developing brain. In contrast, the developing ovaries produce no significant levels of hormones and in the absence of any hormone exposure, the developing brain is feminized.

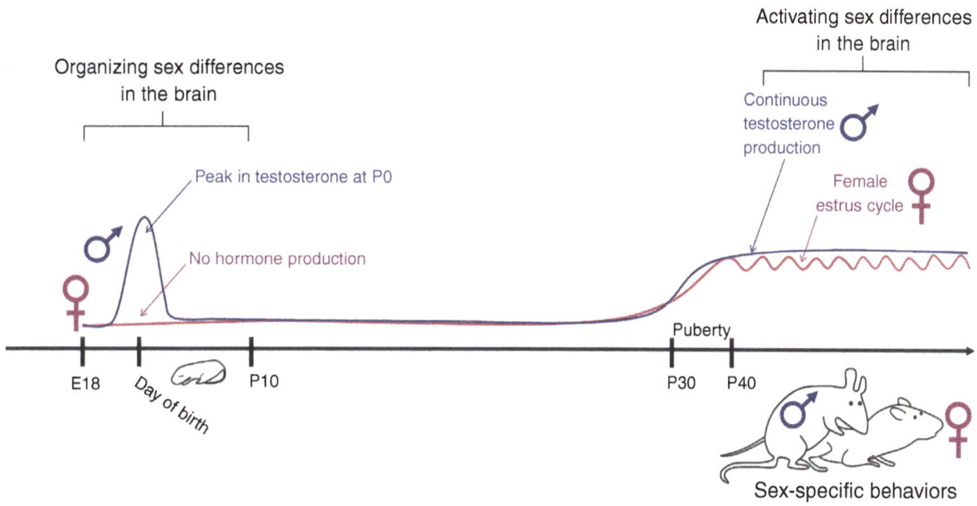


**Figure 10.4** ***The organizational and activational hypothesis of brain sexual differentiation.*** In males, testosterone synthesis peaks around the day of birth in rodents. This testosterone exposure organizes sex differences in the developing brain and permanently masculinizes the brain. After puberty, the testes produce continuous levels of testosterone, which activate the male brain to induce sex-specific behaviors, such as male sex behavior. In the absence of any testosterone exposure during early brain development, the female brain is permanently feminized. After puberty, the ovaries produce estrogens and progestins necessary to activate the feminized brain and induce sex-specific behaviors, specifically female sex behavior.

day 10, exogenous treatment of females with testosterone can override development as a female, and induce a phenotypic male brain (Figure 10.5). After postnatal day 10, the critical period is over and treatment of females with testosterone will have no effect on the organization of the brain (Arnold and Gorski, 1984) (Figure 10.6).

The outcome of sexual differentiation is the distinction between one of two possible pathways, male or female. In the absence of testosterone exposure during the critical period, the default pathway is female. Other developmental processes modified during a critical period, like those mentioned earlier, rely on the correct stimuli to form a functional system, such as in the auditory cortex, visual cortex, or motor system (Hensch, 2004). In the absence of the required stimuli, the proper connections are not formed, and the circuit is not functional. In contrast, in the absence of testosterone during the critical period of sexual differentiation the default female pathway that occurs is a functional circuit regulating the expression of female sexual behavior in adulthood (Baum, 1979). Testosterone exposure changes these pathways during the critical period to allow for the expression of male sexual behavior in adulthood. During this time, normal development of the male rodent brain requires completion of two distinct processes: masculinization and defeminization. Masculinization is the organization of a neural substrate permissive to the adult expression of male sexual behavior. Defeminization is the loss of capacity

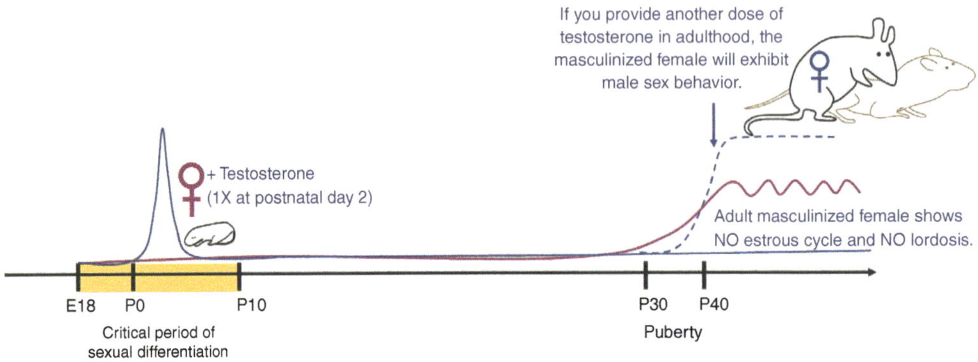


**Figure 10.5** ***The female brain can be masculinized via a single dose of exogenous testosterone during the critical period of sexual differentiation.*** If a neonatal female is injected with testosterone during the critical period of sexual differentiation, for example on postnatal day 2, her brain will be permanently masculinized. As a result, she will show no estrous cycle and no female sex behavior (lordosis) in adulthood. Instead, if you provide the masculinized female with an exogenous dose of testosterone in adulthood, you can activate her masculinized brain and induce the expression of male sex behavior.

as an adult to respond to the activational effects of estradiol and progesterone to induce female sex behavior. Both processes oppose the process of feminization, which occurs in the absence of critical levels of testosterone as a neonate and is the pathway leading to adult female-typical behavior (Baum, 1979; Nordeen and Yahr, 1983). There is no known naturally occurring demasculinization in female rodents, only a default active process of feminization.

An interesting distinction between sexual differentiation and other critical periods involves the closing of the window. In the absence of the appropriate stimulation, many

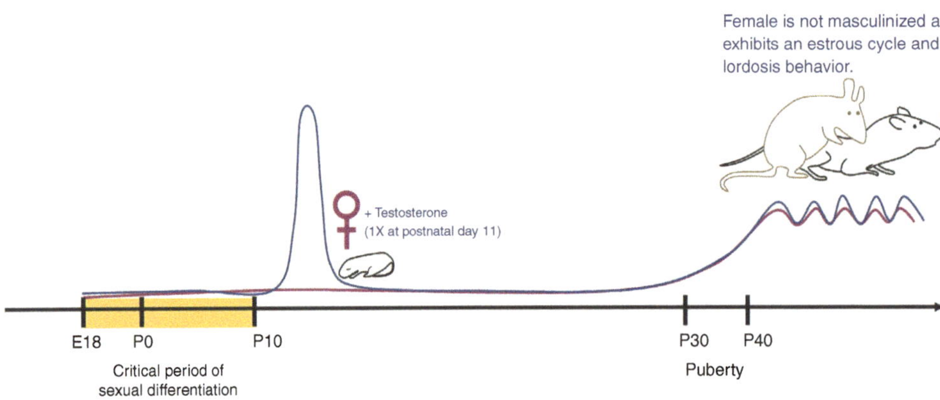


**Figure 10.6** ***Sexual differentiation of the developing brain is constrained to a critical period of development.*** Administration of exogenous testosterone to a female at postnatal day 11, after the critical period of sexual differentiation, will not masculinize her developing brain. As a result, in adulthood, she will exhibit an estrous cycle and female sex behavior.

critical periods, such as that experienced in the visual system, can be extended for a brief time in the expectation of receiving the appropriate input (Hooks and Chen, 2007). In contrast, sexual differentiation ends with the development of the appropriate connections necessary to perform female sexual behavior, or feminization. In this regard, the time frame of sexual differentiation is relatively immutable.

Phoenix et al. (1959) were the first to discover that treatment of female guinea pigs with testosterone could induce permanent masculinization of the brain. Not long after, other groups determined that treatment of female rodents with estradiol could induce masculinization of the brain just as well as testosterone treatment (Booth, 1977; Feder and Whalen, 1965; McEwen et al., 1977a, 1977b). It was hypothesized that testosterone produced by the male gonads was converted locally in the brain to estradiol as a means of permanently differentiating the male brain, and this is referred to as "the aromatization hypothesis." A critical argument in favor of this hypothesis was the discovery that a developing rodent fetus has high levels of circulating alpha-fetoprotein, a protein that potently binds estradiol thereby sequestering the maternal estradiol in the fetal blood stream (Andrews et al., 1982). Coincident with this finding was the discovery that brain nuclei that express sex differences also express high levels of the enzyme P450 aromatase (Naftolin and MacLusky, 1984; Reddy et al., 1974), the enzyme that converts testosterone to estradiol. Blockade of or absence of this neuronal aromatase during the critical period blocks differentiation of the male rodent brain (Bakker et al., 1993, 2004; Lewis et al., 1995). In the rat, testosterone secreted from the testes of the developing male is not bound by alpha-fetoprotein, allowing it to enter the brain where it is locally aromatized to estradiol in specific nuclei. As a result, the concentration of estradiol in the developing male brain is more than twice that of females, particularly in brain regions subject to sexual differentiation (Amateau et al., 2004; Rhoda et al., 1984). Moreover, the estrogen receptor (ER) is expressed at high levels in these same brain regions that undergo sexual differentiation and ER is essential for transducing the steroid signal to achieve the appropriate cellular endpoints of sexual differentiation (McCarthy et al., 1993; Shughrue et al., 1997). Specifically, there are two major ER isoforms, ERα and ERβ, upon which estradiol can act. ERα is the predominant isoform found in the developing hypothalamus and in other brain regions known to be sexually dimorphic (Shughrue et al., 1997). Male mice lacking a functional ERα show low levels of male sex behavior, due to both the organizational and the activational effects of the receptor on sexual behavior (Ogawa et al., 1999; Wersinger et al., 1997). Other work also suggests that the process of estradiol-mediated defeminization is induced exclusively by activation of ERβ (Kudwa et al., 2005, 2006).

It should be noted that while the aromatization hypothesis provides evidence that estradiol, aromatized from testosterone, differentiates the bipotential rodent brain into a male brain, there is no evidence to indicate that the aromatization hypothesis holds true in primates, including humans. Instead, testosterone itself is the hormone that differentiates the primate brain into a male brain during fetal development (Wallen, 2005). In the absence of testosterone exposure, the bipotential primate brain differentiates into a female brain. However, it is important to discuss the aromatization hypothesis and the evidence that estradiol masculinizes the rodent brain because much of what is known regarding the underlying mechanisms of sexual differentiation of the brain is from research performed in rodents. We will next discuss the brain region-specific mechanisms by which the male rat brain is differentiated during the critical period of sexual differentiation. Where relevant, there will also be a discussion of the brain regions that are known to be sexually dimorphic in the human brain as well.

## 4 THE MECHANISMS OF STEROID-INDUCED SEXUAL DIFFERENTIATION IN THE RODENT BRAIN ARE BRAIN REGION SPECIFIC

Many brain regions undergo the process of sexual differentiation in the rodent brain; however, the brain regions that undergo the most pronounced sexual differentiation are those involved in the expression of adult sexual behavior. One fascinating aspect of the research examining the underlying mechanisms of sexual differentiation is that each brain region has a unique mechanism for creating the changes in size and circuitry necessary to make a phenotypic male brain. Understanding the many and complex mechanisms by which hormones can produce long-lasting changes in the circuitry and the function of the developing brain can better inform our understanding of how functional circuits that control behavior are established or constrained by factors produced by the developing brain and environment.

### 4.1 Sex differences in the size of brain nuclei

Sex differences in the size of certain brain nuclei are generally induced by hormone exposure, or the lack thereof, which stimulates or prevents naturally occurring cell death in these areas. The brain regions that show the most robust sex difference in size within the rodent brain include the sexually dimorphic nucleus of the preoptic area (SDN-POA), the spinal nucleus of the bulbocavernosus (SNB), the principal bed nucleus of the stria terminalis (BNSTp), and the anteroventral periventricular nucleus (AVPV). Each of these nuclei is critically involved in the regulation of reproductive behavior or sex-specific physiology. For example, the SDN-POA, which is significantly larger in the male brain than the female brain, is located in the POA of the brain, which is central to the control of male sexual behavior (Ågmo, 1997). It has been implicated in partner preference in sheep, where a larger SDN in the male brain is associated with greater female partner preference (Roselli et al., 2004). The SNB is the motor nucleus of the spinal cord that innervates the muscles controlling the penis and contains many more neurons in males than in females as a result of neonatal testosterone exposure (Freeman et al., 1996; Nordeen et al., 1985). The BNSTp projects to the AVPV and together these nuclei form part of a functional circuit controlling sexually dimorphic gonadotropin secretion from the anterior pituitary (De Vries and Simerly, 2002; Gu and Simerly, 1997; Herbison, 1998). In the SDN,

**Table 10.1** A list of well-studied sexually dimorphic nuclei and their functions

| Brain region | Sex difference | Function | Developmental mechanism establishing the sex difference |
|---|---|---|---|
| SDN-POA | Males ≫ females | Control of sexual behavior (all species) and partner preference in males (sheep). | Testosterone exposure prevents naturally occurring cell death in the male SDN. |
| SNB | Males ≫ females | Control of penile muscles (all species). | Testosterone exposure prevents naturally occurring cell death in the male SDN. |
| BNSTp | Males > females | Control of gonadotropin secretion (all species). | Testosterone exposure prevents naturally occurring cell death in the male SDN. |
| AVPV | Females ≫ males | Control of gonadotropin secretion (all species). | Testosterone exposure increases naturally occurring cell death in the male AVPV. |
| Interstitial nucleus of the anterior hypothalamus-3 (INAH-3) | Males ≫ females | Sexual behavior and preference (humans). | Specific mechanism is unknown, but is differentiated by neonatal testosterone exposure. |

SNB, and BNSTp neonatal testosterone and/or estradiol decreases naturally occurring cell death, resulting in a larger nucleus in males than females (Breedlove and Arnold, 1983; Davis et al., 1996; del Abril et al., 1987; Nordeen et al., 1985) (see Table 10.1). Conversely, in the AVPV testosterone/estradiol increases cell death during the critical period, creating a significantly smaller nucleus in males than in females (Murakami and Arai, 1989; Simerly et al., 1985). These examples provide evidence that one steroid can differentially affect the same cellular process, naturally occurring cell death also known as apoptosis, depending on the brain region analyzed. Two major classes of signaling proteins mediate the process of apoptosis, one being the antiapoptotic proteins such as Bcl-2, the other being the proapoptotic proteins, such as Bax. In transgenic mice that overexpress Bcl-2, the survival-promoting cell signaling protein, the sex difference in the size of the SNB is significantly reduced; suggesting that in the normal male mouse, testosterone might upregulate antiapoptotic proteins such as Bcl-2 during the critical period of sexual differentiation to inhibit cell death in this nucleus (Zup et al., 2003).

Conversely, the AVPV is significantly larger in females than males, with an overall increase in cell number, specifically dopaminergic neurons (Forger et al., 2004; Simerly et al., 1985; Sumida et al., 1993). Reciprocal to this is the number of cells in the

BNSTp, where there are significantly fewer cells in females than in males (Guillamon et al., 1988; Hines et al., 1992). Interestingly, there is also a significant sex difference in the number of synaptic projections from the bed nucleus of the stria terminalis to the AVPV, which contains nearly 10 times the number of fibers in males as in females (Gu and Simerly, 1997). This sex difference is established by testosterone during the critical period. However, rather than modulating the cellular mechanisms of apoptosis, the neurons of the AVPV are impacted by testosterone to induce synthesis of a target-derived trophic factor to promote outgrowth of neurites in order to form the connections originating from the BNST (Ibanez et al., 2001).

## 4.2 The preoptic area

The medial preoptic nucleus (MPN) is the primary brain region controlling male sexual behavior. Lesions of this brain region completely inhibit the expression of male sexual behavior and hormonal, chemical, or electrical stimulation of the MPN can induce or enhance male sexual behavior in adulthood (Ågmo, 1997). Thus, it is a primary target of sexual differentiation during the critical period of sexual differentiation, as this brain region is necessary for the proper expression of male sexual behavior in adulthood. One of the primary sex differences in the MPN is the significant sex difference in the size of the subnuclei described previously, the SDN-POA. In addition to this well-known sex difference, there are also microscopic differences in the neural connections within the MPN that control male sexual behavior. Compared with females, males have two- to three-fold more dendritic spines in the preoptic area (POA), a subregion of the MPN. Dendritic spines are the major sites of excitatory synaptic connections in the brain. The induction of the dendritic spine patterning in the POA is mediated entirely by estradiol aromatized from testosterone, and it occurs during the critical period of sexual differentiation. Thus, administering exogenous estradiol to a newborn female will induce the male pattern of dendritic spines in the POA, which allows for the expression of male sexual behavior in adulthood if the masculinized female is supplied with testosterone again in adulthood (Amateau and McCarthy, 2004).

Estradiol, aromatized from testosterone, permanently alters the dendritic spine connectivity of the POA via prostaglandin E2 (PGE2) (Amateau and McCarthy, 2004). There are at least eight prostanoids, of which PGE2 is one, and synthesis begins with the oxygenative cyclization of arachidonic acid by cyclooxygenase-2 (COX-2). Interestingly, in addition to its important role in male sexual behavior, the POA is also the brain region necessary for fever induction and it relies on PGE2 for this mechanism (Blatteis, 2007). Thus, in hindsight, it may not be surprising that this is also the mechanism by which estradiol sexually differentiates the developing POA during the critical period of sexual differentiation. Specifically, the masculinization of the circuitry in the POA utilizes the pre-existing mechanisms of cellular communication inherent to the POA. COX-2 messenger RNA (mRNA) and protein are higher in the POA of newborn males than females, and treatment of females with estradiol increases COX levels to those seen in the male POA. Blocking COX-2 activity, which inhibits the synthesis of PGE2, in a newborn male will permanently disrupt synaptic patterning of the POA and inhibit the ability of the male to show normal sexual behavior in adulthood (Amateau and McCarthy, 2004). Conversely, administrating PGE2 to newborn females increases the number of dendritic spine synapses within the POA to the levels seen in males, and results in the expression of male sexual behavior in adulthood (Amateau and McCarthy, 2002, 2004). Thus, PGE2 is a critical molecule that is necessary and sufficient to mediate steroid hormone-induced masculinization of the developing rat POA.

The mechanism by which estradiol, via PGE2, induces masculinization of the developing POA requires an orchestration of numerous neural cell types in the developing POA including neurons, astrocytes, and microglia. The evidence indicates that estradiol upregulates COX-2 in neurons, increasing the production and release of PGE2, which can then act on neighboring astrocytes to induce glutamate release (Araque et al., 1999; Bezzi et al., 1998). Glutamate activates glutamate receptors on neighboring neurons in the POA, thereby inducing the formation of dendritic spines (Amateau and McCarthy, 2002), and evidence indicates that the masculinization of the POA requires the activation, phosphorylation, and insertion of new glutamate receptors into the membrane of developing neural cells (Lenz et al., 2011; Wright et al., 2008; Wright and McCarthy, 2009). In addition, PGE2 can induce the production of more PGE2 from neighboring microglia, the immune cells of the brain. Typically, microglia produce PGE2 in the POA as a mechanism of producing fever. In the case of estradiol-induced masculinization, PGE2 enhancement of PGE2 production via microglia produces a feed-forward mechanism by which full masculinization of the neural connectivity in the developing POA can occur (Lenz et al., 2013; Lenz and McCarthy, 2014). This multicellular mechanism of masculinization of the developing POA, which focuses around the induction of PGE2 (see Figure 10.7), permanently changes the response of this brain region and other brain regions in the neural circuitry controlling male sex behavior in response to the presence of estrous cycling females, indicating that the function of this circuitry involved in male sex behavior via the POA has been permanently masculinized during the critical period of sexual differentiation (Nugent et al., 2009).

## 4.3 The ventromedial nucleus of the hypothalamus

During development and in adulthood, the medial basal hypothalamus (MBH) is also a key target for estradiol, with a high concentration of estradiol receptors. A central region in the MBH, the ventromedial nucleus (VMN) is the brain region necessary for female sex behavior (Pfaff and Sakuma, 1979). The VMN is characterized by its well-defined oval shape and sparse thin dendrites (Millhouse, 1973). Similar to the POA, the VMN contains more than three times as many dendritic spine and shaft synapses as females (Matsumoto and Arai, 1983). This sex difference in synaptic

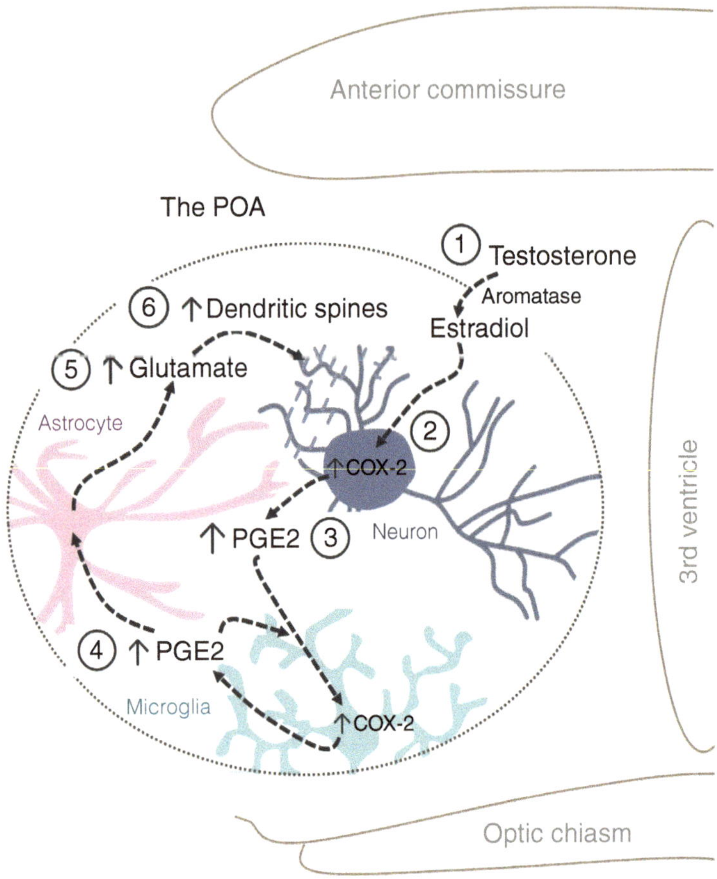


**Figure 10.7** ***The masculinization of the synaptic connectivity in the developing preoptic area (POA) requires estradiol-induced cell-to-cell communication.*** (1) Testosterone is converted to estradiol in the brain via the enzyme aromatase. (2) Estradiol increases the synthesis of cyclooxygenase-2 (COX-2), which in turn increases the synthesis prostaglandin-E2 (PGE2). (3) PGE2 activates neighboring microglia to induce the synthesis of more PGE2, which can (4) also activate neighboring astrocytes. (5) PGE2 induces the release of glutamate from astrocytes, which in turn (6) activates glutamate receptors on neighboring neurons to increase the synthesis of new dendritic spines and synaptic connections in the developing male POA.

connectivity is detected as early as postnatal day 2 and is still present in the rodent brain at postnatal day 100 (Matsumoto and Arai, 1986; Pozzo Miller and Aoki, 1991), consistent with the long-term organizational effects of estradiol in the developing brain. The male pattern of dendritic morphology in the VMN can be induced in females by treatment with either testosterone or estradiol within the first few days of life. In adulthood, the female estrous cycle or exogenous estradiol treatment causes changes in the dendritic spine density and patterning of the VMN, which correlates to changes in female sex receptivity (Frankfurt et al., 1990), indicating that changes in the synaptic connectivity of the VMN is a critical mechanism of inducing or constraining female sex behavior.

Glutamate is the predominant neurotransmitter in the VMN (Ziegler et al., 2002) and estradiol mediates the developmental changes in neuronal morphology in the developing VMN by directly enhancing glutamate release (Schwarz et al., 2008). The classical model of estradiol action involves binding to the nuclear ER and regulation of gene transcription via its interactions at EREs located in the promoter region of genes modulated by estradiol. This is considered the direct mechanism of estradiol on transcription, via an ERE. Estradiol also has the ability to act indirectly via intracellular signaling pathways to act at other promoter elements of genes. Many of these indirect effects of estradiol are initiated at the neuronal membrane and include the activation of protein kinase A (PKA) and CREB (Abraham et al., 2004; Aronica et al., 1994; Zhou et al., 1996), mitogen-activated protein kinase (MAPK) (Singh et al., 1999; Watters et al., 1997; Znamensky et al., 2003), and phosphatidylinositol 3-kinase (PI3 kinase) (Singh et al., 1999; Znamensky et al., 2003). During the critical period of sexual differentiation, estradiol exposure increases the activation of PI3 kinase within only 3 h, which leads to a cascade of events that can induce the defeminization of the synaptic connections in the VMN and subsequent behavior in adulthood. Specifically, activation of PI3 kinase in the developing VMN increases the release of glutamate from presynaptic terminals within the VMN. This, in turn, activates glutamatergic receptors, activating MAPK and masculinizing the dendritic spine number measured in the VMN (Schwarz et al., 2008) (Figure 10.8).

Blocking any one of these steps, including PI3 kinase activation, MAPK activation, or α-amino-3-hydroxy-5-methyl-4-isoxazolepropionic acid receptor/*N*-methyl-D-aspartate receptor glutamate receptor activation, completely prevents in the increase in dendritic spine proteins seen following estradiol treatment of the female rat pup, and ultimately prevents the defeminization of the brain and behavior later in life. Specifically, blocking NMDA receptors during the critical period of sexual differentiation blocks estradiol-induced defeminization of behavior, which results in the production of an adult rat that is capable of expressing both male and female sex behavior (Schwarz and McCarthy, 2008). It should also be noted that estradiol activates PI3 kinase in the developing VMN within 3 h after estradiol exposure, indicating that the effects of estradiol on modulating neurotransmitter function to initiate the process of sexual differentiation with the VMN occurs in a relatively rapid time course. The relatively rapid time course of the estradiol effect suggests that the developing circuitry is extremely sensitive to the organization effects of estradiol exposure during the critical period of sexual differentiation, and most likely this is also the case for other brain regions that undergo sexual differentiation during this critical developmental window.

## 4.4 The arcuate nucleus

The arcuate nucleus is also found in the MBH but serves a different function than the VMN in reproductive physiology. There is no known role for the arcuate nucleus in control of sex behavior per se, but instead it is a major regulator of the anterior pituitary,

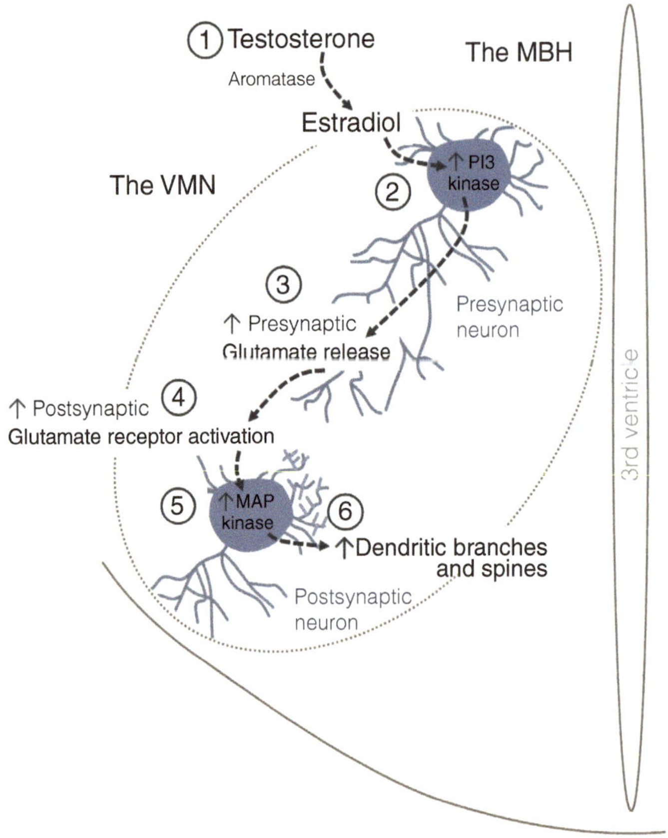


**Figure 10.8** ***The defeminization of the synaptic connectivity in the developing ventromedial nucleus of the hypothalamus requires cell-to-cell communication via estradiol-enhanced glutamate release.*** (1) Testosterone is converted to estradiol in the brain via the enzyme aromatase. (2) Estradiol activates PI3 kinase in developing neurons of the hypothalamus, which in turn (3) enhances the release of glutamate from presynaptic terminals. (4) Glutamate activates receptors on postsynaptic neurons, which (5) activates MAPK and (6) increases the branching and dendritic spines on neurons in the male VMN.

and the subsequent release of luteinizing hormone (LH) and follicle-stimulating hormone (FSH). The LH releasing hormone (LHRH) neurons of the preoptic area project to the anterior pituitary via connections in the arcuate nucleus. They terminate in the median eminence, where LHRH is delivered to the gonadotropic cells of the anterior pituitary containing the gonadotropin hormones, LH and FSH. Once LH and FSH are released into the general circulation these gonadotropin hormones act on their target organ, the gonads, and regulate a variety of cellular activities, including ovulation in females. In particular, ovulation requires a massive surge in LH at the appropriate time in the reproductive cycle. The LH surge is induced by synchronized firing of the LHRH neurons, and this activity is induced by the positive feedback effects of estradiol on the brain (Fink, 1988). Notably, the masculinized brain cannot respond to even exogenous estradiol treatment in adulthood with positive feedback and therefore the masculinized brain is not capable of producing an LH surge.

The arcuate nucleus exerts a modulatory effect on the LH surge. The neurons of this nucleus exhibit striking plasticity in synaptic patterning across the estrus cycle, and this is largely driven by changes in the morphology of local astrocytes (Garcia-Segura et al., 1994; Garcia-Segura and McCarthy, 2004). Exogenous administration of female hormones that recapitulate the estrous cycle to adult males does not induce the same plasticity in astrocytes, and females masculinized at birth by exogenous androgen also do not respond to steroids with changes in astrocyte morphology in adulthood (Barraclough and Gorski, 1961). In neonates, estradiol induces a dramatic morphological shift in astrocytes toward an increased stellation with more primary processes and branches (Mong et al., 1996, 1999). However, there is no evidence of ERs in astrocytes of the arcuate nucleus (Mong et al., 1999). The primary site of estrogen action is neurons, where it increases the synthesis and release of neurotransmitters which then acts on neighboring astrocytes to induce a stellate morphology (McCarthy et al., 2002). In turn, the astrocytes then modulate the formation of dendritic spine synapses by the neurons, reducing their formation in males (Mong et al., 2001) (Figure 10.9). This is an example, similar to that in the POA and the VMN, where crosstalk between neurons and/or astrocytes is an essential component of establishing sex differences in the brain and likely contributes to the ability of the entire population of cells within a nucleus to respond to the steroid, not just those containing ERs.

## 5 THE GENETICS AND EPIGENETICS OF SEX DIFFERENCES IN THE DEVELOPING BRAIN

### 5.1 The sex chromosomes and brain development

Our current understanding of sexual differentiation relies heavily on the organizational hypothesis, which suggests that most sex differences in the brain are determined by the presence or absence of testosterone during early brain development in males and females respectively. However, this view of sexual differentiation does not account for potential sex differences in the brain that are not the direct downstream consequence of sex determination (whether you have testes or ovaries) (Arnold et al., 2013; Arnold, 2012). For example, each cell in the brain has a different set of sex chromosomes – either XY in males, or XX in females. A neural cell that does not express the receptor for a gonadal hormone could still exhibit a sex difference during development either via cell-to-cell communication with cells that do express the ER, or via the direct result of sex differences in the genes expressed by the sex chromosomes themselves in those neural cells that do not express the ER (Davies and Wilkinson, 2006). In fact, sex-specific gene expression has been documented in embryos of many species, including mammals, prior to the onset of gonadogenesis (Deng et al., 2014; Gutierrez-Adan et al., 2006; Kobayashi et al., 2006;

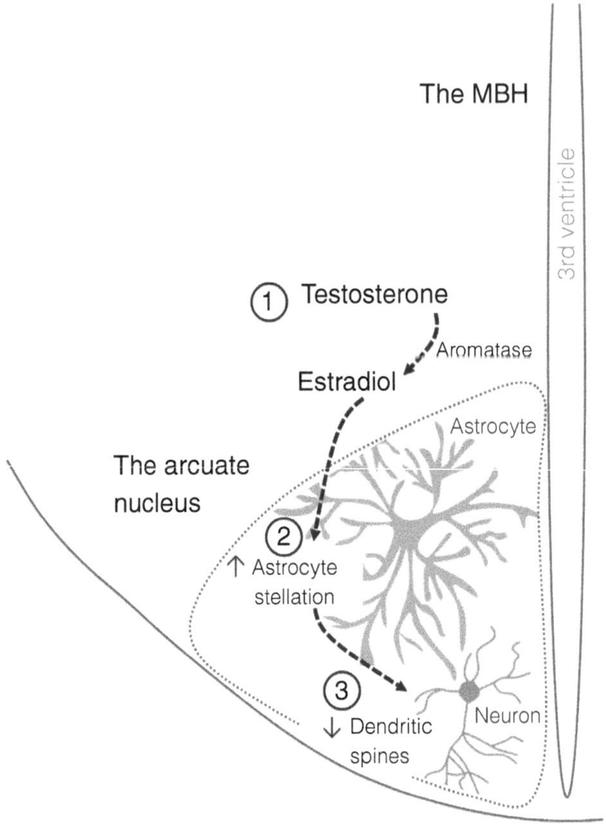


**Figure 10.9** ***The masculinization of the developing arcuate nucleus requires communication between astrocytes and neighboring neurons.*** (1) Testosterone is converted to estradiol in the brain via the enzyme aromatase. (2) Estradiol increases the stellation or the number of processes extending from astrocytes. (3) In turn, via mechanisms not clear, the neurons of the male arcuate nucleus decrease the number of dendritic spines during this process of masculinization.

Tagirov and Rutkowska, 2014; Zhao et al., 2010). As a result, researchers have begun to explore the factors that may be encoded by the sex chromosomes, independent of the gonads.

The first mechanism by which sex differences in gene expression could occur during early brain development, independent of the gonads, is due to the fact that XY neural cells have the Y chromosome and XX neural cells do not (Figure 10.10a). That said, the Y chromosome is small, containing only approximately 27 genes that are transcribed to proteins (Skaletsky et al., 2003), and most of these proteins are expressed exclusively in the male germ cell (not in the brain). In addition, many of the genes expressed on the Y chromosome have a similar gene on the X chromosome (Xu et al., 2002), which could offset any potential sex difference in gene expression caused by the absence of the Y chromosome in females.

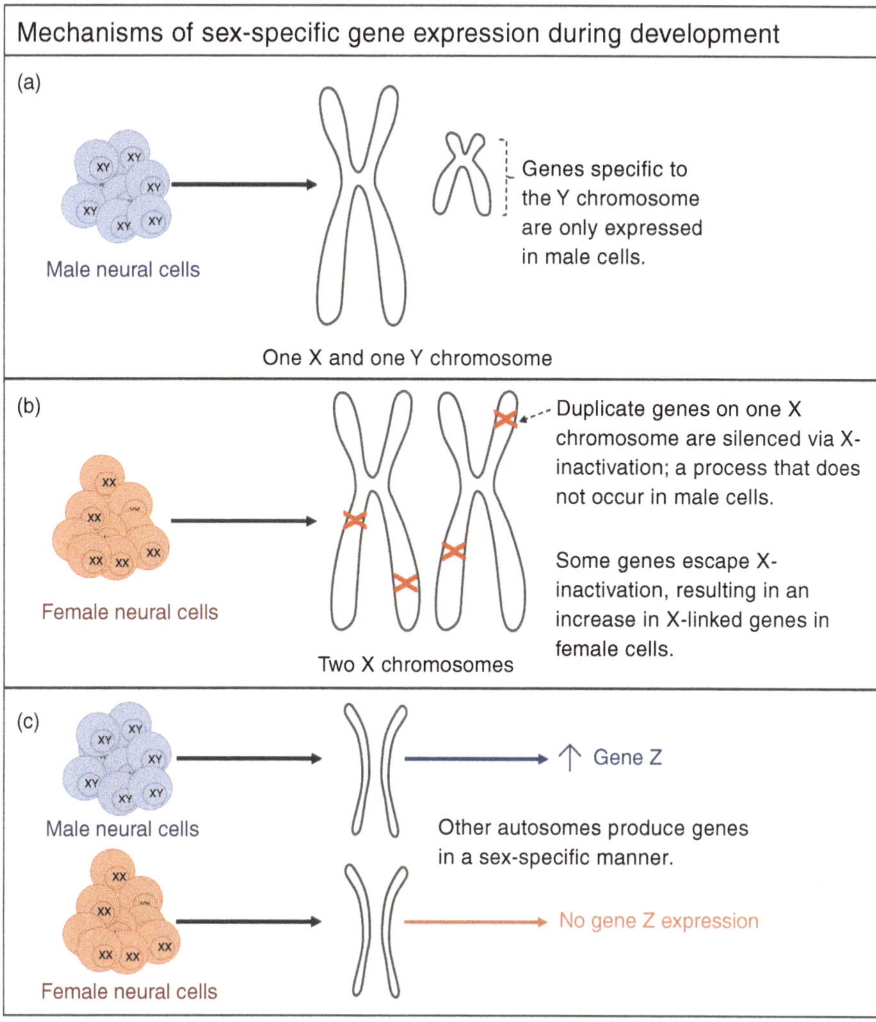


**Figure 10.10** ***There are three primary mechanisms of sex-specific gene expression during early brain development.*** (a) Each male neural cell contains one X and one Y chromosome. As a result, there are genes expressed by the Y chromosome that may influence neural development that would not be expressed in the female neural cells. (b) Each female neural cell contains two X chromosomes. Duplicate genes on the X chromosomes undergo the process of X-inactivation to silence one copy of the duplicate gene. This process does not occur in male cells. In addition, many genes escape X-inactivation, which can result in an increase in the expression X-linked genes in female neural cells relative to the levels of these same genes in male neural cells. (c) Independent of the genes expressed on the sex chromosomes, the autosomes express genes in a sex-specific manner, which can in turn influence brain development in a sex-dependent manner.

A second mechanism by which sex differences in gene expression could occur during early brain development, independent of the gonads, is due to the fact that XX neural cells have two copies of each gene present on the X chromosome. XY neural cells only have one copy of these genes on the X chromosome. As a result, many genes on the X chromosome undergo X-inactivation in female cells, which is a process of silencing one copy of each gene on one copy of the X chromosome (Deng et al., 2014). As one would expect, this process does not occur in the male cells (Figure 10.10b). However, almost 15% of the genes on the X chromosome escape X-inactivation, which would lead to an increase in the expression of certain X chromosome-linked genes in these female cells (Berletch et al., 2011). The differential dosage of certain X-linked genes that result from escaping X-inactivation can also result in sex differences in gene expression on a per cell basis very early in brain development.

This sex difference in gene expression caused by the genes expressed on the X and Y chromosomes is likely important for a number of physiological processes, but most importantly in brain development. The X chromosome is particularly enriched by the number of genes that are important to neurological function (Nguyen and Disteche, 2006a, 2006b; Skuse, 2005; Zechner et al., 2001). In particular, the importance of X-linked genes in neural development is highlighted by the number of X-linked neurodevelopmental disorders. There are more than three times as many mental disorders associated with the X chromosome than other mental disorders (Endele et al., 2010; Ropers, 2006, 2010). As we move forward in this area of research, new tools will likely reveal the sex differences in gene expression that are the result of the mechanisms described previously or even sex differences in gene expression localized to the autosomes (Wolstenholme et al., 2013) (Figure 10.10c) that in turn directly influence the neural development of sex differences and the risk of neurodevelopmental disorders. These advances will allow researchers to move beyond the basic and now incomplete idea that all sex differences in the developing brain are the result of sex hormones.

## 5.2 Epigenetic mechanisms of sex differences in the developing brain

When one considers the genes that regulate neural development, one must also consider the mechanisms that in turn regulate gene expression. Thus, epigenetic regulation of gene transcription has become a mechanism of interest for understanding the biological mechanisms by which gene transcription, neural function, and associated behaviors can be induced or constrained by sexual differentiation that occurs during the critical period. For example, X-inactivation is the direct result of gene silencing via epigenetic mechanisms. Epigenetics refers to the changes on the genome that alter the transcriptional activity of the gene without changing the genomic code. There are three primary mechanisms of epigenetic regulation of the genome, which include: (1) histone modifications such as acetylation and methylation, (2) methylation of specific nucleotides in the DNA, and (3) microRNAs. All three of these mechanisms directly regulate the expression of genes by regulating the availability of the gene to transcriptional machinery or directly interfering with the expression of RNAs already generated via transcription. There have been a few pieces of evidence to suggest that epigenetics may be an important mechanism of sexual differentiation in the developing brain.

In particular, the BNSTp has significantly fewer cells in the female brain than in the male brain (Guillamon et al., 1988; Hines et al., 1992). This striking sex difference is important for reproductive function later in life. This sex difference is established via perinatal testosterone exposure during the critical period of sexual differentiation, which prevents cell death in the male BNSTp. Treatment of male pups with a drug that inhibits histone deacetylation, which in turn increases histone acetylation and thereby inhibits gene transcription, significantly prevents the neuroprotective effects of neonatal testosterone in this region of the brain, such that these males have smaller numbers of surviving neurons in the BNSTp (Murray et al., 2009). As a result, males treated with an inhibitor of histone acetylation have a BNSTp that is similar in size to that of the female. This is one of the first known sex differences in brain structure that is the direct result of epigenetic modifications during the critical period of sexual differentiation.

Around the same time, researchers performed a global analysis of histone acetylation and methylation in the developing mouse brain, including the POA/MBH and the hippocampus and cortex, which revealed dramatic sex differences that were modulated by perinatal testosterone exposure (Tsai et al., 2009), suggesting that histone modifications are an essential component to the sexual differentiation of the developing brain.

Others have sought to examine how sex or neonatal hormone exposure may influence the global methylation of brain regions that are known to be sexually dimorphic or the methylation of specific genes important for sexual differentiation. Estradiol induces differences in DNA methylation on the ERα, ERβ, and progesterone receptor genes in sexually dimorphic brain regions across development (McCarthy and Nugent, 2013; Nugent and McCarthy, 2011; Schwarz et al., 2010). An examination of the entire methylome (methylation on the whole genome) revealed that perinatal testosterone exposure results in significant changes in the methylation of DNA, but that these epigenetic changes emerge in adulthood (Ghahramani et al., 2014). Contrary to the overarching view of DNA methylation as a permanent modification directly tied to gene expression, these data demonstrate that methylation patterns on steroid hormone receptors change across the lifespan and may require time before they are established. However, their role in sexual differentiation of the brain and behavior during development and into adulthood is likely extremely important.

## 6 CONCLUSIONS

As we move forward in this area of research, there are many questions that remain to be answered. For example, what factors influence the development of the female brain? A disproportionate amount of research is directed toward determining the cellular mechanisms establishing the male brain because of the clear advantage of having a known modulator, perinatal testosterone exposure, and thus masculinization of the developing brain can be tested easily using exogenous steroid administration to females. The female brain, in contrast, develops in the absence of any known factor, making the identification of the key underlying mechanisms far more difficult to decipher. Establishing how this occurs is one of the greatest challenges ahead for the field of sex differences in the brain. A second future question in this area of research involves understanding how the genes expressed in a sex-specific manner influence other aspects of brain development and the risk of neurodevelopmental disorders. In addition, how can our understanding of sexual differentiation of the developing brain further our understanding of neurodevelopmental disorders? As we consider the mechanisms that establish sex differences in the developing brain, it is important to understand how these factors may impact the risk of certain neurodevelopmental disorders. Many neuropsychiatric disorders exhibit a striking sex difference in their etiology and prevalence. Males are more likely to be diagnosed with disorders that present at birth or early in childhood. In contrast, females are more often diagnosed with disorders that arise around the onset of adolescence or puberty (see Bao and Swaab (2010) for review). The hypothesis that stems from these epidemiological data is that males are more vulnerable to early life insults than females, and there is some evidence in rodents to support this. This increased vulnerability may be the result of testosterone exposure in males during the critical period of sexual differentiation (Auyeung et al., 2013; Christine-Knickmeyer and Baron-Cohen, 2006), which increases plasticity necessary for altering neural connectivity and function to masculinize the developing brain. Thus, with increased plasticity may come increased vulnerability; however, this hypothesis remains to be determined.

These important questions will require continued basic fundamental research into the mechanisms of sexual differentiation of the brain and requires approaches that consider the hormonal, genetic, and epigenetic processes that regulate sexual differentiation of the developing brain.

## REFERENCES

Abraham, I.M., Todman, M.G., Korach, K.S., Herbison, A.E., 2004. Critical *in vivo* roles for classical estrogen receptors in rapid estrogen actions on intracellular signaling in mouse brain. Endocrinology 145 (7), 3055–3061.

Ågmo, A., 1997. Male rat sexual behavior. Brain Res. Protoc. 1 (2), 203–209.

Amateau, S.K., McCarthy, M.M., 2002. A novel mechanism of dendritic spine plasticity involving estradiol induction of prostaglandin-E2. J. Neurosci. 22 (19), 8586–8596.

Amateau, S.K., McCarthy, M.M., 2004. Induction of PGE2 by estradiol mediates developmental masculinization of sex behavior. Nat. Neurosci. 7 (6), 643–650.

Amateau, S.K., Alt, J.J., Stamps, C.L., McCarthy, M.M., 2004. Brain estradiol content in newborn rats: sex differences, regional heterogeneity, and possible de novo synthesis by the female telencephalon. Endocrinology 145 (6), 2906–2917.

Andrews, G.K., Dziadek, M., Tamaoki, T., 1982. Expression and methylation of the mouse alpha-fetoprotein gene in embryonic, adult, and neoplastic tissues. J. Biol. Chem. 257 (9), 5148–5153.

Araque, A., Sanzgiri, R.P., Parpura, V., Haydon, P.G., 1999. Astrocyte-induced modulation of synaptic transmission. Can. J. Physiol. Pharmacol. 77 (9), 699–706.
Arnold, A.P., 2012. The end of gonad-centric sex determination in mammals. Trends Genet. 28 (2), 55–61.
Arnold, A.P., Breedlove, S.M., 1985. Organizational and activational effects of sex steroids on brain and behavior: a reanalysis. Horm. Behav. 19 (4), 469–498.
Arnold, A.P., Gorski, R.A., 1984. Gonadal steroid induction of structural sex differences in the central nervous system. Annu. Rev. Neurosci. 7, 413–442.
Arnold, A.P., Chen, X., Link, J.C., Itoh, Y., Reue, K., 2013. Cell-autonomous sex determination outside of the gonad. Dev. Dyn. 242 (4), 371–379.
Aronica, S.M., Kraus, W.L., Katzenellenbogen, B.S., 1994. Estrogen action via the cAMP signaling pathway: stimulation of adenylate cyclase and cAMP-regulated gene transcription. Proc. Natl. Acad. Sci. USA 91 (18), 8517–8521.
Auyeung, B., Lombardo, M.V., Baron-Cohen, S., 2013. Prenatal and postnatal hormone effects on the human brain and cognition. Pflugers Arch. 465 (5), 557–571.
Bakker, J., Brand, T., van Ophemert, J., Slob, A.K., 1993. Hormonal regulation of adult partner preference behavior in neonatally ATD-treated male rats. Behav. Neurosci. 107 (3), 480–487.
Bakker, J., Honda, S., Harada, N., Balthazart, J., 2004. Restoration of male sexual behavior by adult exogenous estrogens in male aromatase knockout mice. Horm. Behav. 46 (1), 1–10.
Bao, A.M., Swaab, D.F., 2010. Sex differences in the brain, behavior, and neuropsychiatric disorders. Neuroscientist 16 (5), 550–565.
Barraclough, C.A., Gorski, R.A., 1961. Evidence that the hypothalamus is responsible for androgen-induced sterility in the female rat. Endocrinology 68, 68–79.
Baum, M.J., 1979. Differentiation of coital behavior in mammals: a comparative analysis. Neurosci. Biobehav. Rev. 3 (4), 265–284.
Berletch, J.B., Yang, F., Xu, J., Carrel, L., Disteche, C.M., 2011. Genes that escape from X inactivation. Hum. Genet. 130 (2), 237–245.
Bezzi, P., Carmignoto, G., Pasti, L., Vesce, S., Rossi, D., Rizzini, B.L., Volterra, A., 1998. Prostaglandins stimulate calcium-dependent glutamate release in astrocytes. Nature 391 (6664), 281–285.
Blatteis, C.M., 2007. The onset of fever: new insights into its mechanism. Prog. Brain Res. 162, 3–14.
Booth, J.E., 1977. Sexual behaviour of neonatally castrated rats injected during infancy with oestrogen and dihydrotestosterone. J. Endocrinol. 72 (2), 135–141.
Breedlove, S.M., Arnold, A.P., 1983. Hormonal control of a developing neuromuscular system. II. Sensitive periods for the androgen-induced masculinization of the rat spinal nucleus of the bulbocavernosus. J. Neurosci. 3 (2), 424–432.
Christine-Knickmeyer, R., Baron-Cohen, S., 2006. Fetal testosterone and sex differences. Early Hum. Dev. 82 (12), 755–760.
Davies, W., Wilkinson, L.S., 2006. It is not all hormones: alternative explanations for sexual differentiation of the brain. Brain Res. 1126 (1), 36–45.
Davis, E.C., Popper, P., Gorski, R.A., 1996. The role of apoptosis in sexual differentiation of the rat sexually dimorphic nucleus of the preoptic area. Brain Res. 734 (1–2), 10–18.
De Vries, G.J., Simerly, R.B., 2002. Anatomy, development, and function of sexually dimorphic neural circuits in the mammalian brain. Horm. Brain Behav. 5, 137–192.
del Abril, A., Segovia, S., Guillamon, A., 1987. The bed nucleus of the stria terminalis in the rat: regional sex differences controlled by gonadal steroids early after birth. Brain Res. 429 (2), 295–300.
Deng, X., Berletch, J.B., Nguyen, D.K., Disteche, C.M., 2014. X chromosome regulation: diverse patterns in development, tissues and disease. Nat. Rev. Genet. 15 (6), 367–378.
Desai, N.S., Cudmore, R.H., Nelson, S.B., Turrigiano, G.G., 2002. Critical periods for experience-dependent synaptic scaling in visual cortex. Nat. Neurosci. 5 (8), 783–789.
Endele, S., Rosenberger, G., Geider, K., Popp, B., Tamer, C., Stefanova, I., Kutsche, K., 2010. Mutations in GRIN2A and GRIN2B encoding regulatory subunits of NMDA receptors cause variable neurodevelopmental phenotypes. Nat. Genet. 42 (11), 1021–1026.
Feder, H.H., Whalen, R.E., 1965. Feminine behavior in neonatally castrated and estrogen-treated male rats. Science 147, 306–307.

Fink, G., 1988. Oestrogen and progesterone interactions in the control of gonadotrophin and prolactin secretion. J. Steroid Biochem. 30 (1–6), 169–178.
Forger, N.G., Rosen, G.J., Waters, E.M., Jacob, D., Simerly, R.B., de Vries, G.J., 2004. Deletion of bax eliminates sex differences in the mouse forebrain. Proc. Natl. Acad. Sci. USA 101 (37), 13666–13671.
Frankfurt, M., Gould, E., Woolley, C.S., McEwen, B.S., 1990. Gonadal steroids modify dendritic spine density in ventromedial hypothalamic neurons: a Golgi study in the adult rat. Neuroendocrinology 51 (5), 530–535.
Freeman, L.M., Watson, N.V., Breedlove, S.M., 1996. Androgen spares androgen-insensitive motoneurons from apoptosis in the spinal nucleus of the bulbocavernosus in rats. Horm. Behav. 30 (4), 424–433.
Garcia-Segura, L.M., McCarthy, M.M., 2004. Minireview: role of glia in neuroendocrine function. Endocrinology 145 (3), 1082–1086.
Garcia-Segura, L.M., Chowen, J.A., Duenas, M., Torres-Aleman, I., Naftolin, F., 1994. Gonadal steroids as promoters of neuro-glial plasticity. Psychoneuroendocrinology 19 (5–7), 445–453.
Ghahramani, N., Ngun, T., Chen, P., Tian, Y., Krishnan, S., Muir, S., Vilain, E., 2014. The effects of perinatal testosterone exposure on the DNA methylome of the mouse brain are late-emerging. Biol. Sex Differ. 5 (1), 8.
Gu, G.B., Simerly, R.B., 1997. Projections of the sexually dimorphic anteroventral periventricular nucleus in the female rat. J. Comp. Neurol. 384 (1), 142–164.
Guillamon, A., Segovia, S., del Abril, A., 1988. Early effects of gonadal steroids on the neuron number in the medial posterior region and the lateral division of the bed nucleus of the stria terminalis in the rat. Brain Res. 44 (2), 281–290.
Gutierrez-Adan, A., Perez-Crespo, M., Fernandez-Gonzalez, R., Ramirez, M.A., Moreira, P., Pintado, B., Rizos, D., 2006. Developmental consequences of sexual dimorphism during pre-implantation embryonic development. Reprod. Domest. Anim. 41 (Suppl. 2), 54–62.
Hanson, M.G., Landmesser, L.T., 2004. Normal patterns of spontaneous activity are required for correct motor axon guidance and the expression of specific guidance molecules. Neuron 43 (5), 687–701.
Haverkamp, L.J., Oppenheim, R.W., 1986. Behavioral development in the absence of neural activity: effects of chronic immobilization on amphibian embryos. J. Neurosci. 6 (5), 1332–1337.
Hensch, T.K., 2004. Critical period regulation. Annu. Rev. Neurosci. 27, 549–579.
Herbison, A.E., 1998. Multimodal influence of estrogen upon gonadotropin-releasing hormone neurons. Endocr. Rev. 19 (3), 302–330.
Hines, M., Allen, L.S., Gorski, R.A., 1992. Sex differences in subregions of the medial nucleus of the amygdala and the bed nucleus of the stria terminalis of the rat. Brain Res. 579 (2), 321–326.
Hollier, L.P., Keelan, J.A., Hickey, M., Maybery, M.T., Whitehouse, A.J., 2014. Measurement of androgen and estrogen concentrations in cord blood: accuracy, biological interpretation, and applications to understanding human behavioral development. Front. Endocrinol. 5, 64.
Hooks, B.M., Chen, C., 2007. Critical periods in the visual system: changing views for a model of experience-dependent plasticity. Neuron 56 (2), 312–326.
Ibanez, M.A., Gu, G., Simerly, R.B., 2001. Target-dependent sexual differentiation of a limbic-hypothalamic neural pathway. J. Neurosci. 21 (15), 5652–5659.
Jost, A., 1947. Recherches sur la differenciation sexuelle de l'embryon de lapin. Arch. Microsc. Morph. Exp. (36), 271–315.
Kobayashi, S., Isotani, A., Mise, N., Yamamoto, M., Fujihara, Y., Kaseda, K., Okabe, M., 2006. Comparison of gene expression in male and female mouse blastocysts revealed imprinting of the X-linked gene, Rhox5/pem, at preimplantation stages. Curr. Biol. 16 (2), 166–172.
Koopman, P., Munsterberg, A., Capel, B., Vivian, N., Lovell-Badge, R., 1990. Expression of a candidate sex-determining gene during mouse testis differentiation. Nature 348 (6300), 450–452.
Kudwa, A.E., Bodo, C., Gustafsson, J.A., Rissman, E.F., 2005. A previously uncharacterized role for estrogen receptor beta: defeminization of male brain and behavior. Proc. Natl. Acad. Sci. USA 102 (12), 4608–4612.
Kudwa, A.E., Michopoulos, V., Gatewood, J.D., Rissman, E.F., 2006. Roles of estrogen receptors alpha and beta in differentiation of mouse sexual behavior. Neuroscience 138 (3), 921–928.
Lenz, K.M., McCarthy, M.M., 2014. A starring role for microglia in brain sex differences. Neuroscientist, doi:1073858414536468 [pii].

Lenz, K.M., Wright, C.L., Martin, R.C., McCarthy, M.M., 2011. Prostaglandin E regulates AMPA receptor phosphorylation and promotes membrane insertion in preoptic area neurons and glia during sexual differentiation. PLoS One 6 (4), e18500.

Lenz, K.M., Nugent, B.M., Haliyur, R., McCarthy, M.M., 2013. Microglia are essential to masculinization of brain and behavior. J. Neurosci. 33 (7), 2761–2772.

Lewis, C., McEwen, B.S., Frankfurt, M., 1995. Estrogen-induction of dendritic spines in ventromedial hypothalamus and hippocampus: effects of neonatal aromatase blockade and adult GDX. Brain Res. 87 (1), 91–95.

MacLusky, N.J., Naftolin, F., 1981. Sexual differentiation of the central nervous system. Science 211 (4488), 1294–1302.

Matsumoto, A., Arai, Y., 1983. Sex difference in volume of the ventromedial nucleus of the hypothalamus in the rat. Endocrinol. Jpn. 30 (3), 277–280.

Matsumoto, A., Arai, Y., 1986. Development of sexual dimorphism in synaptic organization in the ventromedial nucleus of the hypothalamus in rats. Neurosci. Lett. 68 (2), 165–168.

McCarthy, M.M., Nugent, B.M., 2013. Epigenetic contributions to hormonally-mediated sexual differentiation of the brain. J. Neuroendocrinol. 25 (11), 1133–1140.

McCarthy, M.M., Schlenker, E.H., Pfaff, D.W., 1993. Enduring consequences of neonatal treatment with antisense oligodeoxynucleotides to estrogen receptor messenger ribonucleic acid on sexual differentiation of rat brain. Endocrinology 133 (2), 433–439.

McCarthy, M.M., Amateau, S.K., Mong, J.A., 2002. Steroid modulation of astrocytes in the neonatal brain: implications for adult reproductive function. Biol. Reprod. 67 (3), 691–698.

McEwen, B.S., Lieberburg, I., Chaptal, C., Krey, L.C., 1977a. Aromatization: important for sexual differentiation of the neonatal rat brain. Horm. Behav. 9 (3), 249–263.

McEwen, B.S., Lieberburg, I., Maclusky, N., Plapinger, L., 1977b. Do estrogen receptors play a role in the sexual differentiation of the rat brain? J. Steroid Biochem. 8 (5), 593–598.

Millhouse, O.E., 1973. The organization of the ventromedial hypothalamic nucleus. Brain Res. 55 (1), 71–87.

Mong, J.A., Kurzweil, R.L., Davis, A.M., Rocca, M.S., McCarthy, M.M., 1996. Evidence for sexual differentiation of glia in rat brain. Horm. Behav. 30 (4), 553–562.

Mong, J.A., Glaser, E., McCarthy, M.M., 1999. Gonadal steroids promote glial differentiation and alter neuronal morphology in the developing hypothalamus in a regionally specific manner. J. Neurosci. 19 (4), 1464–1472.

Mong, J.A., Roberts, R.C., Kelly, J.J., McCarthy, M.M., 2001. Gonadal steroids reduce the density of axospinous synapses in the developing rat arcuate nucleus: an electron microscopy analysis. J. Comp. Neurol. 432 (2), 259–267.

Murakami, S., Arai, Y., 1989. Neuronal death in the developing sexually dimorphic periventricular nucleus of the preoptic area in the female rat: effect of neonatal androgen treatment. Neurosci. Lett. 102 (2–3), 185–190.

Murray, E.K., Hien, A., de Vries, G.J., Forger, N.G., 2009. Epigenetic control of sexual differentiation of the bed nucleus of the stria terminalis. Endocrinology 150 (9), 4241–4247.

Naftolin, F., MacLusky, N., 1984. Aromatization Hypothesis Revisited. Sexual Differentiatioin: Basic and Clinical Aspects. Raven Press, New York, pp. 79.

Nevison, C.M., Brown, G.R., Dixson, A.F., 1997. Effects of altering testosterone in early infancy on social behaviour in captive yearling rhesus monkeys. Physiol. Behav. 62 (6), 1397–1403.

Nguyen, D.K., Disteche, C.M., 2006a. Dosage compensation of the active X chromosome in mammals. Nat. Genet. 38 (1), 47–53.

Nguyen, D.K., Disteche, C.M., 2006b. High expression of the mammalian X chromosome in brain. Brain Res. 1126 (1), 46–49.

Nordeen, E.J., Yahr, P., 1983. A regional analysis of estrogen binding to hypothalamic cell nuclei in relation to masculinization and defeminization. J. Neurosci. 3 (5), 933–941.

Nordeen, E.J., Nordeen, K.W., Sengelaub, D.R., Arnold, A.P., 1985. Androgens prevent normally occurring cell death in a sexually dimorphic spinal nucleus. Science 229 (4714), 671–673.

Nugent, B.M., McCarthy, M.M., 2011. Epigenetic underpinnings of developmental sex differences in the brain. Neuroendocrinology 93 (3), 150–158.

Nugent, B.M., Wright, C.L., Zup, S.L., McCarthy, M.M., 2009. Masculinization induced by neonatal exposure to PGE(2) or estradiol alters c-fos induction by estrous odors in adult rats. Physiol. Behav. 96 (2), 383–388.

Ogawa, S., Chan, J., Chester, A.E., Gustafsson, J.A., Korach, K.S., Pfaff, D.W., 1999. Survival of reproductive behaviors in estrogen receptor beta gene-deficient (betaERKO) male and female mice. Proc. Natl. Acad. Sci. USA 96 (22), 12887–12892.

Pfaff, D.W., Sakuma, Y., 1979. Deficit in the lordosis reflex of female rats caused by lesions in the ventromedial nucleus of the hypothalamus. J. Physiol. 288, 203–210.

Phoenix, C.H., Goy, R.W., Gerall, A.A., Young, W.C., 1959. Organizing action of prenatally administered testosterone propionate on the tissues mediating mating behavior in the female guinea pig. Endocrinology 65, 369–382.

Pozzo Miller, L.D., Aoki, A., 1991. Stereological analysis of the hypothalamic ventromedial nucleus. II. Hormone-induced changes in the synaptogenic pattern. Brain Res. Dev. Brain Res. 61 (2), 189–196.

Reddy, V.V., Naftolin, F., Ryan, K.J., 1974. Conversion of androstenedione to estrone by neural tissues from fetal and neonatal rats. Endocrinology 94 (1), 117–121.

Rhoda, J., Corbier, P., Roffi, J., 1984. Gonadal steroid concentrations in serum and hypothalamus of the rat at birth: aromatization of testosterone to 17 beta-estradiol. Endocrinology 114 (5), 1754–1760.

Ropers, H.H., 2006. X-linked mental retardation: many genes for a complex disorder. Curr. Opin. Genet. Dev. 16 (3), 260–269.

Ropers, H.H., 2010. Genetics of early onset cognitive impairment. Annu. Rev. Genomics Hum. Genet. 11, 161–187.

Roselli, C.E., Larkin, K., Resko, J.A., Stellflug, J.N., Stormshak, F., 2004. The volume of a sexually dimorphic nucleus in the ovine medial preoptic area/anterior hypothalamus varies with sexual partner preference. Endocrinology 145 (2), 478–483.

Ruben, R.J., 1997. A time frame of critical/sensitive periods of language development. Acta Otolaryngol. 117 (2), 202–205.

Schlaggar, B.L., O'Leary, D.D., 1991. Potential of visual cortex to develop an array of functional units unique to somatosensory cortex. Science 252 (5012), 1556–1560.

Schlaggar, B.L., O'Leary, D.D., 1993. Patterning of the barrel field in somatosensory cortex with implications for the specification of neocortical areas. Perspect. Dev. Neurobiol. 1 (2), 81–91.

Schwarz, J.M., McCarthy, M.M., 2008. The role of neonatal NMDA receptor activation in defeminization and masculinization of sex behavior in the rat. Horm. Behav. 54 (5), 662–668.

Schwarz, J.M., Liang, S.L., Thompson, S.M., McCarthy, M.M., 2008. Estradiol induces hypothalamic dendritic spines by enhancing glutamate release: a mechanism for organizational sex differences. Neuron 58 (4), 584–598.

Schwarz, J.M., Nugent, B.M., McCarthy, M.M., 2010. Developmental and hormone-induced epigenetic changes to estrogen and progesterone receptor genes in brain are dynamic across the life span. Endocrinology 151 (10), 4871–4881.

Shughrue, P.J., Lane, M.V., Merchenthaler, I., 1997. Comparative distribution of estrogen receptor-alpha and -beta mRNA in the rat central nervous system. J. Comp. Neurol. 388 (4), 507–525.

Simerly, R.B., Swanson, L.W., Handa, R.J., Gorski, R.A., 1985. Influence of perinatal androgen on the sexually dimorphic distribution of tyrosine hydroxylase-immunoreactive cells and fibers in the anteroventral periventricular nucleus of the rat. Neuroendocrinology 40 (6), 501–510.

Sinclair, A.H., Berta, P., Palmer, M.S., Hawkins, J.R., Griffiths, B.L., Smith, M.J., Goodfellow, P.N., 1990. A gene from the human sex-determining region encodes a protein with homology to a conserved DNA-binding motif. Nature 346 (6281), 240–244.

Singh, M., Sétáló, Jr., G., Guan, X., Warren, M., Toran-Allerand, C.D., 1999. Estrogen-induced activation of mitogen-activated protein kinase in cerebral cortical explants: convergence of estrogen and neurotrophin signaling pathways. J. Neurosci. 19 (4), 1179–1188.

Skaletsky, H., Kuroda-Kawaguchi, T., Minx, P.J., Cordum, H.S., Hillier, L., Brown, L.G., Page, D.C., 2003. The male-specific region of the human Y chromosome is a mosaic of discrete sequence classes. Nature 423 (6942), 825–837.

Skuse, D.H., 2005. X-linked genes and mental functioning. Hum. Mol. Genet. 14 (1), R27–R32.

Sumida, H., Nishizuka, M., Kano, Y., Arai, Y., 1993. Sex differences in the anteroventral periventricular nucleus of the preoptic area and in the related effects of androgen in prenatal rats. Neurosci. Lett. 151 (1), 41–44.

Swaab, D.F., 2004. Sexual differentiation of the human brain: relevance for gender identity, transsexualism and sexual orientation. Gynecol. Endocrinol. 19 (6), 301–312.

Tagirov, M., Rutkowska, J., 2014. Sexual dimorphism in the early embryogenesis in zebra finches. PLoS One 9 (12), e114625.

Tsai, H.W., Grant, P.A., Rissman, E.F., 2009. Sex differences in histone modifications in the neonatal mouse brain. Epigenetics 4 (1), 47–53.

Wallen, K., 2005. Hormonal influences on sexually differentiated behavior in nonhuman primates. Front. Neuroendocrinol. 26 (1), 7–26.

Watters, J.J., Campbell, J.S., Cunningham, M.J., Krebs, E.G., Dorsa, D.M., 1997. Rapid membrane effects of steroids in neuroblastoma cells: effects of estrogen on mitogen activated protein kinase signalling cascade and c-fos immediate early gene transcription. Endocrinology 138 (9), 4030–4033.

Weisz, J., Ward, I.L., 1980. Plasma testosterone and progesterone titers of pregnant rats, their male and female fetuses, and neonatal offspring. Endocrinology 106 (1), 306–316.

Wersinger, S.R., Sannen, K., Villalba, C., Lubahn, D.B., Rissman, E.F., De Vries, G.J., 1997. Masculine sexual behavior is disrupted in male and female mice lacking a functional estrogen receptor alpha gene. Horm. Behav. 32 (3), 176–183.

Wolstenholme, J.T., Rissman, E.F., Bekiranov, S., 2013. Sexual differentiation in the developing mouse brain: contributions of sex chromosome genes. Genes Brain Behav. 12 (2), 166–180.

Wong, R.O., 1999. Retinal waves and visual system development. Annu. Rev. Neurosci. 22, 29–47.

Wright, C.L., McCarthy, M.M., 2009. Prostaglandin E2-induced masculinization of brain and behavior requires protein kinase A, AMPA/kainate, and metabotropic glutamate receptor signaling. J. Neurosci. 29 (42), 13274–13282.

Wright, C.L., Burks, S.R., McCarthy, M.M., 2008. Identification of prostaglandin E2 receptors mediating perinatal masculinization of adult sex behavior and neuroanatomical correlates. Dev. Neurobiol. 68 (12), 1406–1419.

Xu, J., Burgoyne, P.S., Arnold, A.P., 2002. Sex differences in sex chromosome gene expression in mouse brain. Hum. Mol. Genet. 11 (12), 1409–1419.

Zechner, U., Wilda, M., Kehrer-Sawatzki, H., Vogel, W., Fundele, R., Hameister, H., 2001. A high density of X-linked genes for general cognitive ability: a run-away process shaping human evolution? Trends Genet. 17 (12), 697–701.

Zhang, L.I., Poo, M.M., 2001. Electrical activity and development of neural circuits. Nat. Neurosci. 4 (Suppl.), 1207–1214.

Zhao, D., McBride, D., Nandi, S., McQueen, H.A., McGrew, M.J., Hocking, P.M., Clinton, M., 2010. Somatic sex identity is cell autonomous in the chicken. Nature 464 (7286), 237–242.

Zhou, Y., Watters, J.J., Dorsa, D.M., 1996. Estrogen rapidly induces the phosphorylation of the cAMP response element binding protein in rat brain. Endocrinology 137 (5), 2163–2166.

Ziegler, D.R., Cullinan, W.E., Herman, J.P., 2002. Distribution of vesicular glutamate transporter mRNA in rat hypothalamus. J. Comp. Neurol. 448 (3), 217–229.

Znamensky, V., Akama, K.T., McEwen, B.S., Milner, T.A., 2003. Estrogen levels regulate the subcellular distribution of phosphorylated akt in hippocampal CA1 dendrites. J. Neurosci. 23 (6), 2340–2347.

Zup, S.L., Carrier, H., Waters, E.M., Tabor, A., Bengston, L., Rosen, G.J., Forger, N.G., 2003. Overexpression of bcl-2 reduces sex differences in neuron number in the brain and spinal cord. J. Neurosci. 23 (6), 2357–2362.

# CHAPTER 11

# Presence and Absence of Sex Differences in Structure and Function of the Brain Oxytocin System: Implications for Understanding the Regulation of Social Behavior

**Kelly M. Dumais, Alexa H. Veenema**
Boston College, Neurobiology of Social Behavior Laboratory, Department of Psychology, Chestnut Hill, MA, USA

## 1 INTRODUCTION

The neuropeptide oxytocin (OT) and the OT receptor (OTR) have been implicated in the regulation of various social behaviors in vertebrates, including mammals (Heinrichs et al., 2009; Ross and Young, 2009; Goodson and Thompson, 2010; Anacker and Beery, 2013; Hammock, 2015). The classically known function of OT in mammals is its regulation of the maternal response, in which peripheral OT is important in the induction of uterine contractions during parturition and milk ejection during lactation (Fuchs and Poblete, 1970; Belin et al., 1984). Since the late 1970s, researchers have extensively investigated the role of OT in the maternal brain, finding that OT is important in the parturient female via the regulation of both active and passive forms of maternal behavior, such as nesting, retrieving pups, and licking and grooming of pups (Pedersen and Prange, 1979). It was not until the late 1980s and early 1990s that OT's role in other social behaviors, such as social bonding and social recognition, started to be explored. Ironically, when the role of OT in these other forms of social behavior started being investigated, males, rather than females, were predominantly used as subjects. For example, OT was found to be a key regulator of social approach and social recognition in male rats (Dantzer et al., 1987; Popik and van Ree, 1991; Popik et al., 1992; Witt et al., 1992; Benelli et al., 1995; Arletti et al., 1995). With these studies highlighting the role of OT in male social behavior, and others starting to explore the role of OT in female social behaviors other than maternal care (for example, pair bonding in female prairie voles; Williams et al., 1992; Insel and Hulihan, 1995), this classically known "maternal hormone" is now seen as an important regulator of both male and female social behavior. More recently, OT is increasingly recognized as an important modulator of human social behavior (Donaldson and Young, 2008; Heinrichs et al., 2009; Guastella and MacLeod, 2012).

*Sex Differences in the Central Nervous System.* http://dx.doi.org/10.1016/B978-0-12-802114-9.00011-1

As OT became known as a key regulator of social behavior in both males and females, sex differences in the behavioral effects of OT became apparent. Given these sexually dimorphic actions of OT, it is often reported that OT itself is a sexually dimorphic neuropeptide. However, sex differences have been found in the brain OT system (OT peptide and OTR) in some but not all cases, which will be the first topic of discussion of this review (Section 2). We then discuss the implications of these sex differences, or lack thereof, in the OT system for the sex-specific regulation of social behavior in both rodents and humans (Section 3).

In rodents, we focus on sex-specific influences of OT on social behaviors including social recognition, social approach, social play, partner preference formation, mate-guarding, parental care, and alloparental care (Sections 3.1 and 3.2). We have not included a comparative analysis on the role of OT in male and female sexual and aggressive behaviors, mainly because of the differences in the types of behavior (female lordosis vs. male erection/ejaculation, maternal aggression vs. intermale aggression). The reader is referred to the following reviews (although also lacking a comparative analysis) on the role of OT in male and female sexual behavior (Veening et al., 2014), in maternal aggression (Bosch, 2013), and in intermale aggression (Calcagnoli, 2014). In humans, we discuss relevant research describing the influence of OT in both behavior and cognition, including social memory, inferring mental sates of others, cooperation and trust, moral judgments, and couple conflict (Section 3.3). Because of the role of OT in anxiety-related behavior, we also discuss sex-specific influences of OT on anxiety in both rodents and humans (Sections 3.2.2 and 3.3.2). Importantly, the species discussed in this review vary significantly in their social organizations. The social organization of a given species encompasses the organization of social relations within a group, including mating systems (polygamy, monogamy, promiscuity), parental systems (biparental care, maternal care), and sociality systems (the degree to which individuals live in social groups, ranging from solitary to eusocial). We will argue that such species differences should be taken into account as they may help us understand why certain species show sex differences in the brain OT system and OT function while other species do not (Section 2.3). We also discuss whether sex differences in the brain OT system are required for OT to induce sex differences in social behavior (Section 3.4).

The investigation of sex differences in OT parameters and function is especially important given that OT may have therapeutic potential in the treatment of social dysfunction in various neuropsychiatric disorders, including autism spectrum disorder (ASD), schizophrenia, depression, and anxiety disorders (Guastella et al., 2010; Meyer-Lindenberg et al., 2011; Cochran et al., 2013). These disorders show strong sex differences in prevalence, symptom severity, and treatment responses (Fombonne, 2003; Beaudet, 2012). Therefore, a better understanding of whether the brain OT system is sexually dimorphic and whether there are sex-specific actions of OT on brain and behavioral function is imperative if we are to use OT as a potential drug in the treatment of social dysfunction. We will therefore end with a discussion on the advances and precautions of OT as a therapeutic agent in the treatment of social dysfunction in males and females (Section 4).

## 2 SEX DIFFERENCES IN THE BRAIN OT SYSTEM

OT is a highly conserved nine amino acid neuropeptide, which acts as both a hormone in the periphery via the hypothalamic–neurohypophysial system, and as a neuromodulator in the central nervous system. OT-producing magnocellular neurons of the paraventricular nucleus (PVN) and the supraoptic nucleus (SON) of the hypothalamus project to the posterior pituitary, where it is released into the general circulation as a hormone. Moreover, the OT system has evolved to also include an expansion of axonal projections throughout the brain. Research from the late 1970s and early 1980s shows that, in rats, OT synthesized in parvocellular neurons of the PVN projects centrally (Buijs, 1978; Buijs and Swaab, 1979; Sofroniew, 1980, 1983) where it can modulate the activation of many brain regions via binding to the widely distributed OTR (Jard et al., 1987). It was found that, in prairie voles, magnocellular OT neurons of the PVN and SON also project to forebrain regions, namely, the nucleus accumbens (Ross et al., 2009). This was confirmed in rats by Knobloch et al. (2012). They provided important additional details by showing that, although OT projections to forebrain structures come predominantly from the PVN, projections from magnocellular neurons in the SON and in the accessory nuclei (AN) of the hypothalamus also project to a variety of forebrain regions (Knobloch et al., 2012). Notably, early studies used either male rats or did not specify sex, while these latter two studies used females only. A quantitative comparison of OT fiber projections between males and females throughout the entire brain and across species is still missing.

In contrast to the closely related neuropeptide vasopressin (DeVries and Panzica, 2006; Albers, 2014), potential sex differences in brain OT system parameters have been studied less (Dumais and Veenema, 2015). As will be discussed, potential sex differences in the brain OT system have also yielded more inconsistent outcomes, especially across species. Given its important role in maternal behavior, it is sometimes intuitively presumed that OT peptide levels are higher in females compared with males. While this may be true in some species and contexts, it is not always the case. The OT system is greatly evolved in both sexes and this corresponds with a significant role of OT in the regulation of social behavior in both males and females.

### 2.1 Are there sex differences in the brain OT system in rodents?

#### *2.1.1 Sex differences in OT in the rodent brain*

In rats, OT is synthesized in the PVN and SON (Brownstein et al., 1980; Dierickx, 1980; Swanson and Sawchenko, 1983) as well as in groups of magnocellular neurons in the AN that are scattered between the PVN and SON (Rhodes et al., 1981). In other rodent species, including hamsters, gerbils, voles, and naked mole-rats, additional OT synthesizing cells have been reported in the anterior hypothalamus (AH), lateral hypothalamus (LH), medial preoptic area (MPOA), bed nucleus of the stria terminalis (BNST), and medial amygdala (MeA; Wang et al., 1996, 2013; Rosen et al., 2008; Xu et al., 2010).

Importantly, OT synthesis has been found to be sexually dimorphic in some of these areas, but in species- and brain region-specific ways.

In the PVN, sex differences in OT-immunoreactive (ir) neurons largely depend on the species being examined (Table 11.1). For example, there were more OT-ir neurons in the PVN in females in CD mice (Häussler et al., 1990), Brandt's voles (Xu et al., 2010), and mandarin voles (Qiao et al., 2014). A study by Xu et al. (2010) analyzed long-tailed hamsters and Brandt's voles together, and reported more OT-ir in females for both species, but the sex difference seemed to be driven by the Brandt's voles only. In contrast, no sex difference in OT messenger RNA (mRNA) expression in the PVN was found in Wistar rats (Dumais et al., 2013). Likewise, OT-ir neurons in the PVN were similar between males and females in Mongolian gerbils (Wang et al., 2013), Chinese striped hamsters (Wang et al., 2013), naked mole-rats (Rosen et al., 2008), and prairie, pine, meadow, and montane voles (Wang et al., 1996). The PVN, then, is a great example of the species specificity of sex differences in the OT system, with only 3 out of 11 rodent species investigated showing higher OT synthesis in females. As the PVN is a major output region of OT neurons, these species-specific sex differences may correspond with species-specific sex differences in how OT regulates social behavior.

In some species, like the CD mice and mandarin voles, similar sex differences in OT peptide level are seen in additional brain regions, such as the SON (both species) and anterior hypothalamic periventricular nucleus (CD mice) as well as in OT-ir fibers in areas such as the lateral septum (LS), BNST (CD mice), and LH (mandarin voles) (Häussler et al., 1990; Qiao et al., 2014). Likewise, a lack of a sex difference in OT synthesis in the PVN is consistent with a lack of a sex difference in OT synthesis in the SON of Wistar rats (Dumais et al., 2013) and in the MPOA of prairie, pine, meadow, and montane voles (Wang et al., 1996). However, inconsistencies in sex differences within one species are also seen. For example, Mongolian gerbils did not show a sex differences in OT-ir neurons in the PVN, but OT-ir neurons in the MPOA were higher in females than in males (Wang et al., 2013). The same is true for Chinese striped hamsters (Wang et al., 2013). Conversely, Brandt's voles showed a sex difference in OT-ir neurons in the PVN, but not in the MPOA (Xu et al., 2010). These findings indicate that, at least in some species, there may be important brain region-specific sex differences in OT synthesis, which, in turn, may have consequences for OT's sex- and brain region-specific regulation of social behavior (Dumais and Veenema, 2015).

To summarize, of the few rodent species showing sex differences in OT synthesis, it was females who had higher OT-ir compared with males (Table 11.1). The functional significance of these sex differences is still to be determined. Although it may be tempting to assume that such sex differences could reflect the role of OT in female-typical activities and behaviors (such as pregnancy, parturition, nursing, and lordosis), such sex differences would be expected to be found in all species if this assumption was true. The reproductive status (such as mating experience or estrus cycle phase) of the females in the studies described previously is, however, unclear. In order to understand the role of these sex differences it would be important to study whether sex differences, or lack thereof, in OT synthesis change depending on the reproductive state of both males and females and how this relates to the social organization of a given species. For example, could it be that OT synthesis is increased in both lactating females and paternal males compared with virgins? Would such an increase in OT synthesis be absent in nonpaternal males? These are questions that will need to be explored in future studies.

Another important step in determining the functional significance of sex differences, or the lack thereof, in OT synthesis, is the investigation of how levels of OT-ir neurons or OT-ir fiber density correspond to OT release in the brain. In contrast to OT synthesis, OT release is more dynamic and, especially when measured during the display of social behavior (Veenema and Neumann, 2008), can provide information about potential sex differences in local OT release in response to social stimuli. It is possible that such sex differences in dynamic OT release may occur in the absence of sex differences in OT synthesis. Given that OT neurons in the PVN and SON project to the periphery and likely have collaterals that may release OT in the brain (Ross et al., 2009), it will also be important to know how sex differences, or a lack thereof, in OT synthesis relates to peripheral OT levels. The few studies thus far are inconclusive and are not carried out in one design. For example, female mandarin voles were reported to have higher OT-ir (Qiao et al., 2014) as well as plasma OT levels (Cao et al., 2013) than their male counterparts. However, prairie voles did not show sex differences in OT synthesis in the brain (Wang et al., 1996), yet females had higher plasma OT levels than males (Kramer et al., 2004).

With these questions left unanswered, potential correlations between OT synthesis and species-specific social behavior may still give insight into the functional role of sex differences in OT in some rodent species, and is discussed further in Section 2.3. Overall we can conclude that in most rodent species there is a lack of sex differences in OT synthesis. Therefore, studies comparing OT synthesis, OT release in the brain, and OT plasma levels of different species, in different social contexts, and in different reproductive states, will be important to gain a better understanding of whether the OT peptide is or is not sexually dimorphic in rodents.

### 2.1.2 Sex differences in OTR in the rodent brain

Peripheral and central actions of OT are mediated via one type of OTR, which is a G protein-coupled receptor (Gimpl and Fahrenholz, 2001). It is, however, possible for OT to bind to vasopressin receptors (Mouillac et al., 1995; Manning et al., 2012). This is an emerging area of study, with the first functional studies in rodents (Schorscher-Petcu et al., 2010; Song et al., 2014) demonstrating the importance of carefully interpreting behavioral effects mediated by exogenous OT. However, for simplicity, we will only discuss the OTR here. Because the OTR is highly expressed in brain regions involved in the regulation of social behavior (Tribollet et al., 1990; Ophir et al., 2012; Dumais et al., 2013; Zheng et al., 2013), knowledge of sex differences in the expression of OTR (discussed in this section) is important for understanding the role of OT in sex-specific regulation of social behavior (discussed in Section 3).

The first comparative study on OTR binding in the brain was performed in Sprague–Dawley rats and reported that there were no sex differences in OTR binding density in any of the eight forebrain regions analyzed (Tribollet et al., 1990). However, later studies found a robust sex difference in OTR mRNA expression (Bale and Dorsa, 1995) as well as OTR binding (Uhl-Bronner et al., 2005) in the ventromedial hypothalamus (VMH) of Long–Evans and Sprague–Dawley rats, respectively, with higher levels in males than in females. This discrepancy could be due to the use of different ligands across studies: [$^{3}$H] OT in the study by Tribollet et al. (1990) versus a highly specific [$^{125}$I]OTR antagonist in the study by Uhl-Bronner et al. (2005), with the possibility that [$^{3}$H]OT might not exclusively bind OTR (Mouillac et al., 1995; Manning et al., 2012). By using the latter ligand, we recently confirmed and expanded findings of sex differences in OTR binding in the brain of Wistar rats. Specifically, we showed that male rats had higher OTR binding densities than female rats in 9 out of 15 forebrain regions analyzed, including the nucleus accumbens, dorsal caudate putamen, intermediate LS, posterior BNST, MPOA, agranular insular cortex, hippocampal CA1 region, MeA, and VMH (Dumais et al., 2013). Like the sex difference in OTR binding in the MeA of rats, OTR mRNA expression in the MeA in mandarin voles was higher in males compared with females (Cao et al., 2013).

To the best of our knowledge, a comprehensive study comparing OTR binding in the brain between males and females of common laboratory mice (species *Mus musculus*) is lacking. This is quite surprising given the frequent use of these mice to study the role of OT in social behavior. One study reported that male ICR mice had lower OTR binding in the VMH than female ICR mice (Tribollet et al., 2002). A recent study in C57Bl/6J mice found no sex differences in OTR in the hippocampus, LS, and neocortex (Hammock and Levitt, 2013). However, neither of these studies included a comparative analysis in other brain regions showing OTR binding.

In addition, there is a lack of comprehensive research on sex differences in OTR in diverse vole species. Of those species that have been investigated, it was found that female prairie voles show higher OTR binding densities in the medial prefrontal cortex compared with males (Smeltzer et al., 2006); however, no sex differences were found in the ventral pallidum, MeA, central amygdala (CeA), BNST, LS, or cingulate cortex (Bales et al., 2007a). In addition, male mandarin voles showed higher OTR mRNA expression in the MeA, but not the nucleus accumbens (Cao et al., 2013). No other brain regions were investigated (Cao et al., 2013). This overall lack of sex differences in OTR in prairie voles is especially intriguing given the robust sex-specific behavioral effects of exogenous OT and OTR antagonist administrations (see Section 3.1). Although there is a lack of significant sex differences in OTR in prairie voles, the limited regions that were analyzed for OTR in mandarin voles, and the lack of research in other voles species all together, highlight the need for comparative research on sex differences in OTR in a variety of vole species.

Sex differences in OTR binding were compared between two deer mouse species, *Peromyscus maniculatus* and *Peromyscus californicus*. These species differ in mating systems, in which *P. maniculatus* is polygamous and *P. californicus* is monogamous. When the two species were analyzed together, 8 out of 20 brain regions analyzed showed sex differences in OTR binding, including the mitral cell layer and granule cell layer of the olfactory bulb, cingulate cortex, dorsal LS, lateral-, medial anterior-, and medial posterior-BNST, and the CA1 of the hippocampus (Insel et al., 1991). Males had higher OTR binding compared with females in all regions except for the CA1 of the hippocampus, in which females were higher than males. Importantly, the polygamous species *P. maniculatus* seemed to drive the sex differences in OTR binding, especially in areas such as the dorsal LS, the medial anterior BNST, and the CA1. In fact, the only region that showed a sex difference in OTR binding in the monogamous species was the medial posterior BNST (Insel et al., 1991). Furthermore, a lack of sex differences in OTR binding was found for the solitary golden hamster (Dubois-Dauphin et al., 1992). Taken together, these findings clearly indicate that, similar to OT, there are robust species- and brain region-specific sex differences in OTR binding in the brain (Table 11.1).

The sex differences in OTR in brain regions such as the LS, BNST, VMH, and MeA in rats (Dumais et al., 2013), the VMH in some mouse species (Insel et al., 1991; Tribollet et al., 2002), and the MeA in mandarin voles (Cao et al., 2013) may be due to the organizational and/or activational effects of gonadal steroids, with levels of testosterone being higher in males compared with females. Organizational effects of gonadal steroids encompass the permanent effects of hormonal actions on brain development early in life (Arnold, 2009). On the other hand, activational influences of gonadal steroid hormones describe the acute and transitory effects of hormonal actions that occur throughout life. Indeed, it was found that both organizational (Uhl-Bronner et al., 2005) and activational (Tribollet et al., 1990) effects of gonadal hormones influence OTR. Female rats treated neonatally with testosterone show higher levels of OTR binding densities in the VMH, BNST, and MeA compared with control female rats (Uhl-Bronner et al., 2005). However, it is unclear whether these increases in OTR binding are to the level of that seen in male rats in these regions. Furthermore, gonadectomy of adult rats decreased OTR binding in areas such as the VMH, dorsolateral BNST, and dorsal caudate putamen in both males and females (Tribollet et al., 1990). Therefore, sex differences in gonadal hormone levels during the organizational and activational periods may explain sex differences in OTR binding densities in these areas. However, this does not explain the sex differences in OTR in areas that may not be sensitive to gonadal steroids (Tribollet et al., 1990), such as the nucleus accumbens, caudate putamen, and hippocampal CA1 region (Dumais et al., 2013).

Sex differences in OTR may also relate to sex differences in local OT release. Despite the lack of a sex difference in OT mRNA expression in the PVN and SON in Wistar rats (Dumais et al., 2013), it is still possible that there are sex differences in OT peptide and/or OT release due to sex differences in posttranslational mechanisms or release mechanisms, respectively. Several studies have examined OT release in specific forebrain areas of male (Engelmann et al., 1999; Ebner et al., 2000; Waldherr and Neumann, 2007) and female (Nyuyki et al., 2011; Neumann et al., 1993; Bosch et al., 2010, 2004) rats and female prairie voles (Ross et al., 2009), and found significant increases in OT release when these animals were exposed to diverse social stimuli. However, none of these studies included both males and females and therefore a direct comparison of OT release patterns is still lacking. Such a comparison would be an important step in understanding the functional significance of sex differences, or a lack thereof in OTR binding in the brain of rodents.

Together, these findings indicate that, while females of particular species (such as voles, hamsters, and gerbils) show higher OT-ir, males of other species (rats and mice) show higher OTR binding. This suggests that when the OT system shows sexual dimorphisms, it is not always females that show an increase in OT system parameters compared with males (see also Table 11.1).

## 2.2 Are there sex differences in the brain OT system in humans?

Similar to rodents, OT in humans is synthesized in the PVN, SON, and AN (Fliers et al., 1985; Wierda et al., 1991; Meynen et al., 2007). Studies in humans have consistently reported an absence of sex differences in the number of OT neurons in the PVN (Wierda et al., 1991) and in the size of OT neurons in the PVN and SON (Fliers et al., 1985; Ishunina and Swaab, 1999; Table 11.2). This is in agreement with most rodent species, in which OT synthesis (mRNA expression and OT-ir) in the PVN and SON is similar between males and females (see Section 2.1.1; Table 11.1).

Despite this lack of sex differences in OT synthesis in the brain, there could be sex differences in the release of OT. Because of the lack of practicality and feasibility of measuring brain OT directly, plasma OT has served as the next best source for measuring bioactive OT levels in humans, even though it is still unclear as to whether plasma OT levels reflect brain OT release (Evans et al., 2014). Nevertheless, plasma OT levels in men and women may be informative in revealing potential sex differences in circulating OT.

Studies thus far have been inconsistent regarding the presence and the direction of sex differences in plasma OT levels in humans (Table 11.2). For example, preadolescent and adolescent girls were found to have higher plasma OT levels than boys (Miller et al., 2013). However, using a very large cohort ($n$ = 473), Weisman et al. (2013) found that plasma OT levels were higher in adult men compared with adult women. Still, others reported no sex differences in plasma OT (Zhong et al., 2012; Taylor et al., 2010; Grewen et al., 2005; Gordon et al., 2008). There were also no sex differences in plasma

**Table 11.1** Summary of sex differences in the brain OT system and plasma OT in rodents

| OT measure | Species | Region | Sex effect | References |
|---|---|---|---|---|
| OT-ir neurons | Naked mole-rats | PVN, SON, AH, preoptic area, posterior BNST, MeA | = | Rosen et al. (2008) |
| | Prairie voles | PVN, SON, MPOA, BNST, LH preoptic periventricular nucleus, median preoptic nucleus | = | Wang et al. (1996) |
| | Pine voles | PVN, SON, MPOA, BNST, LH preoptic periventricular nucleus, median preoptic nucleus | = | Wang et al. (1996) |
| | Meadow voles | PVN, SON, MPOA, BNST, LH preoptic periventricular nucleus, median preoptic nucleus | = | Wang et al. (1996) |
| | Montane voles | PVN, SON, MPOA, BNST, LH preoptic periventricular nucleus, median preoptic nucleus | = | Wang et al. (1996) |
| | Brandt's voles | PVN | ↑ in females | Xu et al. (2010) |
| | | MPOA, AH, LH, MeA, reticular thalamic nucleus | = | |
| | Long-tailed hamsters, reticular thalamic nucleus | MPOA, AH, LH, MeA | = | Xu et al. (2010) |
| | Mandarin voles | PVN, LH | ↑ in females | Qiao et al. (2014) |
| | | SON | ↑ in dominant females | |
| | | AH | = | |
| | Chinese striped hamsters | Intermediate MPOA | ↑ in females | Wang et al. (2013) |
| | | PVN, AH, LH, anterior MPOA, posterior MPOA | = | |

*(Continued)*

**Table 11.1** Summary of sex differences in the brain OT system and plasma OT in rodents *(cont.)*

| OT measure | Species | Region | Sex effect | References |
|---|---|---|---|---|
| | Mongolian gerbils | Intermediate MPOA | ↑ in females | Wang et al. (2013) |
| | | PVN, AH, LH, anterior MPOA, posterior MPOA | = | |
| | CD mice | PVN, SON, anterior hypothalamic periventricular nucleus | ↑ in females | Häussler et al. (1990) |
| OT-ir fibers | CD mice | LS, BNST | ↑ in females | Häussler et al. (1990) |
| OT mRNA | Wistar rats | PVN, SON | = | Dumais et al. (2013) |
| OTR mRNA | Long–Evans rats | VMH | ↑ in males | Bale and Dorsa (1995) |
| | Mandarin voles | MeA | ↑ in males | Smeltzer et al. (2006) |
| | | Nucleus accumbens | = | |
| OTR binding | *P. maniculatus* and *P. californicus* mice | Mitral cell layer, granule cell layer, cingulate cortex, dorsal LS, lateral-, medial anterior-, and medial posterior-BNST | ↑ in males | Insel et al. (1991) |
| | | When analyzed alone, *P. californicus* showed a sex difference in the medial posterior BNST only | | |
| | Sprague–Dawley rats | Hippocampal CA1 | ↑ in females | |
| | | VMH, spinal cord | ↑ in males | Uhl-Bronner et al. (2005) |
| | | CeA | = | |
| | Wistar rats | Posterior NAc, CPu, LS, BNST, MPOA, agranular insular cortex, hippocampal CA1, MeA, VMH | ↑ in males | Dumais et al. (2013) |
| | | Anterior NAc, ventral LS, PVN, dorsal lateral BNST, CeA | = | |
| | Prairie voles | Medial prefrontal cortex | ↑ in females | Smeltzer et al. (2006) |
| | | Ventral pallidum, MeA, CeA, BNST, LS, or cingulate cortex | = | Bales et al. (2007a) |
| | Mandarin voles | Medial prefrontal cortex | ↑ in females | Smeltzer et al. (2006) |
| Plasma OT | Sprague–Dawley rats | n/a | ↑ in females | Kramer et al. (2004) |
| | Prairie voles | n/a | ↑ in females | Kramer et al. (2004) |
| | Mandarin voles | n/a | ↑ in females | Cao et al. (2013) |

AH, anterior hypothalamus; BNST, bed nucleus of the stria terminalis; CeA, central amygdala; CPu, caudate putamen; LH, lateral hypothalamus; LS, lateral septum; MeA, medial amygdala; MPOA, medial preoptic area; NAc, nucleus accumbens; OT, oxytocin; OTR, oxytocin receptor; PVN, paraventricular nucleus of the hypothalamus; SON, supraoptic nucleus of the hypothalamus; VMH, ventromedial hypothalamus; ↑, higher/more; =, no effect.

**Table 11.2** Summary of sex differences in the brain OT system and plasma OT in humans

| OT measure | Sex effect | References |
|---|---|---|
| PVN-OT neurons | = | Wierda et al. (1991) |
| PVN/SON-OT neuron size | = | Fliers et al. (1985); Ishunina and Swaab (1999) |
| Plasma OT | ↑ in adolescent girls | Miller et al. (2013) |
| | ↑ in men | Weisman et al. (2013) |
| | = | Zhong et al. (2012) |
| | = | Taylor et al. (2010) |
| | = | Grewen et al. (2005) |
| | = | Gordon et al. (2008) |
| | = in controls and patients with social anxiety disorder | Hoge et al. (2008) |
| | = in controls and patients with depression | Ozsoy et al. (2009) |
| | = in fathers and mothers | Gordon et al. (2010) |

OT, oxytocin; PVN, paraventricular nucleus of the hypothalamus; SON, supraoptic nucleus of the hypothalamus; ↑, higher; =, no sex effect.

OT levels when measured in controls and patients with social anxiety disorder (Hoge et al., 2008). These inconsistencies may be due to factors such as age; the development of the OT system across the lifespan, as well as the effects of OT on brain and behavior early in development, is not well known. In addition, these inconsistencies may be due to methodological differences across studies. Indeed, one study has shown that commercially available methods to quantify plasma OT levels are in need of improvement, because of specificity and sensitivity issues as well as differences in OT values before and after extraction and lack of correlation in OT levels between different methods of assay (Szeto et al., 2011). Until methods to measure OT in plasma are proven to be valid and reliable, comparisons between studies remain difficult, if at all possible.

Additionally, sex differences in the human OT system have been improperly cited, adding to the general, but possibly incorrect, idea that OT levels are higher in females than in males. For example, Gordon et al. (2010) and Ozsoy et al. (2009) were cited as showing sex differences in plasma OT levels. However, there were no sex differences in plasma OT levels in mothers and fathers at any time point measured (during the first postpartum weeks and at 6 months postpartum) in Gordon et al. (2010). Instead, this study reported that plasma OT levels in women after parturition are different compared with levels in women before pregnancy and/or parturition (Gordon et al., 2010). Likewise, in Ozsoy et al. (2009), there were no sex differences in plasma OT levels in a cohort of healthy and depressed individuals (although there was a nonsignificant trend toward higher OT levels in females); there was only a sex difference when looking at depressed individuals, in which depressed females had lower plasma OT levels compared with controls, but depressed males were similar to control males (Ozsoy et al., 2009). Interestingly,

Holt-Lunstad et al. (2011) has been listed as a reference supporting both sex differences and no sex differences in plasma OT levels in two separate peer-reviewed publications. In this study, a sex difference in plasma OT levels was driven by a depression factor, and no sex difference was found in plasma OT levels of "low depression" men and women (Holt-Lunstad et al., 2011).

Only one study thus far has addressed potential sex differences in OTR in the human brain. Using receptor autoradiography, Loup et al. (1991) localized human OTR in various brain regions and reported no sex differences, but they had a very low number of female subjects ($n = 4$). Another study using immunohistochemistry and a monoclonal antibody localized human OTR in additional brain regions, but only used females as subjects (Boccia et al., 2013). This technique provides an important opportunity to analyze in more detail potential sex differences in the human OTR system.

To summarize, of the limited data that exist in humans it seems that OT synthesis in the hypothalamus is similar in men and women, that plasma OT levels are rather inconclusive, and that there are few studies that have compared OTR in the brain between men and women (Table 11.2).

## 2.3 Do sex differences in the OT system relate to species-specific social organizations?

The findings in rodents and humans show that there is evidence for both the presence and absence of sex differences in the OT system, depending on the species and the specific OT system parameters investigated. We therefore conclude that we cannot generalize the idea of a "sexually dimorphic brain OT system." To understand why sex differences in OT and OTR are highly species-specific, it may be necessary to take into account species differences in social organization. We propose that the presence or absence of sex differences in the brain OT system depends on the social organization of that particular species (i.e., mating, parental, and sociality systems as well as social behaviors that relate to these systems such as social learning, social memory, and social recognition). Accordingly, we would predict that species in which males and females differ in one or more aspects of social organization (e.g., only females show parental care) will show more sex differences in the OT system, while species in which males and females exhibit very similar social organizations (e.g., both males and females show parental care) will show fewer sex differences in the OT system.

According to these predictions, we would expect to find that polygamous species, which show sex differences in mating and parental systems (Lande and Arnold, 1985), show sex differences in OT system parameters while monogamous species, which show few if any sex differences in mating and parental systems (Kleiman, 1977), show fewer sex differences in OT system parameters. In support of these predictions, sex differences in OT and/or OTR have been found in polygamous species. For example, there are sex differences in OT parameters in CD mice (OT-ir in PVN and SON; Häussler et al., 1990), *P. maniculatus* (OTR binding; Insel et al., 1991), Sprague–Dawley rats (OTR binding; Uhl-Bronner et al., 2005), Wistar rats (OTR binding; Dumais et al., 2013), and Chinese striped hamsters (OT-ir in MPOA; Wang et al., 2013). Similarly, a lack of sex differences in OT system parameters has been found in monogamous species, such as *P. californicus* mice (OTR binding; Insel et al., 1991) and prairie and pine voles (OT-ir in PVN and SON and OTR; Wang et al., 1996; Bales et al., 2007a).

However, in contrast to these predictions, polygamous species such as meadow and montane voles do not show sex differences in OT system parameters (OT-ir in PVN and SON and OTR; Wang et al., 1996). This may indicate the need not only to focus on one or two aspects of the social organization (in this case mating and parental systems), but to consider all aspects of the social organization (including sociality systems). For example, polygamous species can be primarily solitary (such as meadow and montane voles, and hamsters; Madison, 1980; Webster and Brooks, 1981; Gattermann et al., 2001) or can be living in more complex social groups (such as rats and some mouse species; Brett, 1991; Manning et al., 1992). The concept of integrating mating systems with sociality systems has been elegantly proposed by Ophir (2011) to provide a framework that may help us understand species-specific neuropeptide control of social behavior across vertebrate taxa. We believe that a similar approach is necessary to understand the functional significance of sex differences, or a lack thereof, in the OT system.

However, even taking into account all aspects of social organization might not be enough for a full functional understanding of presence or absence of sex differences in the OT system. It will also be important to consider factors such as season and social density, as these might influence changes in the social organization and OT parameters differently in males than in females (Beery and Zucker, 2010; Anacker and Beery, 2013). In addition, it will be important to consider social context and the way we house laboratory animals (Baumans and Van Loo, 2013), as this may or may not represent the social organization of a given species, which likely affects OT parameters. Finally, we should also consider that sex differences in OT parameters may not always function to induce sex differences in behavior, but rather to prevent sex differences in behavior. This dual-function hypothesis of sex differences in the brain proposed by De Vries (2004) may have consequences for how we should interpret the function of the presence or absence of sex differences in the OT system. Translating this to our proposal, the dual-function hypothesis would suggest that males and females with a similar social organization may actually show sex differences in OT parameters in order for them to show similar behaviors.

Given that humans are typically biparental and socially monogamous, our hypothesis that sex differences in the OT system may relate to sex differences in the social organization of a given species lead us to predict that the OT system (OT and OTR) may be similar between men and women. However, given the additional factors as indicated previously as well as the complexity of human social interactions, the vastness of human social networks, and the large scale of human cooperations, men and women may respond or behave differently depending on the specific social context. Therefore, men and women may show significant sex differences, or possibly important individual differences, in OT parameters. The analysis of potential sex differences in the human OT system, however, is very limited and far from conclusive. While a lack of a sex difference in size and amount of OT neurons in the human brain seem to be consistent across studies, reports of sex differences in OT plasma levels in humans are inconsistent. One potential mediator of these inconsistent reports could be differences in the methods of plasma OT measurement and the population of subjects (healthy subjects vs. psychiatric patients, adults vs. adolescents, etc.). Overall, this highlights the need for more comprehensive research into sex differences in the human OT system.

In conclusion, the reported sex differences and/or lack of sex differences in the OT system among rodents and humans indicate that sex differences in the OT system may correlate with variations in social organizations. Surprisingly, there are still large gaps in the literature regarding comprehensive studies comparing OT brain parameters in males and females across different species, making it difficult at this point to draw any firm conclusions regarding the function of sex differences, or a lack thereof, in the brain OT system. However, another way to study how the OT system functionally relates to social behavior, and how this might be different in males and females, is to manipulate the OT system through pharmacological, genetic, or other approaches. This will be discussed in the following section (Section 3).

## 3 SEXUALLY DIMORPHIC ACTIONS OF OT ON BEHAVIOR AND NEURAL RESPONSES

With the extensive investigation into the role of OT in maternal behavior throughout the late 1970s to early 1990s, females were the stars of OT research. However, as researchers started to explore the role of OT in other forms of social bonding in prairie voles, sexually dimorphic effects of OT on partner preference formation started to appear, with a role in females but not males (Winslow et al., 1993; Insel and Hulihan, 1995). However, further research showed that OT does play a role in partner preference formation in male prairie voles (Cho et al., 1999; Liu et al., 2001). Additionally, other social behaviors that were initially being studied in males only (for example, social recognition and social interaction) were also found to be modulated by OT (Popik et al., 1992; Witt et al., 1992). Given these new found roles of OT in both male and female social behavior, a robust literature of sex-specific actions of the OT system in rodents and humans emerged, and is summarized in the following sections and in Tables 11.3–11.8. We have limited the rodent literature to voles, rats, and mice since almost no research exists regarding the role of OT in the regulation of the same social behavior in both sexes in other rodent species such as hamsters (Harmon et al., 2002) and naked mole-rats (Mooney et al., 2014).

**Table 11.3** Summary of sexually dimorphic actions of acute OT administration on adult behavior in prairie voles, rats, and mice

| Species | Route | Dose | Behavioral/brain response | Male effect | Female effect | References |
|---|---|---|---|---|---|---|
| Prairie voles | ICV | 0.5 ng | Partner preference | = | nm | Winslow et al. (1993) |
| | ICV | 0.5 ng | Partner preference | nm | ↑ | Insel and Hulihan (1995) |
| | ICV | 10 and 100 ng | Partner preference | nm | ↑ | Williams et al. (1994) |
| Rats | ICV | 1 ng/4 μL | Social recognition | ↑ | nm | Benelli et al. (1995) |
| | ICV | 1 ng/5 μL | Social recognition | nm | = | Engelmann et al. (1998) |
| | ICV | 0.1 μg/5μL | Social avoidance after maternal defeat | nm | Reversed | Lukas and Neumann (2014) |
| | ICV | 0.1 μg/5μL | Social avoidance after social defeat | Reversed | nm | Lukas et al. (2011) |
| | LS | 400 pg/μL | Social play (in juveniles) | = | ↓ | Bredewold et al. (2014) |
| | Subcutaneous | 1–4 μg/kg | Anxiety (open field) | ↓ | nm | Uvnäs-Moberg et al. (1994) |
| | ICV | 100 ng/h for 5 days | Anxiety (elevated plus maze) | nm | ↓ | Windle et al. (1997) |
| | PVN | 0.02 nmol/μL | Anxiety (elevated plus maze, light–dark box) | ↓ | nm | Blume et al. (2008) |
| | ICV | 10 ng/h for 7 days | Anxiety (light-dark box) | = | ↓ (in highly anxious females) | Slattery and Neumann (2010) |
| | Prelimbic PFC | 1 μg/μL | Anxiety (elevated plus maze, open field, social interaction) | ↓ | ↓ | Sabihi et al. (2014) |
| OT KO mice | ICV | 2 ng | Anxiety (elevated plus maze) | nm | ↓ | Mantella et al. (2003) |
| Mice | i.p., ICV | 10 mg/kg (i.p.) 10 μg/kg (ICV) | Anxiety (four plate test, elevated zero maze) | ↓ | nm | Ring et al. (2006) |

BNST, bed nucleus of the stria terminalis; ERα, estrogen receptor alpha; ICV, intracerebroventricular; i.p., intraperitoneal; KO, knockout; LS, lateral septum; MPOA, medial preoptic area; OT, oxytocin; OTA, oxytocin receptor antagonist; PVN, paraventricular nucleus of the hypothalamus; PFC, prefrontal cortex; V1aR, vasopressin 1a receptor; VMH, ventromedial hypothalamus; ↑, facilitated/increased; ↓ impaired, decreased; =, no effect; nm, not measured.

**Table 11.4** Summary of sexually dimorphic actions of acute OTR antagonist manipulations on adult behavior in prairie voles, rats, and mice

| Species | Route | Dose | Behavioral/brain response | Male effect | Female effect | References |
|---|---|---|---|---|---|---|
| Prairie voles | LS | 0.05 ng or 0.5 ng | Partner preference | ↓ | nm | Liu et al. (2001) |
| | NAc, PFC | 5 ng/μL | Partner preference | nm | ↓ | Young et al. (2001) |
| | NAc | 50 ng (with D1 agonist) | Partner preference | nm | ↓ | Liu and Wang (2003) |
| Rats | ICV | 7.5 ng/h for 7 days | Anxiety (light–dark box) | = | ↑ (in low anxious females) | Slattery and Neumann (2010) |
| | ICV | 1 μg/5μL (with ERβ agonist) | Anxiety (elevated plus maze) | ↑ | ↑ | Kudwa et al. (2014) |
| Mice | ICV | 100 ng | Anxiety (elevated plus maze) | nm | ↑ | Mantella et al. (2003) |
| Rat | ICV | 100 ng/5μL | Social recognition | nm | ↓ | Engelmann et al. (1998) |

ERβ, estrogen receptor beta; ICV, intracerebroventricular; i.p., intraperitoneal; LS, lateral septum; MPOA, medial preoptic area; NAc, nucleus accumbens; OT, oxytocin; OTA, oxytocin receptor antagonist; PVN, paraventricular nucleus of the hypothalamus; PFC, prefrontal cortex; V1aR, vasopressin 1a receptor; ↑, facilitated/increased; ↓ impaired, decreased; =, no effect; nm, not measured.

**Table 11.5** Summary of sexually dimorphic actions of neonatal OT manipulations on juvenile or adult behavior and brain responses in prairie voles

| Dose (μg) | Behavioral/brain response | Age | Male effect | Female effect | References |
|---|---|---|---|---|---|
| 3 and 12 | Partner preference | Adult | ↑ | nm | Bales and Carter (2003a) |
| 3, 6, 12 and 24 | Partner preference | Adult | nm | = (3 μg) | Bales et al. (2007b) |
| | | | nm | ↑ (6 μg) | |
| | | | nm | = (12 μg) | |
| | | | nm | Prefer stranger (24 μg) | |
| 3 | Same-sex aggression | Adult | = | ↑ | Bales and Carter (2003b) |
| 3, 6, 12 and 24 | Alloparental care | Adult | nm | = (3 μg) | Bales et al. (2007b) |
| | | | nm | ↓ (6 μg) | |
| | | | nm | ↑ (12 μg) | |
| | | | nm | = (24 μg) | |
| 3 | Alloparental care | Juvenile | = | = | Bales et al. (2004) |
| 3 | OT-ir | Juvenile | = | ↑ in PVN | Yamamoto et al. (2004) |
| 3 | OTR binding | Adult | = | = | Bales et al. (2007a) |
| 3 | V1aR binding | Adult | ↑ in ventral pallidum and cingulate cortex | ↓ in ventral pallidum cingulate cortex, BNST, LS, and mediodorsal thalamus | Bales et al. (2007a) |
| 3 | ERα-ir | Juvenile | = | ↑ in VMH | Yamamoto et al. (2006) |

BNST, bed nucleus of the stria terminalis; ERα, estrogen receptor alpha; LS, lateral septum; MPOA, medial preoptic area; OT, oxytocin; OTR, oxytocin receptor; PVN, paraventricular nucleus of the hypothalamus; V1aR, vasopressin 1a receptor; VMH, ventromedial hypothalamus; ↑, facilitated/increased; ↓ impaired, decreased; =, no effect; nm, not measured.

**Table 11.6** Summary of sexually dimorphic actions of neonatal OTR antagonist (i.p.) manipulations on juvenile or adult behavior and brain responses in prairie voles and mice

| Species | Dose (μg) | Behavioral/brain response | Age | Male effect | Female effect | References |
|---|---|---|---|---|---|---|
| Prairie voles | 0.3 | Alloparental care | Juvenile | ↓ | = | Bales et al. (2004) |
| Prairie voles | 0.3 | OT-ir | Juvenile | = | ↑ in PVN | Yamamoto et al. (2004) |
| Prairie voles | 0.3 | ERα-ir | Juvenile | = | ↓ in MPOA | Yamamoto et al. (2006) |
| Prairie voles | 0.3 | OTR binding | Adult | = | = | Bales et al. (2007a) |
| Prairie voles | 0.3 | V1aR binding | Adult | ↓ in MPOA, BNST, LS | ↓ in BNST and cingulate cortex | Bales et al. (2007a) |
| Prairie voles | 0.3 | c-Fos after cohabitation | Adult | = | ↑ in CeA | Kramer et al. (2006) |
| Mice | 3 | Social approach and alloparental care | Adult | ↓ alloparental | ↓ social approach | Mogi et al. (2014) |

BNST, bed nucleus of the stria terminalis; CeA, central amygdala; ERα, estrogen receptor alpha; LS, lateral septum; MPOA, medial preoptic area; OT, oxytocin; OTR, oxytocin receptor; PVN, paraventricular nucleus of the hypothalamus; V1aR, vasopressin 1a receptor; ↑, facilitated/increased; ↓ impaired, decreased; =, no effect.

**Table 11.7** Summary of sexually dimorphic actions of intranasal OT on behavior in humans

| Behavior | Male effect | Female effect | References |
|---|---|---|---|
| Anxiety-related behavior | nm | ↓ | Chiodera et al. (1991) |
| Anxiety-related behavior | nm | ↓ | Amico et al. (2004) |
| Anxiety-related behavior | nm | ↓ | Heinrichs et al. (2001) |
| Anxiety-related behavior | ↓ | nm | Heinrichs et al. (2003) |
| Anxiety-related behavior | ↓ | nm | Kirsch et al. (2005) |
| Anxiety-related behavior | ↓ | nm | Radke et al. (2013) |
| Acoustic startle to negative stimuli | ↑ | nm | Striepens et al. (2012) |
| Trust | ↑ | nm | Kosfeld et al. (2005) |
| Generosity | ↑ | nm | Zak et al. (2007) |
| Inferring mental states of others | ↑ | nm | Domes et al. (2007b) |
| Fixations and gaze toward the eyes | ↑ | nm | Guastella et al. (2008) |
| Cooperation with a computer partner | = | nm | Rilling et al. (2012) |
| Cooperation with a computer partner | nm | ↓ | Rilling et al. (2013) |
| Word memory | ↓ | nm | Heinrichs et al. (2004) |
| Facial recognition memory | ↓ | = | Herzmann et al. (2013) |
| Identity recognition memory | ↑ | ↑ | Savaskan et al. (2008) |
| Identification of fearful and negative facial expressions | ↓ | nm | Di Simplicio et al. (2009) |
| Recognition memory of happy and angry faces | ↑ | nm | Schulze et al. (2011) |
| Recognition memory of happy faces | ↑ | nm | Guastella et al. (2008b) |
| Social kinship recognition/competition perception | ↑ competition perception | ↑ social kinship recognition | Fischer-Shofty et al. (2013) |
| Perspective taking ability | ↑ | = | Theodoridou et al. (2013) |
| Behavioral risk assessment | ↓ | = | Lynn et al. (2014) |
| Defensive responding | ↑ | ↑ | Grillon et al. (2013) |
| Positive behavior during couple conflict/cortisol levels | ↑ positive behavior/↓cortisol | ↑ positive behavior/↓cortisol | Ditzen et al. (2009) |
| Sympathetic activity measures during couple conflict | ↑ | ↓ | Ditzen et al. (2013) |
| Self-interest moral judgments | ↑ | ↓ | Scheele et al. (2014) |

↑, Facilitated/increased; ↓, impaired/decreased; =, no effect; nm, not measured.

**Table 11.8** Summary of sexually dimorphic neural activation to intranasal OT in humans using fMRI

| Stimuli/cognitive task | Male effect | Female effect | References |
|---|---|---|---|
| Fear-inducing stimuli | ↓ amygdala activation | nm | Kirsch et al. (2005) |
| Emotional facial stimuli | ↓ amygdala activation | nm | Domes et al. (2007a) |
| Emotional facial stimuli | nm | ↑ amygdala activation | Domes et al. (2010) |
| Reciprocated cooperation | ↑ caudate nucleus and amygdala activation | ↓ amygdala activation | Rilling et al. (2013) |
| Trust game | ↓ amygdala, midbrain, and dorsal striatum activation | nm | Baumgartner et al. (2008) |
| Infant crying labeled as "sick" | nm | ↑ insula and inferior frontal gyrus activation | Riem et al. (2014) |
| Infant crying labeled as "bored" | nm | ↓ insula and inferior frontal gyrus activation | |

↑, Increased; ↓, decreased; =, no effect; nm, not measured.

## 3.1 Sexually dimorphic actions of OT in voles

### *3.1.1 Sexually dimorphic actions of OT on behavior*

Sex-specific actions of the OT system have been studied in a variety of rodent species, including voles, rats, and mice. These rodent species differ largely in their social organization. We will discuss sex-specific roles of OT in light of the social organization of the given species in Section 3.4. Much of the earlier work on sex-specific actions of the OT system was performed on prairie voles. Both sexes show high levels of parental and alloparental care, and paternal care plays an important role in brain and behavior development of the pups (Wang and Novak, 1992, 1994a, 1994b; Ahern and Young, 2009). The research described in the subsequent section provides evidence that, although there are similarities in the expression of social behavior between male and female prairie voles, OT may mediate these social behaviors in sex-specific ways.

Pair bonding in voles can be experimentally tested via the use of the partner preference paradigm. In this paradigm, cohabitation with an opposite-sex partner induces a robust preference to show side-by-side contact with the partner over a stranger (Williams et al., 1992; Winslow et al., 1993). Interestingly, there are sex differences in the temporal parameters of partner preference, wherein a 6 h cohabitation induces partner preference in females, but a 24 h cohabitation is necessary to induce partner preference in males; however, this can be facilitated by mating (DeVries and Carter, 1999; Bales et al., 2013). This suggests that females develop partner preferences more rapidly than males, which may reflect differences in underlying neural mechanisms. With the ability to vary the cohabitation times to suboptimal conditions (i.e., shorter cohabitation time), the partner preference paradigm offers ways to study the mechanisms that facilitate pair bond formation in both males and females, and demonstrate the involvement of OT.

Intracerebroventricular (ICV) OT administration facilitated partner preference formation in female prairie voles (Williams et al., 1994; Insel and Hulihan, 1995), but not in male prairie voles (Winslow et al., 1993). Additional studies indicated that the effects of OT on partner preference in females are located in the nucleus accumbens and prefrontal cortex, but not in the caudate putamen (Young et al., 2001; Liu and Wang, 2003). Interestingly, OTR antagonist administration in the LS blocked partner preference in males after mating (females were not tested; Liu et al., 2001), suggesting a role for septal OTR in pair bonding for males. These local effects of OT and OTR antagonist may point toward different neural circuitries through which OT mediates pair bonding in males versus females.

Additional studies have focused on the developmental and long-lasting effects of OT on pair bonding and other related behaviors in prairie voles. These studies found that a single intraperitoneal (i.p.) injection of OT on postnatal day 1 facilitated partner preference in both males and females, but in dose-dependent ways. Low (3 μg) and high (12 μg) doses of OT, but not an intermediate (6 μg) dose, facilitated partner preference formation in males (Bales and Carter, 2003a; Carter et al., 2008). The opposite was found in females, in which low and high doses of OT had no effect on partner preference, but the intermediate dose facilitated partner preference (Bales et al., 2007b). These dose-dependent differences in males and females may reflect sex differences in the sensitivity to OT on brain and behavior development. This could be due to sex differences in developmental exposure to OT. It would be of interest to determine whether there are variations in exogenous OT transferred via the milk of the mother during lactation or variations in endogenously released OT in response to licking and grooming by the mother.

In pair-bonded prairie voles, aggression often appears in both sexes to serve to guard the mate from same-sex competitors (mate-guarding). After only a 4 h exposure to an animal of the opposite sex (which is too short to establish a pair bond), adult female prairie voles given a low dose of neonatal OT were more aggressive and less social to a stimulus female than control females or females given an OTR antagonist. This effect was not seen in male prairie voles (Bales and Carter, 2003b), suggesting that neonatal exposure to OT facilitates the onset of mate-guarding in female, but not male, prairie voles. Interestingly, as indicated previously, neonatal OT at this same low dose did not have an effect on partner preference in females, but did have an effect on partner preference in males. This suggests that different doses of OT are mediating different aspects of the prairie vole mating system. However, since only one dose of OT was used to test mate-guarding (Bales and Carter, 2003b), it is still possible that a higher dose could have initiated an effect in males. Even so, these findings suggest that there are sex differences in the sensitivity of the brain systems regulating pair bonding and mate-guarding to OT.

Prairie voles of both sexes exhibit alloparental care; as young as 21 days of age, virgin male and female prairie voles will show parental care toward the offspring of others. While a single neonatal injection of a low dose of OT did not alter alloparental care in either sex, a single neonatal injection with an OTR antagonist decreased alloparental care in juvenile males, but not juvenile females (Bales et al., 2004). However, a single neonatal injection with an intermediate dose of OT decreased alloparental care, whereas a high dose of OT increased alloparental care, in adult female prairie voles (males were not tested; Bales et al., 2007b).

These studies not only highlight the dose-dependent sensitivity of the OT system on various mating and parental behaviors in male and female prairie voles, but indicate that just a single neonatal injection of OT can have behavioral consequences later in life that depend on the sex of the animal. For example, female mate-guarding behavior seems to be more sensitive to neonatal OT manipulations, while male partner preference seems to be more sensitive to neonatal OT manipulations. Although it is evident that the OT system is involved in both male and female parental behavior in prairie voles, a direct comparison of OT's role in maternal and paternal care of one's own offspring in both sexes is lacking. Because males do not experience the hormonal changes accompanied by pregnancy and parturition as do females, we can predict that the dose of OT capable of inducing a change in parental care in males may be different than that which produces an effect in maternal care in females. Together, these findings highlight the sex-specific effects of OT on mating and parental behaviors in prairie voles, yet they also reveal the need for more comparative studies of the effects of OT on behavior in both sexes.

### *3.1.2 Sexually dimorphic organizational effects of OT on the brain*

The sexually dimorphic actions of OT on mating and parental behaviors in prairie voles suggest that this may be mediated through differential effects on brain parameters. Indeed, i.p. injections of either a low dose of OT or OTR antagonist on postnatal day 1 had sexually dimorphic organizational effects on the amount of OT-ir in the PVN (Yamamoto et al., 2004), and levels of vasopressin 1a receptor (V1aR; Bales et al., 2007a), which is discussed in the following paragraphs. Whether changes in these systems underlie the sex-specific behavioral changes seen in male and female prairie voles after neonatal OT manipulations is yet unclear.

Neonatally OT-treated females, but not males, showed an increase in OT-ir in the PVN (Yamamoto et al., 2004). Behaviorally, this same dose of OT had an effect on mate-guarding behavior in females only, suggesting a possible association with increased OT-ir in the PVN and mate-guarding in females. Surprisingly, females treated neonatally with an OTR antagonist also showed an increase in OT-ir in the PVN (Yamamoto et al., 2004). Less is known about the implications of neonatal OTR antagonist on female social behavior, but it is interesting to note that neonatal OT and OTR antagonist both increased OT-ir in the PVN.

In addition, neonatal OT increased ERα-ir in the VMH and OTR antagonist decreased ERα-ir in the MPOA of juvenile females, while no effects were seen in males (Yamamoto et al., 2006). Behaviorally, the same dose of neonatal OT increased mate-guarding behavior in adult females (Bales and Carter, 2003b). Given the prominent role of ERα in the VMH in female reproductive behaviors (Musatov et al., 2006), OT could perhaps mediate mate-guarding behavior through its effects on ERα in this brain region, but this remains to be tested. The link between alterations in OT-ir and ERα-ir due to neonatal OT/OTR antagonist in females and later adult social behavior should be interpreted with caution, however, because the behavioral parameters discussed previously (partner preference, mate-guarding) and the analysis of OT-ir and ERα-ir were analyzed at different ages (adulthood and juvenile age, respectively).

Neonatal OT and OTR antagonist did not have an effect on OTR binding in either sex (Yamamoto et al., 2004; Bales et al., 2007a), but did alter V1aR binding in the brains of both adult males and females, albeit in sex-specific ways (Bales et al. 2007a). Activation of the V1aR in the brain is important for the regulation of various social behaviors (Albers, 2014). It is therefore possible that changes in V1aR binding could underly the changes seen in adult social behavior after neonatal OT manipulations. Neonatal OT increased V1aR binding in the ventral pallidum and cingulate cortex of adult males, but decreased V1aR binding in the same regions of adult females (Bales et al., 2007a). Behaviorally, this same dose of OT given neonatally facilitated partner preference formation in males (Bales and Carter, 2003a). Because V1aR in the ventral pallidum is essential for partner preference in males (Lim and Young, 2004), an increase in V1aR binding in this area could be a potential mechanism by which neonatal OT facilitates partner preference. Neonatal OTR antagonist decreased V1aR binding in the MPOA, BNST, and LS of adult males and in the BNST and cingulate cortex of adult females (Bales et al., 2007a). This dose of OTR antagonist had no behavioral effects in females, but decreased alloparental care in juvenile males (Bales et al., 2004), suggesting a potential association between alloparental care and V1aR binding, and possibly vasopressin action, in these particular brain regions in males.

In summary, the sexually dimorphic effects of neonatal OT manipulations on different brain systems is a first step in understanding how neonatal OT manipulations alter social behavior in sexually dimorphic ways (as described in Section 3.1.1). Further research would be required to determine a direct relationship between changes in behavior and changes in brain systems seen after neonatal OT manipulations.

## 3.2 Sexually dimorphic actions of OT on behavioral and brain responses in rats and mice

### 3.2.1 *Sexually dimorphic actions of OT on social behavior*

OT plays a central role in a diverse range of social behaviors in male and female rats and mice (Bielsky and Young, 2004; Veenema and Neumann, 2008; Ross and Young, 2009).

In contrast to prairie voles, rats are typically polygamous, do not form pair bonds, do not readily exhibit alloparental behavior, and pups are mainly raised by the dam (Lonstein and Fleming, 2002). Still, rats live in groups and are social animals, with early social isolation causing behavioral deficits and neurobiological alterations later in life (Veenema, 2009, 2012). These results highlight the importance of social interactions for optimal social functioning in rats. Mice (*Mus musculus*) are also social animals, typically polygamous, and show great variability in their parental responsiveness. Female mice show high levels of maternal behavior, even toward pups of other females, and may form communal nests in which they will nurse each other's pups indiscriminately (Manning et al., 1992). Male mice vary in their expression of paternal care, with male paternal responsiveness often depending on strain and age (Lonstein and De Vries, 2000).

Surprisingly little comparative research has been done on the sex-specific effects of OT in rats and mice. For example, the role of OT in social recognition has been extensively studied in male mice, but much less so in female mice. Social recognition was defined as a decrease in social investigation time of the experimental animal upon re-exposure to the same animal. It was found that male OTR knockout (KO) mice show impaired social recognition (Takayanagi et al., 2005; Lee et al., 2008; Macbeth et al., 2009), but none of these studies tested female OTR KO mice. Moreover, OT in the MeA is necessary and sufficient for social recognition in male mice (Ferguson et al., 2001) and social investigation time correlated positively with OTR mRNA expression in the MeA of male mice (Murakami et al., 2011). Again, none of these studies tested female mice. It is likely, however, that the OT system is also important for social recognition in female mice because not only male (Ferguson et al., 2000; Macbeth et al., 2009) but also female (Choleris et al., 2003, 2006) OT KO mice show impaired social recognition.

Studies on rats suggest that OT may be playing a sex-specific role in social recognition. For example, OT given intracerebroventricularly enhanced social recognition in male rats (Benelli et al., 1995), but not in female rats (Engelmann et al., 1998). However, ICV administration of an OTR antagonist impaired social recognition in both males (Lukas et al., 2011) and females (Engelmann et al., 1998), suggesting that central OT plays a role in social recognition in both sexes. Together, these data indicate that male rats may be more sensitive to the enhancing effects of exogenous ICV OT, possibly due to higher OTR binding densities in various brain regions (Dumais et al., 2013) that are known to be involved in social recognition. Unfortunately, brain region-specific effects of OT injections in the olfactory bulb, septum (Popik et al., 1992), and MPOA (Popik and van Ree, 1991) were only tested in male rats. Although high doses of OT were either not effective or impaired social recognition in male rats (Popik and van Ree, 1991; Popik et al., 1992), these doses were above physiological levels.

By directly comparing male and female rats, we found that OTR binding in the central amygdala correlated negatively with social interest (as reflected by the time the experimental rat investigates a novel juvenile rat) in females, but not males (Dumais et al., 2013), while OTR binding in the MeA correlated positively with social interest in males, but not females (Dumais et al., 2013). Moreover, OT injected ICV reversed social defeat-induced social avoidance in male but not female rats (Lukas et al., 2011; Lukas and Neumann, 2014). These results together may suggest that the OT system is involved in modulating the salience and/or valence of social stimuli differently in males and females, possibly via the activation of different neural circuits.

Finally, OT has been found to regulate juvenile social behavior in sex-specific ways in rats. Here, OT administered into the LS decreased social play behavior in females, but not males (Bredewold et al., 2014). Importantly, sex differences in OTR binding densities in specific brain regions in adult rats (Dumais et al., 2013) have also been found in the same brain regions of juvenile rats (Smith et al., 2014), highlighting that some of the sex differences in the OT system may be independent of maturation processes and may influence the regulation of social behavior differently in males and females across the lifespan. This underlines the importance of looking at the development of the OT system and potential sex-specific influences of OT or other early life manipulations early in development (discussed in Section 3.2.3).

In summary, despite the lack of comparative research, the few studies in rats indicate that social recognition, social interest, social defeat-induced avoidance, and social play may be modulated by OT in sex-specific ways. These findings clearly indicate that further research is warranted, especially regarding the mechanisms by which the OT system regulates these behaviors differently in males and females.

### 3.2.2 Sexually dimorphic actions of OT on anxiety-related behavior

OT plays an important role in the regulation of anxiety-related behaviors and modulation of hypothalamic–pituitary–adrenal (HPA) axis activity (Neumann et al., 2000; Van den Burg and Neumann, 2011), which, in turn, can have indirect effects on OT's role in social behavior. Indeed, social interaction tests have been used as a test of anxiety, with decreased social interaction interpreted as reflecting increased anxiety (File, 1980). However, changes in social interaction do not necessarily reflect changes in anxiety-related behavior, as OT system manipulations have been found to alter social behavior without altering anxiety-related behavior in a nonsocial context (Bredewold et al., 2014). A potential interaction between anxiety and social behavior, however, may still be important to consider when investigating the role of OT in social behavior.

Early studies showed that blood plasma OT levels were higher in response to a variety of stressors in both male (Lang et al., 1983; Gibbs, 1984; Carter and Lightman, 1986; Jezova et al., 1995) and female (Carter and Lightman, 1986; Lang et al., 1983) rats; however, we cannot conclude the specific role of OT in anxiety-related behavior from these studies. Additional studies demonstrated a more functional influence of OT on anxiety, in which it was found that an acute injection of OT decreased anxiety when administered subcutaneously (Uvnäs-Moberg et al., 1994) or into the PVN (Blume et al., 2008) in male rats, or into the prefrontal cortex (Sabihi et al., 2014) in male and female rats. Given that it is less likely that peripherally administered OT will cross the blood–brain barrier (McEwen, 2004), these findings suggest that OT modulates anxiety through different mechanisms, i.e., peripherally as well as centrally. In contrast, blockade of brain OTR in virgin female or in male rats under basal conditions, i.e., in a state of low brain OT activity, did not alter anxiety behavior (Neumann et al., 2000). Furthermore, chronic ICV administration of OT decreased anxiety in normal (Windle et al., 1997) and high-anxious female (Slattery and Neumann, 2010), but not male, rats. In addition, an OTR antagonist increased anxiety-related behavior in low-anxious females only (Slattery and Neumann, 2010).

Investigations in mice found that female OT KO mice showed an increase in anxiety compared with wildtype on the elevated plusmaze (Mantella et al., 2003; Amico et al., 2004), which was subsequently decreased by OT (ICV) injections (Mantella et al., 2003). Further, an OTR antagonist (ICV) increased anxiety on the elevated plusmaze in female wildtype mice (Mantella et al., 2003). In contrast, male OT KO mice show a decrease in anxiety compared with wildtype (Mantella et al., 2003; Winslow et al., 2000). This indicates that a lack of OT during development has opposing effects on anxiety in males and females, suggesting that OT during development may play sex-specific roles in the typical organization of the male and female stress response. However, OT (i.p. or ICV) decreased anxiety in adult male mice (Ring et al., 2006), suggesting that activational and organization effects of OT in male anxiety may be different.

All of these studies strongly support an anxiolytic role of OT in females, but leaves the role of OT in male anxiety less clear. It is evident then, that the role of OT in anxiety-related behavior is complex, and possibly highly situational, depending on the timing of the manipulation (organizational vs. activational influences), the magnitude of the manipulation (acute or chronic), and individual differences (high or low trait anxiety). It may therefore be important to consider the potentially differential role of OT on anxiety-related behaviors in males and females when interpreting the influence of OT on social behaviors that may be affected by anxiety levels.

### *3.2.3 Sexually dimorphic organizational effects of OT on brain and behavior*

Early life experiences can alter central OT system parameters differently in males and females, highlighting potential sex differences in the sensitivity of the OT system to perturbations. For example, it was found that OTR binding was increased in adult female rats that had received high levels of maternal licking and grooming as pups (Francis et al., 2002). Low levels of licking and grooming did not have this effect in females, and no effect was found in males. OT-ir in the PVN was decreased in adult female mice, but not male mice, that were exposed to maternal separation as pups (Veenema et al., 2007). Further, Babb et al. (2014) found that while both male and female offspring of rat dams exposed to chronic social stress show reduced social behavior as juveniles (including social investigation and overall social contact), only juvenile female offspring showed elevated plasma OT levels. These studies indicate that exposure to stress hormones during prenatal development can differentially alter the OT system in males and females, and that females may be more sensitive to variations in maternal care and early life stress compared with males.

In addition, a study using ICR mice found that neonatal manipulations of the OT system induced sex-specific effects on sociability and alloparental care in males and females. In detail, social approach to an unfamiliar conspecific was reduced in females, but not in males, receiving an OTR antagonist on day 1 of birth. In contrast, alloparental care (as measured by pup retrieval) was reduced in males, but not in females, receiving an OTR antagonist on day 1 of birth (Mogi et al., 2014). This is similar to Bales et al. (2004), which reported a decrease in alloparental care in male, but not female, prairie voles after neonatal OTR antagonist treatment. Thus, across two species, blocking the OTR system early in development decreased alloparental care in males only. This may suggest a resilience of the neurobiological mechanisms regulating alloparental care in females, which may reflect the robust evolutionary pressures on females to take care of their young.

## 3.3 Are there sexually dimorphic actions of OT in humans?

Since the discovery that OT can be administered via a nasal spray, research into the role of OT in human social cognition and behavior has been a popular area of interest. Given that OT does not pass the blood–brain barrier, there is still debate over how intranasally applied OT has effects on cognition and behavior, and if OT reaches the brain at all, or has indirect effects on the brain through the periphery (see Evans et al., 2014 for review). With these questions still unanswered, there have been numerous studies reporting effects of intranasal OT on human social cognition and behavior. Notably, studies have also implicated intranasal OT as having therapeutic potential for patients suffering from social dysfunction, which is discussed in the last section of this review.

### *3.3.1 Effects of OT on human social behavior*

The first studies to report intranasal OT effects in humans were done in males only. Among these earlier studies, intranasal OT was found to increase trust by affecting an individual's willingness to accept social risks (Kosfeld et al., 2005), and to increase generosity when given a decision to split a sum of money with a stranger (Zak et al., 2007). In addition, OT increased performance in the Reading the Mind in the Eyes Test (Domes et al., 2007b), suggesting that OT improves the ability to infer the mental state of others from social cues. OT may be improving this ability by affecting the orienting of one's gaze toward social cues, as OT also increased the number of fixations and total gaze toward the eye region of human faces (Guastella et al., 2008a). It was further found that OT enhances the processing of positive versus negative emotional information as evident by a slowed reaction time to correctly identify fearful facial expressions, but a reduction in misclassifications of positive emotions as negative ones (Di Simplicio et al., 2009). Combined, these studies suggest a role for OT in the processing of socially relevant information.

As the importance of using women in research and in clinical trials was increasingly being recognized and promoted, studies started to include both men and women. As some of the first studies looking at the role of OT in men indicated a role of OT in trust, it was later found that plasma OT levels were associated with levels of trust in a sex-specific manner. In men only, subjects with extreme levels of plasma OT (top 20% or bottom 20%) were more likely to be trusting and also appeared more trustworthy to others in a trust game (Zhong et al., 2012).

The importance of OT in human social recognition generated some interesting differences in men and women and across studies. Intranasal OT was found to enhance memory for happy (Guastella et al., 2008b) or happy and angry (Schulze et al., 2011) faces in men. While women were not tested in these studies (Guastella et al., 2008b; Schulze et al., 2011), one study reported that intranasal OT improved identity recognition memory for neutral and angry faces in both men and women (Savaskan et al., 2008). Interestingly, however, another study showed that intranasal OT administration impaired recognition memory of neutral and happy faces in men, but not women (Herzmann et al., 2013). Although not related to social behavior, yet another study reported memory-impairing effect of words in men; women were not tested in this study (Heinrichs et al., 2004). Finally, intranasal OT administration facilitated kinship recognition in women but not in men (Fischer-Shofty et al., 2013).

These discrepancies, especially in men, suggest that OT effects on memory may be specific to the type of memory being examined and the relevance of the different stimuli. Because OT improved recognition memory of angry faces (Savaskan et al., 2008), but impaired recognition in happy faces (Herzmann et al., 2013), the valence of the social stimuli may have an important role in mediating the effects of OT on recognition memory. Interestingly, OT may mediate the perception of emotional valence, as higher plasma OT levels related to perceiving faces as happier in women with or without schizophrenia, but not in men with or without schizophrenia (Rubin et al., 2011). This seems in line with the finding that OT facilitates competition perception in men, but not women (Fischer-Shofty et al., 2013). These studies highlight the importance of emotional valence in the role of OT in perception and recognition of social stimuli.

Given the role of OT in the processing of social stimuli, the role of OT in one's ability to take the perspective of others has also been examined, often revealing a greater influence of exogenous OT in this ability in men compared with women. For example, OT slowed reaction times in an implicit perspective-taking task in men to that of the level of women, but did not have any effects in women (Theodoridou et al., 2013). In addition, OT impaired men's ability, but not women's ability, to adequately perform a task requiring accurate emotion perception in others (Lynn et al., 2014). Specifically, men given OT seemed less able to adapt to changes in risk when shown novel or changing risky environments (angry faces). Finally, OT lowered rates of cooperation with computer partners following deception in women, rendering them to treat computer partners more like humans, an effect that was not found in men (Rilling et al., 2012, 2013). Together, these findings indicate that OT may decrease perspective taking in men, while it has no effects or may increase perspective taking in women. It should be noted that the tasks used in these studies vary significantly, suggesting that the role of OT in this ability may depend on the particular task or component of perspective taking being assessed.

Several studies have found a role of OT in partner bonding in both men and women, which may show some resemblance with the facilitating role of OT on pair bonding in voles. Intranasal OT was found to increase positive behavior during couple conflict and decrease cortisol levels in both male and female partners (Ditzen et al., 2009), and greater partner support as assessed by self-report was related to higher plasma OT in both men and women (Grewen et al., 2005). However, in women, but not men, elevated plasma OT was associated with distress in respect to a pair-bonded relationship (Taylor et al., 2010). Interestingly, OT decreased sympathetic activity measures during couple conflict in women, but increases sympathetic activity measures in men (Ditzen et al., 2013). In line with this, women given intranasal OT reported more distress and anger in response to social evaluation (as assessed with the Trier Social Stress Test), while men given OT reported a less negative affect and had greater vagal rebound (Kubzansky et al., 2012). Elevated OT in women during couple conflict could be serving to decrease the potentially heightened sympathetic responses caused by the stress of couple conflict, or, in the case of social evaluation stress, may help women to face negative social interactions.

In other studies examining the role of OT in social bonding, it was found that OT correlated with parenting behavior and different parenting styles in both men and women. Plasma OT was increased in pregnant and parturient women (Gordon et al., 2008), and increased plasma OT was associated with increased maternal postpartum behavior (Feldman et al., 2007). OT has also been found to be associated with parental behavior in men, as both mothers and fathers showed increased plasma OT compared with nonattached single participants (Gordon et al., 2010). In addition, fathers and mothers who provided high levels of affectionate touch and stimulatory contact with their infants during a play session showed increased salivary OT levels compared with fathers and mothers showing minimal touch (Feldman et al., 2010). Interestingly, OT has also been found to associate with parental behaviors in sex-specific ways. In mothers and fathers, plasma OT levels correlated with affectionate parenting behavior in mothers only, and correlated with stimulatory parenting behavior (proprioceptive contact, tactile stimulation, and object presentation) in fathers only (Gordon et al., 2010). These studies highlight the involvement of the OT system in parental behavior in both mothers and fathers, but also indicate that the specific role of OT in parenting behavior may be different in men and women.

Evidence for the involvement of the OTR in various social behaviors in humans comes from genetic studies of single nucleotide polymorphisms (SNPs) found in the OTR gene (for reviews see Meyer-Lindenberg et al., 2011; Ebstein et al., 2012). Men and women carrying one or two copies of the A allele in the most common SNP, rs53576, show reduced empathy (Rodrigues et al., 2009), whereas reduced positive affect (Lucht et al., 2009) and reward dependence (Tost et al., 2010) were found in men only. Women only were tested for their sensitivity for their toddler's behavior, which was found to be reduced in those carrying rs53576A (Bakermans-Kranenburg and Van IJzendoorn, 2008). Moreover, women, but not men, carrying rs7632287A scored lower on a partner bonding scale and on a relationship quality survey (Walum et al., 2012). A meta-analysis of the most commonly studied SNPs rs53576 and rs2254298 did not reveal any sex effects or any general effects in the social behavior domain (Bakermans-Kranenburg and Van IJzendoorn, 2014), underscoring that it may be unlikely to expect that one SNP in the OTR gene can explain main differences in social behavior.

### 3.3.2 Effects of OT on human anxiety-related behavior

OT also plays a pivotal role in the regulation of human anxiety, but whether it has an anxiolytic or anxiogenic effect, or has similar functions in men and women is still unclear. For example, some studies have found that OT has an anxiolytic effect in men (Heinrichs et al., 2003; Kirsch et al., 2005; Radke et al., 2013) and in postpartum lactating women (Chiodera et al., 1991; Heinrichs et al., 2001), but comparisons between men and women were not made in these studies. In contrast, other studies have reported OT as having an anxiogenic effect in men and women. For example, Striepens et al. (2012) found that OT potentiated acoustic startle responses to negative stimuli in men. Grillon et al. (2013) found that OT increased defensive responding to unpredictable threat in men and women, with no effect of sex. These findings may suggest that OT facilitated context appropriate behavior by showing differential responding to good (positive/rewarding) events versus bad (negative) events.

Other studies have found sex differences in how plasma OT relates to anxiety, indicating that increased plasma OT levels may be related to higher anxiety in women, but not men. It was found that men with higher plasma OT levels had lower trait anxiety, while women with extremely high OT plasma levels had three times the probability of being highly anxious (Weisman et al., 2013). In support, higher OT values were also associated with greater anxiety in adolescent girls, but not boys (Miller et al., 2013). Together with intranasal OT application studies discussed previously, this at least suggests that OT may have opposing roles in anxiety and responses to social stress situations in women compared with men, which deserves future research with regard to potential health benefits of OT (see also Section 4).

### 3.3.3 Sexually dimorphic actions of OT on human brain responses

In addition to sexually dimorphic behavioral effects, intranasally applied OT modulates brain activation differently in men and women, as measured by functional magnetic resonance imaging (fMRI). It was found that intranasal OT reduced amygdala activation in male participants when presented with fear-inducing stimuli (Kirsch et al., 2005). Likewise, intranasal OT reduced activation in the amygdala, dorsal striatum, and midbrain regions in male participants, corresponding with a lack of decrease in trusting behavior when their trust had been breached several times (Baumgartner et al., 2008). Domes et al. (2007a) also found that intranasal OT reduced amygdala activity in men when viewing emotional facial expressions, regardless of valence (fearful, angry, or happy). However, using the same design as in their 2007 study, Domes et al. (2010) found that women given intranasal OT showed an increase in amygdala activity when viewing fearful, but not angry or happy faces, suggesting that OT may increase amygdala-dependent threat processing. Indeed, it was also found that OT increased amygdala activity to social and nonsocial threatening scenes in women (Lischke et al., 2012). This sex difference suggests that OT may be playing different roles in the modulation of social information processing in men and women, which seems to correspond with the findings on behavior as discussed in Sections 3.3.1 and 3.3.2. Interestingly, during cooperative interactions, a different activation pattern was found in which intranasal OT increased neuronal response in the amygdala and caudate nucleus in men, while there was a decreased or no response in the amygdala and caudate nucleus in women, respectively (Rilling et al., 2013).

These findings suggests a complex interaction of the effects of OT on brain activation and the valence of the context or stimuli presentation, illustrating a seemingly opposite effect of OT on amygdala activity in males and females in response to particular social stimuli. Given the important role of the amygdala in fear processing and the sex bias in the prevalence of anxiety-related disorders, these results may indicate the need for sex-specific treatment of anxiety-related disorders.

## 3.4 Do sex differences in the OT system explain sex-specific actions of OT?

The effects of OT on behavior and neuronal responses in male and female rodents and humans described previously support the notion of a sexually dimorphic OT system. There should be caution, however, in the way we interpret a sex-specific effect as equaling a neuropeptide system that shows sex differences in peptide and/or receptor expression. As we discussed in Section 2.1, sex differences in brain OT system parameters are highly species- and brain region-specific. We discussed the hypothesis that a greater difference in OT-modulated social behaviors between males and females in a given species (possibly due to differences in social organization) will predict more sex differences in the OT system. It is now interesting to note that greater sex differences in the OT system do not always predict greater sex differences in the action of exogenous OT on behavior and brain responses. Instead, when combining the outcomes of Sections 2 and 3, it becomes clear that sex differences in the action of OT do not necessarily coincide with sex differences in OT system parameters in the brain. Indeed, sex-specific actions of OT manipulations are seen in species that show sex differences in the OT system, as well as in species that show a lack of sex differences in the OT system.

This illuminates the lack of studies showing causal relationships between sex-specific behavioral effects of OT and sex differences in the OT system. Rats might be a good model system to test this as rats have a sexually dimorphic OTR system (Bale and Dorsa, 1995; Uhl-Bronner et al., 2005; Dumais et al., 2013) and often respond to OT system manipulations in sex-specific ways. For example, OT affects anxiety-related behavior (Slattery and Neumann, 2010), social-defeat induced avoidance (Lukas et al., 2011; Lukas and Neumann, 2014), and social play (Bredewold et al., 2014) differently in males and females. Furthermore, sex differences in OTR in the medial amygdala correlate with social interest in sex-specific ways in rats (Dumais et al., 2013). Though a direct causal relationship remains to be determined, this suggests that sex differences in the OT system may be contributing to the sex-specific effects of OT manipulation.

Interestingly, sex differences in the OT system do not seem to be necessary in order for exogenous OT to induce sex-specific effects on behavior. For example, while there is a lack of sex differences in most OT parameters in prairie voles (OTR in Bales et al., 2007a, OT-ir in Wang et al., 1996), there is an abundance of sex differences in the effects of exogenous OT, especially given early in life, on social behavior, including partner preference formation and parental care (Bales and Carter, 2003a; Bales et al., 2004, 2007b; Carter et al., 2008). These sex-specific effects of OT could be somewhat surprising given the marked similarity between male and female prairie voles in mating and parental systems (with both sexes showing pair bonding, mate guarding, and parental care). This suggests that while the behavioral output is similar, the underlying mechanisms may be different. In support, and described in more detail in Section 3.1.2, exogenous OT influenced OT, estrogen, and vasopressin systems differently in males and females (i.e., OT-ir, ERα-ir, and V1aR; Yamamoto et al., 2004, 2006; Bales et al., 2007a). But clearly, more research is needed to provide evidence that similar behaviors in prairie voles are mediated via different neural mechanisms and circuitries.

Whether there are sexual dimorphisms in the OT system in humans is still unclear and is in need of further research (see Section 2.2). Therefore, since sex differences in plasma OT levels in men and women are inconsistent and sometimes conflicting, it is more difficult to conclude whether greater sex differences in the human OT system predicts greater sex differences in response to OT manipulations. However, when we look at individual studies that indicate sex differences, or a lack thereof, in the OT system, as well as associations of the OT system with behavior, we can conclude that sex differences in the OT system do not necessarily coincide with sex-specific actions of exogenous OT. For example, Zhong et al. (2012) reported no sex differences in plasma OT levels, but found sex-specific correlations of OT with levels of trust. Taylor et al. (2010) also found no sex differences in plasma OT levels, but found sex-specific correlations of OT with distress in social relationships.

The examples discussed previously indicate that the presence or absence of sex differences in the brain OT system may or may not cause sex differences in behavioral output when activated by exogenous OT. This may show some resemblance with the dual-function hypothesis of sex differences in the brain proposed by De Vries (De Vries, 2004). Here he proposed that sex differences in brain structure in both adult and developing brains can either cause sex differences in behavior, or can prevent sex differences in behavior, potentially by compensating for sex differences in gonadal hormone levels.

That leaves us with the question that, if the OT system itself does not necessarily have to be sexually dimorphic to have sexually dimorphic behavioral functions, what could be causing these sexually dimorphic effects of OT? It could be that signaling pathways upon OTR activation are different in males versus females. In support of this idea, Nakajima et al. (2014) found that OTR-expressing interneurons in the medial prefrontal cortex of female mice respond to OT more robustly than those in male mice. Whether this is a common mechanism, or brain region- and species-specific, is not known. Further research into possible sex differences in how OTR-expressing neurons respond to OT will be important to clarify sex differences in the behavioral and brain effects of OT.

Another important future research direction is the investigation into potential sex differences in local OT release patterns in the brain in response to various social stimuli and the functional implications for the regulation of social behavior. Several studies have reported dynamic changes in local release patterns of OT in specific brain areas during the display of different forms of social behavior in males and females (Neumann et al., 1993; Engelmann et al., 1999; Ebner et al., 2000; Bosch et al., 2004; Ross et al., 2009; Waldherr and Neumann, 2007; Nyuyki et al., 2011). However, none of these studies directly compared males and females.

In conclusion, there are numerous examples of sexual dimorphisms of the OT system in a variety of rodent species, and also in humans. Just the same, there are numerous examples of an absence of sex differences in the OT system. This indicates that reference to a "sexually dimorphic OT system" should be specifically linked to particular species. Moreover, social experience and social context are important factors to consider, since both have been found to influence the modulatory action of OT in rodents in a sex-dependent manner (Harmon et al., 2002; Bredewold et al., 2014). Similarly, in humans, context and personality both influence the effects of OT on behavior and cognition (reviewed in Bartz et al., 2011a, 2011b). Therefore, sex differences in the effects of OT are not necessarily hard wired at birth, but may be modulated by social factors. Future research, including more direct comparisons of the behavioral effects of OT in males and females in a variety of species and contexts, is therefore necessary.

## 4 A NOTE ON OT AS A THERAPEUTIC AGENT IN MEN AND WOMEN – ADVANCES AND PRECAUTIONS

Because OT is a key regulator of mammalian social behavior, a role of the OT system in the etiology of neuropsychiatric disorders characterized by social dysfunction has been an important area of research. However, many studies linking the OT system to psychiatric disorder etiology only include one sex, or do not analyze data according to sex. For example, plasma OT levels were found to be lower in children with ASD than in healthy children (Modahl et al., 1998; Green et al., 2001), but only males were sampled in both cases. Links have also been made between SNPs in the OTR gene and autism in a Chinese Han population (Wu et al., 2005) and in a United States Caucasian population (Jacob et al., 2007) and a link between SNPs and Asperger syndrome in a European Caucasian population has been established (Di Napoli et al., 2014). All three studies included males and females; however, an analysis of sex was not reported, likely because of the low number of female subjects. Furthermore, OTR gene variants were found to be associated with specific ASD phenotypes, but again, an analysis of sex was not performed (Campbell et al., 2011). Interestingly, a sex-specific association was found in the expression of the OTR gene and autism (Gregory et al., 2009). Specifically, the OTR gene was found to be hypermethylated in peripheral blood mononuclear cells and in the temporal cortex in male and female patients with autism; however, OTR mRNA expression levels in the temporal cortex were decreased in male patients only.

The OT system has also been implicated in the etiology of other neuropsychiatric disorders characterized by social dysfunction. One study correlated plasma OT levels in men and women with generalized social anxiety disorder, and while there was no effect of sex, higher plasma OT was associated with higher symptom severity and lower satisfaction in social relationships (Hoge et al., 2008). Also, OT levels were found to be decreased in plasma of male schizophrenia patients compared with healthy controls; however, women were not included in this study (Jobst et al., 2014). In addition, polymorphisms in the OTR gene were found to be associated with conduct problems and callous-unemotional traits in children, traits that are a risk for the development of psychopathy in adulthood (Dadds et al., 2014). This association held for both male and female children in their sample.

Initial investigations into the therapeutic potential of OT in improving social dysfunction in patients have been encouraging, but again, direct comparisons of the effects of OT in male and female patients are limited. Hollander et al. (2003) was the first to show that intravenous infusion of OT for 4 h caused a decrease in repetitive behavior in adult patients with autism or Asperger syndrome. A comparison of sex was not performed, however, because this study included only one female. Andari et al. (2010) found that intranasal OT (single dose) increased interactions with a more socially co-operative partner and increased reports of enhanced feelings of trust in individuals with high-functioning autism or Asperger syndrome. This study also had a very low female participant rate ($n = 2$), and an analysis between sexes was not performed. A single dose of intranasal OT also improved emotion recognition in youth with ASD when using the Reading of the Mind in the Eyes Task; however, only males were used as participants (Guastella et al., 2010). In addition, while male patients with Asperger syndrome showed decreased right amygdala activity compared with a control group during facial stimuli processing, intranasal OT increased amygdala activity in response to facial stimuli (Domes et al., 2013). This suggests that OT may promote face processing in individuals with ASD, at least in male patients as no female patients were tested.

OT may also have therapeutic potential in patients with bipolar disorder (BPD), anxiety disorders, schizophrenia, and eating disorders. For example, a single dose of intranasal OT reduced social threat hypersensitivity in women with BPD (Bertsch et al., 2013), but the effect of OT on social threat hypersensitivity in men with BPD was not examined. In addition, while intranasal OT had no effect on amygdala activation to fearful faces in a control group, OT attenuated the heightened amygdala reactivity to fearful faces in patients with generalized social anxiety disorder (Labuschagne et al., 2010), suggesting that OT may have a therapeutic role in those showing exaggerated fear to social stimuli. This study, however, was performed only in men, and a therapeutic role of OT in social anxiety disorder in women is in need of further research. Intranasal OT also improved social cognition in men with schizophrenia; however, women were not included in this study (Woolley et al., 2014). Finally, a 4-week and 8-week intranasal OT treatment was also found to reduce body weight in a group of patients with obesity (Zhang et al., 2013). Both men and women were included in the study, but an analysis of sex was not reported.

Neuropsychiatric disorders characterized by social dysfunction display sex differences in prevalence and symptom severity. For example, ASD is four to eight times more common in boys than in girls (Fombonne, 2003; Beaudet, 2012) while BPD is three times more common in women than in men (Widiger and Trull, 1993). As a result, it may be challenging to find enough participants of both sexes. However, given the sex-specific actions of OT in healthy individuals, it will be important to investigate whether the same is true in male and female patients. There are currently 73 publications that studied the role of OT in ASD and schizophrenia. The studies above, therefore, are certainly not all-inclusive, and are for the purpose of highlighting the potential therapeutic role of OT in a multitude of psychiatric disorders, and highlighting the need for more comparative studies to assess the therapeutic potential of OT in both men and women. For more detailed reviews on the therapeutic potential of OT, see Bakermans-Kranenburg and Van IJzendoorn (2013), Veening and Olivier (2013), and Kapoor (2013).

Importantly, examining how intranasal OT modulates brain functioning and behavior has also shed light on possible side effects of chronic intranasal OT, which is important to consider if using as a drug treatment for particular psychiatric disorders. Long-term effects and adverse side effects of chronic OT administration have to be considered in order to assess the safety of clinical use of OT. Preclinical research in animals and humans has uncovered potential precautions that should be considered when using chronic OT as a therapeutic agent. For example, chronic intranasal OT caused long-term impairments in partner preference formation in male prairie voles (Bales et al., 2013). In addition, while acute intranasal OT slightly increases social interaction, chronic intranasal OT in male mice decreased social interaction toward female mice (Huang et al., 2014). OT was also found to hinder trust and cooperation in patients with BPD (Bartz et al., 2011a,b; Ebert et al., 2013), highlighting the importance of individual variation in the response to OT, and that patients with particular psychiatric disorders and/or in particular social contexts may respond differently to OT than healthy control individuals. Importantly, associations between specific symptoms of BPD and the amount of trust in those given intranasal OT was greater when females were analyzed separately compared with when male and female patients were analyzed together (Ebert et al., 2013). This suggests that OT may be regulating trust differently not only between healthy controls and patients with BPD, but also between men and women with BPD. The study by Bartz et al. (2011a) did include male and female participants, but an analysis of sex was not reported. For more in-depth reviews and commentaries, which bring attention to the importance of understanding more fully the role of intranasal OT on human cognition and behavior, see Young (2013), Guastella and MacLeod (2012), Weisman et al. (2013), Bartz et al. (2011b), MacDonald (2013), Olff et al. (2013), Bartz and Hollander (2006).

There are still unanswered questions at the forefront of research investigating the therapeutic potential of OT in humans. For example, does intranasal OT enter the brain to exert direct effects on behavior, or does intranasal OT primarily enter the bloodstream in which it may elicit indirect effects in the brain via peripheral actions? Does intranasal OT affect levels of endogenous release of OT? Does intranasal OT affect cognition and behavior differently in children versus adults? What are the temporal dynamics and durations of both the behavioral and physiological effects of OT administration?

It has been suggested that intranasal OT reaches the human brain based on findings with vasopressin. Here, an increase in vasopressin in cerebrospinal fluid was observed 30 min after intranasal vasopressin administration in humans (Born et al., 2002). Given the similar structure of OT and vasopressin, similar results for OT are hypothesized. However, this study still does not provide evidence of direct access to the brain, as intranasal vasopressin could have indirectly increased vasopressin levels in the brain via peripheral mechanisms. Moreover, individuals administered intranasal OT showed increased plasma OT 30 min postadministration in men (Gossen et al., 2012), and increased salivary OT up to 7 h postadministration in women (Van IJzendoorn et al., 2012), but only one sex was used in each of these studies. A study looking at intranasal OT administration in both men and women found that individuals administered intranasal OT showed increases in salivary OT across a period of 4 h, reaching a plateau at 45–120 min (Weisman et al., 2012). This study included both men and women, and found no effect of sex. In mice, i.p. OT increased neuronal activity in magnocellular and parvocellular neurons in the PVN, and using *in vitro* models, increased OT release in the PVN in male mice (Zhang and Cai, 2011). Finally, rats and mice given intranasal and peripheral administration of OT show increased brain and plasma OT (Neumann et al., 2013), although it is unclear whether these increases reflect exogenous or endogenous OT. Importantly, these studies suggest that peripheral and/or intranasal OT can stimulate or alter central OT levels; however, the underlying mechanisms are still unknown. Further investigation into how and where intranasal OT exerts its behavioral and physiological effects is necessary for a better understanding of the therapeutic potential of intranasal OT to restore social functioning in both men and women.

## 5 CONCLUSIONS

An abundance of research has implicated OT as a key regulator of a variety of social behaviors. Beginning with the important role of OT in the modulation of maternal behavior in both rodents and humans, OT research was primarily limited to the female brain. With the discovery of the important role of OT in male social behavior, OT research has now extended to examining sex-specific modulatory roles of OT on brain and behavior. Though research has come a long way in terms of determining sex differences in OT system parameters in the brain and sex-specific actions of OT, a surprising few number of studies include both male and female subjects (Zucker and Beery, 2010). Additionally, those studies that do include both sexes often do not report whether males and females were analyzed separately to determine sex-specific effects of OT. This highlights the critical need to include both males and females in both animal and human research. Fortunately, there have been important initiatives that require the inclusion of women in National Institutes of Health-funded clinical research and the inclusion of both sexes in basic, preclinical research (Clayton and Collins, 2014).

Given the evidence of studies discussed previously, it is clear that OT often plays a sex-specific role in the regulation of diverse social behaviors across a wide variety of species. However, sex differences in the OT system are highly species-specific and we have argued that this may, in part, depend on similarities versus differences between males and females regarding the social organization of a species. Further comparative research is needed to test this hypothesis. Because sex differences in various OT system parameters are highly species-specific, and in many cases, are similar between males and females, reference to a "sexually dimorphic OT system" should be specifically referenced with species, context, and brain region.

Finally, investigating sex-specific roles of the OT system in the regulation of social behavior is important given the involvement of the OT system in psychiatric disorders of social dysfunction (Modahl et al., 1998; Green et al., 2001; Wu et al., 2005; Jacob et al., 2007; Campbell et al., 2011; Di Napoli et al., 2014). The sexually dimorphic effects of OT on both brain and behavior, as described in this review, further highlight the importance of understanding the role of OT in both men and women, as this may indicate the need for differential treatment of social dysfunction in males versus females. Knowledge about sex differences in the OT system and the behavioral and neural response to OT will be useful in understanding sex differences in the prevalence of these psychiatric disorders, and may help in the development of proper therapeutic agents in men and women.

## ACKNOWLEDGMENTS

We would like to thank Alex Ophir, Elliott Albers, Rebecca Shansky, John Johnson, and the Veenema lab for giving valuable feedback on earlier versions of this manuscript. We also would like to acknowledge Jim Goodson, a dear friend and colleague who recently passed away, for his support, insights, and inspiration given over the years. This work was supported by NSF IOS 1253386, NIMH R15MH102807, NIMH R01MH102456 to AHV, and NIMH F31MH100891 to KMD.

## REFERENCES

Ahern, T.H., Young, L.J., 2009. The impact of early life family structure on adult social attachment, alloparental behavior, and the neuropeptide systems regulating affiliative behaviors in the monogamous prairie vole (*Microtus ochrogaster*). Front. Behav. Neurosci. 3 (17), 1–19.

Albers, H.E., 2014. Species, sex and individual differences in the vasotocin/vasopressin system: relationship to neurochemical signaling in the social behavior neural network. Front. Neuroendocrinol. 36, 49–71.

Amico, J.A., Mantella, R.C., Vollmer, R.R., Li, X., 2004. Anxiety and stress responses in female oxytocin deficient mice. J. Neuroendocrinol. 16 (4), 319–324.

Anacker, A.M., Beery, A.K., 2013. Life in groups: the roles of oxytocin in mammalian sociality. Front. Behav. Neurosci. 7, 185.

Andari, E., Duhamel, J.R., Zalla, T., Herbrecht, E., Leboyer, M., Sirigu, A., 2010. Promoting social behavior with oxytocin in high-functioning autism spectrum disorders. Proc. Natl. Acad. Sci. USA 107 (9), 4389–4394.

Arletti, R., Benelli, A., Poggioli, R., Luppi, P., Menozzi, B., Bertolini, A., 1995. Aged rats are still responsive to the antidepressant and memory-improving effects of oxytocin. Neuropeptides 29 (3), 177–182.

Arnold, A.P., 2009. The organizational-activational hypothesis as the foundation for a unified theory of sexual differentiation of all mammalian tissues. Horm. Behav. 55 (5), 570–578.

Babb, J.A., Carini, L.M., Spears, S.L., Nephew, B.C., 2014. Transgenerational effects of social stress on social behavior, corticosterone, oxytocin, and prolactin in rats. Horm. Behav. 65, 386–393.

Bakermans-Kranenburg, M.J., Van IJzendoorn, M.H., 2008. Oxytocin receptor (OXTR) and serotonin transporter (5-HTT) genes associated with observed parenting. Soc. Cogn. Affect. Neurosci. 3 (2), 128–134.

Bakermans-Kranenburg, M.J., Van IJzendoorn, M.H., 2013. Sniffing around oxytocin: review and meta-analysis of trials in healthy and clinical groups with implications for pharmacotherapy. Transl. Psychiat. 3 (e258).

Bakermans-Kranenburg, M.J., Van IJzendoorn, M.H., 2014. A sociability gene? Meta-analysis of oxytocin receptor genotype effects in humans. Psychiatr. Genet. 24 (2), 45–51.

Bale, T.L., Dorsa, D.M., 1995. Sex differences in and effects of estrogen on oxytocin receptor messenger ribonucleic acid expression in the ventromedial hypothalamus. Endocrinology 136 (1), 27–32.

Bales, K.L., Carter, C.S., 2003a. Developmental exposure to oxytocin facilitates partner preferences in male prairie voles (*Microtus ochrogaster*). Behav. Neurosci. 117 (4), 854–859.

Bales, K.L., Carter, C.S., 2003b. Sex differences and developmental effects of oxytocin on aggression and social behavior in prairie voles (*Microtus ochrogaster*). Horm. Behav. 44, 178–184.

Bales, K.L., Pfeifer, L.A., Carter, C.S., 2004. Sex differences and developmental effects of manipulations of oxytocin on alloparenting and anxiety in prairie voles. Dev. Psychobiol. 44 (2), 123–131.

Bales, K.L., Plotsky, P.M., Young, L.J., Lim, M.M., Grotte, N., Ferrer, E., Carter, C.S., 2007a. Neonatal oxytocin manipulations have long-lasting, sexually dimorphic effects on vasopressin receptors. Neuroscience 144 (1), 38–45.

Bales, K.L., van Westerhuyzen, J.A., Lewis-Reese, A.D., Grotte, N.D., Lanter, J.A., Carter, C.S., 2007b. Oxytocin has dose-dependent developmental effects on pair bonding and alloparental care in female prairie voles. Horm. Behav. 52, 274–279.

Bales, K.L., Perkeybile, A.M., Conley, O.G., Lee, M.H., Guoynes, C.D., Downing, G.M., Yun, C.R., Solomon, M., Jacob, S., Mendoza, S.P., 2013. Chronic intranasal oxytocin causes long-term impairments in partner preference formation in male prairie voles. Biol. Psychiat. 74, 180–188.

Bartz, J.A., Hollander, E., 2006. The neuroscience of affiliation: forging links between basic and clinical research on neuropeptides and social behavior. Horm. Behav. 50, 518–528.

Bartz, J., Simeon, D., Hamilton, H., Kim, S., Crystal, S., Braun, A., Vicens, V., Hollander, E., 2011a. Oxytocin can hinder trust and cooperation in borderline personality disorder. Soc. Cogn. Affect. Neurosci. 6 (5), 556–563.

Bartz, J.A., Zaki, J., Bolger, N., Ochsner, K.N., 2011b. Social effects of oxytocin in humans: context and person matter. Trends Cogn. Sci. 15 (7), 301–309.

Baumans, V., Van Loo, P.L., 2013. How to improve housing conditions of laboratory animals: the possibilities of environmental refinement. Vet. J. 195 (1), 24–32.

Baumgartner, T., Heinrichs, M., Vonlanthen, A., Fischbacher, U., Fehr, E., 2008. Oxytocin shapes the neural circuitry of trust and trust adaptation in humans. Neuron 58 (4), 639–650.

Beaudet, A.L., 2012. Neuroscience. Preventable forms of autism? Science 338 (6105), 342–343.

Beery, A.K., Zucker, I., 2010. Oxytocin and same-sex social behavior in female meadow voles. Neuroscience 169, 665–673.

Belin, V., Moos, F.C., Richard, P., 1984. Synchronization of oxytocin cells in the hypothalamic paraventricular and supraoptic nucle in suckled rats: direct proof with paired extracellular recordings. Exp. Brain. Res. 57, 201–203.

Benelli, A., Bertolini, A., Poggioli, R., Menozzi, B., Basaglia, R., Arletti, R., 1995. Polymodal dose-response curve for oxytocin in the social recognition test. Neuropeptides 28 (4), 251–255.

Bertsch, K., Gamer, M., Schmidt, B., Schmidinger, I., Walther, S., Kastel, T., Schnell, K., Buchel, C., Domes, G., Herpertz, S.C., 2013. Oxytocin and reduction of social threat hypersensitivity in women with borderline personality disorder. Am. J. Psychiat. 170 (10), 1169–1177.

Bielsky, I.F., Young, L.J., 2004. Oxytocin, vasopressin, and social recognition in mammals. Peptides 25 (9), 1565–1574.

Blume, A., Bosch, O.J., Miklos, S., Torner, L., Wales, L., Waldherr, M., Neumann, I.D., 2008. Oxytocin reduces anxiety via ERK1/2 activation: local effect within the rat hypothalamic paraventricular nucleus. Eur. J. Neurosci. 27, 1947–1956.

Boccia, M.L., Petrusz, P., Suzuki, K., Marson, L., Pedersen, C.A., 2013. Immunohistochemical localization of oxytocin receptors in human brain. Neuroscience 253, 155–164.

Born, J., Lange, T., Kern, W., McGregor, G.P., Bickel, U., Fehm, H.L., 2002. Sniffing neuropeptides: a transnasal approach to the human brain. Nat. Neurosci. 5 (6), 514–516.

Bosch, O.J., Krömer, S.A., Brunton, P.J., Neumann, I.D., 2004. Release of oxytocin in the hypothalamic paraventricular nucleus, but not central amygdala or lateral septum in lactating residents and virgin intruders during maternal defence. Neuroscience 124 (2), 439–448.

Bosch, O.J., Pförtsch, J., Beiderbeck, D.I., Landgraf, R., Neumann, I.D., 2010. Maternal behaviour is associated with vasopressin release in the medial preoptic area and bed nucleus of the stria terminalis in the rat. J. Neuroendocrinol. 22 (5), 420–429.

Bosch, O.J., 2013. Maternal aggression in rodents: brain oxytocin and vasopressin mediate pup defence. Philos. Trans. R Soc. Lond. B Biol. Sci. 368 (1631), 20130085.

Bredewold, R., Smith, C.J., Dumais, K.M., Veenema, A.H., 2014. Sex-specific modulation of juvenile social play behavior by vasopressin and oxytocin depends on social context. Front. Behav. Neurosci. 8 (216), 1–11.

Brett, R.A., 1991. The population structure of naked mole rat colonies. In: Sherman, P.W., Jarvis, J.U.M., Alexander, R.D. (Eds.), The Biology of the Naked Mole Rat. Princeton University Press, Princeton, pp. 97–136.

Brownstein, M.J., Russell, J.T., Gainer, H., 1980. Synthesis, transport, and release of posterior pituitary hormones. Science 207 (4429), 373–378.

Buijs, R.M., 1978. Intra- and extrahypothalamic vasopressin and oxytocin pathways in the rat. Pathways to the limbic system, medulla oblongata and spinal cord. Cell Tissue Res. 192 (3), 423–435.

Buijs, R.M., Swaab, D.F., 1979. Immuno-electron microscopical demonstration of vasopressin and oxytocin synapses in the limbic system of the rat. Cell Tissue Res. 204 (3), 355–365.

Calcagnoli, F., 2014. Oxytocin: the neurochemical mediator of social life (PhD thesis). http://irs.ub.rug.nl/ppn/381337065.

Campbell, D.B., Datta, D., Jones, S.T., Lee, E.B., Sutcliffe, J.S., Hammock, E.A.D., Levitt, P., 2011. Association of oxytocin receptor (OXTR) gene varients with multiple phenotype domains of autism spectrum disorder. J. Neurodev. Disord. 3, 101–112.

Cao, Y., Wu, R., Tai, F., Zhang, X., Yu, P., An, X., Qiao, X., Hao, P., 2013. Neonatal paternal deprivation impairs social recognition and alters levels of oxytocin and estrogen receptor α mRNA expression in the MeA and NAcc, and serum oxytocin in mandarin voles. Horm. Behav. 65 (1), 57–65.

Carter, D.A., Lightman, S.L., 1986. Diurnal patterns of stress-evoked neurohypophyseal hormone secretion: sexual dimorphism in rats. Neurosci. Lett. 71 (2), 252–255.

Carter, C.S., Boone, E.M., Bales, K.L., 2008. Early experience and the developmental programming of oxytocin and vasopressin. Neurobiol. Parental Brain, 417–433.

Chiodera, P., Salvarani, C., Bacchi-Modena, A., Spallanzani, R., Cigarini, C., Alboni, A., Gardini, E., Coiro, V., 1991. Relationship between plasma profiles of oxytoin and adrenocorticotropic hormone during suckling or breast stimulation in women. Horm. Res. 35 (3–4), 119–123.

Cho, M.M., DeVries, A.C., Williams, J.R., Carter, C.S., 1999. The effects of oxytocin and vasopressin on partner preference in male and female prairie voles (*Microtus ochrogaster*). Behav. Neurosci. 113 (5), 1071–1079.

Choleris, E., Gustafsson, J.A., Korach, K.S., Muglia, L.J., Pfaff, D.W., Ogawa, S., 2003. An estrogen-dependent four-gene micronet regulating social recognition: a study with oxytocin and estrogen receptor-alpha and beta-knockout mice. Proc. Natl. Acad. Sci. USA 100 (10), 6192–6197.

Choleris, E., Ogawa, S., Kavaliers, M., Gustafsson, J.A., Korach, K.S., Muglia, L.J., Pfaff, D.W., 2006. Involvement of estrogen receptor alpha, beta and oxytocin in social discrimination: a detailed behavioral analysis with knockout female mice. Genes Brain Behav. 5 (7), 528–539.

Clayton, J.A., Collins, F.S., 2014. Policy: NIH to balance sex in cell and animal studies. Nature 509 (7500), 282–283.

Cochran, D.M., Fallon, D., Hill, M., Frazier, J.A., 2013. The role of oxytoin in psychiatric disorders: a review of biological and therapeutic research findings. Harv. Rev. Psychiat. 21 (5), 219–247.

Dadds, M.R., Moul, C., Cauchi, A., Dobson-Stone, C., Hawes, D.J., Brennan, J., Urwin, R., Ebstein, R.E., 2014. Polymorphisms in the oxytocin receptor gene are associated with the development of psychopathy. Dev. Psychopathol. 26, 21–31.

Dantzer, R., Bluthe, R.M., Koob, G.F., Le Moal, M., 1987. Modulation of social memory in male rats by neurohypophyseal peptides. Psychopharmacology 91 (3), 363–368.

De Vries, G.J., 2004. Minireview: sex differences in adult and developing brains: compensation, compensation, compensation. Endocrinology 145 (3), 1063–1068.

De Vries, G.J., Panzica, G.C., 2006. Sexual differentiation of central vasopressin and vasotocin systems in vertebrates: different mechanisms, similar endpoints. Neuroscience 138 (3), 947–955.

DeVries, A.C., Carter, C.S., 1999. Sex differences in temporal parameters of partner preference in prairie voles (*Microtus ochrogaster*). Can. J. Zool. 77, 885–889.

Di Napoli, A., Warrier, V., Baron-Cohen, S., Chakrabarti, B., 2014. Genetic variation in the oxytocin receptor (OXTR) gene is associated with Asperger syndrome. Mol. Autism 5 (1), 48.

Di Simplicio, M., Massey-Chase, R., Cowen, P.J., Harmer, C.J., 2009. Oxytocin enhances processing of positive versus negative emotional information in healthy male volunteers. J. Psychopharmacol. 23 (3), 241–248.

Dierickx, K., 1980. Immunocytochemical localization of the vertebrate cyclic nonapeptide neurohypophyseal hormones and neuophysins. Int. Rev. Cytol. 62, 119–185.

Ditzen, B., Schaer, M., Gabriel, B., Bodenmann, G., Ehlert, U., Heinrichs, M., 2009. Intranasal oxytocin increases positive communication and reduces cortisol levels during couple conflict. Biol. Psychol. 65, 728–731.

Ditzen, B., Nater, U.M., Schaer, M., Marca, R.L., Bodenmann, G., Ehlert, U., Heinrichs, M., 2013. Sex-specific effects of intranasal oxytocin on autonomic nervous system and emotional responses to couple conflict. SCAN 8, 897–902.

Domes, G., Heinrichs, M., Glascher, J., Muchel, C., Braus, D.F., Herpertz, S.C., 2007a. Oxytocin attenuates amygdala responses to emotional faces regardless of valence. Biol. Psychiatry 62 (10), 1187–1190.

Domes, G., Heinrichs, M., Michel, A., Berger, C., Herpertz, S.C., 2007b. Oxytocin improves "mind-reading" in humans. Biol. Psychiatry 61 (6), 731–733.

Domes, G., Lischke, A., Berger, C., Grossmann, A., Hauenstein, K., Heinrichs, M., Herpertz, S.C., 2010. Effects of intranasal oxytocin on emotional face processing in women. Psychoneuroendocrinology 35 (1), 83–93.

Domes, G., Heinrichs, M., Kumbier, E., Grossmann, A., Hauenstein, K., Herpertz, S.C., 2013. Effects of intranasal oxytocin on the neural basis of face processing in autism spectrum disorder. Biol. Psychiat. 74 (3), 164–171.

Donaldson, Z.R., Young, L.J., 2008. Oxytocin, vasopressin, and the neurogenetics of sociality. Science 322 (5903), 900–904.

Dubois-Dauphin, M., Pévet, P., Barberis, C., Tribollet, E., Dreifuss, J.J., 1992. Localization of binding sites for oxytocin in the brain of the golden hamster. Neuroreport 3 (9), 797–800.

Dumais, K.M., Bredewold, R., Mayer, T.E., Veenema, A.H., 2013. Sex differences in oxytocin receptor binding in forebrain regions: correlations with social interest in brain region- and sex-specific ways. Horm. Behav. 64 (4), 693–701.

Dumais, K.M., Veenema, A.H., 2015. Vasopressin and oxytocin receptor systems in the brain: sex differences and sex-specific regulation of social behavior. Front. Neuroendocrinol., doi:10.1016/j.yfrne.2015.04.003.

Ebert, A., Kolb, M., Heller, J., Edel, M., Roser, P., Brüne, M., 2013. Modulation of interpersonal trust in borderline personality disorder by intranasal oxytocin and childhood trauma. Soc. Neurosci. 8 (4), 305–313.

Ebner, K., Wotjak, C.T., Landrag, R., Engelmann, M., 2000. A single social defeat experience selectively stimulates the release of oxytocin, but not vasopressin, within the septal brain area of male rats. Brain Res. 872 (1–2), 87–92.

Ebstein, R.P., Knafo, A., Mankuta, D., Chew, S.H., Lai, P.S., 2012. The contributions of oxytocin and vasopressin pathway genes to human behavior. Horm. Behav. 61 (3), 359–379.

Engelmann, M., Ebner, K., Wotjak, C.T., Landgraf, R., 1998. Endogenous oxytocin is involved in short-term olfactory memory in female rats. Behavioral. Brain Res. 90, 89–94.

Engelmann, M., Ebner, K., Landgraf, R., Holsboer, F., Wotjak, C.T., 1999. Emotional stress triggers intrahypothalamic but not peripheral release of oxytocin in male rats. J. Neuroendocrinol. 11 (11), 867–872.

Evans, S.L., Dal Monte, O., Noble, P., Averbeck, B.B., 2014. Intranasal oxytocin effects on social cognition: a critique. Brain Res. 1580, 69–77.

Feldman, R., Weller, A., Zagoory-Sharon, O., Levine, A., 2007. Evidence for a neuroendocrinological foundation of human affiliation: plasma oxytocin levels across pregnancy and the postpartum period predict mother-infant bonding. Psychol. Sci. 18, 965–970.

Feldman, R., Gordon, I., Zagoory-Sharon, O., 2010. The cross-generation transmission of oxytocin in humans. Horm. Behav. 58, 669–676.

Ferguson, J.N., Young, L.J., Hearn, E.F., Matzuk, M.M., Insel, T.R., Winslow, J.T., 2000. Social amnesia in mice lacking the oxytocin gene. Nat. Genet. 25 (3), 284–288.

Ferguson, J.N., Aldag, J.M., Insel, T.R., Young, L.J., 2001. Oxytocin in the medial amygdala is essential for social recognition in the mouse. J. Neurosci. 21 (20), 8278–8285.

File, S.E., 1980. The use of social interaction as a method for detecting anxiolytic activity of chlordiazepoxide-like drugs. J. Neurosci. Methods 2 (3), 219–238.

Fischer-Shofty, M., Levkovitz, Y., Shamay-Tsoory, S.G., 2013. Oxytocin facilitates accurate perception of competition in men and kinship in women. Soc. Cogn. Affect. Neurosci. 8, 313–317.

Fliers, E., Swaab, D.F., Pool, C.W., Verwer, R.W.H., 1985. The vasopressin and oxytocin neurons in the human supraoptic and paraventricular nucleus; changes with aging and in senile dementia. Brain Res. 342, 45–53.

Fombonne, E., 2003. The prevalence of autism. JAMA 289 (1), 87–89.

Francis, D.D., Young, L.J., Meaney, M.J., Insel, T.R., 2002. Naturally occurring differences in maternal care are associated with the expression of oxytocin and vasopressin (V1a) receptors: gender differences. J. Neuroendocrinol. 14 (5), 349–353.

Fuchs, A.R., Poblete, Jr., V.F., 1970. Oxytocin and uterine function in pregnant and parturient rats. Biol. Reprod. 2, 387–400.

Gattermann, R., Fritzsche, P., Neumann, K., Al-Hussein, I., Kayser, A., Abiad, M., Yakti, R., 2001. Notes on the current distribution and the ecology of wild golden hamsters (*Mesocricetus auratus*). J. Zool. 254 (3), 359–365.

Gibbs, D.M., 1984. Dissociation of oxytocin, vasopressin and corticotropin secretion during different types of stress. Life Sci. 35 (5), 487–491.

Gimpl, G., Fahrenholz, F., 2001. The oxytocin receptor system: structure, function, and regulation. Physiol. Rev. 81 (2), 630–683.

Goodson, J.L., Thompson, R.R., 2010. Nonapeptide mechanisms of social cognition, behavior and species-specific social systems. Curr. Opin. Neurobiol. 20 (6), 784–794.

Gordon, I., Zagoory-Sharon, O., Schneiderman, I., Leckman, J.F., Weller, A., Feldman, R., 2008. Oxytocin and cortisol in romantically unattached young adults: associations with bonding and psychological distress. Psychophysiology 45, 349–352.

Gordon, I., Zagoory-Sharon, O., Leckman, J.F., Feldman, R., 2010. Oxytocin and the development of parenting in humans. Biol. Psychiat. 68, 377–382.

Gossen, A., Hahn, A., Westphal, L., Prinz, S., Schultz, R.T., Grunder, G., Spreckelmeyer, K.N., 2012. Oxytocin plasma concentrations after single intranasal oxytocin administration – a study in healthy men. Neuropeptides 46 (5), 211–215.

Green, L., Fein, D., Modahl, C., Feinstein, C., Waterhouse, L., Morris, M., 2001. Oxytocin and autistic disorder: alterations in peptide forms. Biol. Psychiat. 50 (8), 609–613.

Gregory, S.G., Connelly, J.J., Towers, A.J., Johnson, J., Biscocho, D., Markunas, C.A., Lintas, C., Abramson, R.K., Wright, H.H., Ellis, P., Langford, C.F., Worley, G., Delong, G.R., Murphy, S.K., Cuccaro, M.L., Persico, A., Persico, A., Pericak-Vance, M.A., 2009. Genomic and epigenetic evidence for oxytocin receptor deficiency in autism. BMC Med. 7, 62.

Grewen, K.M., Girdler, S.S., Amico, J., Light, K.C., 2005. Effects of partner support on resting oxytocin, cortisol, norepinephrine, and blood pressure before and after warm partner contact. Psychosom. Med. 67, 531–538.

Grillon, C., Krimsky, M., Charney, D.R., Vytal, K., Ernst, M., Cornwell, B., 2013. Oxytocin increases anxiety to unpredictable threat. Mol. Psychiat. 18 (9), 958–960.

Guastella, A., MacLeod, C., 2012. A critical review of the influence of oxytocin nasal spray on social cognition in humans: evidence and future directions. Horm. Behav. 61, 410–418.

Guastella, A.J., Mitchell, P.B., Dadds, M.R., 2008a. Oxytocin increases gaze to the eye region of human faces. Biol. Psychiat. 63 (1), 3–5.

Guastella, A.J., Mitchell, P.B., Matthews, F., 2008b. Oxytocin enhances the encoding of positive social memories in humans. Biol. Psychiat. 64 (3), 256–258.

Guastella, A.J., Einfeld, S.L., Gray, K.M., Rinehart, N.J., Tonge, B.J., Lambert, T.J., Hickie, I.B., 2010. Intranasal oxytocin improves emotion recognition for youth with autism spectrum disorders. Biol. Psychiat. 67 (7), 692–694.

Hammock, E.A., 2015. Developmental perspectives on oxytocin and vasopressin. Neuropsychopharmacology 40 (1), 24–42.

Hammock, E.A., Levitt, P., 2013. Oxytocin receptor ligand binding in embryonic tissue and postnatal brain development of the C57BL/6J mouse. Front. Behav. Neurosci. 7, 195.

Harmon, A.C., Moore, T.O., Huhman, K.L., Albers, H.E., 2002. Social experience and social context alter the behavioral response to centrally administered oxytocin in female Syrian hamsters. Neuroscience 109 (4), 767–772.

Häussler, H.U., Jirikowski, G.F., Caldwell, J.D., 1990. Sex differences among oxytocin-immunoreactive neuronal systems in the mouse hypothalamus. J. Chem. Neuroanat. 3, 271–276.

Heinrichs, M., Meinlschmidt, G., Neumann, I., Wagner, S., Kirschbaum, C., Ehlert, U., Hellhammer, D.H., 2001. Effects of suckling on hypothalamic-pituitary-adrenal axis response to psychosocial stress in postpartum lactating women. J. Clin. Endocrinol. Metab. 86 (10), 4798–4804.
Heinrichs, M., Baumgartner, T., Kirschbaum, C., Ehlert, U., 2003. Social support and oxytocin interact to suppress cortisol and subjective responses to psychosocial stress. Biol. Psychiat. 54 (12), 1389–1398.
Heinrichs, M., Meinlschmidt, G., Wippich, W., Ehlert, U., Hellhammer, D.H., 2004. Selective amnesic effects of oxytocin on human memory. Physiol. Behav. 83 (1), 31–38.
Heinrichs, M., von Dawans, B., Domes, G., 2009. Oxytocin, vasopressin, and human social behavior. Front. Neuroendocrinol. 30 (4), 548–557.
Herzmann, G., Bird, C.W., Freeman, M., Curran, T., 2013. Effects of oxytocin on behavioral and ERP measures of recognition memory for own-race and other-race faces in women and men. Psychoneuroendocrinology 38, 2140–2151.
Hoge, E.A., Pollack, M.H., Kaufman, R.E., Zak, P.J., Simon, N.M., 2008. Oxytocin levels in social anxiety disorder. CNS Neurosci. Ther. 14 (3), 165–170.
Hollander, E., Novotny, S., Hanratty, M., Yaffe, R., DeCaria, C.M., Aronowitz, B.R., Mosovich, S., 2003. Neuropsychopharmacology 28 (1), 193–198.
Holt-Lunstad, J., Birmingham, W., Light, K.C., 2011. The influence of depressive symptomology and perceived stress on plasma and salivary oxytocin before, during, and after a support enhancement intervention. Psychoneuroendocrinology 36, 1249–1256.
Huang, H., Michetti, C., Busnelli, M., Managò, F., Sannino, S., Scheggia, D., Giancardo, L., Sona, D., Murino, V., Chini, B., Scattoni, M.L., Papaleo, F., 2014. Chronic and acute intranasal oxytocin produce divergent social effects in mice. Neuropsychopharmacology 39 (5), 1102–1114.
Insel, T.R., Hulihan, T.J., 1995. A gender-specific mechanism for pair bonding: oxytocin and partner preference formation in monogamous voles. Behav. Neurosci. 109 (4), 782–789.
Insel, T.R., Gelhard, R., Shapiro, L.E., 1991. The comparative distribution of forebrain receptors for neurohypophyseal peptides in monogamous and polygamous mice. Neuroscience 43 (2–3), 623–630.
Ishunina, T.A., Swaab, D.F., 1999. Vasopressin and oxytocin neurons of the human supraoptic and paraventricular nucleus; size changes in relation to age and sex. J. Clin. Endocrinol. Metab. 84 (12), 4637–4644.
Jacob, S., Brune, C.W., Carter, C.S., Leventhal, B.L., Lord, C., Cook, Jr., E.H., 2007. Association of the oxytocin receptor gene (OXTR) in Caucasian children and adolescents with autism. Neurosci. Lett. 417 (1), 6–9.
Jard, S., Barberis, C., Audigier, S., Tribollet, E., 1987. Neurohypophyseal hormone receptor systems in brain and periphery. Prog. Brain Res. 72, 173–187.
Jezova, D., Skultetyova, I., Tokarev, D.I., Bakos, P., Vigas, M., 1995. Vasopressin and oxytocin in stress. Ann. NY Acad. Sci. 771, 192–203.
Jobst, A., Dehning, S., Ruf, S., Notz, T., Buchheim, A., Henning-Fast, K., Meißner, D., Meyer, S., Bondy, B., Muller, N., Zill, P., 2014. Oxytocin and vasopressin levels are decreased in the plasma of male schizophrenia patients. Acta Neuropsychiatr. 1–9.
Kapoor, S., 2013. Intranasal oxytocin and its rapidly expanding role in psychiatry. Psychoneuroendocrinology 38 (3), 462–463.
Kirsch, P., Esslinger, C., Chen, Q., Mier, D., Lis, S., Siddhanti, S., Gruppe, H., Mattay, V.S., Gallhofer, B., Meyer-Lindenberg, A., 2005. Oxytocin modulates neural circuitry for social cognition and fear in humans. J. Neurosci. 25 (49), 11489–11493.
Kleiman, D.G., 1977. Monogamy in mammals. Q. Rev. Biol. 52 (1), 39–69.
Knobloch, H.S., Charlet, A., Hoffmann, L.C., Eliava, M., Khrulev, S., Cetin, A.H., Osten, P., Schwarz, M.K., Seeburg, P.H., Stoop, R., Grinevich, V., 2012. Evoked axonal oxytocin release in the central amygdala attenuates fear response. Neuron 73 (3), 553–566.
Kosfeld, M., Heinrichs, M., Zak, P.J., Fischbacher, U., Fehr, E., 2005. Oxytocin increases trust in humans. Nature 435 (7042), 673–676.
Kramer, K.M., Cushing, B.S., Carter, C.S., Wu, J., Ottinger, M.A., 2004. Sex and species differences in plasma oxytocin using an enzyme immunoassay. Can. J. Zool. 82 (8), 1194–1200.
Kramer, K.M., Choe, C., Carter, C.S., Cushing, B.S., 2006. Developmental effects of oxytocin on neural activation and neuropeptide release in response to social stimuli. Horm. Behav. 49, 206–214.
Kubzansky, L.D., Mendes, W.B., Appleton, A.A., Block, J., Adler, G.K., 2012. A heartfelt response: oxytocin effects on response to social stress in men and women. Biol. Psychol. 90, 1–9.

Kudwa, A.E., McGiverm, R.F., Handa, R.J., 2014. Estrogen receptor β and oxytocin interact to modulate anxiety-like behavior and neuroendocrine stress reactivity in adult male and female rats. Physiol. Behav. 129, 287–296.

Labuschagne, I., Phan, K.L., Wood, A., Angstadt, M., Chua, P., Heinrichs, M., Stout, J.C., Nathan, P.J., 2010. Oxytocin attenuates amygdala reactivity to fear in generalized social anxiety disorder. Neuropsychopharmacology 35, 2403–2413.

Lande, R., Arnold, S.J., 1985. Evolution of mating preference and sexual dimorphism. J. Theor. Biol. 117 (4), 651–664.

Lang, R.E., Heil, J.W.E., Ganten, D., Hermann, K., Unger, T., Rascher, W., 1983. Oxytocin unlike vasopressin is a stress hormone in the rat. Neuroendocrinology 37, 314–316.

Lee, H.J., Caldwell, H.K., Macbeth, A.H., Tolu, S.G., Young, III, W.S., 2008. A conditional knockout mice line of the oxytocin receptor. Endocrinology 149 (7), 3256–3263.

Lim, M.M., Young, L.J., 2004. Vasopressin-dependent neural circuits underlying pair bond formation in the monogamous prairie vole. Neuroscience 125 (1), 35–45.

Lischke, A., Gamer, M., Berger, C., Grossmann, A., Hauenstein, K., Heinrichs, M., Herpertz, S.C., Domes, G., 2012. Oxytocin increases amygdala reactivity to threatening scenes in females. Psychoneuroendocrinology 37, 1431–1438.

Liu, Y., Wang, Z.X., 2003. Nucleus accumbens oxytocin and dopamine interact to regulate pair bond formation in female prairie voles. Neuroscience 121, 537–544.

Liu, Y., Curtis, J.T., Wang, Z., 2001. Vasopressin in the lateral septum regulates pair bond formation in male prairie voles (*Microtus ochrogaster*). Behav. Neurosci. 115, 910–919.

Lonstein, J.S., De Vries, G.J., 2000. Sex differences in the parental behavior of rodents. Neurosci. Biobehav. Rev. 24, 669–686.

Lonstein, J.S., Fleming, A.S., 2002. Parental behaviors in rats and mice. Curr. Prot. Neurosci. 8 (15).

Loup, F., Tribollet, E., Dubois-Dauphin, M., Dreifuss, J.J., 1991. Localization of high-affinity binding sites for oxytocin and vasopressin in the human brain. An autoradiographic study. Brain Res. 555 (2), 220–232.

Lucht, M.J., Barnow, S., Sonnenfeld, C., Rosenberger, A., Grabe, H.J., Schroeder, W., Völzke, H., Freyberger, H.J., Herrmann, F.H., Kroemer, H., Rosskopf, D., 2009. Associations between the oxytocin receptor gene (OXTR) and affect, loneliness and intelligence in normal subjects. Prog. Neuropsychopharmacol. Biol. Psychiat. 33 (5), 860–866.

Lukas, M., Neumann, I.D., 2014. Social preference and maternal defeat-induced social avoidance in virgin female rats: sex differences in involvement of brain oxytocin and vasopressin. J. Neurosci. Methods 234, 101–107.

Lukas, M., Toth, I., Reber, S.O., Slattery, D.A., Veenema, A.H., Neumann, I.D., 2011. The neuropeptide oxytocin facilitates pro-social behavior and prevents social avoidance in rats and mice. Neuropsychopharmacology 36 (11), 2159–2168.

Lynn, S.K., Hoge, E.A., Fischer, L.E., Barrett, L.F., Simon, N.M., 2014. Gender differences in oxytocin-associated disruption of decision bias during emotion perception. Psychiat. Res. 14, 1–6.

Macbeth, A.H., Lee, H.J., Edds, J., Young, III, W.S., 2009. Oxytocin and the oxytocin receptor underlie intrastrain, but not interstrain, social recognition. Genes Brain Behav. 8 (5), 558–567.

MacDonald, K.S., 2013. Sex, receptors, and attachment: a review of individual factors influencing response to oxytocin. Front. Neurosci. 6 (194), 1–8.

Madison, D.M., 1980. Space use and social structure in meadow voles, Microtus pennsylvanicus. Behav. Ecol. Sociobiol. 7, 65–71.

Manning, C.J., Wakeland, E.K., Potts, W.K., 1992. Communal nesting patterns in mice implicate MHC genes in kin recognition. Nature 360, 581–583.

Manning, M., Misicka, A., Olma, A., Bankowski, K., Stoev, S., Chini, B., Durroux, T., Mouillac, B., Corbani, M., Guillon, G., 2012. Oxytocin and vasopressin agonists and antagonists as research tools and potential therapeutics. J. Neuroendocrinol. 24 (4), 609–628.

Mantella, R.C., Vollmer, R.R., Li, X., Amico, J.A., 2003. Female oxytocin-deficient mice display enhanced anxiety-related behavior. Endocrinology 144 (6), 2291–2296.

McEwen, B.B., 2004. Brain-fluid barriers: relevance for theoretical controversies regarding vasopressin and oxytocin memory research. Adv. Pharmacol. 50, 655–708.

Meyer-Lindenberg, A., Domes, G., Kirsch, P., Heinrichs, M., 2011. Oxytocin and vasopressin in the human brain: social neuropeptides for translational medicine. Nat. Rev. Neurosci. 12 (9), 524–538.
Meynen, G., Unmehopa, U.A., Hofman, M.A., Swaab, D.F., Hoogendijk, W.J.G., 2007. Hypothalamic oxytocin mRNA expression and melancholic depression. Mol. Psychiat. 12, 118–119.
Miller, M., Bales, K.L., Taylor, S.L., Yoon, J., Hostetler, C.M., Carter, C.S., Solomon, M., 2013. Oxytocin and vasopressin in children and adolescents with autism spectrum disorders: sex differences and associations with symptoms. Autism Res. 6 (2), 91–102.
Modahl, C., Green, L., Fein, D., Morris, M., Waterhouse, L., Feinstein, C., Levin, H., 1998. Plasma oxytocin levels in autistic children. Soc. Biol. Psychiat. 43, 270–277.
Mogi, K., Ooyama, R., Nagasawa, M., Kikusui, T., 2014. Effects of neonatal oxytocin manipulation on development of social behaviors in mice. Physiol. Behav. 133, 68–75.
Mooney, S.J., Douglas, N.R., Holmes, M.M., 2014. Peripheral administration of oxytocin increases social affiliation in the naked mole rat (*Heterocephalus glaber*). Horm. Behav. 65 (4), 380–385.
Mouillac, B., Chini, B., Balestre, M.N., Jard, S., Barberis, C., Manning, M., Tribollet, E., Trumpp-Kallmeyer, S., Hoflack, J., Elands, J., 1995. Identification of agonist binding sites of vasopressin and oxytocin receptors. Adv. Exp. Med. Biol. 395, 301–310.
Murakami, G., Hunter, R.G., Fontaine, C., Ribeiro, A., Pfaff, D., 2011. Relationship among estrogen receptor, oxytocin and vasopressin gene expression and social interaction in male mice. Eur. J. Neurosci. 34 (3), 469–477.
Musatov, S., Chen, W., Pfaff, D.W., Kaplitt, M.G., Ogawa, S., 2006. RNAi-mediated silencing of estrogen receptor α in the ventromedial nucleus of the hypothalamus abolishes female sexual behaviors. Proc. Natl. Acad. Sci. USA 103, 10456–10460.
Nakajima, M., Gorlich, A., Heintz, N., 2014. Oxytocin modulates female sociosexual behavior through a specific class of prefrontal cortical interneurons. Cell 159 (2), 295–305.
Neumann, I., Russell, J.A., Landgraf, R., 1993. Oxytocin and vasopressin release within the supraoptic and paraventricular nuclei of pregnant, parturient and lactating rats: a microdialysis study. Neuroscience 53 (1), 65–75.
Neumann, I.D., Krömer, S.A., Toschi, N., Ebner, K., 2000. Brain oxytocin inhibits the (re)activity of the hypothalamo-pituitary-adrenal exis in male rats: involvement of the hypothalamic and limbic brain regions. Regul. Pept. 96, 31–38.
Neumann, I.D., Maloumby, R., Beiderbeck, D.I., Lukas, M., Landgraf, R., 2013. Increased brain and plasma oxytocin after intranasal and peripheral administration in rats and mice. Psychoneuroendocrinology 38 (10), 1985–1993.
Nyuyki, K.D., Waldherr, M., Baeumi, S., Neumann, I.D., 2011. Yes, I am ready now: differential effects of paced versus unpaced mating on anxiety and central oxytocin release in female rats. PLoS One 6 (8), e23599.
Olff, M., Frijling, J.L., Kubzansky, L.D., Bradley, B., Ellenbogen, M.A., Cardoso, C., Bartz, J.A., Yee, J.R., van Zuiden, M., 2013. The role of oxytocin in social bonding, stress regulation and mental health: an update on the moderating effects of context and interindividual differences. Psychoneuroendocrinology 38, 1883–1894.
Ophir, A.G., 2011. Towards meeting Tingergen's challenge. Horm. Behav. 60 (1), 22–27.
Ophir, A.G., Gessel, A., Zheng, D.J., Phelps, S.M., 2012. Oxytocin receptor density is associated with male mating tactics and social monogamy. Horm. Behav. 61 (3), 445–453.
Ozsoy, S., Esel, E., Kula, M., 2009. Serum oxytocin levels in patients with depression and the effects of gender and antidepressant treatment. Psychiat. Res. 169, 249–252.
Pedersen, C.A., Prange, A.J., 1979. Induction of maternal behavior in virgin rats after intracerebroventricular administration of oxytocin. Proc. Natl. Acad. Sci. USA 76 (12), 6661–6665.
Popik, P., van Ree, J.M., 1991. Oxytocin but not vasopressin facilitates social recognition following injection into the medial preoptic area of the rat brain. Eur. Neuropsychopharmacol. 1 (4), 555–560.
Popik, P., Vetulani, J., van Ree, J.M., 1992. Low doses of oxytocin facilitate social recognition in rats. Psychopharmacology 106 (1), 71–74.
Qiao, X., Yan, Y., Wu, R., Tai, F., Hao, P., Cao, Y., Wang, J., 2014. Sociality and oxytocin and vasopressin in the brain of male and female dominant and subordinate mandarin voles. J. Comp. Physiol. 200, 149–159.
Radke, S., Roelofs, K., de Bruijn, E.R.A., 2013. Acting on anger: social anxiety modulates approach-avoidance tendencies after oxytocin administration. Psychol. Sci. 24, 1573–1578.

Rhodes, C.H., Morrell, J.Y., Pfaff, D.W., 1981. Immunocytochemical analysis of magnocellular elements in rat hypothalamus: distribution and numbers of cells containing neurophysin, oxytocin and vasopressin. J. Comp. Neurol. 198, 45–64.

Riem, M.M., Voorthuis, A., Bakermans-Kranenburg, M.J., van IJzendoorn, M.H., 2014. Pity or peanuts? Oxytocin induces different neural responses to the same infant crying labeled as sick or bored. Dev. Sci. 17 (2), 248–256.

Rilling, J.K., DeMarco, A.C., Hackett, P.D., Thompson, R., Ditzen, B., Patel, R., Pagnoni, G., 2012. Effects of intranasal oxytocin and vasopressin on cooperative behavior and associated brain activity in men. Psychoneuroendocrinology 37 (4), 447–461.

Rilling, J.K., DeMarco, A.C., Hackett, P.D., Chen, X., Gautam, P., Stair, S., Haroon, E., Thompson, R., Ditzen, B., Patel, R., Pagnoni, G., 2013. Sex differences in the neural and behavioral response to intranasal oxytocin and vasopressin during human social interaction. Psychoneuroendocrinology 39, 237–248.

Ring, R.H., Malberg, J.E., Potestio, L., Ping, J., Boikess, S., Luo, B., Schechter, L.E., Rizzo, S., Rahman, Z., Rosenzweig-Lipson, S., 2006. Anxiolytic-like activity of oxytocin in male mice: behavioral and autonomic evidence, therapeutic implications. Psychopharmacology 185, 218–225.

Rodrigues, S.M., Saslow, L.R., Garcia, N., John, O.P., Keltner, D., 2009. Oxytocin receptor genetic variation relates to empathy and stress reactivity in humans. Proc. Natl. Acad. Sci. USA 106 (50), 21437–21441.

Rosen, G.J., De Vries, G.J., Goldman, S.L., Goldman, B.D., Forger, N.G., 2008. Distribution of oxytocin in the brain of a eusocial rodent. Neuroscience 155, 809–817.

Ross, H.E., Young, L.J., 2009. Oxytocin and the neural mechanisms regulating social cognition and affiliative behavior. Front. Neuroendocrinol. 30 (4), 534–547.

Ross, H.E., Cole, C.D., Smith, Y., Neumann, I.D., Landgraf, R., Murphy, A.Z., Young, L.J., 2009. Characterization of the oxytocin system regulating affiliative behavior in female prairie voles. Neuroscience 162 (4), 892–903.

Rubin, L.H., Carter, C.S., Drogos, L., Jamadar, R., Pournajafi-Nazarloo, H., Sweeney, J.A., Maki, P.M., 2011. Sex-specific associations between peripheral oxytocin and emotion perception in schizophrenia. Schizophr. Res. 130 (1–3), 266–270.

Sabihi, S., Durosko, N.E., Dong, S.M., Leuner, B., 2014. Oxytocin in the prelimbic medial prefrontal cortex reduces anxiety-like behavior in female and male rats. Psychoneuroendocrinology 45, 31–42.

Savaskan, E., Ehrhardt, R., Schulz, A., Walter, M., Schächinger, H., 2008. Post-learning intranasal oxytocin modulates human memory for facial identity. Psychoneuroendocrinology 33 (3), 368–374.

Scheele, D., Striepens, N., Kendrick, K.M., Schwering, C., Noelle, J., Wille, A., Schläpfer, T.E., Maier, W., Hurlemann, R., 2014. Opposing effects of oxytocin on moral judgment in males and females. Hum. Brain Mapp. 35 (12), 6067–6076.

Schorscher-Petcu, A., Sotocinal, S., Ciura, S., Dupré, A., Ritchie, J., Sorge, R.E., Crawley, J.N., Hu, S.B., Nishimori, K., Young, L.J., Tribollet, E., Quirion, R., Mogil, J.S., 2010. Oxytocin-induced analgesia and scratching are mediated by the vasopressin-1A receptor in the mouse. J. Neurosci. 30 (24), 8274–8284.

Schulze, L., Lischke, A., Greif, J., Herpertz, S.C., Heinrichs, M., Domes, G., 2011. Oxytocin increases recognition of masked emotional faces. Psychoneuroendocrinology 36 (9), 1378–1382.

Slattery, D.A., Neumann, I.D., 2010. Chronic icv oxytocin attenuates the pathological high anxiety state of selectively bred Wistar rats. Neuropharmacology 58, 56–61.

Smeltzer, M.D., Curtis, J.T., Aragona, B.J., Wang, Z., 2006. Dopamine, oxytocin, and vasopressin receptor binding in the medial prefrontal cortex of monogamous and promiscuous voles. Neurosci. Lett. 394, 146–151.

Smith, C.J.W., Poehlmann, M.L., Wilkins, K.B., Bredewold, R., Veenema, A.H., 2014. Age Differences in the Brain Oxytocin System: Implications for Juvenile Social Motivation. Society for Neuroscience abstract 832.14/NN24, Washington D.C.

Sofroniew, M.V., 1980. Projections from vasopressin, oxytocin, and neurophysin neurons to neural targets in the rat human. J. Histochem. Cytochem. 28 (5), 475–478.

Sofroniew, M.V., 1983. Morphology of vasopressin and oxytocin neurons and their central and vascular projections. Prog. Brain Res. 60, 101–114.

Song, Z., McCann, K.E., McNeill, IV, J.K., Larkin, II, T.E., Huhman, K.L., Albers, H.E., 2014. Oxytocin induces social communication by activating arginine-vasopressin V1a receptors and not oxytocin receptors. Psychoneuroendocrinology 50, 14–19.

Striepens, N., Scheele, D., Kendrick, K.M., Becker, B., Schafer, L., Schwalba, K., Reul, J., Maier, W., Hurlemann, R., 2012. Oxytocin facilitates protective responses to aversive social stimuli in males. Proc. Natl. Acad. Sci. USA 109 (44), 18144–18149.

Swanson, L.W., Sawchenko, P.E., 1983. Hypothalamic integration: organization of the paraventricular and supraoptic nuclei. Annu. Rev. Neurosci. 6, 269–324.

Szeto, A., McCabe, P.M., Nation, D.A., Tabak, B.A., Rossetti, M.A., McCullough, M.E., Schneiderman, N., Mendez, A.J., 2011. Evaluation of enzyme immunoassay and radioimmunoassay methods for the measurement of plasma oxytocin. Psychosom. Med. 73 (5), 393–400.

Takayanagi, Y., Yoshida, M., Bielsky, I.F., Ross, H.E., Kawamata, M., Onaka, T., Yanagisawa, T., Kimura, T., Matzuk, M.M., Young, L.J., Nishimori, K., 2005. Pervasive social deficits, but normal parturition, in oxytocin receptor-deficient mice. Proc. Natl. Acad. Sci. USA 102 (44), 16096–16101.

Taylor, S.E., Saphire-Bernstein, S., Seeman, T.E., 2010. Are plasma oxytocin in women and plasma vasopressin in men biomarkers of distressed pair-bond relationships? Psychol. Sci. 21 (1), 3–7.

Theodoridou, A., Rowe, A.C., Mohr, C., 2013. Men perform comparably to women in a perspective taking task after administration of intranasal oxytocin but not after placebo. Front. Hum. Neurosci. 7 (197), 1–11.

Tost, H., Kolachana, B., Hakimi, S., Lemaitre, H., Verchinski, B.A., Mattay, V.S., Weinberger, D.R., Meyer-Lindenberg, A., 2010. A common allele in the oxytocin receptor gene (OXTR) impacts prosocial temperament and human hypothalamic-limbic structure and function. Proc. Natl. Acad. Sci. USA 107 (31), 13936–13941.

Tribollet, E., Audigier, S., Dubois-Dauphin, M., Dreifuss, J.J., 1990. Gonadal steroids regulate oxytocin receptors but not vasopressin receptors in the brain of male and female rats. An autoradiographical study. Brain Res. 511 (1), 129–140.

Tribollet, E., Ueta, Y., Heitz, F., Marguerat, A., Koizumi, K., Yamashita, H., 2002. Up-regulation of vasopressin and angiotensin II receptors in the thalamus and brainstem of inbred polydipsic mice. Neuroendocrinology 75, 113–123.

Uhl-Bronner, S., Waltisperger, E., Martínez-Lorenzana, G., Condes Lara, M., Freund-Mercier, M.J., 2005. Sexually dimorphic expression of oxytocin binding sites in forebrain and spinal cord of the rat. Neuroscience 135 (1), 147–154.

Uvnäs-Moberg, K., Ahlenius, S., Hillegaart, V., Alster, P., 1994. High doses of oxytocin cause sedation and low doses cause an anxiolytic-like effect in male rats. Pharmacol. Biochem. Behav. 49 (1), 101–106.

Van den Burg, E.H., Neumann, I.D., 2011. Bridging the gap between GPCR activation and behavior: oxytocin and prolactin signaling in the hypothalamus. J. Mol. Neurosci. 43, 200–208.

Van IJzendoorn, M.H., Bhandari, R., van der Veen, R., Grewen, K.M., Bakermans-Kranenburg, M.J., 2012. Elevated salivary levels of oxytoin persist more than 7 h after intranasal administration. Front. Neurosci. 6 (174).

Veenema, A.H., 2009. Early life stress, the development of aggression and neuroendocrine and neurobiological correlates: what can we learn from animal models? Front. Neuroendocrinol. 30 (4), 497–518.

Veenema, A.H., Neumann, I.D., 2008. Central vasopressin and oxytocin release: regulation of complex social behaviours. Prog. Brain Res. 170, 261–276.

Veenema, A.H., Bredewold, R., Neumann, I.D., 2007. Opposite effects of maternal separation on intermale and maternal aggression in C57BL/6 mice: link to hypothalamic vasopressin and oxytocin immunoreactivity. Psychoneuroendocrinology 32 (5), 437–450.

Veenema, A.H., Bredewold, R., De Vries, G.J., 2012. Vasopressin regulates social recognition in juvenile and adult rats of both sexes, but in sex- and age-specific ways. Horm. Behav. 61 (1), 50–56.

Veening, J.G., Olivier, B., 2013. Intranasal administration of oxytocin: behavioral and clinical effects, a review. Neurosci. Biobehav. Rev. 37 (8), 1445–1465.

Veening, J.G., de Jong, T.R., Waldinger, M.D., Korte, S.M., Olivier, B., 2014. The role of oxytocin in male and female reproductive behavior. Eur. J. Pharmacol. 753, 209–228.

Waldherr, M., Neumann, I.D., 2007. Centrally released oxytocin mediates mating-induced anxiolysis in male rats. Proc. Natl. Acad. Sci. USA 104 (42), 16681–16684.

Walum, H., Lichtenstein, P., Neiderhiser, J.M., Reiss, D., Ganiban, J.M., Spotts, E.L., Pedersen, N.L., Anckarsäter, H., Larsson, H., Westberg, L., 2012. Variation in the oxytocin receptor gene is associated with pair-bonding and social behavior. Biol. Psychiat. 71 (5), 419–426.

Wang, Z.X., Novak, M.A., 1992. Influence of the social environment on parental behavior and pup development of meadow voles (*Microtus pennsylvanicus*) and prairie voles (*M. ochrogaster*). J. Comp. Psychol. 106 (2), 163–171.

Wang, Z.X., Novak, M.A., 1994a. Parental care and litter development in primiparous and multiparous prairie voles (*Microtus ochrogaster*). J. Mammal. 75, 18–23.

Wang, Z.X., Novak, M.A., 1994b. Alloparental care and the influence of father presence on juvenile prairie voles (*Microtus ochrogaster*). Anim. Behav. 47, 281–288.

Wang, Z., Zhou, L., Hulihan, T.J., Insel, T.R., 1996. Immunoreactivity of central vasopressin and oxytocin pathways in microtine rodents: a quantitative comparative study. J. Comp. Neurol. 366, 726–737.

Wang, Y., Xu, L., Pan, Y., Wang, Z., Zhang, Z., 2013. Species differences in the immunoreactive expression of oxytocin, vasopressin, tyrosine hydroxylase and estrogen receptor alpha in the brain of Mongolian gerbils (*Meriones unguiculatus*) and Chinese striped hamsters (*Cricetulus barabensis*). PLoS One 8 (6), e65807.

Webster, A.B., Brooks, R.J., 1981. Social behavior of microtus pennsylvanicus in relation to seasonal changes in demography. J. Mammal. 62, 738–751.

Weisman, O., Zagoory-Sharon, O., Feldman, R., 2012. Intranasal oxytocin administration is reflected in human saliva. Psychoneuroendocrinology 37, 1582–1586.

Weisman, O., Zagoory-Sharon, O., Schneiderman, I., Gordon, I., Feldman, R., 2013. Plasma oxytocin distributions in a large cohort of women and men and their gender-specific associations with anxiety. Psychoneuroendocrinology 38, 694–701.

Widiger, T.A., Trull, T.J., 1993. Borderline and narcissistic personality disorders. In: Sutker, P.B., Adams, H.E. (Eds.), Comprehensive Handbook of Psychopathology, second edition. Plenum, New York, pp. 371–394.

Wierda, M., Goudsmit, E., Van der Woude, P.F., Purba, J.S., Hofman, M.A., Bogte, H., Swaab, D.F., 1991. Oxytocin cell number in the human paraventricular nucleus remains constant with aging and in Alzheimer's disease. Neurobiol. Aging 12 (5), 511–516.

Williams, J.R., Carter, C.S., Insel, T., 1992. Partner preference development in female prairie voles in facilitated by mating or the central infusion of oxytocin. Ann. NY Acad. Sci. 652, 487–489.

Williams, J.R., Insel, T.R., Harbaugh, C.R., Carter, C.S., 1994. Oxytocin administered centrally facilitates formation of a partner preference in female prairie voles (*Microtus ochrogaster*). J. Neuroendocrinol. 6 (3), 247–250.

Windle, R.J., Shanks, N., Lightman, S.L., Ingram, C.D., 1997. Central oxytocin administration recues stress-induced corticosterone release an anxiety behavior in rats. Endocrinology 138 (7), 2829–2834.

Winslow, J.T., Hastings, N., Carter, C.S., Harbaugh, C.R., Insel, T.R., 1993. A role for central vasopressin in pair bonding in monogamous prairie voles. Nature 365, 545–548.

Winslow, J.T., Hearn, E.F., Ferguson, J., Young, L.J., Matzuk, M.M., Insel, T.R., 2000. Infant vocalization, adult aggression, and fear behavior of an oxytocin null mutant mouse. Horm. Behav. 37 (2), 145–155.

Witt, D.M., Winslow, J.T., Insel, T.R., 1992. Enhanced social interactions in rats following chronic, centrally infused oxytocin. Pharmacol. Biochem. Behav. 43 (3), 855–861.

Woolley, J.D., Chuang, B., Lam, O., Lai, W., O'Donovan, A., Rankin, K.P., Mathalon, D.H., Vinogradov, S., 2014. Oxytocin administration enhances controlled social cognition in patients with schizophrenia. Psychoneuroendocrinology 47, 116–125.

Wu, S., Jia, M., Ruan, Y., Liu, J., Guo, Y., Shuang, M., Gong, X., Zhang, Y., Yang, X., Zhang, D., 2005. Positive association of the oxytocin receptor gene (OXTR) with autism in the Chinese Han population. Biol. Psychiat. 58 (1), 74–77.

Xu, L., Pan, Y., Young, K.A., Wang, Z., Zhang, Z., 2010. Oxytocin and vasopressin immunoreactive staining in the brains of Brandt's voles (*Lasiopodomys brandtii*) and greater long-tailed hamsters (*Tscherskia triton*). Neuroscience 169 (3), 1235–1247.

Yamamoto, Y., Cushing, B.S., Kramer, K.M., Epperson, P.D., Hoffman, G.E., Carter, C.S., 2004. Neonatal manipulations of oxytocin alter expression of oxytocin and vasopressin immunoreactive cells in the paraventricular nucleus of the hypothalamus in a gender-specific manner. Neuroscience 125, 947–955.

Yamamoto, Y., Carter, C.S., Cushing, B.S., 2006. Neonatal manipulation of oxytocin affects expression of estrogen receptor alpha. Neuroscience 137, 157–164.

Young, L.J., 2013. When too much of a good thing is bad: chronic oxytocin, development, and social impairments. Biol. Psychiat. 74 (3), 160–161.

Young, L.J., Lim, M.M., Gingrich, B., Insel, T.R., 2001. Cellular mechanisms of social attachment. Horm. Behav. 40 (2), 133–138.
Zak, P.J., Stanton, A.A., Ahmadi, S., 2007. Oxytocin increases generosity in humans. PLoS One 2 (11), e1128.
Zhang, G., Cai, D., 2011. Circadian intervention of obesity development via resting-stage feeding manipulation or oxytocin treatment. Am. J. Physiol. Endocrinol. Metab. 301 (5), 1004–1012.
Zhang, H., Wu, C., Chen, Q., Chen, X., Xu, Z., Wu, J., Cai, D., 2013. Treatment of obesity and diabetes using oxytocin or analogs in patients and mouse models. PLoS One 8, 5.
Zheng, D.J., Larsson, B., Phelps, S.M., Ophir, A.G., 2013. Female alternative mating tactics, reproductive success and nonapeptide receptor expression in the social decision-making network. Behav. Brain Res. 246, 139–147.
Zhong, S., Monakhov, M., Mok, H.P., Tong, T., Lai, P.S., Chew, S.H., Ebstein, R.P., 2012. U-shaped relation between plasma oxytocin levels and behavior in the trust game. PLoS One 7 (12), e51095.
Zucker, I., Beery, A.K., 2010. Males still dominate animal studies. Nat. Opin. 465, 690.

## CHAPTER 12

# Sex Differences in Neurological Diseases

**Farida Sohrabji*, C. Jane Welsh**, Doodipala Samba Reddy***
*TAMHSC College of Medicine, Department of Neuroscience and Experimental Therapeutics, Bryan, TX, USA
**Texas A&M University, Department of Veterinary Integrative Biosciences, College Station, TX, USA

## 1 INTRODUCTION

The major causes of adult onset brain disorders in the United States include traumatic brain injury, stroke, epilepsy, Alzheimer's disease, Parkinson's disease, multiple sclerosis (MS), and Huntington's disease. In some of these diseases, such as Huntington's, there are no sex differences in the incidence of this disease. In other cases, there is a marked preponderance of the disease in one sex or the other, for example, more males suffer from Parkinson's disease, while more women suffer from MS. In this chapter, we will examine the evidence for sex differences and the role of gonadal steroids in three of major disorders: stroke, MS, and epilepsy.

## 2 ISCHEMIC STROKE

### 2.1 Introduction

In the United States, stroke is the fourth leading cause of death and the leading nonmartial cause of disability. Unlike multiple sclerosis and epilepsy, stroke is an acute disease, caused by disruption of normal blood supply to the brain, and resulting in a range of neurologic deficits, including loss of sensory-motor function, language, and executive function depending on the brain area that is affected. There are two broad types of stroke: ischemic and hemorrhagic. In an ischemic stroke, the brain is deprived of blood and, consequently, glucose and oxygen. Ischemic stroke is the most common type of stroke (87%). The second type is a hemorrhagic stroke, where a blood vessel ruptures, causing blood to flood into the brain. Hemorrhagic strokes are more rare (13%) but they are also more severe and likely to result in death. Sex differences have been observed in a specific type of hemorrhagic stroke called the subarachnoid hemorrhage (Park et al., 2008); however, sex differences are much more prevalent in ischemic stroke. Thus, the primary focus here will be ischemic stroke, where sex differences have been reported for a number of stroke-related measures including stroke incidence, severity, and pathophysiology.

*Sex Differences in the Central Nervous System*. http://dx.doi.org/10.1016/B978-0-12-802114-9.00012-3

Ischemic stroke is a complex disease with an acute onset. Loss of blood supply to the brain, due to either a clot or atherosclerotic deposits in a cerebral vessel, results in a progressive loss of oxygen and glucose to neurons. The ischemic environment of the brain causes cell death and impaired function of critical support cells. Numerous factors can alter the risk for stroke, including demographic factors such as age and sex, and health conditions such as hypertension, diabetes, and hypercholesterolemia (Allen and Bayraktutan, 2008). In women, additional factors may apply, related to specific hormonal events (reviewed in Bushnell, 2008). For example, pregnancy-related changes in angiogenesis have been shown to increase the risk for stroke. During perimenopause, where ovarian hormone levels begin to decline, women are more likely to develop diabetes, which is a significant risk factor for stroke and cardiovascular disease. In addition, at menopause, women may consider hormone therapy, which has also been shown to increase the risk for stroke. In young women, hormone-based birth control medication has also been shown to elevate the risk for stroke especially when combined with cigarette smoking. Atrial fibrillation, which doubles the risk of stroke in men, increases the risk of stroke in women fivefold, especially when patients are taken off anticoagulation therapy (Fang et al., 2005).

Women live longer than men and, as a result of this longevity, projected stroke-related disability and institutionalization is likely to affect women more than men (Lai et al., 2005). Although earlier studies indicated that the increased incidence of stroke in women did not occur until age 80 (Rosamond et al., 2007), newer studies suggest that the risk for stroke in middle-aged women doubles that of men (Towfighi et al., 2007). Furthermore, women account for 60% of stroke-related deaths (Lloyd-Jones et al., 2010), even after normalization for age. A Canadian stroke registry study reported that 10% of female stroke patients were discharged to long-term care compared with 5% of males (Kapral et al., 2005), despite the observation that stroke size tends not to be different in males and females (Silva et al., 2010). Finally, 5-year stroke recurrence is disproportionately higher in females (20%) compared with males (10%) in the 45–64 age range (Roger et al., 2011). In the Danish National Registry analysis, women were reported to have more severe strokes than men although they exhibited a survival advantage compared with men, especially at advanced ages (Olsen and Andersen, 2010).

## 2.2 Sex differences in preclinical models

Sex differences in stroke outcome are also well recognized in preclinical models. In animals, ischemic stroke can be modeled in several ways usually involving reversible or nonreversible blockage of cerebral vessels, such as occlusion of the middle cerebral artery (MCAo). Stroke severity or outcomes are typically assessed by histological analysis of the volume of infarction and neurological scores of sensory motor performance. Such studies have shown that young females have a smaller infarct and better cerebral blood flow than age-matched males both in normoglycemic (Alkayed et al., 1998) and diabetic

(Toung et al., 2000) animals, although female obese diabetic mice (KKAy strain) fare worse than male counterparts (Sakata et al., 2011). Young females sustain a smaller infarct compared with young males or aged female mice (Liu et al., 2009); however, aging reverses this sex difference, such that aged females showed significantly more mortality and poorer stroke outcomes compared with older males (Manwani et al., 2011). Sex differences have also been noted at middle age, where adult female rats have smaller infarcts than middle-aged females (Selvamani and Sohrabji, 2010a).

The female advantage seen in young human demographics as well as animal models may be associated with either chromosomal sex or sex differences in gonadal hormones. While the separate effects of chromosomal sex and gonadal sex are difficult to dissect, a recent animal model called the 4-core genotype provides an animal model where the chromosomal contribution and gonadal contribution can be evaluated in a particular disease (Arnold and Chen, 2009). Using XX and Sry-deficient $XY^-$ female mice and $XX^-$Sry and $XY^-$Sry male mice to examine the effects of chromosome complement without any sex hormones, one study showed that sex differences in stroke are influenced by sex hormones and not by chromosomes (Manwani et al., 2014). This is consistent with a large body of experimental research that estrogen, the major ovarian hormone, may improve stroke outcomes.

## 2.3 Role of estrogen in stroke outcomes

Using natural variations in circulating estrogen levels across the estrous cycle Liao et al. (2001) showed that the extent of ischemic damage was inversely related to circulating levels of estrogen. In fact, replacement with 17β-estradiol and its inactive stereoisomer 17α-estradiol (Simpkins et al., 1997) as well as the conjugate equine estrogen preparation (McCullough et al., 2001) all reduce infarct volume in female animals. Interestingly, estrogen also reduces infarct volume in males, while the precursor steroid testosterone increases infarct volume in this group (Hawk et al., 1998). Exogenous estrogen replacement is neuroprotective when given prior (Dubal et al., 1998) or subsequent to the injury (Liu et al., 2007; Yang et al., 2003; Jover-Mengual et al., 2010), or even to the diabetic KKAy mouse, where intact females have worse outcomes than males (Sakata et al., 2011). Estrogen treatment is also effective in a variety of ischemic stroke models including the occlusive suture model (Dubal and Wise, 2001), vasoconstrictive endothelin-1 model (Selvamani and Sohrabji, 2010a; Glendenning et al., 2008), and the carotid occlusion model (Brann et al., 2007).

However, the aging process complicates hormone effects on stroke outcomes. Hormone therapy at menopause is a specific concern for stroke outcome as this middle-aged demographic is at increased risk for stroke. In the Women's Health Initiative, a placebo-controlled multicenter randomized study, the group of postmenopausal women who received estrogen + rogestins had an increased risk of ischemic stroke compared with placebo controls. This increased risk due to hormones occurred irrespective of age and other baseline health conditions (Wassertheil-Smoller et al., 2003). In animal models, the impact of estrogen treatment is usually studied in younger animals whose ovaries are removed bilaterally, creating a surgical menopause. In these studies estrogen treatment is usually neuroprotective. In alternate models, where middle-aged acyclic females or aged females are studied, the effects of estrogen are ambiguous. Research from Sohrabji's laboratory has specifically addressed this reproductive transition phase by studying acyclic, middle-aged female animals (Jezierski and Sohrabji, 2000; Johnson and Sohrabji, 2005; Selvamani and Sohrabji, 2010a, 2010b). Reproductive senescent rats are characterized by reproductive criteria including multiple previous successful pregnancies, current reproductive failure, and constant diestrus (a low estrogen state) as determined by vaginal smears. Plasma levels of estradiol are low and FSH levels are elevated in this group, consistent with a menopausal pattern. MCAo in this middle-aged female results in a significantly larger infarct as compared with young females (Selvamani and Sohrabji, 2010a), suggesting that the loss of ovarian hormones may accelerate age-related stroke severity. However, estrogen treatment to ovariectomized reproductive senescent females paradoxically increases infarct volume, indicating that the aging brain may be refractory to hormone-mediated neuroprotection (Selvamani and Sohrabji, 2010a, 2010b). Similarly, estrogen also failed to show neuroprotection to aged females subject to a clot-based ischemic injury (Leon et al., 2012). In middle-aged gerbils, estrogen is also not neuroprotective in a carotid artery occlusion model (De Butte-Smith et al., 2007). One possible explanation for the loss of estrogen's neuroprotective capacity may be the age-related loss of other endogenous growth factors, especially those that work cooperatively with the estrogen receptor system. One such example is insulin-like growth factor (IGF)-1, a peptide hormone that promotes cell survival and decreases with age. In middle-aged females with ischemia, estrogen treatment is neuroprotective when combined with IGF-1 (Selvamani and Sohrabji, 2010b), indicating that age-related changes in other growth factors may alter the impact of estrogen.

## 2.4 Progestins and stroke

Although not as widely studied in stroke as estradiol, progesterone, another major ovarian hormone, has also been shown to promote recovery in other neurologic diseases such as traumatic brain injury (Stein, 2011). Progesterone has also been shown to be beneficial in experimental stroke models. In a transient intraluminal MCAo followed by reperfusion, DMSO-dissolved, but not water-soluble, progesterone, delivered either pre- or 2 h postischemia, effectively reduced infarct volume and improved neurological outcomes (Jiang et al., 1996). Further work has shown that the effects of progesterone also persist long term, resulting in reduced impairment of spatial learning and memory (Moralí et al., 2011). In a permanent ischemic model where the MCA was cauterized, animals that received either intraperitoneal injections of progesterone followed by subcutaneous injections 6, 48, and 72 h later had a significant reduction in infarct volume and improvement in the rotorad task and grip strength task (Ishrat et al., 2009). Progesterone was shown to reduce cell death in the CA1 and CA2 fields of the hippocampus of cats subject to acute global cerebral ischemia, compared with vehicle treated controls (González-Vidal et al., 1998).

Progesterone treatment was also an effective neuroprotectant in the focal ischemic model when delivered in the late stage of ischemia (2–7 days poststroke) in spontaneously hypertensive male rats (Kumon et al., 2000). Interestingly, progesterone pretreatment of female mice did not show a protective effect in stroke. Ovariectomized females implanted with progesterone 7 days prior to stroke surgery did not show an improvement in infarct volume or edema as compared with controls (Coomber and Gibson, 2010), suggesting that in females, estrogen was a more effective hormone for stroke.

## 2.5 Mechanisms of hormone action

Estrogen affects many aspects of stroke pathophysiology and may directly affect neuronal survival. Consistent with this hypothesis, estrogen treatment activates the survival kinase Akt, as well as reduces DNA fragmentation and elevates the antiapoptotic protein Bcl2 (Choi et al., 2004), and directly promotes cell survival *in vitro*. Several studies have shown that estrogen attenuates the production of reactive oxygen species and reduces oxidative stress in the ischemic brain (reviewed in Brann et al., 2012). Estrogen and estrogen homologues have been shown to enhance or preserve mitochondrial function during pathologic circumstances such as oxidative stress (Simpkins et al., 2010). Other neuroprotective mechanisms include estrogen's ability to modulate cell death pathways and stimulate neurogenesis poststroke (Shao et al., 2012). Similar to estrogen, progesterone treatment has also been shown to impact apoptosis, poststroke neurogenesis, inflammation, and oxidative stress (reviewed in Singh and Su, 2013; Espinosa-García et al., 2013). There are similarities and subtle differences in the neuroprotective actions of estrogen and progesterone. Pretreatment of ovariectomized females with either 17β-estradiol, progesterone, or 17α-estradiol all resulted in reduced markers of oxidative stress such as malondialdehyde and increased glutathione levels following ischemia (Ozacmak and Sayan, 2009). However, while estrogen increases poststroke neurogenesis, progesterone treatment attenuates neuronal proliferation but increases the long-term survival of newborn neurons (Zhang et al., 2010).

## 2.6 The inflammatory response to stroke and the role of estrogen

An important target of estrogen, common to many neurologic diseases, is the immune system. The rapid loss of neurons following ischemia triggers an inflammatory response that activates local microglia and astrocytes, resulting in the recruitment of peripheral immune cells to the ischemic brain. During the initial phase of stroke, secretions from activated microglia cause a change in the vascular endothelium, weaken the blood–brain barrier, thus increasing adhesion and recruitment of immune cells. Once peripheral immune cells gain access to the ischemic brain, the inflammatory response is amplified.

One way in which estrogen may protect against ischemic damage is by suppressing microglial activation. This is supported by evidence from studies on primary cultures of microglia or glial cell lines, where estrogen pretreatment attenuates lipopolysaccharide (LPS)-induced superoxide release, phagocytic activity, and an increase in inducible nitric oxide synthase (iNOS), which reduces activation of local microglia and peripheral monocytes following *in vivo* intracerebral LPS injections and inhibits the proinflammatory COX-2 pathway in cerebral microvessels (summarized in Sohrabji and Bake, 2006). On the other hand, 17β-estradiol fails to suppress cytokine release in activated N9 microglial cells *in vitro* (Suuronen et al., 2005) and stimulates cytokine messenger RNA in the brain in animals injected with LPS systemically (Soucy et al., 2005). Furthermore, long-term treatment of ovariectomized animals with estrogen (8 weeks) failed to attenuate microglial activation *in vivo* when animals were infused with LPS into the ventricles (Marriott et al., 2002). Interestingly, intact females had an attenuated microglial response as compared with ovariectomized animals, suggesting that endogenous estrogens or a combination of endogenous ovarian hormones may regulate the activation of microglia.

Estrogen's effects on the inflammatory response also appear to be dependent on the "reproductive age" of the individual. In an excitotoxic forebrain injury model, which activates microglial and inflammatory cytokine synthesis, estrogen replacement to ovariectomized young adult animals reduced the expression of the proinflammatory cytokine interleukin (IL)-1β. However, estrogen replacement increased the expression of this cytokine when administered to acyclic middle-aged females with forebrain injury (Nordell et al., 2003). IL-1β or similar cytokines in the brain can result from activation of local immune cells or from increased traffic of circulating immune cells into the brain. Microglia from adult and middle-aged females responded to LPS with an increase in IL-1β; however, estrogen failed to attenuate cytokine expression at either age (Johnson and Sohrabji, 2005). However, in whole blood cultures obtained from adult females, estrogen attenuated LPS-induced TNFα expression, and paradoxically increased expression of this cytokine in whole blood cultures obtained from reproductive senescent animals (Johnson and Sohrabji, 2005). Thus, estrogen's effects on the immune response are age- and target-specific.

In view of the evidence that the somatic response to stroke shapes the evolving infarct, estrogen's actions on peripheral immune cells may be critical for its neuroprotective effects. The splenic response to ischemic stroke, specifically, amplifies neurodegeneration and, conversely, splenectomy reduces infarct volume (Ajmo et al., 2008). In the acute phase after stroke, spleen volume shrinks rapidly and this shrinkage is attributed to the migration of cells into circulation, and many of these cells can be tracked to the brain vasculature (Seifert et al., 2012). Poststroke, diverse immune cells are seen in the brain, including neutrophils, macrophages, and dendritic cells. Estrogen treatment has been shown to reduce poststroke splenic contraction, but did not reduce inflammatory cytokines in the brain (Zhang et al., 2010), suggesting that estrogen may impact peripheral immune responses. In a comparison of 5–6, 14–15, and 20–22-month-old males and females, aged females and males had greater poststroke splenic contraction compared with females who were cyclic (5–6 and 14–15 months), with a corresponding increase in the influx of T cells into the brain (Manwani et al., 2013). These data suggest that the peripheral immune response and immune recruitment is blunting cycling females (with normal levels of estrogen), which have the best outcome from stroke. Reduced trafficking of leukocytes into the brain may occur due to an efficient blood–brain barrier or decreased capacity of the endothelium to bind leukocytes. Estrogen has been shown to affect both targets. In adult females, estrogen reduces the permeability of the blood–brain barrier constitutively (Bake and Sohrabji, 2004; Cipolla et al., 2009) and estrogen treatment, administered shortly before ischemia, attenuates blood–brain barrier disruption as well as inhibits the action of proteases such as MMP-9 and MMP-2 (Liu et al., 2005) and reduces the expression of adhesion molecule ICAM-1 on endothelial cells (Galea et al., 2002).

Collectively, these studies indicate that estrogen may improve stroke outcomes by modulating the immune response, and that the increased stroke severity seen with aging may reflect both the loss of endogenous hormones and an aging immune system.

### 2.7 Summary

While males are more likely to sustain a stroke in younger demographics, this is reversed in aging, where more women will suffer strokes and have more severe strokes. The protection seen in younger women is attributed to ovarian hormones, but paradoxically, hormone therapy to older women increases the incidence of stroke. In preclinical studies, estrogen treatment has been shown to reliably improve stroke outcomes in young animals, but not in older animals, suggesting that there may be a critical window for the effectiveness of this hormone.

## 3 MULTIPLE SCLEROSIS

### 3.1 Introduction

MS is the most common demyelinating disease of the central nervous system (CNS) occurring at an incidence of approximately 350,000 in the United States and 2.5 million worldwide (Anderson et al., 1992; Noonan et al., 2002; Compston and Coles, 2002). MS is characterized by acute, focal inflammatory demyelination, and neurodegeneration of the CNS. The lesions arise as a result of immune-mediated destruction of myelin and subsequent axonal damage (Noseworthy et al., 2000; Wingerchuk et al., 2001). Activated macrophages, T cells, and some plasma cells are present in MS lesions and are thought to contribute to the pathology (Lucchinetti et al., 2000). The symptoms of MS reflect its pathological changes, and may include, but are not limited to: optic neuritis, clumsiness, gait ataxia, limb weakness, paralysis, cognitive impairment, sexual dysfunction, fatigue, pain, incontinence, and depression (Noseworthy et al., 2000).

Because MS is a chronic disease, often with an insidious onset, the actual exposure age is difficult to determine. Migration studies point to an environmental factor experienced before the age of 15 (Kurtzke, 1975). The onset of MS typically occurs in late adolescence, followed by a relapsing remitting course that eventually develops into a secondary progressive disease (Kantarci and Weinshenker, 2005). The transition to secondary progressive MS is thought to be mediated by accumulative neurodegeneration and cerebral atrophy (Kantarci and Weinshenker, 2005).

## 3.2 Multiple sclerosis and sex differences

Multiple sclerosis occurs two to three times more frequently in females (Whitacre, 2001) although in one study this female preponderance has been shown to have increased to 4:1 (Orton et al., 2006). The increased prevalence of MS in women occurs irrespective of racial ethnicity (Confavreux et al., 1998). When men develop MS, they tend to suffer more severe clinical signs and deteriorate faster than women (Cottrell et al., 1999). Males tend to develop MS later in life (30–40 years of age), when testosterone levels begin to decline (Weinshenker, 1994).

While being female increases the likelihood of developing MS, female sex hormones appear to have a beneficial effect. This is especially noted in the third trimester of pregnancy in patients with MS when estriol increases to high levels. Sex hormones may have beneficial effects on MS through various mechanisms: by modulating the immune system and/or affecting the nervous system. During pregnancy, the immune system shifts responses from T helper 1 (Th1)/Th17 toward Th2/regulatory T cell (Treg), which could account for the disease remission seen in MS during pregnancy (Confavreux et al., 1998).

MS was originally thought to result from autoimmune Th1 cell activation and subsequent trafficking of encephalitogenic T cells and macrophages into the CNS (Sospedra and Martin, 2005). However, Th17 cells have also been shown to be crucial in MS immunopathogenesis and other autoimmune diseases: collagen-induced arthritis, and experimental autoimmune myocarditis, rheumatoid arthritis, and autoimmune myocarditis (Bettelli et al., 2006; Cua et al., 2003; Chen et al., 2006; Komiyama et al., 2006; Sutton et al., 2006; Rangachari et al., 2006). Th1 and Th17 cytokines: TNF-$\alpha$ and IFN-$\gamma$ and IL-17 generally increase inflammation in MS whereas the Th2/Treg cytokines IL-4 and IL-10 reduce inflammation (Fletcher et al, 2010). CD4+CD25+Foxp3+ Tregs are nonresponsive *in vitro* and suppress other T cells mainly via the production of immunosuppressive cytokines such as IL-10 (reviewed in Thompson and Powrie, 2004). Decreased numbers of regulatory T cells have been reported in MS patients (Viglietta et al., 2004) and increased Tregs have been reported in both pregnant and estrogen-treated mice (Polanczyk et al., 2005).

## 3.3 Sex differences and hormone effects in animal models of MS

There are two main types of animal models for MS research: experimental autoimmune encephalomyelitis (EAE) and virus-induced demyelination. EAE involves immunizing experimental animals against myelin components, either spinal cord homogenate or encephalitogenic proteins or peptides. These preparations are injected into susceptible strains of animals with complete Freund's adjuvant (an emulsion of oil and heat-killed *Mycobacterium tuberculosis*). In some EAE models, this is accompanied by injection of *Bordetella pertussis*, which is thought to open the blood–brain barrier. The viral models of MS are arguably more naturalistic than EAE and include Theiler's virus-induced demyelination (Welsh et al., 1990).

### *3.3.1 Sex differences and hormone effects in EAE*

The role of sex in susceptibility of animal models of MS is complicated. Female SJL mice are more susceptible to EAE, similar to MS in humans (Bebo et al., 1996, 1998, 1999; Voskuhl et al., 1996). In other strains of mice such as B10.PL and PL/J, males are more susceptible to EAE (Papenfuss et al., 2004) and in C57BL/6 mice there are no sex differences (Okuda et al., 2002).

The EAE model has been utilized to examine the role of estrogen and estrogen receptor agonists (Bebo et al., 2001; Morales et al., 2006; Tiwari-Woodruff et al., 2007). Experimentally, 17β-estradiol has been shown to reduce EAE in mice by inhibiting both activation of encephalitogenic T cells and cell migration into the CNS and increasing the activity of CD4+CD25+ regulatory T cells (Offner and Polanczyk, 2006). 17β-estradiol has also been shown to promote axon and myelin survival (Offner and Polanczyk, 2006). Similarly, pregnancy also plays a protective role in EAE through immunoregulatory cytokine production in particular, increases in IL-10 and decreases in IL-17, TNFα (McClain et al., 2007).

Physiological levels of estradiol and estriol administered prior to the onset of EAE are protective, whereas if given at the onset of active disease, these hormones do not affect disease progression (Bebo et al., 2001). This difference may be explained by the fact that naïve immune cells are more sensitive to sex hormones than differentiated cells (Bebo et al., 2001). Estriol has been shown to be protective in studies with human MS patients and in the nonovariectomized EAE model of MS (Palaszynski et al., 2004; Sicotte et al., 2002; Spence and Voskuhl, 2012).

Another pregnancy-related hormone, progesterone, has been shown to reduce inflammatory cell infiltration, demyelination, microglia/astroglial activation and axonal loss in EAE but again only if administered predisease onset (Garay et al., 2007). In another study, progesterone has been show to attenuate EAE symptoms and increase levels of protective cytokines, notably IL-10 when administered at pregnancy levels to intact female C57BL/6 mice with the myelin oligodendrocyte glycoprotein model of disease (Yates et al., 2010).

Androgens have been shown to have a protective role in EAE (Palaszynski et al., 2004). Testosterone has been shown to have a protective role by causing a shift toward a Th2 response, and both testosterone and estriol have been shown to have neuroprotective roles (Gold and Voskuhl, 2009). Significantly, testosterone has been shown to reduce focal gray matter loss in a small clinical trial of 10 men with relapsing remitting MS (Kurth et al., 2014). In fact the authors note a significant increase in gray matter in the right frontal cortex. This is of great significance since the loss of gray matter has become recognized as an important factor in MS progression.

### 3.3.2 Theiler's virus-induced demyelination as a model for MS and epilepsy

Theiler's murine encephalomyelitis virus (TMEV) is a natural pathogen of mice, which belongs to the Picornaviridae family (Theiler, 1934; Lipton, 1975). Infection of SJL mice with TMEV causes an autoimmune-mediated demyelination of the central nervous system resulting in inflammatory lesions with similarities to human MS (Lipton, 1975; Lucchinetti et al., 2000). TMEV infection of C57BL/6 mice induces an epilepsy-like disease (Kirkman et al., 2010) and so the virus can be used to mimic both diseases in different mouse strains.

Studies of the effects of sex on susceptibility to TMEV-induced demyelination are contradictory (Hill et al., 1998; Alley et al., 2003; Sieve et al., 2004; Fuller et al., 2007). Hill et al. (1998) showed that female mice were more susceptible to early Theiler's virus disease. Whereas another study showed that male SJL/J mice had higher viral loads and lower virus-specific antibody levels during the early phase of the disease and then developed worse demyelinating disease compared with females (Alley et al., 2003). Sieve et al. (2004) reported that both male and female SJL/J mice developed similar levels of CNS inflammation but males suffered more axonal loss compared with female mice suggesting a neuroprotective role for estrogen (Young et al., 2010). In Theiler's virus infection of a unique strain, C57L/J mice, estrogen had a protective role since castration increased the susceptibility to Theiler's virus-induced demyelination and prolonged treatment of castrated mice with estrogen restored the resistance to the level of control mice. Estrogen-treated mice showed significantly decreased level of virus-specific Th1 responses, both in the periphery and in the CNS. In addition, *in vitro* estrogen treatment inhibited viral replication directly in macrophages, consistent with the lower level of viral RNA in microglia/macrophages in the CNS from castrated estrogen-treated mice compared with controls. Also, estrogen treatment inhibited VCAM-1 expression induced by TNFα in cerebrovascular endothelial cells via inhibition of nuclear factor-kappa B (NFkB) (Fuller et al., 2007). Similar to EAE, estrogen treatment in the TMEV model appears to exert its effects on the induction of immune responses, as well as infiltration of activated immune cells into the CNS via inhibition of NFkB function. It also reduces viral replication, which, although not relevant in the EAE model which is purely autoimmune, may be very relevant to MS.

Future studies on hormone therapies for both the epilepsy and MS models induced by TMEV may provide important therapeutic insights.

## 3.4 The effects of estrogens in the CNS

While estrogens have many effects on the immune system, they also impact many cells within the CNS. Astrocytes are essential components of the blood–brain barrier, and they are important in cell-to-cell communication, not only with CNS cells, but also with inflammatory cells (Kipp and Beyer, 2009). Estradiol has been shown to prevent astrocytes from expressing major histocompatibility complex class II and to enhance the induction of T cell apoptosis (Kipp and Beyer, 2009). Expression of ERα is necessary in astrocytes to prevent axonal loss and gliosis, and diminish immune cell infiltration into the CNS (Spence et al., 2011).

Both estriol and estradiol can bind to estrogen receptors α and β (ERα and ERβ, respectively) although estradiol has a higher affinity for ERα while estriol has a higher affinity for ERβ (Spence and Voskuhl, 2012). ERα has been implicated in EAE as having a potent anti-inflammatory effect with regard to Th1/Th17 responses (Lélu et al., 2011). At a cellular level binding of estradiol to ERα inhibits the transport of the pro-inflammatory transcription factor NFkB into the nucleus via interaction with PI3K, most notably in microglia, the resident macrophages of the CNS (Spence et al., 2011; Vegeto et al., 2008). *In vitro* experiments have shown that stimulation of ERα by estradiol or estriol agonists suppresses transcription of NFkB in astrocytes and thereby decreases the TNFα-mediated production of the pro-inflammatory chemokine CCL2 (Giraud et al., 2010). This reduced levels of CCL2 and resulted in the inhibiting of the infiltration of peripheral lymphocytes into the CNS (Spence and Voskuhl, 2012). In addition, treatment with ERα ligand reduced spinal cord inflammation, whereas ERβ ligand treatment only showed protective effects in the long term, but no reduction in inflammation (Tiwari-Woodruff and Voskuhl, 2009). ERα expression on astrocytes but not neurons has been shown to account for neuroprotection in EAE and decrease both macrophage inflammation and axonal loss (Spence et al., 2011). ERα appears to decrease the expression of the chemokines CCL2 and CCL7 on astrocytes. Although ERβ ligand treatment is neuroprotective in EAE, it did not act via receptors on astrocytes or neurons (Spence et al., 2013).

Finally, estradiol increases the ability of astrocytes to resorb excess glutamate, thus preventing neuronal loss due to excitotoxicity, which may explain the relative effectiveness of estradiol versus estriol's therapeutic effects on treated mice (Spence and Voskuhl, 2012).

Sex hormones have also been shown to control neural growth factors such as brain-derived neurotrophic factor, glial cell line-derived neurotrophic factor, IGF-1, and vascular endothelial growth factor (Kipp et al., 2012), which may also contribute to their neuroprotective effects. Progesterone has a neuroprotective effect on oligodendrocyte differentiation (Crawford et al., 2010), and promotes myelination (Chesik and De Kesyer, 2010). Taken together, it is likely that the therapeutic effects of pregnancy on MS are due to a combination of sex hormones since these hormones are present at high concentrations during the third trimester.

## 3.5 Hormone treatment in MS patients

The Pregnancy in Multiple Sclerosis European multicenter study enrolled 254 pregnant women with MS. During pregnancy, the relapse rate decreased especially in the third trimester, coinciding with the highest levels of estrogens and then relapses increased markedly upon delivery when levels of estrogens fell (Abramsky, 1994; Confavreux et al., 1998; Gold and Voskuhl, 2009; Rinta et al., 2010; Stuart and Bergstrom, 2011). As endogenous estrogen levels decrease in postmenopausal patients, the relapse rate increases (Smith and Studd, 1992). Clinical trials have shown that estriol may be a promising treatment option for MS and suggests that hormones may indeed be playing a role in disease modulation in MS patients (Sicotte et al., 2002; Soldan et al., 2003). In addition, oral contraceptives have been shown to decrease the number of relapses (Stuart and Bergstrom, 2011), the severity of symptoms, and the incidence of MS. Disease severity was reported to be 40% lower if the oral contraceptives contained ethinyl estradiol and a progestagen (Alonso et al., 2005; Nicot, 2009).

Estradiol has been shown to enhance the generation of Tregs in MS patients (Valor et al., 2011; Yates et al., 2010) and increase perforin expression (Valor et al., 2011). Perforin causes cell death via the granzyme-perforin pathway found in cytotoxic T cells but has also been postulated as a mechanism of Treg suppression. At higher pregnancy levels of estradiol, Treg perforin expression is diminished in accordance with the recruitment of Tregs to the fetal/maternal interface (Valor et al., 2011).

Hormones appear to play a role in sexual dimorphism seen in autoimmune diseases, but chromosomal sex also has an effect. Smith-Bouvier et al. (2008) examined the influence of chromosomal sex on the susceptibility to the autoimmune diseases MS and systemic lupus erythematosus. The group tested this hypothesis using XX and Sry-deficient $XY^-$ female mice and XX-Sry and $XY^-$Sry male mice that were gonadectomized to examine the effects of chromosome complement without any sex hormones (Smith-Bouvier et al., 2008). They found that the XX sex chromosome complement rendered SJL mice more susceptible to both EAE and the pristane-induced lupus model (Smith-Bouvier et al., 2008). In addition, the immune response is altered by XX complement, with $XY^-$ females producing anti-inflammatory Th2 cytokines such as IL-13 and IL-5 (Smith-Bouvier et al., 2008). This may help explain why females may be more susceptible to autoimmune diseases, and why they are more often afflicted.

Eighty percent of MS patients develop relapsing remitting disease and approximately 20% develop primary progressive disease (PPMS). Interestingly, there is no association with sex and PPMS, with a male-to-female ratio of 1:1 (Cottrell et al., 1999). As patients with this rare form of MS age, the ratio begins to switch to the traditional 1:2 seen in relapsing-remitting MS that may indicate an association with sex hormones, as female hormones decline with menopause (Cottrell et al., 1999).

### 3.6 Summary

Sex hormones are clearly implicated in MS and estrogens modulate cells of both the immune and nervous system thereby mediating therapeutic effects on inflammation and neurodegeneration. Further studies on the effects of sex hormones on animal models of neurological diseases will provide insights into the mechanisms of action of hormones and may allow the development of novel therapeutics with relevance to many conditions.

## 4 EPILEPSY

### 4.1 Introduction

The incidence of epilepsy is strongly sex-specific, with more men than women affected by epilepsy. Steroid hormones and neurosteroids play a key role in sex-specific forms of epilepsies such as catamenial epilepsy, a menstrual cycle-related seizure disorder in women. Neuroinflammation is emerging as a key factor in the neuropathology of epilepsy, including sex-specific forms of epilepsies.

### 4.2 Epilepsy and sex differences in epilepsies

Epilepsy is a chronic neurological disorder characterized by the unpredictable occurrence of seizures. Epilepsy affects an estimated 3 million Americans and about 65 million people worldwide in a variety of ways (Jacobs et al., 2009; Hesdorffer and Begley, 2013). About 150,000 new cases of epilepsy are diagnosed in the United States annually (Hesdorffer and Begley, 2013). Epileptic seizures are classified into partial (simple and complex partial seizures) and generalized (absence, tonic–clonic, myoclonic, and atonic seizures). Epilepsy may develop because of an abnormality in neural connectivity, an imbalance in inhibitory and excitatory neurotransmitters or some combination of these factors. Epileptogenesis is a process whereby a brain becomes progressively epileptic due to an initial precipitating event of diverse origin such as brain injury, stroke, infection, or prolonged seizures. Epileptogenesis involves three distinct stages: (1) the initial precipitating event, (2) the latent period, and (3) the chronic period with spontaneous seizures. Neuroinflammation and neurodegeneration appear to trigger epileptogenesis but there are other factors that play critical roles in the neuropathology (Pitkänen et al., 2009; Reddy, 2013a, b).

The incidence of epilepsy is sex specific as men exhibit relatively greater seizure susceptibility than women (Herzog, 2009; Verrotti et al., 2007; Reddy, 2010a, 2013a). Although epilepsy affects more men than women, women exhibit seizure syndromes that are much more complex and often intractable. Epilepsy affects an estimated 25 million women worldwide. There are about 1.2 million women with epilepsy of childbearing age in the United States (El-Sayed, 1998; Kaplan et al., 2007; Harden et al., 2009). Epilepsy is associated with special concerns for women of childbearing age (Morrell, 2003). There is abundant evidence of hormone effects on seizures since many women report that their seizure patterns change at puberty, menstrual cycle, and with menopause. The natural fluctuation in steroid hormones that occurs during the menstrual cycle significantly influences seizure patterns and potentially influences drug therapy. An estimated 10 million women in the United States and 100 million women worldwide use hormonal contraceptives. This is associated with risks including drug interactions and seizure exacerbations (Reddy, 2010b).

Sex difference in seizure susceptibility is one of the long-standing issues of epilepsy. Clinical evidence shows gender and age-related expression in many seizure syndromes. The incidence of epilepsy is generally higher in males than in females (Hauser, 1997; Christensen et al., 2005). More women than men are diagnosed with idiopathic generalized epilepsy, but localization-related symptomatic epilepsies are more frequent in men, and cryptogenic localization-related epilepsies are more frequent in women (Hauser, 1997; Christensen et al., 2005). Overall, there is considerable evidence indicating that men exhibit greater seizure susceptibility than females, while many females exhibit greater fluctuations in susceptibility to seizures, including menstrual cycle-related changes in seizure activity. Sex-based differences in seizure sensitivity may arise from variations between men and women in factors such as body weight, steroid hormones, cytochrome P450 activity, and biologic differences in neuronal networks in the brain. Steroid hormones play a key role in the neuroendocrine control of neuronal excitability and seizure susceptibility in men and women with epilepsy (Herzog, 2009; Verrotti et al., 2007; Reddy, 2010a, 2013a). Changes in seizure sensitivity are also evident at puberty, which is associated with rigorous changes in reproductive hormones and behavioural patterns (Reddy, 2009b). The relationship between menstrual cycle and seizure sensitivity in females is well known and is greatly influenced by hormonal fluctuations associated with menstrual cycle phases.

## 4.3 Catamenial epilepsy

Epilepsy is characterized by the unpredictable occurrence of seizures. However, seizures do not occur randomly in many women with epilepsy. Seizure clusters occur with a temporal periodicity following circadian or lunar cycles. In women with epilepsy, seizure periodicity may conform to the menstrual cycle according to a "menstrual clock" provided by a common phase marker of the onset of menses. This condition is referred as catamenial epilepsy. Catamenial epilepsy, derived from the Greek word kataminios, meaning "monthly," is characterized by seizures that cluster around specific points in the menstrual cycle. Catamenial epilepsy is a neuroendocrine condition in which seizures are most often clustered around perimenstrual or periovulatory period.

Catamenial epilepsy affects up to 70% of women with epilepsy (Herzog et al., 2004; Bazan et al., 2005; Gilad et al., 2008; Reddy, 2009a, 2013b). Overall, these studies support the prevailing notion that at least one in every three women with epilepsy shows catamenial seizure exacerbation.

Three types of catamenial seizures, perimenstrual (C1), periovulatory (C2), and inadequate luteal phase (C3), have been identified (Herzog et al., 1997). The perimenstrual is the most common clinical type. In perimenstrual catamenial epilepsy (C1), women with epilepsy experience an increase in seizure activity before, during, or after the onset of menstruation (Reddy, 2009a). Catamenial epilepsy is observed in women with both ovulatory and anovulatory cycles. Women with ovulatory cycles can experience either the perimenstrual or periovulatory catamenial types or even both within a single cycle. The diagnosis of ovulatory or anovulatory cycles is often made by estimating the midluteal phase progesterone levels. Progesterone levels lower than 5 ng/mL during day 20–22 of the cycle would certainly indicate inadequate luteal phase. In women with anovulatory cycles, inadequate luteal pattern is identified as a third type of catamenial epilepsy (C3).

## 4.4 Neuroendocrine mechanisms of catamenial epilepsy

Catamenial epilepsy is a multifaceted condition attributed to numerous causes. Catamenial epilepsy is an acquired disorder and currently there is no clear evidence of genetic components (Herzog, 2009; Quigg et al., 2009). A variety of mechanisms such as fluctuations in antiepileptic drug levels, changes in water and electrolyte balance, and physiological variation in ovarian hormone secretion have been proposed as causes for catamenial epilepsy (Reddy, 2013a, 2013b).

### 4.4.1 Estrogens

Estradiol has been known to play a role in the exacerbation of seizures in women with epilepsy (Logothetis et al., 1959; Bäckström, 1976; Jacono and Robinson, 1987). Plasma estradiol levels are found to increase during both the follicular and luteal phase of the normal menstrual cycle. Bäckström, 1976 characterized the relationship between seizures and steroid hormones, finding that in women with epilepsy, a positive correlation between seizure susceptibility and the estrogen-to-progesterone ratio was observed, peaking in the premenstrual and preovulatory periods and declining during the midluteal phase. Logothetis et al. (1959) have demonstrated that intravenous infusions of estrogen were associated with rapid interictal epileptiform activity in women with epilepsy and seizures were exacerbated when estrogen was given premenstrually. Therefore, it is thought that estrogens may facilitate some forms of catamenial seizures observed during these phases. The periovulatory catamenial exacerbation has been attributed to the midcycle surge of estrogen that is relatively unopposed by progesterone until early luteal phase (Logothetis et al., 1959). An increase in the ratio of estrogen-to-progesterone levels during the perimenstrual period (described in the subsequent section) might at least partly contribute to the development of perimenstrual seizure exacerbation (Bonuccelli et al., 1989; Herzog et al., 1997).

### 4.4.2 Progesterone

Progesterone plays a key role in catamenial epilepsy. Progesterone has long been known to have antiseizure activity in a variety of animal models of epilepsy (Craig, 1966; Landgren et al., 1978; Reddy, 2009a). Progesterone also has antiepileptogenic actions. The antiseizure actions of progesterone are mostly mediated by its metabolic conversion into neurosteroids (Reddy et al., 2004a, 2004b, 2010; Reddy and Mohan, 2011; Reddy and Ramanathan, 2012). In women with epilepsy, natural cyclic variations in progesterone during the menstrual cycle could influence catamenial seizure susceptibility. Seizures decrease in the midluteal phase when serum progesterone levels are high and increase premenstrually when progesterone levels fall and there is a decrease in the serum progesterone-to-estrogen ratio (Bäckström, 1976; Bonuccelli et al., 1989; Herzog et al., 2001). Changes in progesterone levels have been directly correlated with catamenial seizures (Tuveri et al., 2008; El-Khayat et al., 2008). Despite some limitations, these findings provide evidence that disruption in ovarian cycle-related fluctuations in progesterone can be correlated to catamenial seizure exacerbation. The emerging evidence clearly indicates that catamenial seizures are associated with a rapid decline in progesterone around menstruation. In clinical studies progesterone has been found to reduce seizures (Bäckström et al., 1984; Herzog, 2009). Herzog et al. (2012) assessed the short-term efficacy and safety of adjunctive cyclic progesterone therapy in the treatment of intractable seizures in 462 women with partial epilepsy. There was no significant difference in the proportions of responders for all seizures combined between progesterone and placebo in women with catamenial and noncatamenial epilepsy. However, the prespecified exploratory findings suggest that the level of perimenstrual seizure exacerbation is a significant predictor of the responder rate with progesterone therapy. Therefore, progesterone therapy may provide a clinically significant benefit for many women with perimenstrual catamenial epilepsy.

### 4.4.3 Neurosteroids

A variety of neurosteroids are synthesized in the brain. The most widely studied are allopregnanolone (AP), allotetrahydro-deoxycorticosterone, and androstanediol (Reddy, 2010a). Neurosteroids rapidly alter neuronal excitability through direct interaction with gamma-aminobutyric acid (GABA)-A receptors (Harrison et al., 1987; Hosie et al., 2007; Carver and Reddy, 2013). AP and other structurally related neurosteroids act as positive allosteric modulators and direct activators of GABA-A receptors. AP and related neurosteroids are powerful anticonvulsants (Reddy, 2010a; Reddy and Jian, 2010). Therefore, neurosteroids play a critical role in catamenial seizure susceptibility in women with epilepsy. When neurosteroid levels fluctuate, loss of seizure control can occur. Neurosteroids allopregnanolone and tetrahydrodeoxycorticosterone have been implicated in perimenstrual seizure exacerbations in women with normal menstrual cycle. It is hypothesized that withdrawal of progesterone-derived neurosteroids leads to enhanced excitability predisposing to seizures (Reddy et al., 2001, 2012; Reddy, 2013b). In addition, plasticity in GABA-A receptor subunits could play a role in the enhanced seizure susceptibility in perimenstrual catamenial epilepsy. Animal studies have shown that prolonged exposure to allopregnanolone followed by withdrawal such as that which occurs during menstruation causes a marked increase in expression of $\alpha$4-subunit, a key subunit linked to enhanced neuronal excitability, seizure susceptibility, and benzodiazepine resistance (Smith et al., 2007; Gangisetty and Reddy, 2010; Wu et al., 2013). These neuroendocrine changes can result in reduced inhibition resulting in enhanced excitability, which, among other effects, predisposes to seizures.

It is thought that catamenial seizures may occur due to multiple mechanisms (Reddy, 2014). Perimenstrual type (C1) occurs in women with ovulatory cycle possibly due to a sharp decline ("withdrawal") in the serum level of progesterone and, consequently, of the level of progesterone-derived anticonvulsant neurosteroids in the brain around the perimenstrual period. The estradiol/neurosteroid ratio is highest during menstruation. Because neurosteroids potentiate GABA-A receptor-mediated inhibition, the rapid loss of neurosteroid-mediated inhibition, such as that which occurs before, during or after the onset of menses, could exacerbate seizures in many women with catamenial epilepsy. Periovulatory type (C2) occurs in women with ovulatory cycles, possibly due to estradiol surge just before ovulation, and low neurosteroid levels do not offset the estradiol-induced excitation because the rise of anticonvulsant neurosteroid levels would not occur until after ovulation. The relatively low neurosteroid inhibition and marked estradiol excitation could lead to periovulatory seizures. Inadequate luteal type (C3) occurs in women with anovulatory cycles possibly due to a loss of neurosteroid-mediated inhibition during luteal phase for a prolonged time and also due to elevated estrogen levels. Progesterone secretion that occurs normally during the luteal phase is markedly decreased during anovulatory cycles, resulting in abnormally low levels of neurosteroid in the brain (Tuveri et al., 2008; Reddy, 2014).

Based on neurosteroid physiology, Reddy and colleagues have developed multiple rodent models of catamenial epilepsy (Reddy et al., 2001, 2012; Reddy and Zeng, 2007). Rodents have a 4–5 day estrous cycle and studies of fluctuations in seizure susceptibility in cycling female rodents have not led to results that are relevant to the human menstrual cycle. In order to provide a model that more closely mimics the human situation, a condition of elevated progesterone was created in rats by gonadotropin treatment. This resulted in prolonged high circulating levels of estrogen and progesterone similar to those that occur in the luteal phase of the menstrual cycle. Then, to simulate the withdrawal of allopregnanolone that occurs at the time of menstruation, the animals were treated with finasteride (a $5\alpha$-reductase and neurosteroid synthesis inhibitor) 11 days after the initiation of gonadotropin treatment. Withdrawal of neurosteroids led to decreased seizure threshold and increased seizures (Reddy et al., 2001, 2012; Reddy and Zeng, 2007), suggesting that endogenous neurosteroids play a critical role in catamenial seizures.

These animal models are used to investigate therapies for catamenial epilepsy (Reddy and Rogawski, 2000a, 2000b, 2001; Reddy et al., 2012). A key result is that conventional AEDs, including benzodiazepines and valproate, are less potent in protecting against seizures during the period of enhanced seizure susceptibility following neurosteroid withdrawal. This pharmacoresistance appears to mimic the situation in women with catamenial epilepsy, in whom breakthrough seizures occur despite treatment with antiepileptic drugs. In contrast, with the results with conventional antiepileptic drugs, neurosteroids such as AP and ganaxolone had enhanced activity in the catamenial epilepsy models (Reddy and Rogawski, 2001; Reddy et al., 2012). Consequently, they proposed a "neurosteroid replacement" approach to treat catamenial seizure exacerbations (Reddy and Rogawski, 2009). A neurosteroid could be administered in a "pulse" prior to menstruation and then withdrawn, or continuously administered throughout the month. The neurosteroid would be administered at low doses to avoid sedative side effects. Such low doses are expected to contribute little anticonvulsant activity during most of the menstrual cycle, but may prevent the occurrence of perimenstrual catamenial seizures.

## 4.5 Neuroinflammation and epilepsy

Neuroinflammation is a common consequence of seizures and related neuronal injury events. Brain inflammation is emerging to play a central role in the pathogenesis of acquired epilepsy (Vezzani et al., 2013; Rojas et al., 2014). Acute seizures and status epilepticus cause neuroinflammation by activating microglia, astrocytes, and induction and enhancement of inflammatory cytokines such as IL-1β, IL-6, and TNF in key brain regions such as the hippocampus (Reddy and Kuruba, 2013). Further, seizure-related expression of inflammatory cytokines occurs in brain regions that may undergo neuronal damage. Moreover, it is likely that seizure-induced release of inflammatory cytokines like IL-1β from astrocytes may cause brain inflammation and damage the blood–brain barrier. All currently used animal models, including chemical, electrical, and mechanical paradigms, can trigger a robust neuroinflammation, which appears to play a central role in the progression of epileptogenesis resulting in recurrent seizures (Reddy and Kuruba, 2013). A prolonged seizure episode is generally considered as a trigger to initiate the inflammatory cascade that leads to activation of astrocytes and microglia and related secondary autocrine inflammatory markers. Like ischemic stroke, neuronal injury, and autoimmune flare-ups, prolonged seizures can lead to significant cell necrosis, neuroinflammation, and neurodegeneration as hallmark neuropathology features. There are many components of neuroinflammation that are linked to sex differences including hormones, immunological changes, and neuronal network organization. Therefore, sex differences in such factors that control inflammation could likely contribute to sex-based differences in the incidence of epilepsy.

## 5 CONCLUSIONS

Steroid hormone and neurosteroids are believed to play a key role in the sex-specific forms of epilepsies and gender-related seizure susceptibility. Menstrual and stress-related fluctuations in seizures may be related to alterations in brain neurosteroid levels. In addition to neurosteroid fluctuations, there is emerging evidence that plasticity in GABA-A receptor subunits could play a role in the enhanced seizure susceptibility in gender-specific forms of epilepsy such as catamenial epilepsy. Such neuroendocrine changes can result in reduced inhibition, resulting in enhanced excitability, which, among other effects, predisposes to catamenial seizures. A neurosteroid replacement therapy has been suggested as a rational treatment strategy for management of catamenial epilepsy.

## REFERENCES

Abramsky, O., 1994. Pregnancy and multiple sclerosis. Ann. Neurol. 36, S38–41.

Ajmo, Jr., C.T., Vernon, D.O., Collier, L., Hall, A.A., Garbuzova-Davis, S., Willing, A., Pennypacker, K.R., 2008. The spleen contributes to stroke-induced neurodegeneration. J. Neurosci. Res. 86, 2227–2234.

Alkayed, N.J., Harukuni, I., Kimes, A.S., London, E.D., Traystman, R.J., Hurn, P.D., 1998. Gender-linked brain injury in experimental stroke. Stroke 29, 159–165, discussion 166.

Allen, C.L., Bayraktutan, U., 2008. Risk factors for ischaemic stroke. Int. J. Stroke 3, 105–116.

Alley, J., Khasabov, S., Simone, D., Beitz, A., Rodriguez, M., Njenga, M.K., 2003. More severe neurologic deficits in SJL/J male than female mice following Theiler's virus-induced CNS demyelination. Exp. Neurol. 180, 14–24.

Alonso, A., Jick, S., Olek, M.J., Ascherio, A., Jick, H., Hernan, M.A., 2005. Recent use of oral contraceptives and the risk of multiple sclerosis. Arch. Neurol. 62, 1362–1365.

Anderson, D.W., Ellenberg, J.H., Leventhal, C.M., Reingold, S.C., Rodriguez, M., Silberberg, D.H., 1992. Revised estimate of the prevalence of multiple sclerosis in the United States. Ann. Neurol. 31, 333–336.

Arnold, A.P., Chen, X., 2009. What does the "four core genotypes" mouse model tell us about sex differences in the brain and other tissues? Front. Neuroendocrinol. 30, 1–9.

Bäckström, T., 1976. Epileptic seizures in women related to plasma estrogen and progesterone during the menstrual cycle. Acta Neurol. Scand. 54, 321–347.

Bäckström, T., Zetterlund, B., Blom, S., Romano, M., 1984. Effect of intravenous progesterone infusions on the epileptic discharge frequency in women with partial epilepsy. Acta Neurol. Scand. 69, 240–248.

Bake, S., Sohrabji, F., 2004. 17β-Estradiol differentially regulates blood-brain barrier permeability in young and aging female rats. Endocrinol. 145, 5471–5475.

Bazan, A.C., Montenegro, M.A., Cendes, F., Min, L.L., Guerreiro, C.A., 2005. Menstrual cycle worsening of epileptic seizures in women with symptomatic focal epilepsy. Arq. Neuropsiquiatr. 63 (3B), 751–756.

Bebo, B.F., Vandenbark, A.A., Offner, H., 1996. Male SJL mice do not relapse after induction of EAE with PLP 139-151. J. Neuroci. Res. 45, 680–689.

Bebo, B.F., Schuster, J.C., Vandenbark, A.A., Offner, H., 1998. Gender differences in experimental autoimmune encephalomyelitis develop during the induction of the immune response to encephalitogenic peptides. J. Neurosci. Res. 52, 420–426.

Bebo, B.F., Adlard, K., Schuster, J.C., Unsicker, L., Vanderbark, A.A., Offner, H., 1999. Gender differences in protection from EAE induced by oral tolerance with a peptide analogue of MBP-Ac1-11. J. Neurosci. Res. 55, 432–440.

Bebo, B.F., Fyfe-Johnson, A., Adlard, K., Beam, A.G., Vandenbark, A.A., Offner, H., 2001. Low-dose estrogen therapy ameliorates experimental autoimmune encephalomyelitis in two different inbred mouse strains. J. Immunol. 166, 2080–2089.

Bettelli, E., Carrier, Y., Gao, W., Korn, T., Strom, T.B., Oukka, M., Weiner, H.L., Kuchroo, V.K., 2006. Reciprocal developmental pathways for the generation of pathogenic effector TH17 and regulatory T cells. Nature 441, 235–238.

Bonuccelli, U., Melis, G.B., Paoletti, A.M., Fioretti, P., Murri, L., Muratorio, A., 1989. Unbalanced progesterone and estradiol secretion in catamenial epilepsy. Epilepsy Res. 3, 100–106.

Brann, D.W., Dhandapani, K., Wakade, C., Mahesh, V.B., Khan, M.M., 2007. Neurotrophic and neuroprotective actions of estrogen: basic mechanisms and clinical implications. Steroids 72, 381–405.

Brann, D., Raz, L., Wang, R., Vadlamudi, R., Zhang, Q., 2012. Oestrogen signalling and neuroprotection in cerebral ischaemia. J. Neuroendocrinol. 24, 34–47.

Bushnell, C.D., 2008. Stroke in women: risk and prevention throughout the lifespan. Neurol. Clin. 26 (4), 1161–1176.

Carver, C.M., Reddy, D.S., 2013. Neurosteroid interactions with synaptic and extrasynaptic GABA-A receptors: regulation of subunit plasticity, phasic and tonic inhibition, and neuronal network excitability. Psychopharmacology 230 (2), 151–188.

Chen, Y., Langrish, C.L., Mckenzie, B., Joyce-Shaikh, B., Stumhofer, J.S., McClanahan, T., Blumenschein, W., Churakovsa, T., Low, J., Presta, L., Hunter, C.A., Kastelein, R.A., Cua, D.J., 2006. Anti-IL-23 therapy inhibits multiple inflammatory pathways and ameliorates autoimmune encephalomyelitis. J. Clin. Invest. 116, 1317–1326.

Chesik, D., De Kesyer, J., 2010. Progesterone and dexamethasone differentially regulate the IGF-system in glial cells. Neurosci. Lett. 468, 178–182.

Choi, Y.C., Lee, J.H., Hong, K.W., Lee, K.S., 2004. 17 Beta-estradiol prevents focal cerebral ischemic damages via activation of Akt and CREB in association with reduced PTEN phosphorylation in rats. Fundam. Clin. Pharmacol. 18, 547–557.

Christensen, J., Kjeldsen, M.J., Anderson, H., Friis, M.L., Sidenius, P., 2005. Gender differences in epilepsy. Epilepsia 46, 956–960.

Cipolla, M., Godfrey, J., Wiegman, M., 2009. The effect of ovariectomy and estrogen on penetrating brain arterioles and blood-brain barrier permeability. Microcirculation 16, 685–693.

Compston, A., Coles, A., 2002. Multiple sclerosis. Lancet 359, 1221–1231.

Confavreux, C., Hutchinson, M., Hours, M.M., Cortinovis-Tourniaire, P., Moreau, T., 1998. Rate of pregnancy-related relapse in multiple sclerosis. N. Engl. J. Med. 339, 285–291.

Coomber, B., Gibson, C.L., 2010. Sustained levels of progesterone prior to the onset of cerebral ischemia are not beneficial to female mice. Brain Res. 1361, 124–132.

Cottrell, D.A., Kremenchutzky, M., Rice, G.P.A., Koopman, W.J., Hader, W., Bakerville, J., Ebers, G.C., 1999. The natural history of multiple sclerosis: a geographically based study 5. The clinical features and natural history of primary progressive multiple sclerosis. Brain 122, 625–639.

Cottrell, D.A., Kremenchutzky, M., Rice, G.P.A., Koopman, W.J., Hader, W., Baskerville, J., Ebers, G.C., 1999. The natural history of multiple sclerosis: a geographically based study. Brain 122, 625–639.

Craig, C.R., 1966. Anticonvulsant activity of steroids: separability of anticonvulsant from hormonal effects. J. Pharmacol. Exp. Ther. 153, 337–343.

Crawford, D.K., Mangiardi, M., Song, B., Patel, R., Du, S., Sofroniew, M.V., Voskhul, R.R., Tiwari-Woodruff, S.K., 2010. Estrogen receptor beta ligand: a novel treatment to enhance endogenous functional remyelination. Brain 133, 2999–3016.

Cua, D.J., Sherlock, J., Chen, Y., Murphy, C.A., Joyce, B., Seymour, B., Lucian, L., To, W., Kwan, S., Churakova, T., Zurawski, S., Wiekowski, M., Lira, S.A., Gorman, D., Kastelein, R.A., Sedgwick, J.D., 2003. Interleukin-23 rather than interleukin-12 is the critical cytokine for autoimmune inflammation of the brain. Nature 421, 744–748.

De Butte-Smith, M., Nguyen, A.P., Zukin, R.S., Etgen, A.M., Colbourne, F., 2007. Failure of estradiol to ameliorate global ischemia-induced CA1 sector injury in middle-aged female gerbils. Brain Res. 1153, 214–220.

Dubal, D.B., Wise, P.M., 2001. Neuroprotective effects of estradiol in middle-aged female rats. Endocrinology 142, 43–48.

Dubal, D., Kashon, M., Pettigrew, L., Ren, J., Finklestein, S., Rau, S., Wise, P., 1998. Estradiol protects against ischemic injury. J. Cereb. Blood Flow Metab. 18, 1253–1258.

El-Khayat, H.A., Soliman, N.A., Tomoum, H.Y., Omran, M.A., El-Wakad, A.S., Shatla, R., 2008. Reproductive hormonal changes and catamenial pattern in adolescent females with epilepsy. Epilepsia 49, 1619–1626.

El-Sayed, Y.Y., 1998. Obstetric and gynecologic care of women with epilepsy. Epilepsia 39 (Suppl. 8), S17–S25.

Espinosa-García, C., Vigueras-Villaseñor, R.M., Rojas-Castañeda, J.C., Aguilar-Hernández, A., Monfil, T., Cervantes, M., Moralí, G., 2013. Post-ischemic administration of progesterone reduces caspase-3 activation and DNA fragmentation in the hippocampus following global cerebral ischemia. Neurosci. Lett. 550, 98–103.

Fang, M.C., Singer, D.E., Chang, Y., Hylek, E.M., Henault, L.E., Jensvold, N.G., Go, A.S., 2005. Gender differences in the risk of ischemic stroke and peripheral embolism in atrial fibrillation: the AnTicoagulation and Risk factors In Atrial fibrillation (ATRIA) study. Circulation 112, 1687–1691.

Fletcher, J.M., Lalor, S.J., Sweeney, C.M., Tubridy, N., Mills, K.H., 2010. T cells in multiple sclerosis and experimental autoimmune encephalomyelitis. Clin. Exp. Immunol. 162, 1–11.

Fuller, A., Yahikozawa, H., So, E.Y., Dal Canto, M., Koh, C.S., Welsh, C.J., Kim, B.S., 2007. Castration of male C57L/J mice increases susceptibility and estrogen treatment restores resistance to Theiler's virus-induced demyelinating disease. J. Neurosci. Res. 85, 871–881.

Galea, E., Santizo, R., Feinstein, D.L., Adamsom, P., Greenwood, J., Koenig, H.M., Pelligrino, D.A., 2002. Estrogen inhibits NF kappa B-dependent inflammation in brain endothelium without interfering with I kappa B degradation. Neuroreport 13 (11), 1469–1472.

Gangisetty, O., Reddy, D.S., 2010. Neurosteroid withdrawal regulates GABA-A receptor α4-subunit expression and seizure susceptibility by activation of progesterone receptor-independent early growth response factor-3 pathway. Neuroscience 170, 865–880.

Garay, L., Deniselle, M.C.G., Lima, A., Roig, P., De Nicola, A.F., 2007. Effects of progesterone in the spinal cord of a mouse model of multiple sclerosis. J. Steroid Biochem. Mol. Biol. 107, 228–237.

Gilad, R., Sadeh, M., Rapoport, A., Dabby, R., Lampl, Y., 2008. Lamotrigine and catamenial epilepsy. Seizure 17, 531–534.

Giraud, S.N., Caron, C.M., Pham-Dinh, D., Kitabgi, P., Nicot, A.B., 2010. Estradiol inhibits ongoing autoimmune neuroinflammation and NFkB-dependent CCL2 expression in reactive astrocytes. Proc. Natl. Acad. Sci. USA 107, 8416–8421.

Glendenning, M.L., Lovekamp-Swan, T., Schreihofer, D.A., 2008. Protective effect of estrogen in endothelin-induced middle cerebral artery occlusion in female rats. Neurosci. Lett. 445, 188–192.

Gold, S.M., Voskuhl, R.R., 2009. Estrogen and testosterone therapies in multiple sclerosis. In: Verhaagen, J., Hol, E.M., Huitenga, I., Wijnholds, J., Bergen, A.B., Boer, G.J., Dick, F.S. (Eds.), Progress in Brain Research. Elsevier, New York, pp. 239–251.

González-Vidal, M.D., Cervera-Gaviria, M., Ruelas, R., Escobar, A., Moralí, G., Cervantes, M., 1998. Progesterone: protective effects on the cat hippocampal neuronal damage due to acute global cerebral ischemia. Arch. Med. Res. 29, 117–124.

Harden, C.L., Pennell, P.B., Koppel, B.S., et al., 2009. Management issues for women with epilepsy – focus on pregnancy (an evidence-based review): III. Vitamin K, folic acid, blood levels, and breast-feeding: Report of the Quality Standards Subcommittee and Therapeutics and Technology Assessment Subcommittee of the American Academy of Neurology and the American Epilepsy Society. Epilepsia 50, 1247–1255.

Harrison, N.L., Majewska, M.D., Harrington, J.W., Barker, J.L., 1987. Structure-activity relationships for steroid interactions with the γ-aminobutyric acidA receptor complex. J. Pharmacol. Exp. Ther. 241, 346–353.

Hauser, W.A., 1997. Incidence and prevalence. In: Engel, Jr., J., Pedley, T.A. (Eds.), Epilepsy: A Comprehensive Textbook. Lippincott-Raven Publishers, Philadelphia, pp. 47–57.

Hawk, T., Zhang, Y.Q., Rajakumar, G., Day, A.L., Simpkins, J.W., 1998. Testosterone increases and estradiol decreases middle cerebral artery occlusion lesion size in male rats. Brain Res. 796, 296–298.

Herzog, A.G., 2009. Hormonal therapies: progesterone. Neurotherapeutics 6, 383–391.

Herzog, A.G., Klein, P., Ransil, B.J., 1997. Three patterns of catamenial epilepsy. Epilepsia 38, 1082–1088.

Herzog, A.G., Friedman, M.N., Freund, S., Pascual-Leone, A., 2001. Transcranial magnetic stimulation evidence of a potential role for progesterone in the modulation of premenstrual corticocortical inhibition in a woman with catamenial seizure exacerbation. Epilepsy Behav. 2, 367–369.

Herzog, A.G., Harden, C.L., Liporace, J., Pennell, P., Schomer, D.L., Sperling, M., Fowler, K., Nikolov, B., Shuman, S., Newman, M., 2004. Frequency of catamenial seizure exacerbation in women with localization-related epilepsy. Ann. Neurol. 56, 431–434.

Herzog, A.G., Fowler, K.M., Smithson, S.D., Kalayjian, L.A., Heck, C.N., Sperling, M.R., Liporace, J.D., Harden, C.L., Dworetzky, B.A., Pennell, P.B., Massaro, J.M., Progesterone Trial Study Group, 2012.

Progesterone vs placebo therapy for women with epilepsy: a randomized clinical trial. Neurology 78 (24), 1959–1966.

Hesdorffer, D.C., Begley, C.E., 2013. Surveillance of epilepsy and prevention of epilepsy and its sequelae: lessons from the Institute of Medicine report. Curr. Opin. Neurol. 26 (2), 168–173.

Hill, K.E., Pigmans, M., Fujinami, R.S., Rose, J.W., 1998. Gender variations in early Theiler's virus induced demyelinating disease: differential susceptibility and effects of IL-4, IL-10 and combined IL-4 with IL-10. J. Neuroimmunol. 85, 44–51.

Hosie, A.M., Wilkins, M.E., Smart, T.G., 2007. Neurosteroid binding sites on GABAA receptors. Pharmacol. Ther. 116, 7–19.

Ishrat, T., Sayeed, I., Atif, F., Stein, D.G., 2009. Effects of progesterone administration on infarct volume and functional deficits following permanent focal cerebral ischemia in rats. Brain Res. 1257, 94–101.

Jacobs, M.P., Leblanc, G.G., Brooks-Kayal, A., Jensen, F.E., Lowenstein, D.H., Noebels, J.L., 2009. Curing epilepsy: progress and future directions. Epilepsy Behav. 14, 438–445.

Jacono, J.J., Robinson, J., 1987. The effects of estrogen, progesterone, and ionized calcium on seizures during the menstrual cycle in epileptic women. Epilepsia 28, 571–577.

Jezierski, M.K., Sohrabji, F., 2000. Region- and peptide-specific regulation of the neurotrophins by estrogen. Mol. Brain Res. 85, 77–84.

Jiang, N., Chopp, M., Stein, D., Feit, H., 1996. Progesterone is neuroprotective after transient middle cerebral artery occlusion in male rats. Brain Res. 735, 101–107.

Johnson, A.B., Sohrabji, F., 2005. Estrogen's effects on central and circulating immune cells vary with reproductive age. Neurobiol. Aging 26, 1365–1374.

Jover-Mengual, T., Miyawaki, T., Latuszek, A., Alborch, E., Zukin, R.S., Etgen, A.M., 2010. Acute estradiol protects CA1 neurons from ischemia-induced apoptotic cell death via the PI3K/Akt pathway. Brain Res. 1321, 1–12.

Kantarci, O.H., Weinshenker, B.G., 2005. Natural history of multiple sclerosis. Neurol. Clin. 23, 17–38.

Kaplan, P.W., Norwitz, E.R., Ben Menachem, E., Pennell, P.B., Druzin, M., Robinson, J.N., Gordon, J.C., 2007. Obstetric risks for women with epilepsy during pregnancy. Epilepsy Behav. 11, 283–291.

Kapral, M.K., Fang, J., Hill, M.D., Silver, F., Richards, J., Jaigobin, C., Cheung, A.M., 2005. Sex differences in stroke care and outcomes: results from the Registry of the Canadian Stroke Network. Stroke 36, 809–814.

Kipp, M., Amor, S., Krauth, R., Beyer, C., 2012. Multiple sclerosis: neuroprotective alliance of estrogen-progesterone and gender. Front. Neuroendocrinol. 33, 1–16.

Kipp, M., Beyer, C., 2009. Impact of sex steroids on neuroinflammatory processes and experimental multiple sclerosis. Front. Neuroendocrinol. 30, 188–200.

Kirkman, N.J., Libbey, J.E., Wilcox, K.S., White, H.S., Fujinami, R.S., 2010. Innate but not adaptive immune responses contribute to behavioral seizures following viral infection. Epilepsia 51 (3), 454–464.

Komiyama, Y., Nakae, S., Matsuki, T., Nambu, A., Ishigame, H., Kakuta, S., Sudo, K., Iwakura, Y., 2006. IL-17 plays an important role in the development of experimental autoimmune encephalomyelitis. J. Immunol. 177, 566–573.

Kumon, Y., Kim, S.C., Tompkins, P., Stevens, A., Sakaki, S., Loftus, C.M., 2000. Neuroprotective effect of postischemic administration of progesterone in spontaneously hypertensive rats with focal cerebral ischemia. J. Neurosurg. 92, 848–852.

Kurth, F., Luders, E., Sicotte, N.L., Gaser, C., Giesser, B.S., Swerdloff, R.S., Montag, M.J., Voskuhl, R.R., Mackenzie-Graham, A., 2014. Neuroprotective effects of testosterone treatment in men with multiple sclerosis. Neuroimage Clin. 4, 454–460.

Lai, S.M., Duncan, P.W., Dew, P., Keighley, J., 2005. Sex differences in stroke recovery. Prev. Chronic Dis. 2, A13.

Landgren, S., Bäckström, T., Kalistratov, G, 1978. The effect of progesterone on the spontaneous interictal spike evoked by the application of penicillin to the cat's cerebral cortex. J. Neurol. Sci. 36, 119–133.

Lélu, K., Laffont, S., Delpy, L., Paulet, P.E., Périnat, T., Tschanz, S.A., Pelletier, L., Engelhardt, B., Guéry, J.C., 2011. Estrogen receptor α signaling in T lymphocytes is required for estradiol-mediated inhibition of Th1 and Th17 cell differentiation and protection against experimental autoimmune encephalomyelitis. J. Immunol. 187, 2386–2393.

Leon, R.L., Li, X., Huber, J.D., Rosen, C.L., 2012. Worsened outcome from middle cerebral artery occlusion in aged rats receiving 17b-estradiol. Endocrinology 153, 3386–3393.

Liao, S., Chen, W., Kuo, J., Chen, C., 2001. Association of serum estrogen level and ischemic neuroprotection in female rats. Neurosci. Lett. 297, 159–162.

Lipton, H.L., 1975. Theiler's virus infection in mice: an unusual biphasic disease process leading to demyelination. Infect. Immun. 11, 1147–1155.

Liu, R1, Wen, Y., Perez, E., Wang, X., Day, A.L., Simpkins, J.W., Yang, S.H., 2005. 17beta-Estradiol attenuates blood-brain barrier disruption induced by cerebral ischemia-reperfusion injury in female rats. Brain Res. 1060, 55–61.

Liu, R., Wang, X., Liu, Q., Yang, S.H., Simpkins, J.W., 2007. Dose dependence and therapeutic window for the neuroprotective effects of 17beta-estradiol when administered after cerebral ischemia. Neurosci. Lett. 415, 237–241.

Liu, F., Yuan, R., Benashski, S.E., McCullough, L.D., 2009. Changes in experimental stroke outcome across the life span. J. Cereb. Blood Flow Metab. 29, 792–802.

Lloyd-Jones, D., Adams, R.J., Brown, T.M., et al., 2010. Heart disease and stroke statistics – 2010 update: a report from the American Heart Association. Circulation 121, e46–e215.

Logothetis, J., Harner, R., Morrel, F., 1959. The role of estrogens in catamenial exacerbation of epilepsy. Neurology 9, 352–360.

Lucchinetti, C., Bruck, W., Parisi, J., Scheithauer, B., Rodriguez, M., Lassmann, H., 2000. Heterogeneity of multiple sclerosis lesion: implications for the pathogenesis of demyelination. Ann. Neurol. 47, 707–717.

Manwani, B., Liu, F., Xu, Y., Persky, R., Li, J., McCullough, L.D., 2011. Functional recovery in aging mice after experimental stroke. Brain. Behav. Immun. 25, 1689–1700.

Manwani, B., Liu, F., Scranton, V., Hammond, M.D., Sansing, L.H., McCullough, L.D., 2013. Differential effects of aging and sex on stroke induced inflammation across the lifespan. Exp. Neurol. 249, 120–131.

Manwani, B., Bentivegna, K., Benashski, S.E., Venna, V.R., Xu, Y., Arnold, A.P., McCullough LD, 2014. Sex differences in ischemic stroke sensitivity are influenced by gonadal hormones, not by sex chromosome complement. J. Cereb. Blood Flow Metab. 35 (2), 221–229.

Marriott, L.K., Hauss-Wegrzyniak, B., Benton, R.S., Vraniak, P.D., Wenk, G.L., 2002. Long-term estrogen therapy worsens the behavioral and neuropathological consequences of chronic brain inflammation. Behav. Neurosci. 116, 902–911.

McClain, M.A., Gatson, N.N., Powell, N.D., Papenfuss, T.L., Gienapp, I.E., Song, F., Shawler, T.M., Kithcart, A., Whitacre, C.C., 2007. Pregnancy suppresses experimental autoimmune encephalomyelitis through immunoregulatory cytokine production. J. Immunol. 179, 8146–8152.

McCullough, L.D., Alkayed, N.J., Traystman, R.J., Williams, M.J., Hurn, P.D., 2001. Postischemic estrogen reduces hypoperfusion and secondary ischemia after experimental stroke. Stroke 32, 796–802.

Morales, L.B.J., Loo, K.K., Liu, H., Peterson, C., Tiwari-Woodruff, S., Voskuhl, R.R., 2006. Treatment with an estrogen receptor – a ligand is neuroprotective in experimental autoimmune encephalomyelitis. Neurobiol. Dis. 26, 6823–6833.

Moralí, G., Montes, P., Hernández-Morales, L., Monfil, T., Espinosa-García, C., Cervantes, M., 2011. Neuroprotective effects of progesterone and allopregnanolone on long-term cognitive outcome after global-cerebral ischemia. Restor. Neurol. Neurosci. 29, 1–15.

Morrell, M.J., 2003. Reproductive and metabolic disorders in women with epilepsy. Epilepsia 44 (Suppl. 4), 11–20.

Nicot, A., 2009. Gender and sex hormones in multiple sclerosis pathology and therapy. Front. Biosci. 14, 4477–4515.

Noonan, C.W., Kathman, S.J., White, M.C., 2002. Prevalence estimates for MS in the United States and evidence of an increasing trend for women. Neurology 58 (1), 136–138.

Nordell, V.L., Scarborough, M.M., Buchanan, A.K., Sohrabji, F., 2003. Differential effects of estrogen in the injured forebrain of young adult and reproductive senescent animals. Neurobiol. Aging 24, 733–743.

Noseworthy, J.H., Lucchinetti, C., Rodriguez, M., Weinshenker, B.G., 2000. Multiple sclerosis. N. Engl. J. Med. 343, 938–952.

Offner, H., Polanczyk, M., 2006. A potential role for estrogen in experimental autoimmune encephalomyelitis and multiple sclerosis. Ann. NY Acad. Sci. 1089, 343–372.

Okuda, Y., Okuda, M., Bernard, C.C., 2002. Gender does not influence the susceptibility of C57Bl/6 mice to develop chronic experimental autoimmune encephalomyelitis induced by myelin oligodendrocyte glycoprotein. Immunol. Lett. 81, 25–29.

Olsen, T.S., Andersen, K.K., 2010. Female survival advantage relates to male inferiority rather than female superiority: a hypothesis based on the impact of age and stroke severity on 1-week to 1-year case fatality in 40,155 men and women. Gend. Med. 7, 284–295.
Orton, S.M., Herrera, B.M., Yee, I.M., Valdar, W., Ramagopalan, S.V., Sadovnick, A.D., Ebers, G.C., 2006. Sex ratio of multiple sclerosis in Canada: a longitudinal study. Lancet Neurol. 5, 932–936.
Ozacmak, V.H., Sayan, H., 2009. The effects of 17beta estradiol, 17alpha estradiol and progesterone on oxidative stress biomarkers in ovariectomized female rat brain subjected to global cerebral ischemia. Physiol. Res. 58, 909–912.
Palaszynski, K.M., Liu, H., Loo, K.K., Voskuhl, R.R., 2004. Estriol treatment ameliorates disease in males with experimental autoimmune encephalomyelitis: implications for multiple sclerosis. J. Neuroimmunol. 149, 84–89.
Palaszynski, K.M., Loo, K.K., Ashouri, J.F., Liu, H., Voskuhl, R.R., 2004. Androgens are protective in experimental autoimmune encephalomyelitis: implications for multiple sclerosis. J. Neuroimmunol. 146, 144–152.
Papenfuss, T.L., Roger, C.J., Glenapp, I., Yurris, M., McClain, M., Damico, N., Valo, J., Song, F., Whitacre, C.C., 2004. Sex differences on experimental autoimmune encephalomyelitis in multiple mouse strains. J. Neuroimmunol. 150, 59–69.
Park, S.K., Kim, J.M., Kim, J.H., Cheong, J.H., Bak, K.H., Kim, C.H., 2008. Aneurysmal subarachnoid hemorrhage in young adults: a gender comparison study. J Clin. Neurosci. 15, 389–392.
Pitkänen, A., Immonen, R.J., Gröhn, O.H., Kharatishvili, I., 2009. From traumatic brain injury to posttraumatic epilepsy, what animal models tell us about the process and treatment options. Epilepsia 50 (Suppl. 2), 21–29.
Polanczyk, M.J., Hopke, C., Huan, J., Vandenbark, A.A., Offner, H., 2005. Enhanced FoxP3 expression and Treg cell function in pregnant and estrogen-treated mice. J. Neuroimmunol. 170, 85–92.
Quigg, M., Smithson, S.D., Fowler, K.M., Sursal, T., Herzog, A.G. and the Progesterone Trial Study Group, 2009. Laterality and location influence catamenial seizure expression in women with partial epilepsy. Neurology 73, 223–227.
Rangachari, M., Nauermann, N., Marty, R.R., Dirnhofer, S., Kurrer, M.O., Komnenovic, V., Penninger, J.M., Eriksson, U., 2006. T-bet negatively regulates autoimmune myocarditis by suppressing local production of interleukin 17. J. Exp. Med. 203, 2009–2019.
Reddy, D.S., 2004a. Anticonvulsant activity of the testosterone-derived neurosteroid 3α-androstanediol. Neuroreport 15, 515–518.
Reddy, D.S., 2004b. Testosterone modulation of seizure susceptibility is mediated by neurosteroids 3α-androstanediol and 17β-estradiol. Neuroscience 129, 195–207.
Reddy, D.S., 2009a. The role of neurosteroids in the pathophysiology and treatment of catamenial epilepsy. Epilepsy Res. 85, 1–30.
Reddy D.S., 2009. Steroid hormones and sex differences in seizure susceptibility. In: Schwartzkroin, P. (Ed.), Encyclopedia of Basic Epilepsy Research, vol. 1. Academic Press, Oxford, pp. 526–533.
Reddy, D.S., 2010a. Neurosteroids: endogenous role in the human brain and therapeutic potentials. Prog. Brain Res. 186, 113–137.
Reddy, D.S., 2010b. Clinical pharmacokinetic interactions between antiepileptic drugs and hormonal contraceptives. Expert Rev. Clin. Pharmacol. 3 (2), 183–192.
Reddy, D.S., 2013a. Neuroendocrine aspects of catamenial epilepsy. Horm. Behav. 63, 254–266.
Reddy, D.S., 2013b. Role of hormones and neurosteroids in epileptogenesis. Front. Cell. Neurosci. 7 (115), 1–20.
Reddy, D.S., 2014. Neurosteroids and their role in sex-specific epilepsies. Neurobiol. Dis. 72, 198–209.
Reddy, D.S., Jian, K., 2010. The testosterone-derived neurosteroid androstanediol is a positive allosteric modulator of GABAA receptors. J. Pharmacol. Exp. Ther. 334, 1031–1041.
Reddy, D.S., Kuruba, R., 2013. Experimental models of status epilepticus and neuronal injury for evaluation of therapeutic interventions. Int. J. Mol. Sci. 14 (9), 18284–18318.
Reddy, D.S., Mohan, A., 2011. Development and persistence of limbic epileptogenesis are impaired in mice lacking progesterone receptors. J. Neurosci. 31, 650–658.
Reddy, D.S., Ramanathan, G., 2012. Finasteride inhibits the disease-modifying activity of progesterone in the hippocampus kindling model of epileptogenesis. Epilepsy Behav. 25, 92–97.

Reddy, D.S., Rogawski, M.A., 2000a. Enhanced anticonvulsant activity of ganaxolone after neurosteroid withdrawal in a rat model of catamenial epilepsy. J. Pharmacol. Exp. Ther. 294, 909–915.

Reddy, D.S., Rogawski, M.A., 2000b. Chronic treatment with the neuroactive steroid ganaxolone in the rat induces anticonvulsant tolerance to diazepam but not to itself. J. Pharmacol. Exp. Ther. 295, 1241–1248.

Reddy, D.S., Rogawski, M.A., 2001. Enhanced anticonvulsant activity of neuroactive steroids in a rat model of catamenial epilepsy. Epilepsia 42, 303–310.

Reddy, D.S., Rogawski, M.A., 2009. Neurosteroid replacement therapy for catamenial epilepsy. Neurotherapeutics 6, 392–401.

Reddy, D.S., Zeng, Y.C., 2007. Effect of neurosteroid withdrawal on spontaneous recurrent seizures in a rat model of catamenial epilepsy. FASEB J. 21 (6), A1179–A11179.

Reddy, D.S., Kim, H.Y., Rogawski, M.A., 2001. Neurosteroid withdrawal model of perimenstrual catamenial epilepsy. Epilepsia 42, 328–336.

Reddy, D.S., Castaneda, D.C., O'Malley, B.W., Rogawski, M.A., 2004. Anticonvulsant activity of progesterone and neurosteroids in progesterone receptor knockout mice. J. Pharmacol. Exp. Ther. 310 (1), 230–239.

Reddy, D.S., Gangisetty, O., Briyal, S., 2010. Disease-modifying activity of progesterone in the hippocampus kindling model of epileptogenesis. Neuropharmacology 59, 573–581.

Reddy, D.S., Gould, J., Gangisetty, O., 2012. A mouse kindling model of perimenstrual catamenial epilepsy. J. Pharmacol. Exp. Ther. 341, 784–793.

Rinta, S., Airas, L., Elovaara, I., 2010. Is the modulatory effect of pregnancy in multiple sclerosis associated with changes in blood apoptotic molecules? Acta Neurol. Scand. 122, 168–174.

Roger, V.L., Go, A.S., Lloyd-Jones, D.M., 2011. Heart disease and stroke statistics – 2011 update: a report from the American Heart Association. Circulation 123, e18–e209.

Rojas, A., Jiang, J., Ganesh, T., Yang, M.S., Lelutiu, N., Gueorguieva, P., Dingledine, R., 2014. Cyclooxygenase-2 in epilepsy. Epilepsia 55, 17–25.

Rosamond, W., Flegal, K., Friday, G., Furie, K., Go, A., Greenlund, K., Haase, N., Ho, M., Howard, V., Kissela, B., Kittner, S., Lloyd-Jones, D., McDermott, M., Meigs, J., Moy, C., Nichol, G., O'Donnell, C.J., Roger, V., Rumsfeld, J., Sorlie, P., Steinberger, J., Thom, T., Wasserthiel-Smoller, S., Hong, Y., 2007. Heart disease and stroke statistics – 2007 update: a report from the American Heart Association Statistics Committee and Stroke Statistics Subcommittee. Circulation 115, e69–e171.

Sakata, A., Mogi, M., Iwanami, J., Tsukuda, K., Min, L.J., Jing, F., Ohshima, K., Ito, M., Horiuchi, M., 2011. Female type 2 diabetes mellitus mice exhibit severe ischemic brain damage. J. Am. Soc. Hypertens. 5, 7–11.

Seifert, H.A., Hall, A.A., Chapman, C.B., Collier, L.A., Willing, A.E., Pennypacker, K.R., 2012. A transient decrease in spleen size following stroke corresponds to splenocyte. J. Neuroimmune Pharmacol. 7 (4), 1017–1024.

Selvamani, A., Sohrabji, F., 2010a. Reproductive age modulates the impact of focal ischemia on the forebrain as well as the effects of estrogen treatment in female rats. Neurobiol. Aging 31, 1618–1628.

Selvamani, A., Sohrabji, F., 2010b. The neurotoxic effects of estrogen on ischemic stroke in older female rats is associated with age-dependent loss of insulin-like growth factor-1. J. Neurosci. 30, 6852–6861.

Shao, B., Cheng, Y., Jin, K., 2012. Estrogen, neuroprotection and neurogenesis after ischemic stroke. Curr. Drug Targets 13, 188–198.

Sicotte, N.L., Liva, S.M., Klutch, R., et al., 2002. Treatment of multiple sclerosis with the pregnancy hormone estriol. Ann. Neurol. 52, 421–428.

Sieve, A.N., Steelman, A.J., Young, C.R., Storts, R., Welsh, T.H., Welsh, C.J.R., Meagher, M.W., 2004. Chronic restraint stress during early Theiler's virus infection exacerbates the subsequent demyelinating disease in SJL mice. J. Neuroimmunol. 155, 103–118.

Silva, G.S., Lima, F.O., Camargo, E.C., Smith, W.S., Lev, M.H., Harris, G.J., Halpern, E.F., Koroshetz, W., Furie, K.L., 2010. Gender differences in outcomes after ischemic stroke: role of ischemic lesion volume and intracranial large-artery occlusion. Cerebrovasc. Dis. 30, 470–475.

Simpkins, J.W., Rajakumar, G., Zhang, Y.Q., Simpkins, C.E., Greenwald, D., Yu, C.J., Bodor, N., Day, A.L., 1997. Estrogens may reduce mortality and ischemic damage caused by middle cerebral artery occlusion in the female rat. J. Neurosurg. 87, 724–730.

Simpkins, J.W., Yi, K.D., Yang, S.H., Dykens, J.A., 2010. Mitochondrial mechanisms of estrogen neuroprotection. Biochim. Biophys. Acta 1800, 1113–1120.

Singh, M., Su, C., 2013. Progesterone-induced neuroprotection: factors that may predict therapeutic efficacy. Brain Res. 1514, 98–106.

Smith, R., Studd, J.M., 1992. A pilot study of the effect upon multiple sclerosis of the menopause, hormone replacement therapy and menstrual cycle. J. R. Soc. Med. 85, 612–613.

Smith, S.S., Shen, H., Gong, Q.H., Zhou, X., 2007. Neurosteroid regulation of GABA-A receptors: focus on the alpha4 and delta subunits. Pharmacol. Ther. 116, 58–76.

Smith-Bouvier, D.L., Divekar, A.A., Sasidhar, M., Du, S., Tiwari-Woodruff, S.K., King, J.K., Arnold, A.P., Singh, R.R., Voskuhl, R.R., 2008. A role for sex chromosome complement in the female bias in autoimmune disease. J. Exp. Med. 205, 1099–1108.

Sohrabji, F., Bake, S., 2006. Age-related changes in neuroprotection: is estrogen pro-inflammatory for the reproductive senescent brain? Endocrine 29, 191–197.

Soldan, S., Alvarez Retuerto, A., Sicotte, N., Voskuhl, R., 2003. Immune modulation in multiple sclerosis patients treated with the pregnancy hormone estriol. J. Immunol. 171, 6267–6274.

Sospedra, M., Martin, R., 2005. Immunology of multiple sclerosis. Annu. Rev. Immunol. 23, 683–747.

Soucy, G., Boivin, G., Labrie, F., Rivest, S., 2005. Estradiol is required for a proper immune response to bacterial and viral pathogens in the female brain. J. Immunol. 174, 6391–6398.

Spence, R.D., Voskuhl, R.R., 2012. Neuroprotective effects of estrogens and androgens in CNS inflammation and neurodegeneration. Front. Neuroendocrinol. 33, 105–115.

Spence, R.D., Hamby, M.E., Umeda, E., Itoh, N., Du, S., Wisdom, A.J., Cao, Y., Bondar, G., Lam, J., Ao, Y., Sandoval, F., Suriany, S., Sofroniew, M.V., Voskuhl, R.R., 2011. Neuroprotection mediated through estrogen receptor-α in astrocytes. Proc. Natl. Acad. Sci. 108, 8867–8872.

Spence, R.D., Wisdom, A.J., Cao, Y., Hill, H.M., Mongerson, C.R.L., Stapornkul, B., Itoh, N., Sofroniew, M.V., Voskuhl, R.R., 2013. Estrogen mediates neuroprotection and anti-inflammatory effects during EAE through ERα signaling on astrocytes but not through ERβ signaling on astrocytes or neurones. J. Neurosci. 33, 10924–10933.

Stein, D.G., 2011. Progesterone in the treatment of acute traumatic brain injury: a clinical perspective and update. Neuroscience 191, 101–106.

Stuart, M., Bergstrom, L., 2011. Pregnancy and multiple sclerosis. J. Midwifery Womens Health 56, 41–47.

Sutton, C., Brereton, C., Keogh, B., Mills, K.H.G., Lavelle, E.C., 2006. A crucial role for interleukin (IL)-1 in the induction of IL-17-producing T cells that mediate autoimmune encephalomyelitis. J. Exp. Med. 203, 1685–1691.

Suuronen, T., Nuutinen, T., Huuskonen, J., Ojala, J., Thornell, A., Salminen, A., 2005. Anti-inflammatory effect of selective estrogen receptor modulators (SERMs) in microglial cells. Inflamm. Res. 54, 194–203.

Theiler, M., 1934. Spontaneous encephalomyelitis of mice – a new virus. Science 80, 122.

Thompson, C., Powrie, F., 2004. Regulatory T cells. Curr. Opin. Pharmacol. 4, 408–414.

Tiwari-Woodruff, S., Voskuhl, R., 2009. Neuroprotective and anti-inflammatory effects of estrogen receptor ligand treatment in mice. J. Neurol. Sci. 286, 81–85.

Tiwari-Woodruff, S., Morales, L.B.J., Lee, R., Voskuhl, R.R., 2007. Differential neuroprotective and anti-flammatory effects of estrogen receptor (ER) alpha and ER beta ligand treatment. Proc. Natl. Acacd. Sci. 104, 14813–14818.

Toung, T.K., Hurn, P.D., Traystman, R.J., Sieber, F.E., 2000. Estrogen decreases infarct size after temporary focal ischemia in a genetic model of type 1 diabetes mellitus. Stroke 31, 2701–2706.

Towfighi, A., Saver, J.L., Engelhardt, R., Ovbiagele, B., 2007. A midlife stroke surge among women in the United States. Neurol. 69, 1898–1904.

Tuveri, A., Paoletti, A.M., Orrù, M., Melis, G.B., Marotto, M.F., Zedda, P., Marrosu, F., Sogliano, C., Marra, C., Biggio, G., Concas, A., 2008. Reduced serum level of THDOC, an anticonvulsant steroid, in women with perimenstrual catamenial epilepsy. Epilepsia 49, 1221–1229.

Valor, L., Teijeiro, R., Aristimuño, C., Faure, F., Alonso, B., de Andres, C., Tejera, M., Lopez-Lazareno, N., Fernandez-Cruz, E., Sanchez-Ramon, S., 2011. Estradiol-dependent perforin expression by human regulatory T-cells. Eur. J. Clin. Invest. 41, 357–364.

Vegeto, E., Benedusi, V., Maggi, A., 2008. Estrogen anti-inflammatory activity in brain: a therapeutic opportunity for menopause and neurodegenerative diseases. Front. Neuroendocrinol. 29, 507–519.

Verrotti, A., Latini, G., Manco, R., De Simone, M., Chiarelli, F., 2007. Influence of sex hormones on brain excitability and epilepsy. J. Endocrinol. Invest. 30 (9), 797–803.

Vezzani, A., Aronica, E., Mazarati, A., Pittman, Q.J., 2013. Epilepsy and brain inflammation. Exp. Neurol. 244, 11–21.

Viglietta, V., Baechder-Allan, C., Weiner, H.L., Hafler, D.A., 2004. Loss of functional suppression by CD4+ CD25+ regulatory T cells in patients with multiple sclerosis. J. Exp. Med. 199, 1285–1291.

Voskuhl, R.R., Pitchekian-Halabi, H., MacKenzie-Graham, A., McFarland, H.F., Raine, C.S., 1996. Gender differences in autoimmune demyelination in the mouse: implications for multiple sclerosis. Ann. Neurol. 39, 724–733.

Wassertheil-Smoller, S., Hendrix, S.L., Limacher, M., Heiss, G., Kooperberg, C., Baird, A., Kotchen, T., Curb, J.D., Black, H., Rossouw, J.E., Aragaki, A., Safford, M., Stein, E., Laowattana, S., Mysiw, W.J., WHI Investigators, 2003. Effect of estrogen plus progestin on stroke in postmenopausal women: the Women's Health Initiative: a randomized trial. JAMA 289, 2673–2684.

Weinshenker, B.G., 1994. Natural history of multiple sclerosis. Ann. Neurol. 36 (Suppl.), S6–S11.

Welsh, C.J.R., Tonks, P., Borrow, P., Nash, A.A., 1990. Theiler's virus: an experimental model of virus-induced demyelination. Autoimmunity 6, 105–112.

Whitacre, C.C., 2001. Sex differences in autoimmune disease. Nat. Immunol. 2, 777–780.

Wingerchuk, D.M., Lucchinetti, C.F., Noseworthy, J.H., 2001. Multiple sclerosis: current pathophysiological concepts. Lab. Invest. 81, 263–281.

Wu, X., Gangisetty, O., Carver, C.M., Reddy, D.S., 2013. Estrous cycle regulation of extrasynaptic δ-containing GABAA receptor-mediated tonic inhibition and limbic epileptogenesis. J. Pharmacol. Exp. Ther. 346, 146–160.

Yang, S.H., Liu, R., Wu, S.S., Simpkins, J.W., 2003. The use of estrogens and related compounds in the treatment of damage from cerebral ischemia. Ann. N.Y. Acad. Sci. 1007, 101–107.

Yates, M.A., Li, Y., Chlebeck, P., Proctor, T., Vandenbark, A.A., Offner, H., 2010. Progesterone treatment reduces disease severity and increases IL-10 in experimental autoimmune encephalomyelitis. J. Neuroimmunol. 220, 136–139.

Young, E.E., Sieve, A.N., Vichaya, E.G., Carcoba, L.M., Young, C.R., Ambrus, A., Storts, R., Welsh, C.J.R., Meagher, M.W., 2010. Chronic restraint stress during early Theiler's virus infection exacerbates the subsequent demyelinating disease in SJL mice: II. CNS disease severity. J. Neuroimmunol. 220, 79–89.

Zhang, B., Subramanian, S., Dziennis, S., Jia, J., Uchida, M., Akiyoshi, K., Migliati, E., Lewis, A.D., Vandenbark, A.A., Offner, H., Hurn, P.D., 2010. Estradiol and G1 reduce infarct size and improve immunosuppression after experimental stroke. J. Immunol. 184, 4087–4094.

Zhang, Z., Yang, R., Cai, W., Bai, Y., Sokabe, M., Chen, L., 2010. Treatment with progesterone after focal cerebral ischemia suppresses proliferation of progenitor cells but enhances survival of newborn neurons in adult male mice. Neuropharmacology 58 (6), 930–939.

CHAPTER 13

# Steroid Hormone Signaling Pathways and Sex Differences in Neuroendocrine and Behavioral Responses to Stress

**Mario G. Oyola*, Anna M. Malysz**, Shailaja K. Mani*,**, Robert J. Handa†**
*Baylor College of Medicine, Department of Neuroscience, Houston, TX, USA
**Baylor College of Medicine, Department of Molecular and Cellular Biology, Houston, TX, USA
†University of Arizona College of Medicine, Department Basic Medical Sciences, Phoenix, AZ, USA

## 1 INTRODUCTION

### 1.1 Overview of sex differences in neuroendocrine function

Organisms do not live in isolation from the outside environment. Their survival depends upon their ability to sense diverse external stimuli, which are integrated within the hypothalamus and ultimately trigger behavioral, autonomic, or neuroendocrine responses that allow for the organism's adaptation to an ever-changing world. Stressful events, whether real or perceived, redirect the physiology and behavior of the organism toward immediate survival and maintenance of homeostasis by a coordinated and well-orchestrated activation of the neuroendocrine and autonomic nervous systems. Activation of these pathways is crucial for animal survival during stressful situations, as they stimulate the "fight-or-flight" response that heightens the animal's sensory and motor abilities and mobilizes stored energy sources, thereby allowing it to cope with threatening situations. During an acute stressful episode, mobilization of energy stores and cognitive processes are enhanced, whereas appetite, digestion, bodily repair, pain perception, and reproduction are suppressed. Although this system is highly capable of homeostatic recovery following the cessation of the stressor, chronic activation of the stress circuitry can lead to detrimental outcomes. The inability to reinstate homeostasis following exposure to stress can ultimately lead to disease states such as immunodeficiency, memory impairment, and obesity, among others (Sapolsky et al., 2000).

Hypothalamic homeostatic responses are evolutionarily conserved and many of them are sex specific. Males and females respond differentially to the same physical or psychological threats. This may in part be due to differential sensory processing in males and females. Ultimately, the sexually dimorphic responses are controlled by differing concentrations of circulating hormones occurring either in adulthood or during development, and are driven by steroid-specific actions on the neuroendocrine stress-response system, the hypothalamo–pituitary–adrenal (HPA) axis, a brain region that integrates neural

*Sex Differences in the Central Nervous System.* http://dx.doi.org/10.1016/B978-0-12-802114-9.00013-5

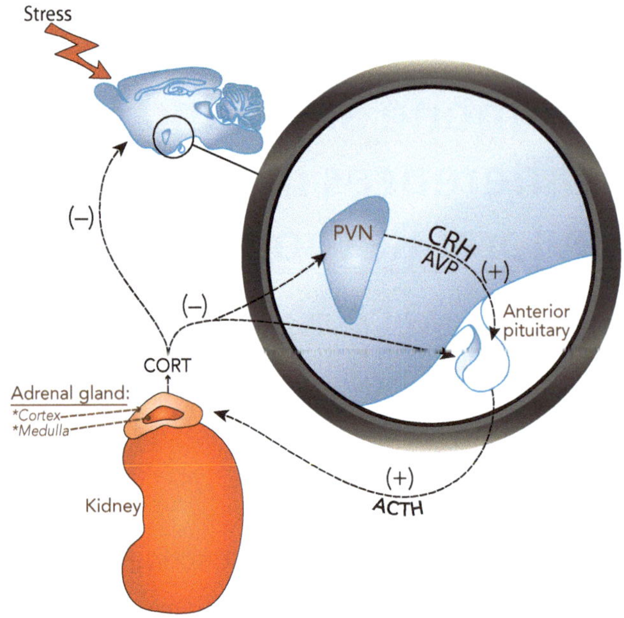


**Figure 13.1** ***Schematic diagram of the hypothalamo–pituitary–adrenal (HPA) axis response to stress.*** Stressful events activate the HPA axis by relaying information through neurons residing in the PVN of the hypothalamus, an integratory node for multiple positive and negative inputs. In response, the neurons of the PVN synthesize and secrete corticotropin releasing hormone (CRH) into the hypophyseal portal system. CRH action in the pituitary is enhanced by other hypothalamic factors, such as arginine vasopressin (AVP). CRH secretion enhances the synthesis and release of adrenocorticotropic hormone (ACTH) from the anterior pituitary, which acts on the adrenal cortex to stimulate the synthesis and secretion of corticosterone (CORT). Circulating CORT subsequently acts at the level of the pituitary, hypothalamus (PVN), and limbic–cortical brain areas to limit further hormone secretion. PVN, paraventricular nucleus of the hypothalamus; CRH, corticotropin releasing hormone; AVP, arginine vasopressin; ACTH, adrenocorticotropic hormone; CORT, corticosterone.

and humoral signals from both the environment and the circadian pacemaker. Stressful events activate the HPA axis by relaying information through neurons residing in the paraventricular nucleus of the hypothalamus (PVN), an integratory node for multiple positive and negative inputs (Figure 13.1). In response, the neurons of the parvocellular region of the PVN synthesize and secrete corticotropin-releasing hormone (CRH) into the hypophyseal portal system. This secretion, in turn, enhances the synthesis and release of adrenocorticotropic hormone (ACTH) from corticotrophs of the anterior pituitary. CRH action in the pituitary is enhanced by other hypothalamic factors such as arginine vasopressin (AVP) and oxytocin (OT). ACTH acts on the adrenal cortex to stimulate the synthesis and secretion of glucocorticoids (corticosterone (CORT) in rodents or cortisol in humans). Circulating CORT subsequently acts at the level of the pituitary, hypothalamus, and limbic–cortical brain areas (hippocampus, prefrontal cortex, and amygdala) to limit further hormone secretion. Glucocorticoids have profound effects on all tissues and organs throughout the body and influence fat metabolism and muscle catabolism, induce lipolysis, proteolysis, immune-redistribution, and immunosuppression, thus aiding survival during stressful episodes and limiting nonessential activity to only the bare minimum (Buchanan, 2000; Husak and Moore, 2008).

## 1.2 Activational and organizational actions of steroid hormones

Gonadal steroids are instrumental in influencing sex-specific differences in the stress response. Fluctuations in the levels of these steroids during a critical developmental period organize the brain by regulating gene expression, epigenetics, protein synthesis, and cellular excitability (Bingham et al., 2011; McCormick and Mahoney, 1999), resulting in permanent changes in male and female adult brain function. In addition, activational effects, those direct actions of hormones on transitory cellular function that occur only when the hormone is present, can be influenced by the underlying organizational differences that lead to sex-specific hormone sensitivity, physiology, and behavior. The organizational effects of steroids on the brain are driven by gonadal hormone fluctuations that take place during a critical period or window of susceptibility, usually during the period of early brain development. These effects are permanent, and the physiological and behavioral effects that follow in adulthood cannot be simply reversed by hormone manipulation. Gonadal steroid hormones have pronounced effects on the HPA axis in adulthood, and the neonatal steroid environment may also influence some of these effects. It has been shown that estrogens increase and androgens decrease reactivity of the HPA axis (Burgess and Handa, 1992) in an activational manner, whereas organizational effects on the HPA axis have also been reported (Seale et al., 2004a,b; Seale et al., 2005a,b).

## 1.3 Basic mechanisms of sexual differentiation of the brain

During development, gonadal hormones serve as regulatory elements for the sexual differentiation of the brain. Evidence for this phenomenon arises from early studies by Phoenix et al., who demonstrated that sex differences in behavior, and by expansion, in the brain, are driven by testosterone (T), the primary androgen secreted by the testes (Phoenix et al., 1959). The presence of the liver-derived estrogen-binding protein, α-fetoprotein, in the bloodstream of pregnant dams not only prevents masculinization of the females from their mother's estrogens, but also allows testis-derived T to enter neurons of male brains, where it is locally converted to estradiol within distinct regions, by the enzyme aromatase. Thus, aromatization, the process by which the enzyme aromatase converts T to estradiol, also plays an important role in sexual differentiation during the end of gestation and shortly after birth (Kudwa et al., 2006). Neural aromatase responsiveness not only varies in different regions of the brain, but also is sexually differentiated (Roselli et al., 2009). These observations laid the framework for subsequent studies determining how changes in gonadal hormone levels contribute to sex differences in the phenotype of diverse, nongonadal tissues and at all levels of analysis from behavioral to cellular and molecular (Tsai et al., 2009; McCarthy and Nugent, 2013). Similar to their effects on perinatal organization, gonadal hormone surges during puberty maintain and accentuate sexually dimorphic behaviors during adolescence and into adulthood (Ahmed et al., 2008). Increasing evidence supports the notion that sex differences are not exclusively driven by gonadal hormones, and can occur prior to gonadal differentiation and even before plasma T levels are found at sexually dimorphic levels (Arnold, 2009). Studies indicate that these nongonadal hormone-driven factors contributing to phenotypical sex differences in the male and female brain can be attributed to the primary actions of genes encoded by the X and Y sex chromosomes (Arnold, 1996; Wade and Arnold, 1996; McCarthy and Arnold, 2011). Sex-specific effects of the environment also play a critical factor in gender-based phenotypes, especially in humans. Thus, steroid hormones, factors encoded by sex chromosomes, nongonadal factors downstream of sex chromosome genes, and sex-specific environments, interact with each other to cause sex differences in the brain. The specific actions of steroid hormones and the enzymes involved in the biosynthesis of steroid hormones in the response to stress are discussed in later sections.

# 2 MECHANISMS OF STEROID HORMONE ACTION

## 2.1 Overview of steroid hormone action

Steroid hormones, androgens, estrogens, progestins, and corticosteroids regulate a wide variety of functions in living organisms resulting in changes in physiology and behaviors. In addition to their effects on reproduction, these hydrophobic molecules regulate diverse biological processes, including development, metabolism, homeostasis, immune responses, reproductive and nonreproductive behaviors, etc. Steroid hormones exert their physiological effects in a tissue- and region-specific manner primarily by binding to their cognate receptors specific for each class of hormones. Upon hormone binding in the cytoplasm, the receptors dissociate from chaperone, and heat shock proteins (HSP) undergo conformational changes, leading to dimerization, nuclear translocation, DNA binding (direct or indirect), and interactions with the general transcriptional machinery assisted by coactivator molecules to initiate gene transcription (Beato, 1989; Beato and Klug, 2000; Tyagi et al., 2000) (see Figure 13.2).

Steroid hormone receptors belong to a superfamily of nuclear receptor proteins and are classified based on their mechanisms of action (types I to IV) and the DNA sequences to which they bind (subfamilies 0–6). The steroid receptor superfamily contains receptors for thyroid hormones, vitamin D, retinoic acid, nerve growth factor 1B, and steroidogenic factor 1, among others, as well as for androgens, estrogens, corticosteroids, and progestins. Of importance for this discussion, the steroid hormone receptors for androgens, estrogens, corticosteroids, and progestins have been placed into nuclear receptor subfamily three (estrogen receptor-like), which regulates transcription through interactions with DNA response elements consisting of two inverted repeats, a nucleotide sequence that is an inverted complement of another downstream sequence, both of which are separated by a variable number of nucleotides.

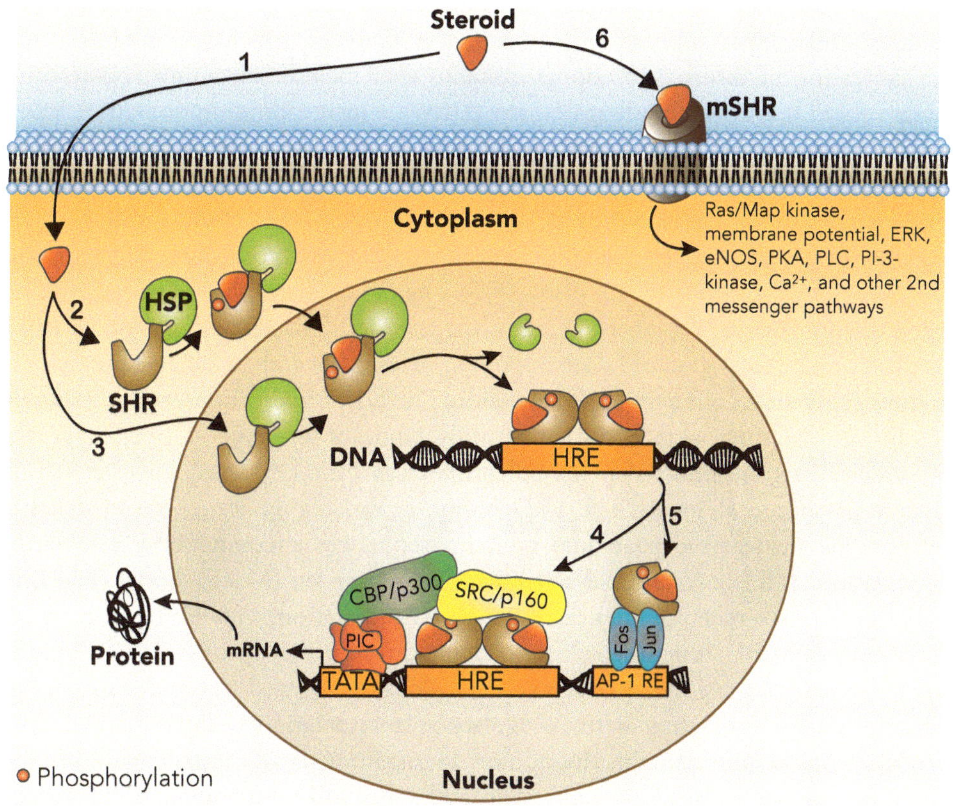


**Figure 13.2** ***Schematic representation of steroid hormone action in target cells.*** In the classical steroid hormone action model (1) steroid hormones freely diffuse through the plasma membrane lipid bilayer of the cell and (2) bind to the inactive steroid hormone receptor (SHR)-heat shock protein (HSP) complex in the cytoplasm, or (3) bind to SHR complex within the nucleus. The binding of the steroid hormone to SHR induces a conformational change and dissociates the HSP from the complex, leading to SHR phosphorylation and nuclear translocation. (4) Within the nucleus, the liganded SHRs undergo dimerization, bind to the hormone regulatory element (HRE) in the target DNA and recruit steroid receptor coactivator (SRC/160) family of coactivators, cointegrator cyclic adenosine monophosphate response element binding protein (CBP), and other general transcription factors to produce a transcriptionally active, protein initiation complex (PIC). PIC binds to the TATA box of the promoter DNA to initiate transcription (mRNA) and translation (protein). (5) Gene expression can also be regulated by the steroid-bound SHR, which can act as a coactivator or corepressor, by its interaction with other DNA-bound transcription factors (e.g., Fos, Jun) at the AP-1 response element (AP-1 RE). (6) Steroid hormones can also elicit actions through membrane-bound SHR (mSHR). This mechanism alters membrane potential and secondary messenger pathways by interacting with SHR tethered to a membrane, G-protein coupled receptors (GPCRs), or ion channels.

The organization of the structural and functional domains is similar between the members of the superfamily of nuclear receptors. The receptors have a modular structure consisting of distinct functional domains that include an amino (N)-terminal domain (NTD), a DNA binding domain (DBD), a ligand-binding domain (LBD), a flexible "hinge region" separating the LBD and DBD, and a protease-sensitive carboxyl (C)-terminal domain (Green et al., 1986; Kumar et al., 1986; Warnmark et al., 2003). Each of the domains has specialized subdomains controlling transactivation, nuclear localization, gene transcription, heat shock protein (HSP) interactions, and transcriptional silencing (Lavery and McEwan, 2005). The NTD, which varies in sequence between various nuclear receptors, contains the transactivation function (AF-1), which recruits coregulators and core transcriptional machinery and modulates the cell and promoter specificity of target gene activation. The DBD, highly conserved among the members of the nuclear receptor superfamily, contains two type II zinc finger motifs, which recognize and target the receptors to specific DNA sequences called the hormone response elements. The LBD region is also highly conserved in structure between the nuclear receptors and contains sequences encoding HSP association, dimerization, and transcriptional activation and repression (Maeda, 2001). The LBD is able to act as a receptor silencer through its interaction with HSP, masking and unmasking the LBD functional domain depending on the ligand-receptor binding state. A second activation function (AF-2) is located in the LBD and assists robust transcription by interacting with coregulators, in association with AF-1. The small, variable, hinge region contains the nuclear localization signal sequence and plays an important role in intracellular trafficking and subcellular distribution. The variable C-terminal domain has specialized functions and contains retention signals for protein sorting.

While early studies suggested that the steroid receptors are found in their unbound state in the cytoplasm, later reports indicate that receptors for estrogens, androgens, and progestins can also be found in the nucleus in the unliganded state (Price et al., 2001; King and Greene, 1984; Perrot-Applanat et al., 1986; Husmann et al., 1990). Nonetheless, glucocorticoid and mineralocorticoid receptors (GRs and MRs) have been localized in the unbound state in the cytoplasm stabilized by the HSP complex, which blocks the LBD and nuclear localization signal, rendering the receptor complexes inactive (Nishi and Kawata, 2006; Pratt et al., 2006). The receptor–DNA interaction is critical for the activation or repression of gene transcription and is conserved throughout the nuclear receptor family 3 group C (NR3C) subgroup. Of note, coregulatory proteins and histone acetylation/deacetylation play a key role in the regulation of gene transcription. While acetylation disassembles and exposes DNA, deacetylation tightly coils the DNA, making it inaccessible to the transcription machinery, and hence, serves as a transcription repressor. Furthermore, transcription factors can selectively interact with steroid receptors to influence gene activity in a cell-specific manner (Diamond et al., 1990). Finally, steroid receptor splice variants (Price et al., 2001) and posttranslational modifications, like phosphorylation (Orti et al., 1992), can alter the receptor's function and further gene transcription. Nuclear receptors, their targets, ligands and coregulators, and signaling pathway databases are accessible at the Nuclear Receptor Atlas consortium: http://www.nursa.org/nursa/index.jsf.

The "classical" view that intracellular steroid receptors function as ligand-dependent transcription factors to activate or repress target gene expression has undergone modifications to include rapid, nonclassical mechanisms of hormone action. Secondary messengers and cytosolic signaling cascades mediate these effects, initiated at the membrane. Studies describing fast actions of steroid hormones depict rapid changes in intracellular calcium levels (Chaban et al., 2011) and second messenger pathways (Kelly and Ronnekleiv, 2012), and have been shown to ultimately participate in gene transcription (Vasudevan et al., 2005). Rapid actions have been demonstrated for all the members of the nuclear receptor subfamily (Foradori et al., 2008; Tabori et al., 2005) and could also involve cross-talk between various signaling transduction pathways (Kousteni et al., 2001; Wu et al., 2005) In addition, a number of studies in the last few decades have also demonstrated that intracellular nuclear receptors can be activated by factors other than their cognate ligands. Although this mechanism, known as "ligand-independent activation," has been reported for the activation of nuclear receptors (androgen, progestin, and estrogen) by cyclic nucleotides, neurotransmitters, and growth factors (Mani and Portillo, 2010), its role in modulation of the HPA axis has not yet been demonstrated.

## 2.2 Androgen receptors

Androgen receptors (ARs), members of the ligand-activated transcription factors of the nuclear receptor superfamily, are involved in development, brain differentiation, and physiological responses to a variety of stimuli (Evans, 1988). Although there are two ligands for the AR, only one receptor has been identified and cloned (Brinkmann, 2011). T and 5α-dihydrotestosterone (DHT), the main ligands, have inhibitory actions on the HPA axis. While both bind to the AR, DHT binds to AR with twofold greater affinity, and functions as a more potent transcriptional activator (Brinkmann, 2011; Grino et al., 1990). Furthermore, the disassociation rate of T from the AR is fivefold greater than that of DHT (Grino et al., 1990). Following synthesis from cholesterol in the Leydig cells of the testis, T reaches its target cell and binds the AR directly or indirectly through its conversion to DHT by the 5α-reductase enzyme. Of importance, unliganded ARs are mostly found in the cytoplasm (Tyagi et al., 2000). Studies, however, have detected AR in extranuclear regions of neurons and glial processes, where they could play an important role in the rapid effects of androgens (Sarkey et al., 2008; DonCarlos et al., 2006).

### 2.2.1 Androgen receptor location in the brain

Androgens are critical for the expression of male reproductive phenotypes, and ARs are present in several brain regions involved in the regulation of these male-specific behaviors, such as the medial preoptic area (POA), ventromedial nucleus (VMN), and arcuate nucleus (Sar and Stumpf, 1973, 1977; Handa et al., 1986; McGinnis et al., 1983; Roselli, 1991). Moreover, the presence of ARs has also been reported in limbic regions, which are involved in the behavioral response to stress (Simerly et al., 1990). Important for studies examining androgen regulation of stress responses, ARs are minimally present in the PVN, suggesting that androgenic signaling does not work directly in this brain region for the regulation of the HPA axis. ARs have been detected proximal to the PVN of the hypothalamus, surrounding CRH-expressing cells, where implants of DHT have been shown to decrease the activation of the HPA axis triggered by orchiectomy (Handa et al., 2008; Lund et al., 2006, 2004). Therefore, inhibition of the HPA axis might be due to the actions of androgens indirectly on AR-positive, extra-hypothalamic regions (i.e., bed nucleus of the stria terminalis (BnST), lateral septum, medial amygdala, hippocampus, and cortex).

### 2.2.2 Androgen effects on gene expression

In the early 1970s, Gaskin and Kitay reported that prepubertal castration of male rats and hamsters increases the level of the adrenal hormone, CORT, in response to stressors, suggesting T as a potential modulator of the stress response (Gaskin and Kitay, 1970). It is known that androgens can interact with the regulatory elements of the HPA axis, as studies in adult rats have shown that orchiectomy results in higher levels of ACTH and CORT, and that T replacement decreases this response (Handa et al., 1994b; McCormick et al., 1998; Mitsushima et al., 2008; Viau and Meaney, 1996). Furthermore, a 30 min restraint stressor significantly increases the peak and prolongs the return to baseline of plasma levels of CORT and ACTH, and the expression levels of CRH and AVP in the PVN in gonadectomized male rats, compared with those of intact control males (Seale et al., 2004a; Viau et al., 2003) (Williamson and Viau, 2008). In addition to T, the nonaromatizable androgen, DHT, also reduces stress-induced c-Fos messenger (m)RNA expression in the PVN of gonadectomized rats, suggesting that inhibitory effects of androgens on the HPA axis are not due to its aromatization to estradiol (Lund et al., 2006, 2004; Viau et al., 2003; Kerr et al., 1996). Interestingly, the magnitude and duration of stress response also changes during development, with prepubertal males exhibiting a prolonged stress response (ACTH and CORT levels) compared with adults. In addition, expression of CRH in the PVN in response to acute and chronic stress remains higher in prepubertal males compared with adults, suggesting that HPA axis plasticity is significantly modulated by androgens during pubertal development (Romeo et al., 2006; Viau et al., 2005).

CRH and AVP neurons are highly involved in the activation of the HPA axis; however, studies using immunohistochemistry have shown that ARs are not present in PVN neuroendocrine neurons (Bingaman et al., 1994). ARs, however, have been found in nonneuroendocrine neurons of the dorsal and ventral medial parvocellular region of the rat PVN, which have terminal projections into the spinal cord and brain stem autonomic nuclei (Bingaman et al., 1994). These observations suggest that a transsynaptic component in the androgenic regulation of the PVN, and further the HPA axis, is plausible (Viau and Meaney, 1996). Furthermore, a series of elegant studies by Viau's group have identified the medial POA as a critical part of the neurocircuitry mediating the inhibitory effects of androgens on the HPA axis response to stress (Viau and Meaney, 1996; Williamson and Viau, 2008). Evidence supporting the extra-PVN modulation of the HPA axis by androgens comes from studies demonstrating that direct androgenic activation of the medial POA and the BnST reduces CORT response to acute stress (Viau and Meaney, 1996). Thus, both the medial POA and BnST could play a key role in the androgenic inhibition of the PVN. In contrast, Lund et al. (2006) were able to show an inhibition of the HPA axis response to restraint stress when DHT pellets were placed near the PVN, suggesting the potential for a direct action of androgens on HPA regulatory neurons in the PVN (Lund et al., 2006). Ultimately, this effect was postulated to be a response mediated by the metabolism of DHT to 5α-androstane-3β,17β-diol (3β-diol), a weak androgen that selectively binds and activates estrogen receptor (ER) β.

Although the mechanism by which androgens inhibit the HPA axis is still unclear, it is possible that ARs modulate descending projections to the PVN, by regulating the release of gamma-aminobutyric acid (GABA). This idea is supported by data showing the increased sensitivity of GABAergic cells that release GABA in response to stress or orchiectomy (Williamson et al., 2005). Moreover, AR distribution overlaps to a great extend with glutamic acid decarboxylase (an enzyme necessary for the production of GABA) in the medial POA and BnST.

On the other hand, studies by Lund et al. (2006) show a direct effect of androgen on the PVN to inhibit HPA responses, an observation that may be explained by the fact that DHT can be metabolized to 3α-androstanediol (3α-diol), a metabolite with high affinity for the GABA-A receptors, which could also result in direct inhibition of stress-related pathways (Lund et al., 2006). Furthermore, DHT can also be unidirectionally metabolized to 3β-diol, which can act as an ERβ agonist, preferentially binding to and activating transcription of ERβ regulated genes to modulate the HPA axis. Collectively, these data suggest potential indirect mechanisms for the inhibitory actions of androgens, mediated by AR, in the stress circuitry.

### 2.2.3 Androgen effects on behavior

In addition to modulating changes in the levels of stress hormones, androgens are known to be involved in regulating reproductive and nonreproductive behaviors, such as anxiety and depressive-like behaviors. Androgen-insensitive male mice and rats with testicular feminization mutation (a single base deletion spawning an immature stop codon downstream of the mutation, resulting in a truncated version of the AR protein (He et al., 1991) showed deficits in learning and memory tasks (Rizk et al., 2005) and displayed higher levels of anxiety-like behaviors than females carrying a functional copy of the AR gene (Hamson et al., 2014; Zuloaga et al., 2008a, 2008b, 2011), providing further evidence of AR involvement.

## 2.3 Estrogen receptors

Estrogens influence many neurophysiological and behavioral actions, including reproductive and nonreproductive functions. All vertebrates synthetize estrogens, suggestive of a well-conserved evolutionary molecule with a common origin (Eick and Thornton, 2011). While their role in mediating reproductive functions has been well established, studies on their effects on nonreproductive functions remain incomplete. In addition to their actions on bone density and strength, blood lipid levels, fat deposition, and water and salt balance (Bondesson et al., 2014), estrogens are also involved in modulating neuroendocrine, autonomic, and behavioral responses to stress. Among these actions, the physiologically relevant ones are the regulation of inflammatory processes, pain, anxiety, depressive-like behaviors, and cognitive functions (Sarvari et al., 2011; Hunter et al., 2011; Diz-Chaves et al., 2012; Kiss et al., 2012). In this section, we focus on estrogen signaling, and ERs as regulators of stress and stress-related behaviors.

### 2.3.1 *Two forms of estrogen receptors*

The classical actions of estrogens are mediated through two distinct intracellular receptors: ERα (NR3A1) and ERβ (NR3A2), which are part of the superfamily of ligand-activated transcription factors. While estradiol binds with almost equivalent affinity to both ERs, a 56% homology exists between the LBDs of ERα and ERβ (Kuiper et al., 1996). The DBD is 96% homologous and both receptors bind to the same DNA response element (Kuiper et al., 1996). Furthermore, the AF-1 function in the N-terminal region shows the most heterogeneity between the ERα and ERβ, and is thought to recruit distinct coregulators, which could lead to the activation of different sets of target genes (Charn et al., 2010; Grober et al., 2011; Zhao et al., 2010). Such differential downstream regulation is exemplified in the HPA axis, where activation of ERβ has been shown to reduce stress-induced c-Fos expression in the PVN and plasma CORT and ACTH levels (Lund et al., 2006; Weiser and Handa, 2009; Weiser et al., 2009, 2010; Oyola et al., 2012), while selective activation of ERα exacerbates these levels (Lund et al., 2005; Weiser and Handa, 2009). The distinct pattern of expression of each receptor type could also contribute to these differences and will be discussed in the subsequent section.

### 2.3.2 *Estrogen receptor localization in the brain*

The expression pattern of ERs varies in the brain in an age- and sex-dependent manner. Additionally, the expression patterns differ between ER types throughout the rostral-caudal range of the brain and spinal cord (Simerly and Young, 1991). For example, in the rat hypothalamus, ERα expression is predominant in the medial POA, arcuate nucleus, and the VMN of the hypothalamus, whereas ERβ expression is abundant in the supraoptic nucleus (SON), PVN, and suprachiasmatic nucleus (SCN) (Chu and Fuller, 1997; Laflamme et al., 1998; Shughrue et al., 1996). Coexpression of ERα and ERβ has been observed in brain nuclei involved in HPA axis circuitry, such as the BnST, POA, arcuate nucleus, and medial amygdala, among others. Studies using the Esr2 bacterial artificial chromosome mouse line, in which ERβ-expressing neurons could be identified by the green fluorescent protein (ERβ-eGFP), have also corroborated the expression patterns of ERβ (Milner et al., 2010). The overlapping, yet distinct expressions of these two ERs in limbic structures suggest their active role in conserved animal behaviors and physiological responses. Furthermore, ERα and ERβ coexpression has also been reported in astrocytes and oligodendrocytes (Azcoitia et al., 1999; Platania et al., 2003; Mhyre and Dorsa, 2006; Santagati et al., 1994; Zhang et al., 2004). Although more research is necessary to decipher the function of these receptors in nonneuronal cells, it has been postulated that ERβ could modulate anti-inflammatory properties (Saijo et al., 2011; Wu et al., 2013; Zuloaga et al., 2012).

Expression levels of ERα and ERβ mRNA levels have been found to vary not only throughout development, but also between sexes (Cao and Patisaul, 2011). Studies in developing female rats demonstrate that while ERα mRNA levels remain high in the anteroventral PVN, the VMN, and the medial POA regions, ERβ mRNA levels remain high only in the VMN until postnatal day 19. On the other hand, male rats express higher levels of ERβ mRNA at birth until postnatal day 19, from which point females show higher expression. In a study using ERβ-eGFP mice, Zuloaga et al. (2014) describe a developmentally dynamic expression pattern of ERβ in the cortex, hippocampus, arcuate, VMN, and PVN (Zuloaga et al., 2014). Furthermore, a sex difference was noted in ERβ-eGFP expression in the BnST and anteroventral periventricular nucleus (AVPV). The AVPV shows greater clustering of ERβ-eGFP-positive cells near the third ventricle in females, on postnatal day 21, but not days 0 and 4. The BnST shows greater concentrations of ERβ neurons in the male versus the female on postnatal days 4 and 21, but not day 0, but with dramatic developmental increases occurring in both sexes between ages 4 and 21. The significance of postnatal day 21 in mice roughly translates to the early stage of puberty onset, the hormonal fluctuations of which may explain the alterations in ER expression beginning around this time. While these findings shed light on transcriptional levels of the receptors, very little is known about the protein levels of ERβ, as dependable antibodies for the ERβ protein do not currently exist (Snyder et al., 2010).

Despite their unique localization pattern, ERs also overlap with other steroid receptors in the brain, particularly in the circuitry controlling the HPA axis. For instance, early studies found that AR and ER mRNAs are coexpressed on the same neurons in regions

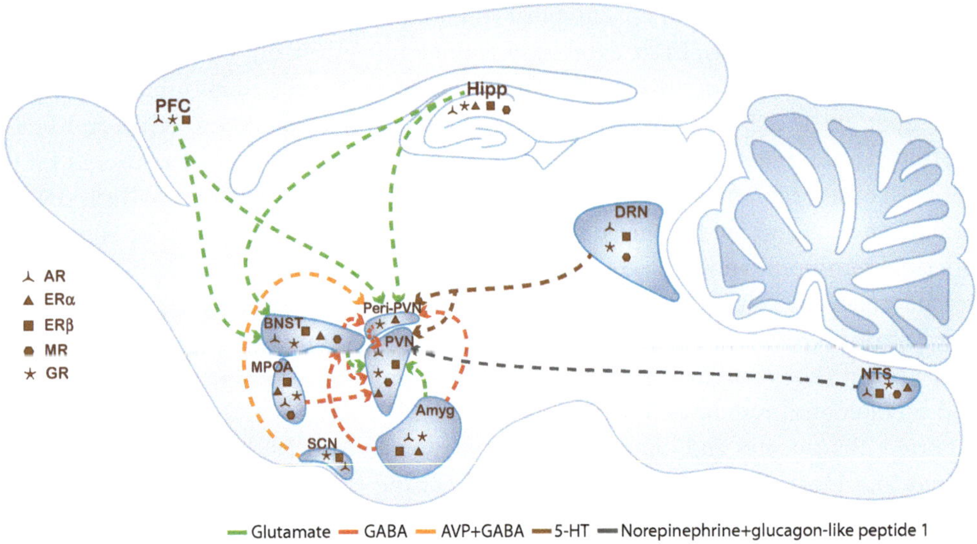


**Figure 13.3** ***Steroid hormone receptors in neuroendocrine stress circuitry.*** The paraventricular nucleus (PVN) and the peri-PVN regions of the hypothalamus receive inputs from several brain regions involved in the modulation of the stress response. These brain regions have a considerable amount of overlap among the various steroid hormone receptors that are involved in mediating the activation or inhibition of the HPA axis. The degree of activation of the HPA axis is highly dependent on the type of stress, which recruits different brain regions to determine the final outcome of the response. Arrow colors: stimulatory glutamate connections (green), inhibitory GABAergic connections (red), AVP and GABA mixed connections (yellow), serotonergic connections (brown), and norepinephrine and glucagon-like peptide 1 (gray). PFC, prefrontal cortex; Hipp, hippocampus; BNST, bed nucleus of stria terminalis; MPOA, medial preoptic area; SCN, suprachiasmatic nucleus; AMYG, amygdala; DRN, dorsal raphe nucleus; AVP, arginine vasopressin; NTS, nucleus of the solitary tract, AR, androgen receptors; ERα, estrogen receptor alpha; ERβ, estrogen receptor beta; MR, mineralocorticoid receptor; GR, glucocorticoid receptor.

involved in stress responses, namely the hypothalamus, the BnST, and the hippocampus (Simerly et al., 1990) (Figure 13.3)

### 2.3.3 Estrogen effects on gene expression

Estrogen-regulated gene expression has been well studied in reproductive organs and in the brain. Estrogens refer to a class of hormones, with estradiol being the most common circulating estrogen. These compounds are well known for their organizational effects on the sexual differentiation of the rodent brain. Exposure to estrogens before and immediately after birth leads to permanent changes in the brain that ultimately result in differing responses to stress in adulthood (McCormick and Mathews, 2007). Estradiol in the brain of males arises predominantly from the intracellular aromatization of circulating T to estradiol. This is demonstrated by studies that inhibited aromatization within the first 12 h of neonatal life in male rodents and found increased plasma CORT levels and c-Fos expression in the PVN in response to stress during adulthood (Bingham et al., 2012). Interestingly, the HPA axis response in these animals resembles that of an adult female, suggesting that estradiol acts perinatally to masculinize the stress response of adult rodents. Conversely, administering high levels of estradiol to neonatal female rats for the first week of life defeminizes their typical adult HPA axis response (Patchev et al., 1995) and they show a more male-like stress response.

Taken together, these findings imply a critical role for estrogens during neonatal life in sculpting sex differences in adult HPA axis responses of the rodent. However, such hormonal manipulations, when performed during adulthood, bring about differential results. Removal of endogenous sex steroids by way of ovariectomy in adult female rodents results in lower plasma CORT and ACTH levels following exposure to a stressor, an effect that could be reversed by exogenous estradiol supplementation (Burgess and Handa, 1992; Handa et al., 1994a; Serova et al., 2010; Weiser and Handa, 2009). Conversely, orchiectomy of adult male rats causes an increase in the HPA axis response to stress presumably by relieving the androgenic inhibition imposed by circulating gonadal steroids (Handa et al., 2013). These latter studies demonstrate a role for activational effects of estrogens and androgens on the HPA axis, in addition to the organizational effects previously described.

### 2.3.4 Estrogen effects on behavior

Estrogens have been shown to regulate mood, affect, anxiety, fear, locomotion (Mora et al., 1996; Sell et al., 2000; Walf and Frye, 2007), and vulnerability to drug addiction (Jackson et al., 2006). Furthermore, estrogens' effects on these behaviors are often reported to be sex-specific. For example, behavioral responses that hijack the limbic system, such as addiction, are influenced by the fluctuations of estradiol that occur throughout the estrous cycle in rodents (Jackson et al., 2006). Estradiol administration to ovariectomized female rats increases susceptibility to self-administer cocaine, compared with orchiectomized and intact male rats (Hu et al., 2004), further supporting the idea that estrogens work in a sex-specific manner.

As mentioned earlier, the effects of estrogens can be attributed to actions on specific ER-positive brain areas. For instance, selective activation of the medial POA by estradiol implants results in increased activity on the running wheel (Fahrbach et al., 1985). In contrast, estrogen implants in the PVN can regulate HPA axis response to restraint stress (Lund et al., 2006), and glucocorticoid implants in the amygdala can control anxiety-like behaviors (Handa and Weiser, 2014), both of which can be modulated by estrogen levels, providing further examples of ER-sensitive regions regulating distinct behaviors.

Reports on estrogen-induced behavioral responses vary in the literature. These behavioral discrepancies can be attributed to estrogens' actions on the two distinct receptor systems: ER$\alpha$ and ER$\beta$ (Handa et al., 2012). While ER$\alpha$ activation influences reproductive behaviors (Lindzey et al., 1998; Wersinger et al., 1997), ER$\beta$ activation affects nonreproductive behaviors, such as learning and memory (Jacome et al., 2010), as well as anxiety-related behaviors (Donner and Handa, 2009; Lund et al., 2005; Oyola et al., 2012; Weiser et al., 2009). Although estrogens have been reported to exhibit both anxiogenic and anxiolytic properties (Oyola et al., 2012; Burgess and Handa, 1992; Leret et al., 1994; Nomikos and Spyraki, 1988), the critical ERα and ERβ targets that mediate estrogenic effects on anxiety and the stress circuitry have not been yet identified. We have recently confirmed that selective activation of ERβ decreases anxiety-like behaviors in wild-type mice, in a similar fashion to that which has been reported in rats. Furthermore, administration of a selective ERβ agonist reduces anxiety-like behaviors and corresponding stress hormone (CORT and ACTH) levels in wild-type mice, but is ineffective in mice with targeted null mutation of the ERβ gene (Oyola et al., 2012). We also identified the anterodorsal medial amygdala and the bed nucleus of the stria terminalis (BnST) as potential anatomical targets for ERβ mediation of these anxiolytic actions, as these brain areas showed elevated c-Fos expression following behavioral testing when treated with an ERβ agonist, compared with vehicle-treated controls. The increased level of c-Fos expression during a test where the ERβ agonist increases anxiolytic behaviors suggests involvement of an ERβ-mediated pathway that activates anxiolytic or inhibits anxiogenic behaviors.

## 2.4 Corticosteroid receptors

Corticosteroids, released from the adrenal cortex, regulate the functions of the central nervous system (CNS) by binding two different corticosteroid receptors: the type II or GRs (NR3C1) and the type I or MRs (NR3C2). As with previously described receptor systems, sex differences in physiological actions also exist between the corticosteroid receptor types. These differences may be due to differences in physical structure, binding affinity, and distribution patterns within the CNS (Evans and Arriza, 1989; Canny, 1990; de Kloet et al., 1993). While MRs and GRs share common ligands, the downstream effects differ between the receptors. Similarly, the binding affinities of the ligands for MR and GR are not the same. Both aldosterone and CORT preferentially bind MRs over GRs (Reul and de Kloet, 1985; Ratka et al., 1989; Coirini et al., 1985). Not surprisingly, the expression patterns of GRs and MRs in the brain are different, which could also be responsible for their distinct functions. GRs and MRs are often thought of as having opposing actions; whereas GRs show neuroendangering properties, MRs show neuroprotective properties *in vivo* (Almeida et al., 2000) and *in vitro* (Crochemore et al., 2005) alike. Furthermore, GRs are widely expressed throughout the brain, often showing coexpression with MRs in some brain regions, including the hypothalamus and hippocampus (Ahima and Harlan, 1990; Ahima et al., 1991). Of all the hypothalamic nuclei, the PVN shows the highest density of GRs. GRs are also coexpressed by CRH neurons, though not exclusively (Ceccatelli et al., 1989; Liposits et al., 1987). In summary, the actions of CORT are highly dependent on the type of receptor, the brain region, and sex.

### *2.4.1 Glucocorticoid receptors*

The structural organization of GRs is similar to those in the other members of the nuclear receptor superfamily. Upon ligand binding and nuclear translocation, GRs form homodimers to activate or repress target gene transcription. In addition, several studies have also reported the formation of GR–MR heterodimers (Trapp et al., 1994), which result in different transactivation. Studies comparing intact and adrenalectomized animals report translocation of ligand-bound GR receptors into the nucleus in intact animals, whereas unliganded GRs in adrenalectomized subjects remain in the cytoplasm (Ahima and Harlan, 1990; Morimoto et al., 1996). In addition to this classical mechanism, rapid actions of GRs have been reported in the cardiovascular, immune, and neuroendocrine systems (Ayroldi et al., 2012; Bellavance and Rivest, 2012; Lee et al., 2012). Moreover, nongenomic signaling has shown to activate signal transduction pathways, such as mitogen-activated protein kinase pathways, through cytoplasmic or membrane-bound GR receptors (Ayroldi et al., 2012; Busillo and Cidlowski, 2013).

In addition to subcellular localization, regional localization of GRs is also implicated in the actions of CORT. In the rat brain, GR mRNA expression has been identified in virtually every brain region (e.g., cortex, hippocampus, thalamus, hypothalamus, amygdala, septum, striatum, and cerebellum) (Ahima and Harlan, 1990; Cintra et al., 1994; McGimsey et al., 1991; Morimoto et al., 1996) with highest concentrations in the cortex and hippocampus (Ahima and Harlan, 1990; Aronsson et al., 1988; Cintra et al., 1994; Morimoto et al., 1996). Within the hypothalamus, the PVN shows the highest expression of GR mRNA (Ahima et al., 1991). Notably, these same neurons also coexpress other HPA-modulating proteins (i.e., CRH, AVP) (Ceccatelli et al., 1989; Liposits et al., 1987; Sawchenko, 1987a; Uht et al., 1988), supporting the functional importance of CORT in HPA axis modulation. Consistent with these findings, specific deletion of GRs in the forebrain or amygdala of adult mice has uncovered their role in HPA axis regulation under stressful conditions, and in memory acquisition and fear conditioning in the two structures, respectively (Arnett et al., 2011).

### *2.4.2 Mineralocorticoid receptors*

Structurally and mechanistically, the MR resembles the GR and they both belong to group C of the subfamily 3 of nuclear receptor proteins. However, functional differences between these two receptors exist, which allow for their selective modulation of different downstream targets (Datson et al., 2001). While both GRs and MRs are involved in the response to homeostatic perturbances, the extent of their involvement differs, due to their differing affinities for CORT. Due to their greater affinity for CORT, MRs are preferentially occupied compared with GRs, which have a relatively lower binding affinity and are little occupied during periods of low CORT (i.e., basal). In contrast, GRs show increased occupancy by ligand during elevated levels of circulating CORT, either after stress-induced HPA axis activation or with the circadian rise of CORT that occurs just prior to the active part of the circadian cycle (Joels and de Kloet, 1994; Ratka et al., 1989; Reul and de Kloet, 1985). Hence, due to these receptors' differential affinities for ligand (Coirini et al., 1985; Ratka et al., 1989; Reul and de Kloet, 1985), the HPA axis is able to respond to a wide range of glucocorticoid levels as exemplified by negative feedback that occurs during basal hormone secretion and following stress-induced changes in circulating hormone levels.

Significant overlap exists between the expression patterns of MRs and GRs in the limbic system. Brain structures expressing MRs include the cortex, hypothalamus, hippocampus, cingulate cortex, and subcortical regions. Whereas GR mRNA is uniformly distributed throughout these regions, MR mRNA is highly expressed by hippocampal neurons (Chao et al., 1989) and some hypothalamic nuclei. Using receptor binding studies of adult rat brains, it has been shown that GR exhibits the greatest expression in the hippocampal cornus ammonis region (CA) 1 and CA2 regions, whereas MRs predominate over the CA2 region (Coirini et al., 1985). Similarly, though MRs and GRs are coexpressed in the cingulate cortex, a dichotomous regional separation exists within the structure. Van Steensel et al. (1996) demonstrated that, in addition to regional overlap of the two receptors, coexpression exists within individual cells.

### *2.4.3 Corticosteroid effects on gene expression*

Although the signaling mechanisms for GRs and MRs are well studied, the downstream targets of each receptor are unclear. Given the common use of steroid-based therapies in modern medicine, knowledge of the specific functional target genes with which they interact is imperative for our understanding and improvement of drugs that target these genes. In an effort to understand specific GR gene targets, Phuc Le and colleagues performed a genome-wide analysis that identified 300 GR-bound promoters. This approach revealed multiple new target genes, and confirmed previously identified ones. The identified genes include the homeobox genes (i.e., *Hoxa13a13*, *Hoxb4*, *Hoxc5*, *Hoxc9*, *Hex1*), transcription factors (i.e., AP-4, 21, 15), transformation related protein 53, cyclic adenosine monophosphate response element binding protein1, and Fos (Phuc Le et al., 2005). The target genes for MRs included sodium channel and $Na^+/K^+$ ATP pumps, serine/threonine kinase gene coding for cell signaling, serum glucocorticoid regulated kinase-1, a mediator of renal sodium balance, and corticosteroid hormone induced factor and K-Ras2, sodium channel activity regulators (Brennan and Fuller, 1999, 2006; Soundararajan et al., 2005). However, further research is needed to examine specific GR and MR targeted gene activation in the brain, as it is possible that different genes can be regulated in different brain regions based on the region-specific localization of coregulators and transcription factors.

#### 2.4.4 Corticosteroid effects on behavior

The balance of GR- and MR-mediated actions in the limbic system has important implications on the HPA axis' ability to adapt to homeostatic challenges. Responses to threats can be accentuated by reduced GR expression, and diminished with increased MR expression (Harris et al., 2013). Moreover, imbalance of GR and MR activity is thought to increase susceptibility to developing affective disorders (De Kloet et al., 1998). Indeed, in support of this hypothesis, in humans, MR and GR mutations have been linked with HPA axis dysregulation and depression.

Supporting the physiological importance of GR-MR balance, Oitzl et al. (2001) show that a point mutation that prevents GR dimerization and binding to DNA impairs GR signaling in transgenic mice, resulting in spatial memory impairments in the Morris water maze task (Oitzl et al., 2001). Reduced GR gene expression in GR-heterozygous mutant mice results in HPA axis hypoactivation, while mice with GR gene overexpression show augmented negative feedback signaling (Ridder et al., 2005) and decreased anxiety-like behaviors in both sexes (Rozeboom et al., 2007). Similarly, mice with a forebrain-specific reduction in GRs show HPA axis hyperactivation, negative feedback impairments, and depressive-like behaviors (Boyle et al., 2005). Interestingly, while deletion of MRs in the limbic system results in impaired spatial and working memory (Berger et al., 2006), MR overexpression increases spatial memory retention, reduces anxiety-like behaviors, and results in neuroprotection from neuronal insults (Lai et al., 2007). In summary, imbalances in GR and MR signaling activity are central to many neurological disorders.

## 3 THE HYPOTHALAMO–PITUITARY–ADRENAL AXIS

### 3.1 Anatomy of the HPA

One of the most conserved survival mechanisms throughout evolution involves an organism's ability to respond to a real or perceived threat. The principal system mediating this stress response is the HPA axis. When activated by stressors, the hypothalamus initiates the secretion of a series of hormones into the bloodstream, which ultimately trigger the release of glucocorticoids from the adrenal cortex into the systemic circulation. In humans, the main glucocorticoid is cortisol, while in rats and mice it is CORT. These hormones travel throughout the body and regulate gene transcription in all cells, including neurons of the brain that express MRs or GRs (Reul and de Kloet, 1985). The brain perceives the rise of the stress hormones as a feedback signal to turn off the stress-response pathway and return to homeostasis, as well as to activate behaviors that may help deal with the homeostatic challenge.

Of importance for this discussion, the hypothalamic PVN initiates the stress response cascade. The PVN is a wing-shaped structure adjacent to the dorsal part of the third ventricle in the rostral part of the hypothalamus. Gurdjian (1927) first described the parcellation of the PVN based on Nissl staining and identified a medial parvocellular region, characterized by cells with small soma, and a lateral magnocellular region, characterized by cells with large, dense soma (Gurdjian, 1927). Subsequently, these neurons were further classified by their output into long-projecting neurons, neurosecretory magnocellular cells, and the neurosecretory parvocellular cells. The long-projecting cells send their axons to the brain and spinal cord, and are involved in controlling autonomic motor function. In the rat, the magnocellular neurons of the PVN project their axons to the posterior pituitary gland, where they secrete OT and AVP directly into general circulation. The parvocellular neurons found in the parvocellular region of the PVN secrete neuropeptide releasing factors (e.g., CRH, AVP, thyrotropin releasing hormone) into the hypothalamo-hypophyseal portal vasculature. From here, these peptides travel to the anterior pituitary gland where they activate specific anterior pituitary cells to secrete hormones such as ACTH, thyroid stimulating hormone, and growth hormone into the general circulation. Of importance, in the rat, there is substantial overlap between the chemical phenotype of parvocellular neurons, with up to 40% of CRH neurons coexpressing and coreleasing AVP and OT (Sawchenko et al., 1984). Such overlap does not appear to be the case in the mouse, where there are far fewer CRH neurons that coexpress AVP or OT (Biag et al., 2012).

Most relevant to the context of the HPA axis are the hormones, CRH and ACTH. CRH, released from hypothalamic neurons of the PVN, stimulates the secretion of ACTH from corticotrophs of the anterior pituitary. Subsequently, upon reaching the adrenal glands, ACTH prompts the release of CORT from the adrenal cortex. Glucocorticoids circulate throughout the body and upon reaching the brain, particularly the hippocampus, hypothalamus, and the anterior pituitary (Walker et al., 2012), signal the termination of the HPA axis response to stressors by binding to MRs and GRs (Reul and de Kloet, 1985; Sutanto and De Kloet, 1987). Primary sites for GR and MR binding include the PVN, BnST, and the amygdala. The BnST and amygdala lie upstream of the PVN and communicate with neurons of the PVN, either directly or indirectly, to terminate the cascade (Dong et al., 2001; Prewitt and Herman, 1998). The termination of the HPA axis reactivity by glucocorticoids is dose- and duration-dependent (Abe and Critchlow, 1980; Sapolsky et al., 2000), and the correct response of this negative feedback system is crucial for avoiding the deleterious effects of chronic HPA axis activation and glucocorticoid stimulation. It should also be noted, however, that the termination of the HPA axis response to stress is not the only function of circulating glucocorticoids, which have been shown to regulate behaviors by acting upon a number of other brain areas (Goel et al., 2014).

## 4 HORMONE SECRETORY PATTERNS

### 4.1 Circadian patterns of secretion

Basal activity of the HPA axis is heavily influenced by the endogenous circadian rhythm of a species (Kalsbeek et al., 2012). The circadian rhythm of glucocorticoid secretion is characterized by an elevation just prior to waking, which predicts the onset of the active period and prepares physiological systems for greater activity. Thus, the glucocorticoid rhythm is vital for synchronization between the internal and external environment and aids an organism in adaptation. Early research by Pincus (1943) discovered a diurnal variation in urine ketosteroid secretion, which was attributed to a circadian rhythmicity of the endocrine system (Pincus, 1943). Ultimately, the circadian rhythm of glucocorticoids was shown to be regulated by the circadian pacemaker, which is said to reside in the SCN of the anterior hypothalamus (Hendrickson et al., 1972). It receives retinal inputs and sends projections to the parvocellular neurons in the PVN and other hypothalamic nuclei, as well as autonomic areas of the brain, thus influencing the tone of the HPA axis (Kalsbeek et al., 2010b; Watts et al., 1987). Experiments in which the SCN was lesioned have confirmed its central role by completely disrupting the circadian pattern of CORT levels in the adrenal cortex (Moore and Eichler, 1972). Interestingly, while this hormonal circadian rhythm is innate, it does not present itself until 3 weeks of age in rats (Moore-Ede et al., 1982).

The basal fluctuations of CORT in an animal's system are linked to its activity (Kaneko et al., 1980). Diurnal species, such as humans, show peak CORT levels in the morning and trough levels at night, while nocturnal species (e.g., rodents) demonstrate the reverse pattern. The rise in glucocorticoids predicts the onset of the active period in the diurnal cycle and is used to synchronize the central circadian clock with peripheral clocks throughout the body. In this way, glucocorticoids are essential for circadian timing and allow for rhythms in metabolism, immune function, cardiovascular function, and others to be synchronized.

In the HPA axis, there also exists an ultradian, or pulsatile, pattern of adrenal glucocorticoid secretory activity (Moore-Ede et al., 1982). Current thought suggests that the ultradian rhythm is generated through a feedforward:feedback system. The pulsatile pattern of CORT arises from a delay in CORT secretion following a stimulus as a consequence of the requirement for *de novo* synthesis of steroid in the adrenal cortex. This synthesis is coupled with a rapid nongenomic effect of CORT at the pituitary level to inhibit CRH-induced ACTH secretion (Walker et al., 2010, 2012). Indeed, even under constant CRH infusion, an ultradian pattern of ACTH and CORT secretion has been detected with ACTH pulses that precede the CORT pulses. Moreover, when high levels of CRH were infused, the pattern was disrupted and CORT oscillations were dampened (Rankin et al., 2012).

A diurnal rhythm of ACTH is also present in both humans and rodents (Kaneko et al., 1980). Furthermore, ACTH circadian patterns mirror the CORT circadian function, but to a lesser extent (Akana et al., 1986). This decreased amplitude is thought to be caused by similarly rhythmic adrenal sensitivity to ACTH (Kaneko et al., 1980). As this sensitivity increases, the negative feedback effects of the corresponding rise in CORT act on the anterior pituitary to diminish further ACTH release. Moreover, the pulsatile patterns of CORT secretion have been shown to be sexually dimorphic, with females having higher amplitude, frequency, and number of CORT pulses compared with males (Seale et al., 2004b).

AVP has also been shown to contribute significantly to the regulation of the HPA axis. Early studies indicated that AVP alone could stimulate ACTH secretion (Krieger and Zimmerman, 1978;Yates and Maran, 1974). In addition, AVP acts through the V1b receptors of the pituitary to enhance CRH-driven secretion of ACTH (Armario, 2010). In the rat, AVP is highly expressed by neurons of the PVN, where its distribution fits with its role as a neuroendocrine regulator. In the mouse, this overlap is not as robust (Biag et al., 2012). The distribution of AVP throughout the hypothalamus has been classified into several systems: (1) the magnocellular neurons of the PVN and SON, which project to the posterior pituitary; (2) a sexually dimorphic system in the rostral forebrain, with efferent projections from the BnST and the medial amygdala; (3) an autonomic and neuroendocrine system, with efferent projections from parvocellular neurons of the PVN that overlap with CRH neurons; and (4) a circadian system, with efferent projections from neurons in the SCN (Kalsbeek et al., 2010a). The significant coexpression of AVP in CRF-expressing neurons of the rat PVN further gives way to the intricate relationship between AVP and the HPA axis (Watts, 2005).

In the rat, AVP neurotransmission from the SCN has an inhibitory effect on CORT release by the adrenal cortex that coincides with the diurnal CORT fluctuations (Kalsbeek et al., 1992, 1996). However, AVP's inhibitory tone does not coincide with the entire circadian CORT pattern. In search of an alternative stimulatory tone from the SCN, Kalsbeek et al. disinhibited the negative outputs to the PVN by administering an antagonist of the V1 receptor at various points throughout a 24-h cycle. Correspondingly, these subjects showed higher CORT and ACTH levels than vehicle-injected controls. Though still largely unexplored, the SCN's role in inducing the diurnal rise in ACTH has been demonstrated by Cascio (1987), who found that adrenalectomized mice implanted with CORT pellets do not demonstrate the typical evening ACTH surge after bilateral SCN lesions (Cascio et al., 1987). Finally, a strong correlation between lower AVP production in the SCN and poor sleep-wake cycles in aging humans supports the AVP's postulated role in circadian regulation (Hofman and Swaab, 1994).

Besides regulating the secretion of releasing factors, like CRH, the SCN also regulates the HPA axis through autonomic innervation of the adrenal glands (Engeland, 1998; Edwards and Jones, 1993). Transneuronal retrograde tracer studies have identified second order neuronal connections from the PVN to the preganglionic spinal cord neurons in the thoracolumbar region and even third order neurons traced back to the SCN (Buijs et al., 1999). In another series of studies, it was found that electric stimulation of the splanchnic nerve increases adrenal sensitivity to ACTH (Edwards and Jones, 1993).

## 5 REGULATION OF HORMONE SECRETION BY STEROID HORMONES

### 5.1 Negative feedback by glucocorticoids

The HPA axis is a closed loop neuroendocrine system that governs its own regulation through dose- and duration-dependent glucocorticoid negative feedback action (Sapolsky et al., 2000; Abe and Critchlow, 1980). Disrupting this negative feedback circuit by adrenalectomy leads to higher basal and stress-induced PVN neuropeptide expression and elevated levels of circulating CORT and ACTH (Imaki et al., 1991; Kovacs et al., 1986; Sawchenko, 1987b). Thus, this negative feedback action is imperative for avoidance of deleterious effects of prolonged high levels of glucocorticoids in systemic circulation.

Circulating glucocorticoids can inhibit the HPA axis at several different levels. Inhibition occurs either directly by CORT binding to MRs and GRs in neurons responsible for the origin of the cascade (i.e., PVN, CRF, and AVP neurons or anterior pituitary corticotrophs), or indirectly by CORT targeting MRs and GRs in other limbic structures (i.e., hippocampus) projecting to the PVN (Akana et al., 1986; Dallman et al., 1987; Bradbury et al., 1994; Sawchenko, 1987a; Kovacs et al., 2000). CORT's effect on the HPA axis is highly dependent on the differential sensitivity of MRs and GRs (Reul and de Kloet, 1985; Reul et al., 1990). This differential sensitivity allows for selective actions on the HPA axis. For example, under basal CORT levels, MR binding predominates and therefore the available pool of MRs is largely occupied. In contrast, following exposure to a stressor, CORT levels rise quickly and the elevated levels saturate any available MRs and then occupy the lower affinity GRs. Initial studies have shown MR's high affinity for CORT (Dallman et al., 1987, 1985; Rabadan-Diehl et al., 1997). Consistent with this finding, following adrenalectomy of rats, CRH and ACTH levels rise; however, upon readministration of CORT at MR-selective low doses, the HPA axis restores homeostatic function. Conversely, antagonism of hippocampal MRs increases both ACTH and CORT basal levels, similar to adrenalectomy (van Haarst et al., 1997). Not surprisingly, the hippocampus contains the highest expression of MRs in the brain, which could render it the primary site of glucocorticoid negative feedback through MRs. Moreover, transgenic mice overexpressing MR in the forebrain have a reduced response to stressors as seen by lower CORT levels in both males and females (Rozeboom et al., 2007). Such data also implicate the MR in feedback regulation following stress-induced rises in CORT. Finally, PVN-specific deletion of GR results in increased CRH immunoreactivity in the PVN (Jeanneteau et al., 2012). This GR imbalance also augments downstream ACTH and CORT responses to stress, implicating the participation of GRs in the negative feedback response. In summary, glucocorticoids take advantage of distinct GR and MR sensitivities and their differential expression patterns to modulate negative feedback actions on the HPA axis.

# 6 STEROID METABOLISM AND THE REGULATION OF THE HPA AXIS

## 6.1 Biosynthesis of steroid hormones

Steroids encompass a large group of hormones that obtain their name from their common cholesterol derivation. They share a typical ring structure and are categorized into two main classes by their tissue of origin: the adrenal steroids (also called adrenocorticoids), synthesized in the adrenal cortex, and the gonadal steroids, synthesized in the ovaries and testes (Nelson, 2000). In pregnant women, the placenta also contributes to the production of sex steroids (i.e., estrogen and progesterone). While the adrenal cortex is the primary producer of glucocorticoids (e.g., cortisol) and mineralocorticoids (e.g., aldosterone), and the gonads remain the principal producers of sex steroids, these glands are not mutually exclusive in function and also produce alternate metabolites of these steroids, though to a much lesser extent. Since there is no mechanism for steroid storage within a cell, all steroid hormones are immediately released upon production. Hence, circulating steroid hormone levels largely represent synthesis of the hormone *de novo*.

Adrenocorticoids are subdivided into the glucocorticoid and mineralocorticoid classes based on the type of receptor each binds to and the effect each achieves. Glucocorticoids are produced in the zona fasciculata of the adrenal cortex and receive their name from their prominent role in glucose metabolism via binding to GRs. These receptors are present in almost every cell in the body and additionally mediate immune, cardiovascular, metabolic, and general homeostatic functions (Kadmiel and Cidlowski, 2013). The chief glucocorticoid in humans is cortisol, which is synthesized from its inactive precursor, cortisone, by the action of the enzyme 11β-hydroxysteroid dehydrogenase (HSD). Their relationship to other steroid hormones is shown in Figure 13.4.

While the impact of glucocorticoids on HPA axis homeostasis has already been addressed in detail, their homeostatic role extends to the immune system where they also work to decrease inflammation (Silverman and Sternberg, 2012). Mineralocorticoids, however, target MRs to balance an organism's fluid and salt content. The body's principal mineralocorticoid is aldosterone, which is synthesized from 18-OH-corticosterone by the enzyme CYP11B2 in the zona glomerulosa of the adrenal cortex (Condon et al., 2002). Aldosterone's major action is peripheral; it maintains blood pressure homeostasis by stimulating salt reabsorption in the nephron's collecting duct, thus triggering the passive reabsorption of water. Importantly, research shows that cortisol binds preferentially to MRs over GRs and that it is this receptor preference, by way of its specific brain distribution, results in its actions on the HPA axis (Reul and de Kloet, 1985; Sutanto and De Kloet, 1987). However, the MR is not selective for aldosterone versus CORT, and selectivity of MRs arises not from the receptor, but from the tissue. Tissues that respond to aldosterone possess the enzyme 11β-HSD, which rapidly metabolizes glucocorticoids to 11-keto metabolites, rendering them incapable of activating the MR, thereby allowing those tissues to respond specifically to fluctuations in aldosterone. This mechanism

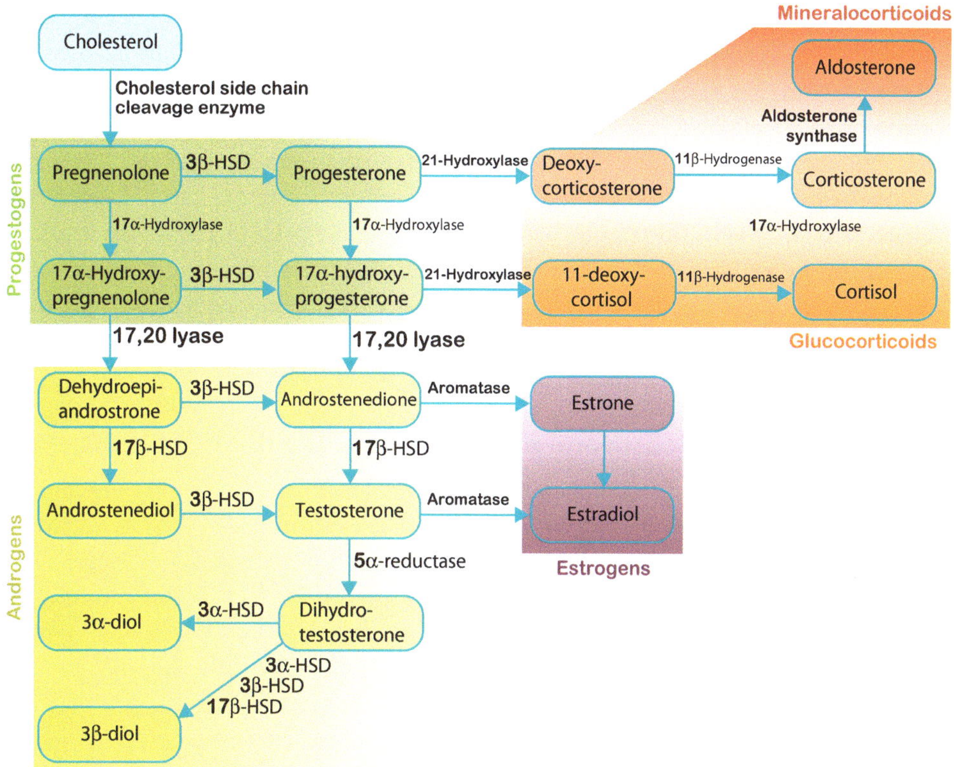


**Figure 13.4** ***Diagram depicting major pathways in the biosynthesis of progestogens, androgens, estrogens, and corticosteroids from cholesterol precursor (grouped by color).*** Steroidogenic enzymes that are involved in the metabolic conversion of steroids are indicated in bold.

presents an interesting prereceptor regulation of the ligand through targeted metabolism of interfering ligands.

Gonadal steroid hormones can be further divided into androgens, estrogens, and progestogens. In males, androgen production and secretion is dominated by the testes, and in both sexes, to a lesser extent, produced and secreted by the adrenal cortex Nelson, 2000; see Figure 13.4. The principal androgen is T; however, its more potent metabolite in terms of AR binding is DHT, produced in tissues that express the enzyme 5α-reductase (Lephart et al., 2001). While T is functionally responsible for the development of male secondary sex characteristics and the maintenance of reproductive function (Shifren et al., 2000; Goldstat et al., 2003), it also has anabolic functions of muscle mass development (Ferrando et al., 2002) and red blood cell production (Gordon et al., 1970).

Similarly, estrogens and progesterone are produced and secreted by the granulosa cells of the ovaries, and, to a much lesser extent, by the zona reticularis of the adrenal cortex in both sexes (Nelson, 2000). Males also produce estrogens and females androgens, though in comparably smaller amounts. Estradiol is the most potent estrogen and is produced from the metabolism of androgens (T or androstenedione) by the aromatase enzyme (Roselli et al., 1985). Estrogens also prompt the development of secondary female sex characteristics, and enhance sperm maturation in males (Nelson, 2000). Progesterone is a precursor of androgens, estrogens, and corticosteroids. It is derived from cholesterol by the P450 side chain cleavage enzyme and 3β-HSD. It exhibits its role in reproduction by maintaining the endometrial lining for pregnancy. It is also secreted by the ovaries and adrenal glands, as well as by the placenta during pregnancy. Importantly, progestins, androgens, and estrogens share the same biosynthetic pathway, originating with the uptake and metabolism of cholesterol (Figure 13.4).

## 6.2 Androgen metabolism and prereceptor regulation of the HPA axis

Regulation of the HPA axis is dependent upon several factors, including the types of steroid in circulation, their levels, and the availability and distribution of their respective receptors. In addition to these regulatory factors, the metabolic conversion and availability of active steroid ligands can also modulate HPA axis reactivity. Enzymatic steps taken in order to synthesize an active ligand are classified as prereceptor modulations. Enzymatic conversions were first identified in the activation of glucocorticoids by types I and II, 11β-HSD1 and 11β-HSD2, respectively (Funder et al., 1988; White, 2001; Whorwood and Stewart, 1996). Moreover, androgen metabolism is central to the prereceptor regulation of the HPA axis (Handa et al., 2009).

Steroid hormone metabolism has been well studied in peripheral and CNS tissues. The main gonadal steroids regulating the HPA axis, estrogens and androgens, are synthetized from T. Testosterone can be intracellularly metabolized into 17β-estradiol through the actions of aromatase (CYP19A1) enzyme (Roselli et al., 1985), or reduced to DHT by the enzyme 5α-reductase (Lephart et al., 2001). The actions of T in modulation of the HPA axis are highly regulated through metabolism to its more potent form, DHT. DHT is not a substrate for aromatization and thus for years has been used as a prototypical androgen, thought to be capable of only activating the AR. In addition, 5α-reductase levels could regulate ligand availability and serve as a crucial step for prereceptor regulation in the HPA axis (Handa et al., 2009). The complexities of this system are only beginning to come to light in the last decade.

Estradiol also mediates HPA axis activity in a dichotomous manner, perhaps due to its ability to bind to ERs, ERα, and ERβ, with subnanomolar affinity (Kuiper et al., 1998). Functioning through ERα, estradiol enhances stress-induced CORT secretion and CRH mRNA expression, whereas the activation of ERβ has the opposite effect, reducing CORT and ACTH and inhibiting CRH mRNA induction. This "yin-yang" mechanism of regulation is important for adjusting the gain of the axis in response to changes in reproduction. Of importance, it has recently been shown that the metabolism of DHT into other androgens, such as 3α-diol or 3β-diol, by three enzymes in the aldo-keto reductase superfamily: 3α-, 3β-, and 17β-HSD (Gangloff et al., 2003; Jin and Penning, 2001; Steckelbroeck et al., 2004; Torn et al., 2003; Weihua et al., 2002) can also have an impact on the HPA axis. 3β-diol reportedly has a significantly higher selectivity for ERβ than ERα (Kuiper et al., 1998) and works to decrease HPA axis activity by binding to ERβ (Lund et al., 2005; Oyola et al., 2012). A role for 3α-diol in the regulation of the HPA axis has not yet been demonstrated. In summary, the metabolism of T in the brain may involve three sites for HPA axis regulation: (1) through 5α-reduction and the production of the androgen, DHT; (2) through aromatization of androgens for the production of estradiol; and (3) through 3α-, 3β-, and 17β-HSD in the production of 3β-diol. The final step of 3β-diol metabolism, or termination, takes place by the enzymatic actions of CYP7B1, which converts 3β-diol into inactive 6α- or 7α-triols, providing yet another prereceptor regulation of the overall amount of ERβ-binding compounds on the HPA axis (Weihua et al., 2002; Sundin et al., 1987). Moreover, another compound, 3α-diol, can be converted back into DHT (Bauman et al., 2006); therefore, it can serve as a sink for further androgen production. Importantly, none of these estrogenic metabolites are able to efficiently bind to the AR (Handa et al., 1987), and analogously, ERs do not bind DHT (Kuiper et al., 1997; Handa et al., 1987), thus rendering direct estrogenic actions on androgenic pathways, or T or DHT regulation through ER pathways on the HPA axis unlikely.

The type of steroid that modulates the HPA axis, and its availability, which is dependent upon spatial-specific expression of enzymes in the brain, are necessary for HPA axis regulation by steroid metabolites. Pattern-specific expressions of the enzymes involved in prereceptor regulation are critical for the sensitivity to circulating glucocorticoids and further regulation of the HPA axis. The two enzymes involved in the metabolism of glucocorticoids are 11β-HSD1 and 11β-HSD2. While 11β-HSD1 converts biologically inactive cortisone into active cortisol, 11β-HSD2 does the reverse (Tomlinson et al., 2004). In rodents, these same enzymes also metabolize CORT to its inactive form and back. Furthermore, since MRs have the highest affinity for CORT and appear to be nearly perpetually occupied by corticosteroids, 11β-HSD2 prevents the activation of MRs in the brain regions in which it is expressed by converting CORT to its biologically inactive metabolite (de Kloet et al., 2000). Playing the reverse role, 11β-HSD1 increases glucocorticoid signal, and thus enhances MR binding by increasing local CORT levels.

Rodent studies have identified 11β-HSD1 as a prereceptor modulator of HPA axis activity (Harris et al., 2001). This regulatory enzyme is mainly found in the pituitary gland, the PVN, and the hippocampus (Yard and Kadowitz, 1972; Alzamora et al., 2000; Brem et al., 1995). Not surprisingly, these are also regions where the highest expression of corticosteroid receptors is found. Supporting these findings, mice lacking 11β-HSD1 show both high plasma CORT and ACTH levels with extended CORT peaks (Harris et al., 2001). Additionally, upon exposure to an acute stressor, 11β-HSD1-null mice present exacerbated CORT and ACTH levels and protracted homeostatic recovery. Furthermore, their negative feedback response to exogenous glucocorticoid treatment is also impaired (Souness et al., 2002). Upstream metabolic disruption of 11β-HSD1 by disabling cofactors vital for its production also results in negative feedback impairments (Muto et al., 1998). Of note, these deleterious effects have been shown to be strain-dependent (Whitworth et al., 2002). Finally, aldosterone's actions on MRs are highly mediated by prereceptor modulation. CORT's and aldosterone's affinities for MRs are similar, with CORT circulating at greater concentrations than aldosterone. Mineralocorticoid tissue-specificity is mainly determined by the actions of the enzyme 11β-HSD2, which renders the glucocorticoids inactive, thus allowing MRs to be available for binding by aldosterone (Funder, 1990). In summary, enzymatic steps involved in the regulation of steroid receptor ligands are critical prereceptor modulatory steps for the regulation of HPA axis homeostasis.

## 6.3 Evidence for organizational sex differences in HPA

The organization of the HPA axis, and consequential postnatal neuroendocrine responses to perturbations in homeostasis, are directed by gonadal steroids before and after birth (McCormick and Mathews, 2007). Exposure to different levels of steroid hormones at critical pre- and perinatal stages of life results in abnormalities in the HPA axis response to stress in adulthood. For example, removal of gonads in the male rat immediately after birth results in aberrantly elevated levels of ACTH and CORT in response to restraint stress, in addition to hyperactivation of the PVN during adulthood (McCormick et al., 1998). Interestingly, when these castrated animals are treated with T as adults, the detrimental effects are reversed (Bingham and Viau, 2008). Females treated with testosterone propionate 24 h after birth results in the neuroendocrine response to stress mimicking that of the male HPA axis, characterized by a reduced CORT response and PVN CRH and AVP expression levels, and greater PVN GR mRNA expression (Seale et al., 2005a). Readministration of exogenous estradiol in adulthood, further increases the response to stress (activational actions). Similar effects are also observed when T is injected to females on postnatal day 1 and continued for 1 week. T administration to newly born females after the first week of life does not result in defeminization of the HPA axis upon stress (Goel and Bale, 2008), and these animals grow up with an HPA axis reactivity that is similar to that of normal females. As previously mentioned, the organizational sex effects of T are due, in part, to its aromatization to estradiol, since inhibition of aromatase during perinatal stages results in increased CORT and PVN activation following restraint stress (Bingham et al., 2012). In a series of elegant studies, Seale et al. not only reiterate the perinatal organizational effects of T, specifically on CORT pulse patterns in adulthood, but also suggest a similar organizational importance of estrogen exposure during the initial 24-h window (Seale et al., 2004a, 2004b, 2005a, 2005b). Female rats that received T treatment on postnatal day 1 had overall lower amplitude CORT peaks, similar to males, as compared to adult females that were injected with vehicle during postnatal life (Seale et al., 2005a). Furthermore, the masculinized CORT pulse patterns of adult females that were injected with T at birth could be reversed by estradiol pellet implantation in adulthood, indicating that they were still responsive to the activational effects of estradiol. Similarly, males administered with either flutamide (an AR antagonist) or 1,4,6-androstatriene-3,17-dione (an aromatase inhibitor) 24 h after birth exhibited female-typical CORT pulse patterns in adulthood, showing that T's organizational actions may take place through both the estrogen and androgen pathways (Seale et al., 2005b). Taken together, these data support the notion that regulation of gonadal steroids plays an important role in the brain's organizational sex differences.

## 7 THE ROLE OF STEROID HORMONES IN MEDIATING SEX DIFFERENCES IN ANXIETY AND DEPRESSION

### 7.1 Limbic circuitry underlying anxiety and depression

Increasing data indicate that steroid hormones are involved in the modulation of not only the HPA axis, but also the behavioral responses to stress. Estrogens elicit dichotomous behavioral responses, depending on the receptor they bind to. Activation of ERβ decreases responsiveness of the HPA axis and correspondingly increases anxiolytic behaviors. This was posited by Kretzel et al. (2001) who showed that global deletion of ERβ increases anxiety-like behaviors in open field activity (OFA) and the elevated plus maze (EPM) (Krezel et al., 2001). It has been proposed that ERβ works in conjunction with the brain's serotonergic system to modify anxiety and depressive-like behaviors. This suggestion is supported by the studies of Levkovitz and Segal (1997) that showed an increased expression of 5-HT1a receptor, which primarily regulates GABA function (Levkovitz and Segal, 1997). Changes in serotonin (5-HT) in ERβ-null mice were also noted in the BnST, POA, and hippocampus following exposure to the EPM (Imwalle et al., 2005). Direct agonism of ERβ in the main 5-HT-synthesizing brain nuclei, the dorsal raphe, increased the expression of the brain-specific, rate-limiting enzyme for the production of 5-HT, tryptophan hydroxylase 2 (TPH2) (Donner and Handa, 2009). Furthermore, the THP2 enzyme is coexpressed by more than 90% of ERβ-positive cells in the dorsal raphe and periaqueductal gray matter of the mouse (Gundlah et al., 2001), guinea pig (Lu et al., 1999), and primate (Nomura et al., 2005). Collectively, these studies indicate that estradiol works via ERβ to enhance 5-HT actions, part of which has been shown to control HPA axis function and behaviors.

In addition to anxiety-like behaviors, ERβ has been also implicated in mediating depression-like behaviors. In rodents depressive-like behaviors are measured by the amount of time the animal struggles during the tail suspension test or time they remain immobile in the forced swim test (Porsolt et al., 1977). Rocha et al. (2005) demonstrated the importance of ERβ as a regulator of depressive-like behaviors using estradiol administration to ERβ global knockout and wild-type mice, the former of which showed lower antidepressive-like behaviors. ERβ activation also reduces depressive-like behaviors in the forced swim test in rats (Clark et al., 2012; Donner and Handa, 2009).

Further support that ERβ activation decreases anxiety-like behaviors comes from pharmacological experiments in rodents in which subtype-selective ER agonists were used. Unlike its effect in wild-type mice, the administration of an ERβ agonist to ERβ-null mice did not affect anxiolytic behaviors (Oyola et al., 2012), as measured by time spent in the perimeter of the open arena and closed arms in the OFA and EPM tests, respectively. This finding has also been observed in female ovariectomized mice and rats treated with the same agonist (Lund et al., 2005; Walf and Frye, 2005; Weiser et al., 2009). The results of these experiments are attributed to ERβ being a key mediator of anxiolytic behaviors, although the brain site(s) mediating its actions remain unknown. Conversely, selective activation of ERα using the agonist propyl-pyrazole-triol not only enhances HPA axis reactivity (Lund et al., 2006), but also increases anxiety-like behaviors in the EPM and forced swim test, respectively (Lund et al., 2005; Walf and Frye, 2006). Collectively, these studies indicate that the opposing actions of these two receptor subtypes may account for estradiol's dichotomous actions on both the HPA axis and behavioral responses to stress.

## 8 CONCLUSIONS

In summary, current studies indicate the indispensable nature of gonadal steroid hormones in the sex-dependent organization of the brain and its consequent activational actions which regulate stress responses to environmental threats by the HPA axis. Several checkpoints exist for the regulation of this stress circuitry, some of which are sex-specific. Type of steroid, steroid receptor expression, spatial organization of these receptors, prereceptor modulation of the enzymes involved in the synthesis of steroid receptor ligands, and their spatial organization work together to carefully maintain the system's homeostatic state. Changes to any of the above variables for a period of time can result in disease states, including stress-related psychiatric disorders (i.e., anxiety, posttraumatic stress disorder, depression), compromised immune, cardiovascular, and excretory systems, and memory impairments. Due to the existing gaps in knowledge on the HPA axis microcircuitry, further research is needed to identify the subtype-specific components in both developmental programming and adult-specific effects of steroid hormone signaling. Such information would advance the development of target-specific therapeutics to combat the above-mentioned diseases.

## REFERENCES


Abe, K., Critchlow, V., 1980. Delayed feedback inhibition of stress-induced activation of pituitary-adrenal function: effects of varying dose, rate and duration of corticosterone administration and of telencephalon removal. Neuroendocrinology 31 (5), 349–354.

Ahima, R.S., Harlan, R.E., 1990. Charting of type II glucocorticoid receptor-like immunoreactivity in the rat central nervous system. Neuroscience 39 (3), 579–604.

Ahima, R., Krozowski, Z., Harlan, R., 1991. Type I corticosteroid receptor-like immunoreactivity in the rat CNS: distribution and regulation by corticosteroids. J. Comp. Neurol. 313 (3), 522–538.

Ahmed, E.I., Zehr, J.L., Schulz, K.M., Lorenz, B.H., DonCarlos, L.L., Sisk, C.L., 2008. Pubertal hormones modulate the addition of new cells to sexually dimorphic brain regions. Nat. Neurosci. 11 (9), 995–997.

Akana, S.F., Cascio, C.S., Du, J.Z., Levin, N., Dallman, M.F., 1986. Reset of feedback in the adrenocortical system: an apparent shift in sensitivity of adrenocorticotropin to inhibition by corticosterone between morning and evening. Endocrinology 119 (5), 2325–2332.

Almeida, O.F., Conde, G.L., Crochemore, C., Demeneix, B.A., Fischer, D., Hassan, A.H., Meyer, M., Holsboer, F., Michaelidis, T.M., 2000. Subtle shifts in the ratio between pro- and antiapoptotic molecules after activation of corticosteroid receptors decide neuronal fate. FASEB J. 14 (5), 779–790.

Alzamora, R., Michea, L., Marusic, E.T., 2000. Role of 11beta-hydroxysteroid dehydrogenase in nongenomic aldosterone effects in human arteries. Hypertension 35 (5), 1099–1104.

Armario, A., 2010. Activation of the hypothalamic–pituitary–adrenal axis by addictive drugs: different pathways, common outcome. Trends Pharmacol. Sci. 31 (7), 318–325.

Arnett, M.G., Kolber, B.J., Boyle, M.P., Muglia, L.J., 2011. Behavioral insights from mouse models of forebrain-and amygdala-specific glucocorticoid receptor genetic disruption. Mol. Cell Endocrinol. 336, 1–2, 2–5.

Arnold, A.P., 1996. Genetically triggered sexual differentiation of brain and behavior. Horm. Behav. 30 (4), 495–505.

Arnold, A.P., 2009. The organizational-activational hypothesis as the foundation for a unified theory of sexual differentiation of all mammalian tissues. Horm. Behav. 55 (5), 570–578.

Aronsson, M., Fuxe, K., Dong, Y., Agnati, L.F., Okret, S., Gustafsson, J.A., 1988. Localization of glucocorticoid receptor mRNA in the male rat brain by *in situ* hybridization. Proc. Natl. Acad. Sci. USA 85 (23), 9331–9335.

Ayroldi, E., Cannarile, L., Migliorati, G., Nocentini, G., Delfino, D.V., Riccardi, C., 2012. Mechanisms of the anti-inflammatory effects of glucocorticoids: genomic and nongenomic interference with MAPK signaling pathways. FASEB J. 26 (12), 4805–4820.

Azcoitia, I., Sierra, A., Garcia-Segura, L.M., 1999. Localization of estrogen receptor beta-immunoreactivity in astrocytes of the adult rat brain. Glia 26 (3), 260–267.

Bauman, D.R., Steckelbroeck, S., Williams, M.V., Peehl, D.M., Penning, T.M., 2006. Identification of the major oxidative 3alpha-hydroxysteroid dehydrogenase in human prostate that converts 5alpha-androstane-3alpha, 17beta-diol to 5alpha-dihydrotestosterone: a potential therapeutic target for androgen-dependent disease. Mol. Endocrinol. 20 (2), 444–458.

Beato, M., 1989. Gene regulation by steroid hormones. Cell 56 (3), 335–344.

Beato, M., Klug, J., 2000. Steroid hormone receptors: an update. Hum. Reprod. Update 6 (3), 225–236.

Bellavance, M.A., Rivest, S., 2012. The neuroendocrine control of the innate immune system in health and brain diseases. Immunol. Rev. 248 (1), 36–55.

Berger, S., Wolfer, D.P., Selbach, O., Alter, H., Erdmann, G., Reichardt, H.M., Chepkova, A.N., Welzl, H., Haas, H.L., Lipp, H.P., Schutz, G., 2006. Loss of the limbic mineralocorticoid receptor impairs behavioral plasticity. Proc. Natl. Acad. Sci. USA 103 (1), 195–200.

Biag, J., Huang, Y., Gou, L., Hintiryan, H., Askarinam, A., Hahn, J.D., Toga, A.W., Dong, H.W., 2012. Cyto- and chemoarchitecture of the hypothalamic paraventricular nucleus in the C57BL/6j male mouse: a study of immunostaining and multiple fluorescent tract tracing. J. Comp. Neurol. 520 (1), 6–33.

Bingaman, E.W., Magnuson, D.J., Gray, T.S., Handa, R.J., 1994. Androgen inhibits the increases in hypothalamic corticotropin-releasing hormone (CRH) and CRH-immunoreactivity following gonadectomy. Neuroendocrinology 59 (3), 228–234.

Bingham, B., Viau, V., 2008. Neonatal gonadectomy and adult testosterone replacement suggest an involvement of limbic arginine vasopressin and androgen receptors in the organization of the hypothalamic-pituitary-adrenal axis. Endocrinology 149 (7), 3581–3591.

Bingham, B., Gray, M., Sun, T., Viau, V., 2011. Postnatal blockade of androgen receptors or aromatase impair the expression of stress hypothalamic-pituitary-adrenal axis habituation in adult male rats. Psychoneuroendocrinology 36 (2), 249–257.

Bingham, B., Wang, N.X., Innala, L., Viau, V., 2012. Postnatal aromatase blockade increases c-fos mRNA responses to acute restraint stress in adult male rats. Endocrinology 153 (4), 1603–1608.

Bondesson, M., Hao, R., Lin, C.Y., Williams, C., Gustafsson, J.A., 2014. Estrogen receptor signaling during vertebrate development. Biochim. Biophys. Acta.

Boyle, M.P., Brewer, J.A., Funatsu, M., Wozniak, D.F., Tsien, J.Z., Izumi, Y., Muglia, L.J., 2005. Acquired deficit of forebrain glucocorticoid receptor produces depression-like changes in adrenal axis regulation and behavior. Proc. Natl. Acad. Sci. USA 102 (2), 473–478.

Bradbury, M.J., Akana, S.F., Dallman, M.F., 1994. Roles of type I and II corticosteroid receptors in regulation of basal activity in the hypothalamo-pituitary-adrenal axis during the diurnal trough and the peak: evidence for a nonadditive effect of combined receptor occupation. Endocrinology 134 (3), 1286–1296.

Brem, A.S., Bina, R.B., King, T., Morris, D.J., 1995. Bidirectional activity of 11 beta-hydroxysteroid dehydrogenase in vascular smooth muscle cells. Steroids 60 (5), 406–410.

Brennan, F.E., Fuller, P.J., 1999. Acute regulation by corticosteroids of channel-inducing factor gene messenger ribonucleic acid in the distal colon. Endocrinology 140 (3), 1213–1218.

Brennan, F.E., Fuller, P.J., 2006. Mammalian K-Ras2 is a corticosteroid-induced gene *in vivo*. Endocrinology 147 (6), 2809–2816.

Brinkmann, A.O., 2011. Molecular mechanisms of androgen action – a historical perspective. Methods Mol. Biol. 776, 3–24.

Buchanan, K.L., 2000. Stress and the evolution of condition-dependent signals. Trends Ecol. Evol. 15 (4), 156–160.

Buijs, R.M., Wortel, J., Van Heerikhuize, J.J., Feenstra, M.G., Ter Horst, G.J., Romijn, H.J., Kalsbeek, A., 1999. Anatomical and functional demonstration of a multisynaptic suprachiasmatic nucleus adrenal (cortex) pathway. Eur. J. Neurosci. 11 (5), 1535–1544.

Burgess, L.H., Handa, R.J., 1992. Chronic estrogen-induced alterations in adrenocorticotropin and corticosterone secretion, and glucocorticoid receptor-mediated functions in female rats. Endocrinology 131 (3), 1261–1269.

Busillo, J.M., Cidlowski, J.A., 2013. The five Rs of glucocorticoid action during inflammation: ready, reinforce, repress, resolve, and restore. Trends Endocrinol. Metab. 24 (3), 109–119.

Canny, B.J., 1990. Hippocampal glucocorticoid receptors and the regulation of ACTH secretion. Mol. Cell Endocrinol. 71 (1), C35–38.

Cao, J., Patisaul, H.B., 2011. Sexually dimorphic expression of hypothalamic estrogen receptors alpha and beta and kiss1 in neonatal male and female rats. J. Comp. Neurol. 519 (15), 2954–2977.

Cascio, C.S., Shinsako, J., Dallman, M.F., 1987. The suprachiasmatic nuclei stimulate evening ACTH secretion in the rat. Brain Res. 423, 1–2, 173–178.

Ceccatelli, S., Cintra, A., Hokfelt, T., Fuxe, K., Wikstrom, A.C., Gustafsson, J.A., 1989. Coexistence of glucocorticoid receptor-like immunoreactivity with neuropeptides in the hypothalamic paraventricular nucleus. Exp. Brain Res. 78 (1), 33–42.

Chaban, V., Li, J., McDonald, J.S., Rapkin, A., Micevych, P., 2011. Estradiol attenuates the adenosine triphosphate-induced increase of intracellular calcium through group ii metabotropic glutamate receptors in rat dorsal root ganglion neurons. J. Neurosci. Res. 89 (11), 1707–1710.

Chao, H.M., Choo, P.H., McEwen, B.S., 1989. Glucocorticoid and mineralocorticoid receptor mRNA expression in rat brain. Neuroendocrinology 50 (4), 365–371.

Charn, T.H., Liu, E.T., Chang, E.C., Lee, Y.K., Katzenellenbogen, J.A., Katzenellenbogen, B.S., 2010. Genome-wide dynamics of chromatin binding of estrogen receptors alpha and beta: mutual restriction and competitive site selection. Mol. Endocrinol. 24 (1), 47–59.

Chu, S., Fuller, P.J., 1997. Identification of a splice variant of the rat estrogen receptor beta gene. Mol. Cell Endocrinol. 132, 1–2, 195–199.

Cintra, A., Bhatnagar, M., Chadi, G., Tinner, B., Lindberg, J., Gustafsson, J.A., Agnati, L.F., Fuxe, K., 1994. Glial and neuronal glucocorticoid receptor immunoreactive cell populations in developing, adult, and aging brain. Ann. NY Acad. Sci. 746, 42–61, discussion 61–63.

Clark, J.A., Alves, S., Gundlah, C., Rocha, B., Birzin, E.T., Cai, S.J., Flick, R., Hayes, E., Ho, K., Warrier, S., Pai, L., Yudkovitz, J., Fleischer, R., Colwell, L., Li, S., Wilkinson, H., Schaeffer, J., Wilkening, R., Mattingly, E., Hammond, M., Rohrer, S.P., 2012. Selective estrogen receptor-beta (SERM-beta) compounds modulate raphe nuclei tryptophan hydroxylase-1 (Tph-1) mRNA expression and cause antidepressant-like effects in the forced swim test. Neuropharmacology 63 (6), 1051–1063.

Coirini, H., Magarinos, A.M., De Nicola, A.F., Rainbow, T.C., McEwen, B.S., 1985. Further studies of brain aldosterone binding sites employing new mineralocorticoid and glucocorticoid receptor markers *in vitro*. Brain Res. 361, 1–2, 212–216.

Condon, J.C., Pezzi, V., Drummond, B.M., Yin, S., Rainey, W.E., 2002. Calmodulin-dependent kinase I regulates adrenal cell expression of aldosterone synthase. Endocrinology 143 (9), 3651–3657.

Crochemore, C., Lu, J., Wu, Y., Liposits, Z., Sousa, N., Holsboer, F., Almeida, O.F., 2005. Direct targeting of hippocampal neurons for apoptosis by glucocorticoids is reversible by mineralocorticoid receptor activation. Mol. Psychiat. 10 (8), 790–798.

Dallman, M.F., Makara, G.B., Roberts, J.L., Levin, N., Blum, M., 1985. Corticotrope response to removal of releasing factors and corticosteroids *in vivo*. Endocrinology 117 (5), 2190–2197.

Dallman, M.F., Akana, S.F., Cascio, C.S., Darlington, D.N., Jacobson, L., Levin, N., 1987. Regulation of ACTH secretion: variations on a theme of B. Recent Prog. Horm. Res. (43), 113–173.

Datson, N.A., van der Perk, J., de Kloet, E.R., Vreugdenhil, E., 2001. Identification of corticosteroid-responsive genes in rat hippocampus using serial analysis of gene expression. Eur. J. Neurosci. 14 (4), 675–689.

de Kloet, E.R., Oitzl, M.S., Joels, M., 1993. Functional implications of brain corticosteroid receptor diversity. Cell Mol. Neurobiol. 13 (4), 433–455.

De Kloet, E.R., Vreugdenhil, E., Oitzl, M.S., Joels, M., 1998. Brain corticosteroid receptor balance in health and disease. Endocr. Rev. 19 (3), 269–301.

de Kloet, E.R., Van Acker, S.A., Sibug, R.M., Oitzl, M.S., Meijer, O.C., Rahmouni, K., de Jong, W., 2000. Brain mineralocorticoid receptors and centrally regulated functions. Kidney Int. 57 (4), 1329–1336.

Diamond, M.I., Miner, J.N., Yoshinaga, S.K., Yamamoto, K.R., 1990. Transcription factor interactions: selectors of positive or negative regulation from a single DNA element. Science 249 (4974), 1266–1272.

Diz-Chaves, Y., Kwiatkowska-Naqvi, A., Von Hulst, H., Pernia, O., Carrero, P., Garcia-Segura, L.M., 2012. Behavioral effects of estradiol therapy in ovariectomized rats depend on the age when the treatment is initiated. Exp. Gerontol. 47 (1), 93–99.

DonCarlos, L.L., Sarkey, S., Lorenz, B., Azcoitia, I., Garcia-Ovejero, D., Huppenbauer, C., Garcia-Segura, L.M., 2006. Novel cellular phenotypes and subcellular sites for androgen action in the forebrain. Neuroscience 138 (3), 801–807.

Dong, H.W., Petrovich, G.D., Swanson, L.W., 2001. Topography of projections from amygdala to bed nuclei of the stria terminalis. Brain Res. Brain Res. Rev. 38, 1–2, 192–246.

Donner, N., Handa, R.J., 2009. Estrogen receptor beta regulates the expression of tryptophan-hydroxylase 2 mRNA within serotonergic neurons of the rat dorsal raphe nuclei. Neuroscience 163 (2), 705–718.

Edwards, A.V., Jones, C.T., 1993. Autonomic control of adrenal function. J. Anat. 183 (Pt 2), 291–307.

Eick, G.N., Thornton, J.W., 2011. Evolution of steroid receptors from an estrogen-sensitive ancestral receptor. Mol. Cell Endocrinol. 334, 1–2, 31–38.

Engeland, W.C., 1998. Functional innervation of the adrenal cortex by the splanchnic nerve. Horm. Metab. Res. 30, 6–7, 311–314.

Evans, R.M., 1988. The steroid and thyroid hormone receptor superfamily. Science 240 (4854), 889–895.

Evans, R.M., Arriza, J.L., 1989. A molecular framework for the actions of glucocorticoid hormones in the nervous system. Neuron 2 (2), 1105–1112.

Fahrbach, S.E., Meisel, R.L., Pfaff, D.W., 1985. Preoptic implants of estradiol increase wheel running but not the open field activity of female rats. Physiol. Behav. 35 (6), 985–992.

Ferrando, A.A., Sheffield-Moore, M., Yeckel, C.W., Gilkison, C., Jiang, J., Achacosa, A., Lieberman, S.A., Tipton, K., Wolfe, R.R., Urban, R.J., 2002. Testosterone administration to older men improves muscle function: molecular and physiological mechanisms. Am. J. Physiol. Endocrinol. Metab. 282 (3), E601–E607.

Foradori, C.D., Weiser, M.J., Handa, R.J., 2008. Non-genomic actions of androgens. Front. Neuroendocrinol. 29 (2), 169–181.

Funder, J.W., 1990. Target tissue specificity of mineralocorticoids. Trends Endocrinol. Metab. 1 (3), 145–148.

Funder, J.W., Pearce, P.T., Smith, R., Smith, A.I., 1988. Mineralocorticoid action: target tissue specificity is enzyme, not receptor, mediated. Science 242 (4878), 583–585.

Gangloff, A., Shi, R., Nahoum, V., Lin, S.X., 2003. Pseudo-symmetry of C19 steroids, alternative binding orientations, and multispecificity in human estrogenic 17beta-hydroxysteroid dehydrogenase. FASEB J. 17 (2), 274–276.

Gaskin, J.H., Kitay, J.I., 1970. Adrenocortical function in the hamster. Sex differences and effects of gonadal hormones. Endocrinology 87 (4), 779–786.

Goel, N., Bale, T.L., 2008. Organizational and activational effects of testosterone on masculinization of female physiological and behavioral stress responses. Endocrinology 149 (12), 6399–6405.

Goel, N., Workman, J.L., Lee, T.T., Innala, L., Viau, V., 2014. Sex differences in the HPA axis. Compr. Physiol. 4 (3), 1121–1255.

Goldstat, R., Briganti, E., Tran, J., Wolfe, R., Davis, S.R., 2003. Transdermal testosterone therapy improves well-being, mood, and sexual function in premenopausal women. Menopause 10 (5), 390–398.

Gordon, A.S., Zanjani, E.D., Levere, R.D., Kappas, A., 1970. Stimulation of mammalian erythropoiesis by 5beta-H steroid metabolites. Proc. Natl. Acad. Sci. USA 65 (4), 919–924.

Green, S., Walter, P., Kumar, V., Krust, A., Bornert, J.M., Argos, P., Chambon, P., 1986. Human oestrogen receptor cDNA: sequence, expression and homology to V-Erb-A. Nature 320 (6058), 134–139.

Grino, P.B., Griffin, J.E., Wilson, J.D., 1990. Testosterone at high concentrations interacts with the human androgen receptor similarly to dihydrotestosterone. Endocrinology 126 (2), 1165–1172.

Grober, O.M., Mutarelli, M., Giurato, G., Ravo, M., Cicatiello, L., De Filippo, M.R., Ferraro, L., Nassa, G., Papa, M.F., Paris, O., Tarallo, R., Luo, S., Schroth, G.P., Benes, V., Weisz, A., 2011. Global analysis of estrogen receptor beta binding to breast cancer cell genome reveals an extensive interplay with estrogen receptor alpha for target gene regulation. BMC Genomics 12, 36.

Gundlah, C., Lu, N.Z., Mirkes, S.J., Bethea, C.L., 2001. Estrogen receptor beta (ERbeta) mRNA and protein in serotonin neurons of macaques. Brain. Res. Mol. Brain Res. 91, 1–2, 14–22.

Gurdjian, E.S., 1927. The diencephalon of the albino rat: studies on the brain of the rat. J. Comp. Neurol. 43, 1–114.

Hamson, D.K., Jones, B.A., Csupity, A.S., Ali, F.M., Watson, N.V., 2014. Androgen insensitive male rats display increased anxiety-like behavior on the elevated plus maze. Behav. Brain Res. (259), 158–163.

Handa, R.J., Weiser, M.J., 2014. Gonadal steroid hormones and the hypothalamo–pituitary–adrenal axis. Front. Neuroendocrinol. 35 (2), 197–220.

Handa, R.J., Reid, D.L., Resko, J.A., 1986. Androgen receptors in brain and pituitary of female rats: cyclic changes and comparisons with the male. Biol. Reprod. 34 (2), 293–303.

Handa, R.J., Stadelman, H.L., Resko, J.A., 1987. Effect of estrogen on androgen receptor dynamics in female rat pituitary. Endocrinology 121 (1), 84–89.

Handa, R.J., Burgess, L.H., Kerr, J.E., O'Keefe, J.A., 1994a. Gonadal steroid hormone receptors and sex differences in the hypothalamo–pituitary–adrenal axis. Horm. Behav. 28 (4), 464–476.

Handa, R.J., Nunley, K.M., Lorens, S.A., Louie, J.P., McGivern, R.F., Bollnow, M.R., 1994b. Androgen regulation of adrenocorticotropin and corticosterone secretion in the male rat following novelty and foot shock stressors. Physiol. Behav. 55 (1), 117–124.

Handa, R.J., Pak, T.R., Kudwa, A.E., Lund, T.D., Hinds, L., 2008. An alternate pathway for androgen regulation of brain function: activation of estrogen receptor beta by the metabolite of dihydrotestosterone, 5alpha-androstane-3beta,17beta-diol. Horm. Behav. 53 (5), 741–752.

Handa, R.J., Weiser, M.J., Zuloaga, D.G., 2009. A role for the androgen metabolite, 5α-androstane-3β,17β-diol, in modulating oestrogen receptor β-mediated regulation of hormonal stress reactivity. J. Neuroendocrinol. 21 (4), 351–358.

Handa, R.J., Mani, S.K., Uht, R.M., 2012. Estrogen receptors and the regulation of neural stress responses. Neuroendocrinology 96 (2), 111–118.

Handa, R.J., Kudwa, A.E., Donner, N.C., McGivern, R.F., Brown, R., 2013. Central 5-alpha reduction of testosterone is required for testosterone's inhibition of the hypothalamo-pituitary-adrenal axis response to restraint stress in adult male rats. Brain Res. (1529), 74–82.

Harris, H.J., Kotelevtsev, Y., Mullins, J.J., Seckl, J.R., Holmes, M.C., 2001. Intracellular regeneration of glucocorticoids by 11beta-hydroxysteroid dehydrogenase (11beta-HSD)-1 plays a key role in regulation of the hypothalamic-pituitary-adrenal axis: analysis of 11beta-HSD-1-deficient mice. Endocrinology 142 (1), 114–120.

Harris, A.P., Holmes, M.C., de Kloet, E.R., Chapman, K.E., Seckl, J.R., 2013. Mineralocorticoid and glucocorticoid receptor balance in control of HPA axis and behaviour. Psychoneuroendocrinology 38 (5), 648–658.

He, W.W., Kumar, M.V., Tindall, D.J., 1991. A frame-shift mutation in the androgen receptor gene causes complete androgen insensitivity in the testicular-feminized mouse. Nucleic Acids Res. 19 (9), 2373–2378.

Hendrickson, A.E., Wagoner, N., Cowan, W.M., 1972. An autoradiographic and electron microscopic study of retino-hypothalamic connections. Z. Zellforsch Mikrosk Anat. 135 (1), 1–26.

Hofman, M.A., Swaab, D.F., 1994. Alterations in circadian rhythmicity of the vasopressin-producing neurons of the human suprachiasmatic nucleus (SCN) with aging. Brain Res. 651, 1–2, 134–142.

Hu, M., Crombag, H.S., Robinson, T.E., Becker, J.B., 2004. Biological basis of sex differences in the propensity to self-administer cocaine. Neuropsychopharmacology 29 (1), 81–85.

Hunter, D.A., Barr, G.A., Amador, N., Shivers, K.Y., Kemen, L., Kreiter, C.M., Jenab, S., Inturrisi, C.E., Quinones-Jenab, V., 2011. Estradiol-induced antinociceptive responses on formalin-induced nociception are independent of cox and HPA activation. Synapse 65 (7), 643–651.

Husak, J.F., Moore, I.T., 2008. Stress hormones and mate choice. Trends Ecol. Evol. 23 (10), 532–534.

Husmann, D.A., Wilson, C.M., McPhaul, M.J., Tilley, W.D., Wilson, J.D., 1990. Antipeptide antibodies to two distinct regions of the androgen receptor localize the receptor protein to the nuclei of target cells in the rat and human prostate. Endocrinology 126 (5), 2359–2368.

Imaki, T., Nahan, J.L., Rivier, C., Sawchenko, P.E., Vale, W., 1991. Differential regulation of corticotropin-releasing factor mRNA in rat brain regions by glucocorticoids and stress. J. Neurosci. 11 (3), 585–599.

Imwalle, D.B., Gustafsson, J.A., Rissman, E.F., 2005. Lack of functional estrogen receptor beta influences anxiety behavior and serotonin content in female mice. Physiol. Behav. 84 (1), 157–163.

Jackson, L.R., Robinson, T.E., Becker, J.B., 2006. Sex differences and hormonal influences on acquisition of cocaine self-administration in rats. Neuropsychopharmacology 31 (1), 129–138.

Jacome, L.F., Gautreaux, C., Inagaki, T., Mohan, G., Alves, S., Lubbers, L.S., Luine, V., 2010. Estradiol and ERbeta agonists enhance recognition memory, and DPN, an ERbeta agonist, alters brain monoamines. Neurobiol. Learn Mem. 94 (4), 488–498.

Jeanneteau, F.D., Lambert, W.M., Ismaili, N., Bath, K.G., Lee, F.S., Garabedian, M.J., Chao, M.V., 2012. BDNF and glucocorticoids regulate corticotrophin-releasing hormone (CRH) homeostasis in the hypothalamus. Proc. Natl. Acad. Sci. USA 109 (4), 1305–1310.

Jin, Y., Penning, T.M., 2001. Steroid 5alpha-reductases and 3alpha-hydroxysteroid dehydrogenases: key enzymes in androgen metabolism. Best Pract. Res. Clin. Endocrinol. Metab. 15 (1), 79–94.

Joels, M., de Kloet, E.R., 1994. Mineralocorticoid and glucocorticoid receptors in the brain. Implications for ion permeability and transmitter systems. Prog. Neurobiol. 43 (1), 1–36.

Kadmiel, M., Cidlowski, J.A., 2013. Glucocorticoid receptor signaling in health and disease. Trends Pharmacol. Sci. 34 (9), 518–530.

Kalsbeek, A., Buijs, R.M., van Heerikhuize, J.J., Arts, M., van der Woude, T.P., 1992. Vasopressin-containing neurons of the suprachiasmatic nuclei inhibit corticosterone release. Brain Res. 580, 1–2, 62–67.

Kalsbeek, A., van der Vliet, J., Buijs, R.M., 1996. Decrease of endogenous vasopressin release necessary for expression of the circadian rise in plasma corticosterone: a reverse microdialysis study. J. Neuroendocrinol. 8 (4), 299–307.

Kalsbeek, A., Fliers, E., Hofman, M.A., Swaab, D.F., Buijs, R.M., 2010a. Vasopressin and the output of the hypothalamic biological clock. J. Neuroendocrinol. 22 (5), 362–372.

Kalsbeek, A., Yi, C.X., la Fleur, S.E., Buijs, R.M., Fliers, E., 2010b. Suprachiasmatic nucleus and autonomic nervous system influences on awakening from sleep. Int. Rev. Neurobiol. 93, 91–107.

Kalsbeek, A., van der Spek, R., Lei, J., Endert, E., Buijs, R.M., Fliers, E., 2012. Circadian rhythms in the hypothalamo–pituitary–adrenal (HPA) axis. Mol. Cell Endocrinol. 349 (1), 20–29.

Kaneko, M., Hiroshige, T., Shinsako, J., Dallman, M.F., 1980. Diurnal changes in amplification of hormone rhythms in the adrenocortical system. Am. J. Physiol. 239 (3), R309–R316.

Kelly, M.J., Ronnekleiv, O.K., 2012. Membrane-initiated actions of estradiol that regulate reproduction, energy balance and body temperature. Front. Neuroendocrinol. 33 (4), 376–387.

Kerr, J.E., Beck, S.G., Handa, R.J., 1996. Androgens selectively modulate c-Fos messenger RNA induction in the rat hippocampus following novelty. Neuroscience 74 (3), 757–766.

King, W.J., Greene, G.L., 1984. Monoclonal antibodies localize oestrogen receptor in the nuclei of target cells. Nature 307 (5953), 745–747.

Kiss, A., Delattre, A.M., Pereira, S.I., Carolino, R.G., Szawka, R.E., Anselmo-Franci, J.A., Zanata, S.M., Ferraz, A.C., 2012. 17beta-estradiol replacement in young, adult and middle-aged female ovariectomized rats promotes improvement of spatial reference memory and an antidepressant effect and alters monoamines and BDNF levels in memory- and depression-related brain areas. Behav. Brain Res. 227 (1), 100–108.

Kousteni, S., Bellido, T., Plotkin, L.I., O'Brien, C.A., Bodenner, D.L., Han, L., Han, K., DiGregorio, G.B., Katzenellenbogen, J.A., Katzenellenbogen, B.S., Roberson, P.K., Weinstein, R.S., Jilka, R.L., Manolagas, S.C., 2001. Nongenotropic, sex-nonspecific signaling through the estrogen or androgen receptors: dissociation from transcriptional activity. Cell 104 (5), 719–730.

Kovacs, K., Kiss, J.Z., Makara, G.B., 1986. Glucocorticoid implants around the hypothalamic paraventricular nucleus prevent the increase of corticotropin-releasing factor and arginine vasopressin immunostaining induced by adrenalectomy. Neuroendocrinology 44 (2), 229–234.

Kovacs, K.J., Foldes, A., Sawchenko, P.E., 2000. Glucocorticoid negative feedback selectively targets vasopressin transcription in parvocellular neurosecretory neurons. J. Neurosci. 20 (10), 3843–3852.

Krezel, W., Dupont, S., Krust, A., Chambon, P., Chapman, P.F., 2001. Increased anxiety and synaptic plasticity in estrogen receptor beta-deficient mice. Proc. Natl. Acad. Sci. USA 98 (21), 12278–12282.

Krieger, D.T., Zimmerman, E.A., 1978. The nature of CRF and its relationship to vasopressin. Clinical Neuroendocrinology. In: Besser, M., Martini, L. (Eds.), Clinical Neuroendocrinology, Academic Press, New York, pp. 363–391.

Kudwa, A.E., Michopoulos, V., Gatewood, J.D., Rissman, E.F., 2006. Roles of estrogen receptors alpha and beta in differentiation of mouse sexual behavior. Neuroscience 138 (3), 921–928.

Kuiper, G.G., Enmark, E., Pelto-Huikko, M., Nilsson, S., Gustafsson, J.A., 1996. Cloning of a novel receptor expressed in rat prostate and ovary. Proc. Nat. Acad. Sci. USA 93 (12), 5925–5930.

Kuiper, G.G., Carlsson, B., Grandien, K., Enmark, E., Haggblad, J., Nilsson, S., Gustafsson, J.A., 1997. Comparison of the ligand binding specificity and transcript tissue distribution of estrogen receptors alpha and beta. Endocrinology 138 (3), 863–870.

Kuiper, G.G., Lemmen, J.G., Carlsson, B., Corton, J.C., Safe, S.H., van der Saag, P.T., van der Burg, B., Gustafsson, J.A., 1998. Interaction of estrogenic chemicals and phytoestrogens with estrogen receptor beta. Endocrinology 139 (10), 4252–4263.

Kumar, V., Green, S., Staub, A., Chambon, P., 1986. Localisation of the oestradiol-binding and putative DNA-binding domains of the human oestrogen receptor. EMBO J. 5 (9), 2231–2236.

Laflamme, N., Nappi, R.E., Drolet, G., Labrie, C., Rivest, S., 1998. Expression and neuropeptidergic characterization of estrogen receptors (ERalpha and ERbeta) throughout the rat brain: anatomical evidence of distinct roles of each subtype. J. Neurobiol. 36 (3), 357–378.

Lai, M., Horsburgh, K., Bae, S.E., Carter, R.N., Stenvers, D.J., Fowler, J.H., Yau, J.L., Gomez-Sanchez, C.E., Holmes, M.C., Kenyon, C.J., Seckl, J.R., Macleod, M.R., 2007. Forebrain mineralocorticoid receptor overexpression enhances memory, reduces anxiety and attenuates neuronal loss in cerebral ischaemia. Eur. J. Neurosci. 25 (6), 1832–1842.

Lavery, D.N., McEwan, I.J., 2005. Structure and function of steroid receptor Af1 transactivation domains: induction of active conformations. Biochem. J. 391 (Pt 3), 449–464.

Lee, S.R., Kim, H.K., Youm, J.B., Dizon, L.A., Song, I.S., Jeong, S.H., Seo, D.Y., Ko, K.S., Rhee, B.D., Kim, N., Han, J., 2012. Non-genomic effect of glucocorticoids on cardiovascular system. Pflugers Arch. 464 (6), 549–559.

Lephart, E.D., Lund, T.D., Horvath, T.L., 2001. Brain androgen and progesterone metabolizing enzymes: biosynthesis, distribution and function. Brain Res. Brain Res. Rev. 37, 1–3, 25–37.

Leret, M.L., Molina-Holgado, F., González, M.I., 1994. The effect of perinatal exposure to estrogens on the sexually dimorphic response to novelty. Physiol. Behav. 55 (2), 371–373.

Levkovitz, Y., Segal, M., 1997. Serotonin 5-HT1a receptors modulate hippocampal reactivity to afferent stimulation. J. Neurosci. 17 (14), 5591–5598.

Lindzey, J., Wetsel, W.C., Couse, J.F., Stoker, T., Cooper, R., Korach, K.S., 1998. Effects of castration and chronic steroid treatments on hypothalamic gonadotropin-releasing hormone content and pituitary gonadotropins in male wild-type and estrogen receptor-alpha knockout mice. Endocrinology 139 (10), 4092–4101.

Liposits, Z., Uht, R.M., Harrison, R.W., Gibbs, F.P., Paull, W.K., Bohn, M.C., 1987. Ultrastructural localization of glucocorticoid receptor (GR) in hypothalamic paraventricular neurons synthesizing corticotropin releasing factor (CRF). Histochemistry 87 (5), 407–412.

Lu, N.Z., Shlaes, T.A., Gundlah, C., Dziennis, S.E., Lyle, R.E., Bethea, C.L., 1999. Ovarian steroid action on tryptophan hydroxylase protein and serotonin compared to localization of ovarian steroid receptors in midbrain of guinea pigs. Endocrine 11 (3), 257–267.

Lund, T.D., Munson, D.J., Haldy, M.E., Handa, R.J., 2004. Androgen inhibits, while oestrogen enhances, restraint-induced activation of neuropeptide neurones in the paraventricular nucleus of the hypothalamus. J. Neuroendocrinol. 16 (3), 272–278.

Lund, T.D., Rovis, T., Chung, W.C.J., Handa, R.J., 2005. Novel actions of estrogen receptor-beta on anxiety-related behaviors. Endocrinology 146 (2), 797–807.

Lund, T.D., Hinds, L.R., Handa, R.J., 2006. The androgen 5alpha-dihydrotestosterone and its metabolite 5alpha-androstan-3beta, 17beta-diol inhibit the hypothalamo–pituitary–adrenal response to stress by acting through estrogen receptor beta-expressing neurons in the hypothalamus. J. Neurosci. 26 (5), 1448–1456.

Maeda, M., 2001. The conserved residues of the ligand-binding domains of steroid receptors are located in the core of the molecules. J. Mol. Graph Model 19 (6), 543–551, 601–606.

Mani, S., Portillo, W., 2010. Activation of progestin receptors in female reproductive behavior: interactions with neurotransmitters. Front. Neuroendocrinol. 31 (2), 157–171.

McCarthy, M.M., Arnold, A.P., 2011. Reframing sexual differentiation of the brain. Nat. Neurosci. 14 (6), 677–683.

McCarthy, M.M., Nugent, B.M., 2013. Epigenetic contributions to hormonally-mediated sexual differentiation of the brain. J. Neuroendocrinol. 25 (11), 1133–1140.

McCormick, C.M., Mahoney, E., 1999. Persistent effects of prenatal, neonatal, or adult treatment with flutamide on the hypothalamic–pituitary–adrenal stress response of adult male rats. Horm. Behav. 35 (1), 90–101.

McCormick, C.M., Mathews, I.Z., 2007. HPA function in adolescence: role of sex hormones in its regulation and the enduring consequences of exposure to stressors. Pharmacol. Biochem. Behav. 86 (2), 220–233.

McCormick, C.M., Furey, B.F., Child, M., Sawyer, M.J., Donohue, S.M., 1998. Neonatal sex hormones have "organizational" effects on the hypothalamic–pituitary–adrenal axis of male rats. Brain Res. Dev. Brain Res. 105 (2), 295–307.

McGimsey, W.C., Cidlowski, J.A., Stumpf, W.E., Sar, M., 1991. Immunocytochemical localization of the glucocorticoid receptor in rat brain, pituitary, liver, and thymus with two new polyclonal antipeptide antibodies. Endocrinology 129 (6), 3064–3072.

McGinnis, M.Y., Davis, P.G., Meaney, M.J., Singer, M., McEwen, B.S., 1983. *In vitro* measurement of cytosol and cell nuclear androgen receptors in male rat brain and pituitary. Brain Res. 275 (1), 75–82.

Mhyre, A.J., Dorsa, D.M., 2006. Estrogen activates rapid signaling in the brain: role of estrogen receptor alpha and estrogen receptor beta in neurons and glia. Neuroscience 138 (3), 851–858.

Milner, T.A., Thompson, L.I., Wang, G., Kievits, J.A., Martin, E., Zhou, P., McEwen, B.S., Pfaff, D.W., Waters, E.M., 2010. Distribution of estrogen receptor B containing cells in the brains of bacterial artificial chromosome transgenic mice. Brain Res. 1351, 74–96.

Mitsushima, D., Takase, K., Funabashi, T., Kimura, F., 2008. Gonadal steroid hormones maintain the stress-induced acetylcholine release in the hippocampus: simultaneous measurements of the extracellular acetylcholine and serum corticosterone levels in the same subjects. Endocrinology 149 (2), 802–811.

Moore, R.Y., Eichler, V.B., 1972. Loss of a circadian adrenal corticosterone rhythm following suprachiasmatic lesions in the rat. Brain Res. 42 (1), 201–206.

Moore-Ede, M.C., Sulzman, F.M., Fuller, C.A., 1982. Circadian timing of physiological systems. The Clocks that Time Us. Physiology of the Circadian Timing System. Harvard University Press, 201–294.

Mora, S., Dussaubat, N., Diaz-Veliz, G., 1996. Effects of the estrous cycle and ovarian hormones on behavioral indices of anxiety in female rats. Psychoneuroendocrinology 21 (7), 609–620.

Morimoto, M., Morita, N., Ozawa, H., Yokoyama, K., Kawata, M., 1996. Distribution of glucocorticoid receptor immunoreactivity and mRNA in the rat brain: an immunohistochemical and *in situ* hybridization study. Neurosci. Res. 26 (3), 235–269.

Muto, S., Nemoto, J., Ebata, S., Kawakami, K., Asano, Y., 1998. Corticosterone and 11-dehydrocorticosterone stimulate Na,K-ATPase gene expression in vascular smooth muscle cells. Kidney Int. 54 (2), 492–508.

Nelson, R.J., 2000. An Introduction to Behavioral Endocrinology, second ed. Sinauer Associates, Sunderland, Mass.

Nishi, M., Kawata, M., 2006. Brain corticosteroid receptor dynamics and trafficking: implications from live cell imaging. Neuroscientist 12 (2), 119–133.

Nomikos, G.G., Spyraki, C., 1988. Influence of oestrogen on spontaneous and diazepam-induced exploration of rats in an elevated plus maze. Neuropharmacology 27 (7), 691–696.

Nomura, M., Higuchi, T., Ishida, Y., Ohta, S., Komari, T., Imaizumi, N., Miyao-Tokutomi, M., Matsuoka, M., Tajima, S., 2005. Differential expression pattern of C4 bundle sheath expression genes in rice, a C3 plant. Plant Cell Physiol. 46 (5), 754–761.

Oitzl, M.S., Reichardt, H.M., Joels, M., de Kloet, E.R., 2001. Point mutation in the mouse glucocorticoid receptor preventing DNA binding impairs spatial memory. Proc. Natl. Acad. Sci. USA 98 (22), 12790–12795.

Orti, E., Bodwell, J.E., Munck, A., 1992. Phosphorylation of steroid hormone receptors. Endocr. Rev. 13 (1), 105–128.

Oyola, M.G., Portillo, W., Reyna, A., Foradori, C.D., Kudwa, A., Hinds, L., Handa, R.J., Mani, S.K., 2012. Anxiolytic effects and neuroanatomical targets of estrogen receptor-beta (ERbeta) activation by a selective ERbeta agonist in female mice. Endocrinology 153 (2), 837–846.

Patchev, V.K., Hayashi, S., Orikasa, C., Almeida, O.F., 1995. Implications of estrogen-dependent brain organization for gender differences in hypothalamo–pituitary–adrenal regulation. FASEB J. 9 (5), 419–423.

Perrot-Applanat, M., Groyer-Picard, M.T., Logeat, F., Milgrom, E., 1986. Ultrastructural localization of the progesterone receptor by an immunogold method: effect of hormone administration. J. Cell Biol. 102 (4), 1191–1199.

Phoenix, C.H., Goy, R.W., Gerall, A.A., Young, W.C., 1959. Organizing action of prenatally administered testosterone propionate on the tissues mediating mating behavior in the female guinea pig. Endocrinology 65, 369–382.

Phuc Le, P., Friedman, J.R., Schug, J., Brestelli, J.E., Parker, J.B., Bochkis, I.M., Kaestner, K.H., 2005. Glucocorticoid receptor-dependent gene regulatory networks. PLoS Genet. 1 (2), e16.

Pincus, G., 1943. A diurnal rhythm in the excretion of urinary ketosteroids by young men. J. Clin. Endocrinol. 3, 195–199.

Platania, P., Laureanti, F., Bellomo, M., Giuffrida, R., Giuffrida-Stella, A.M., Catania, M.V., Sortino, M.A., 2003. Differential expression of estrogen receptors alpha and beta in the spinal cord during postnatal development: localization in glial cells. Neuroendocrinology 77 (5), 334–340.

Porsolt, R.D., Le Pichon, M., Jalfre, M., 1977. Depression: a new animal model sensitive to antidepressant treatments. Nature 266 (5604), 730–732.

Pratt, W.B., Morishima, Y., Murphy, M., Harrell, M., 2006. Chaperoning of glucocorticoid receptors. Handb. Exp. Pharmacol. 172, 111–138.

Prewitt, C.M., Herman, J.P., 1998. Anatomical interactions between the central amygdaloid nucleus and the hypothalamic paraventricular nucleus of the rat: a dual tract-tracing analysis. J. Chem. Neuroanat. 15 (3), 173–185.

Price, Jr., R.H., Butler, C.A., Webb, P., Uht, R., Kushner, P., Handa, R.J., 2001. A splice variant of estrogen receptor beta missing exon 3 displays altered subnuclear localization and capacity for transcriptional activation. Endocrinology 142 (5), 2039–2049.

Rabadan-Diehl, C., Makara, G., Kiss, A., Zelena, D., Aguilera, G., 1997. Regulation of pituitary corticotropin releasing hormone (CRH) receptor mRNA and CRH binding during adrenalectomy: role of glucocorticoids and hypothalamic factors. J. Neuroendocrinol. 9 (9), 689–697.

Rankin, J., Walker, J.J., Windle, R., Lightman, S.L., Terry, J.R., 2012. Characterizing dynamic interactions between ultradian glucocorticoid rhythmicity and acute stress using the phase response curve. PLoS One 7 (2), e30978.

Ratka, A., Sutanto, W., Bloemers, M., de Kloet, E.R., 1989. On the role of brain mineralocorticoid (type I) and glucocorticoid (type II) receptors in neuroendocrine regulation. Neuroendocrinology 50 (2), 117–123.

Reul, J.M., de Kloet, E.R., 1985. Two receptor systems for corticosterone in rat brain: microdistribution and differential occupation. Endocrinology 117 (6), 2505–2511.
Reul, J.M., de Kloet, E.R., van Sluijs, F.J., Rijnberk, A., Rothuizen, J., 1990. Binding characteristics of mineralocorticoid and glucocorticoid receptors in dog brain and pituitary. Endocrinology 127 (2), 907–915.
Ridder, S., Chourbaji, S., Hellweg, R., Urani, A., Zacher, C., Schmid, W., Zink, M., Hortnagl, H., Flor, H., Henn, F.A., Schutz, G., Gass, P., 2005. Mice with genetically altered glucocorticoid receptor expression show altered sensitivity for stress-induced depressive reactions. J. Neurosci. 25 (26), 6243–6250.
Rizk, A., Robertson, J., Raber, J., 2005. Behavioral performance of TFM mice supports the beneficial role of androgen receptors in spatial learning and memory. Brain Res. 1034, 1–2, 132–138.
Rocha, B.A., Fleischer, R., Schaeffer, J.M., Rohrer, S.P., Hickey, G.J., 2005. 17 beta-estradiol-induced antidepressant-like effect in the forced swim test is absent in estrogen receptor-beta knockout (BERKO) mice. Psychopharmacology (Berl.) 179 (3), 637–643.
Romeo, R.D., Bellani, R., Karatsoreos, I.N., Chhua, N., Vernov, M., Conrad, C.D., McEwen, B.S., 2006. Stress history and pubertal development interact to shape hypothalamic–pituitary–adrenal axis plasticity. Endocrinology 147 (4), 1664–1674.
Roselli, C.E., 1991. Sex differences in androgen receptors and aromatase activity in microdissected regions of the rat brain. Endocrinology 128 (3), 1310–1316.
Roselli, C.E., Horton, L.E., Resko, J.A., 1985. Distribution and regulation of aromatase activity in the rat hypothalamus and limbic system. Endocrinology 117 (6), 2471–2477.
Roselli, C.E., Liu, M., Hurn, P.D., 2009. Brain aromatization: classic roles and new perspectives. Semin. Reprod. Med. 27 (3), 207–217.
Rozeboom, A.M., Akil, H., Seasholtz, A.F., 2007. Mineralocorticoid receptor overexpression in forebrain decreases anxiety-like behavior and alters the stress response in mice. Proc. Natl. Acad. Sci. USA 104 (11), 4688–4693.
Saijo, K., Collier, J.G., Li, A.C., Katzenellenbogen, J.A., Glass, C.K., 2011. An adiol-ERbeta-CTBP transrepression pathway negatively regulates microglia-mediated inflammation. Cell 145 (4), 584–595.
Santagati, S., Melcangi, R.C., Celotti, F., Martini, L., Maggi, A., 1994. Estrogen receptor is expressed in different types of glial cells in culture. J. Neurochem. 63 (6), 2058–2064.
Sapolsky, R.M., Romero, L.M., Munck, A.U., 2000. How do glucocorticoids influence stress responses? Integrating permissive, suppressive, stimulatory, and preparative actions. Endocr. Rev. 21 (1), 55–89.
Sar, M., Stumpf, W.E., 1973. Autoradiographic localization of radioactivity in the rat brain after the injection of 1,2-3H-testosterone. Endocrinology 92 (1), 251–256.
Sar, M., Stumpf, W.E., 1977. Distribution of androgen target cells in rat forebrain and pituitary after [3H]-dihydrotestosterone administration. J. Steroid Biochem. 8 (11), 1131–1135.
Sarkey, S., Azcoitia, I., Garcia-Segura, L.M., Garcia-Ovejero, D., DonCarlos, L.L., 2008. Classical androgen receptors in non-classical sites in the brain. Horm. Behav. 53 (5), 753–764.
Sarvari, M., Hrabovszky, E., Kallo, I., Solymosi, N., Toth, K., Liko, I., Szeles, J., Maho, S., Molnar, B., Liposits, Z., 2011. Estrogens regulate neuroinflammatory genes via estrogen receptors alpha and beta in the frontal cortex of middle-aged female rats. J. Neuroinflammation 8, 82.
Sawchenko, P.E., 1987a. Evidence for a local site of action for glucocorticoids in inhibiting CRF and vasopressin expression in the paraventricular nucleus. Brain Res. 403 (2), 213–223.
Sawchenko, P.E., 1987b. Evidence for differential regulation of corticotropin-releasing factor and vasopressin immunoreactivities in parvocellular neurosecretory and autonomic-related projections of the paraventricular nucleus. Brain Res. 437 (2), 253–263.
Sawchenko, P.E., Swanson, L.W., Vale, W.W., 1984. Corticotropin-releasing factor: co-expression within distinct subsets of oxytocin-, vasopressin-, and neurotensin-immunoreactive neurons in the hypothalamus of the male rat. J. Neurosci. 4 (4), 1118–1129.
Seale, J.V., Wood, S.A., Atkinson, H.C., Bate, E., Lightman, S.L., Ingram, C.D., Jessop, D.S., Harbuz, M.S., 2004a. Gonadectomy reverses the sexually diergic patterns of circadian and stress-induced hypothalamic–pituitary–adrenal axis activity in male and female rats. J. Neuroendocrinol. 16 (6), 516–524.
Seale, J.V., Wood, S.A., Atkinson, H.C., Harbuz, M.S., Lightman, S.L., 2004b. Gonadal steroid replacement reverses gonadectomy-induced changes in the corticosterone pulse profile and stress-induced hypothalamic-pituitary-adrenal axis activity of male and female rats. J. Neuroendocrinol. 16 (12), 989–998.

Seale, J.V., Wood, S.A., Atkinson, H.C., Harbuz, M.S., Lightman, S.L., 2005a. Postnatal masculinization alters the HPA axis phenotype in the adult female rat. J. Physiol. 563 (Pt 1), 265–274.
Seale, J.V., Wood, S.A., Atkinson, H.C., Lightman, S.L., Harbuz, M.S., 2005b. Organizational role for testosterone and estrogen on adult hypothalamic–pituitary–adrenal axis activity in the male rat. Endocrinology 146 (4), 1973–1982.
Sell, S.L., Scalzitti, J.M., Thomas, M.L., Cunningham, K.A., 2000. Influence of ovarian hormones and estrous cycle on the behavioral response to cocaine in female rats. J. Pharmacol. Exp. Ther. 293 (3), 879–886.
Serova, L.I., Harris, H.A., Maharjan, S., Sabban, E.L., 2010. Modulation of responses to stress by estradiol benzoate and selective estrogen receptor agonists. J. Endocrinol. 205 (3), 253–262.
Shifren, J.L., Braunstein, G.D., Simon, J.A., Casson, P.R., Buster, J.E., Redmond, G.P., Burki, R.E., Ginsburg, E.S., Rosen, R.C., Leiblum, S.R., Caramelli, K.E., Mazer, N.A., 2000. Transdermal testosterone treatment in women with impaired sexual function after oophorectomy. N. Engl. J. Med. 343 (10), 682–688.
Shughrue, P.J., Komm, B., Merchenthaler, I., 1996. The distribution of estrogen receptor-beta mRNA in the rat hypothalamus. Steroids 61 (12), 678–681.
Silverman, M.N., Sternberg, E.M., 2012. Glucocorticoid regulation of inflammation and its functional correlates: from HPA axis to glucocorticoid receptor dysfunction. Ann. NY Acad. Sci. 1261, 55–63.
Simerly, R.B., Young, B.J., 1991. Regulation of estrogen receptor messenger ribonucleic acid in rat hypothalamus by sex steroid hormones. Mol. Endocrinol. 5 (3), 424–432.
Simerly, R.B., Chang, C., Muramatsu, M., Swanson, L.W., 1990. Distribution of androgen and estrogen receptor mRNA-containing cells in the rat brain: an *in situ* hybridization study. J. Comp. Neurol. 294 (1), 76–95.
Snyder, M.A., Smejkalova, T., Forlano, P.M., Woolley, C.S., 2010. Multiple ERbeta antisera label in ERbeta knockout and null mouse tissues. J. Neurosci. Methods 188 (2), 226–234.
Soundararajan, R., Zhang, T.T., Wang, J., Vandewalle, A., Pearce, D., 2005. A novel role for glucocorticoid-induced leucine zipper protein in epithelial sodium channel-mediated sodium transport. J. Biol. Chem. 280 (48), 39970–39981.
Souness, G.W., Brem, A.S., Morris, D.J., 2002. 11 beta-hydroxysteroid dehydrogenase antisense affects vascular contractile response and glucocorticoid metabolism. Steroids 67, 3–4, 195–201.
Steckelbroeck, S., Jin, Y., Gopishetty, S., Oyesanmi, B., Penning, T.M., 2004. Human cytosolic 3alpha-hydroxysteroid dehydrogenases of the aldo-keto reductase superfamily display significant 3beta-hydroxysteroid dehydrogenase activity: implications for steroid hormone metabolism and action. J. Biol. Chem. 279 (11), 10784–10795.
Sundin, M., Warner, M., Haaparanta, T., Gustafsson, J.A., 1987. Isolation and catalytic activity of cytochrome P-450 from ventral prostate of control rats. J. Biol. Chem. 262 (25), 12293–12297.
Sutanto, W., De Kloet, E.R., 1987. Species-specificity of corticosteroid receptors in hamster and rat brains. Endocrinology 121 (4), 1405–1411.
Tabori, N.E., Stewart, L.S., Znamensky, V., Romeo, R.D., Alves, S.E., McEwen, B.S., Milner, T.A., 2005. Ultrastructural evidence that androgen receptors are located at extranuclear sites in the rat hippocampal formation. Neuroscience 130 (1), 151–163.
Tomlinson, J.W., Walker, E.A., Bujalska, I.J., Draper, N., Lavery, G.G., Cooper, M.S., Hewison, M., Stewart, P.M., 2004. 11beta-hydroxysteroid dehydrogenase type 1: a tissue-specific regulator of glucocorticoid response. Endocr. Rev. 25 (5), 831–866.
Torn, S., Nokelainen, P., Kurkela, R., Pulkka, A., Menjivar, M., Ghosh, S., Coca-Prados, M., Peltoketo, H., Isomaa, V., Vihko, P., 2003. Production, purification, and functional analysis of recombinant human and mouse 17beta-hydroxysteroid dehydrogenase type 7. Biochem. Biophys. Res. Commun. 305 (1), 37–45.
Trapp, T., Rupprecht, R., Castren, M., Reul, J.M., Holsboer, F., 1994. Heterodimerization between mineralocorticoid and glucocorticoid receptor: a new principle of glucocorticoid action in the CNS. Neuron 13 (6), 1457–1462.
Tsai, H.W., Grant, P.A., Rissman, E.F., 2009. Sex differences in histone modifications in the neonatal mouse brain. Epigenetics 4 (1), 47–53.
Tyagi, R.K., Lavrovsky, Y., Ahn, S.C., Song, C.S., Chatterjee, B., Roy, A.K., 2000. Dynamics of intracellular movement and nucleocytoplasmic recycling of the ligand-activated androgen receptor in living cells. Mol. Endocrinol. 14 (8), 1162–1174.

Uht, R.M., McKelvy, J.F., Harrison, R.W., Bohn, M.C., 1988. Demonstration of glucocorticoid receptor-like immunoreactivity in glucocorticoid-sensitive vasopressin and corticotropin-releasing factor neurons in the hypothalamic paraventricular nucleus. J. Neurosci. Res. 19 (4), 405–411, 68–69.
van Haarst, A.D., Oitzl, M.S., de Kloet, E.R., 1997. Facilitation of feedback inhibition through blockade of glucocorticoid receptors in the hippocampus. Neurochem. Res. 22 (11), 1323–1328.
van Steensel, B., van Binnendijk, E.P., Hornsby, C.D., van der Voort, H.T., Krozowski, Z.S., de Kloet, E.R., van Driel, R., 1996. Partial colocalization of glucocorticoid and mineralocorticoid receptors in discrete compartments in nuclei of rat hippocampus neurons. J. Cell Sci. 109 (Pt 4), 787–792.
Vasudevan, N., Kow, L.M., Pfaff, D., 2005. Integration of steroid hormone initiated membrane action to genomic function in the brain. Steroids 70, 5–7, 388–396.
Viau, V., Meaney, M.J., 1996. The inhibitory effect of testosterone on hypothalamic-pituitary-adrenal responses to stress is mediated by the medial preoptic area. J. Neurosci. 16 (5), 1866–1876.
Viau, V., Lee, P., Sampson, J., Wu, J., 2003. A testicular influence on restraint-induced activation of medial parvocellular neurons in the paraventricular nucleus in the male rat. Endocrinology 144 (7), 3067–3075.
Viau, V., Bingham, B., Davis, J., Lee, P., Wong, M., 2005. Gender and puberty interact on the stress-induced activation of parvocellular neurosecretory neurons and corticotropin-releasing hormone messenger ribonucleic acid expression in the rat. Endocrinology 146 (1), 137–146.
Wade, J., Arnold, A.P., 1996. Functional testicular tissue does not masculinize development of the zebra finch song system. Proc. Natl. Acad. Sci. USA 93 (11), 5264–5268.
Walf, Alicia A., Frye, Cheryl A., 2005. ERbeta-selective estrogen receptor modulators produce antianxiety behavior when administered systemically to ovariectomized rats. Neuropsychopharmacology 30 (9), 1598–1609.
Walf, A.A., Frye, C.A., 2006. A review and update of mechanisms of estrogen in the hippocampus and amygdala for anxiety and depression behavior. Neuropsychopharmacology 31 (6), 1097–1111.
Walf, A.A., Frye, C.A., 2007. The use of the elevated plus maze as an assay of anxiety-related behavior in rodents. Nat. Protoc. 2 (2), 322–328.
Walker, J.J., Terry, J.R., Lightman, S.L., 2010. Origin of ultradian pulsatility in the hypothalamic–pituitary–adrenal axis. Proc. Biol. Sci. 277 (1688), 1627–1633.
Walker, J.J., Spiga, F., Waite, E., Zhao, Z., Kershaw, Y., Terry, J.R., Lightman, S.L., 2012. The origin of glucocorticoid hormone oscillations. PLoS Biol. 10 (6), e1001341.
Warnmark, A., Treuter, E., Wright, A.P., Gustafsson, J.A., 2003. Activation functions 1 and 2 of nuclear receptors: molecular strategies for transcriptional activation. Mol. Endocrinol. 17 (10), 1901–1909.
Watts, A.G., 2005. Glucocorticoid regulation of peptide genes in neuroendocrine crh neurons: a complexity beyond negative feedback. Front. Neuroendocrinol. 26, 3–4, 109–130.
Watts, A.G., Swanson, L.W., Sanchez-Watts, G., 1987. Efferent projections of the suprachiasmatic nucleus: I. Studies using anterograde transport of phaseolus vulgaris leucoagglutinin in the rat. J. Comp. Neurol. 258 (2), 204–229.
Weihua, Z., Lathe, R., Warner, M., Gustafsson, J.A., 2002. An endocrine pathway in the prostate, ERbeta, AR, 5α-androstane-3β,17β-diol, and CYP7b1, regulates prostate growth. Proc. Natl. Acad. Sci. USA 99 (21), 13589–13594.
Weiser, M.J., Handa, R.J., 2009. Estrogen impairs glucocorticoid dependent negative feedback on the hypothalamic–pituitary–adrenal axis via estrogen receptor alpha within the hypothalamus. Neuroscience 159 (2), 883–895.
Weiser, M.J., Wu, T.J., Handa, R.J., 2009. Estrogen receptor-beta agonist diarylpropionitrile: biological activities of R- and S-enantiomers on behavior and hormonal response to stress. Endocrinology 150 (4), 1817–1825.
Weiser, M.J., Foradori, C.D, Handa, R.J., 2010. Estrogen receptor beta activation prevents glucocorticoid receptor-dependent effects of the central nucleus of the amygdala on behavior and neuroendocrine function. Brain Res. 1336, 78–88.
Wersinger, S.R., Sannen, K., Villalba, C., Lubahn, D.B., Rissman, E.F., De Vries, G.J., 1997. Masculine sexual behavior is disrupted in male and female mice lacking a functional estrogen receptor alpha gene. Horm. Behav. 32 (3), 176–183.
White, P.C., 2001. 11beta-hydroxysteroid dehydrogenase and its role in the syndrome of apparent mineralocorticoid excess. Am. J. Med. Sci. 322 (6), 308–315.

Whitworth, J.A., Schyvens, C.G., Zhang, Y., Andrews, M.C., Mangos, G.J., Kelly, J.J., 2002. The nitric oxide system in glucocorticoid-induced hypertension. J. Hypertens. 20 (6), 1035–1043.

Whorwood, C.B., Stewart, P.M., 1996. Human hypertension caused by mutations in the 11 beta-hydroxysteroid dehydrogenase gene: a molecular analysis of apparent mineralocorticoid excess. J. Hypertens. Suppl. 14 (5), S19–24.

Williamson, M., Viau, V., 2008. Selective contributions of the medial preoptic nucleus to testosterone-dependent regulation of the paraventricular nucleus of the hypothalamus and the HPA axis. Am. J. Physiol. Regul. Integr. Comp. Physiol. 295 (4), R1020–1030.

Williamson, M., Bingham, B., Viau, V., 2005. Central organization of androgen-sensitive pathways to the hypothalamic-pituitary-adrenal axis: implications for individual differences in responses to homeostatic threat and predisposition to disease. Prog. Neuropsychopharmacol. Biol. Psychiat. 29 (8), 1239–1248.

Wu, T.W., Wang, J.M., Chen, S., Brinton, R.D., 2005. 17beta-estradiol induced Ca2+ influx via l-type calcium channels activates the SRC/ERK/cyclic-AMP response element binding protein signal pathway and BCL-2 expression in rat hippocampal neurons: a potential initiation mechanism for estrogen-induced neuroprotection. Neuroscience 135 (1), 59–72.

Wu, W.F., Tan, X.J., Dai, Y.B., Krishnan, V., Warner, M., Gustafsson, J.A., 2013. Targeting estrogen receptor beta in microglia and T cells to treat experimental autoimmune encephalomyelitis. Proc. Natl. Acad. Sci. USA 110 (9), 3543–3548.

Yard, A.C., Kadowitz, P.J., 1972. Studies on the mechanism of hydrocortisone potentiation of vasoconstrictor responses to epinephrine in the anesthetized animal. Eur. J. Pharmacol. 20 (1), 1–9.

Yates, E.E., Maran, J.W., 1974. Stimulation and inhibition of adrenocorticotropin release. Handb. Physiol. Sect. 7 (4), 367–404.

Zhang, Z., Cerghet, M., Mullins, C., Williamson, M., Bessert, D., Skoff, R., 2004. Comparison of *in vivo* and *in vitro* subcellular localization of estrogen receptors alpha and beta in oligodendrocytes. J. Neurochem. 89 (3), 674–684.

Zhao, C., Gao, H., Liu, Y., Papoutsi, Z., Jaffrey, S., Gustafsson, J.A., Dahlman-Wright, K., 2010. Genome-wide mapping of estrogen receptor-beta-binding regions reveals extensive cross-talk with transcription factor activator protein-1. Cancer Res. 70 (12), 5174–5183.

Zuloaga, D.G., Morris, J.A., Jordan, C.L., Breedlove, S.M., 2008a. Mice with the testicular feminization mutation demonstrate a role for androgen receptors in the regulation of anxiety-related behaviors and the hypothalamic–pituitary–adrenal axis. Horm. Behav. 54 (5), 758–766.

Zuloaga, D.G., Puts, D.A., Jordan, C.L., Breedlove, S.M., 2008b. The role of androgen receptors in the masculinization of brain and behavior: what we've learned from the testicular feminization mutation. Horm. Behav. 53 (5), 613–626.

Zuloaga, D.G., Poort, J.E., Jordan, C.L., Breedlove, S.M., 2011. Male rats with the testicular feminization mutation of the androgen receptor display elevated anxiety-related behavior and corticosterone response to mild stress. Horm. Behav. 60 (4), 380–388.

Zuloaga, K.L., O'Connor, D.T., Handa, R.J., Gonzales, R.J., 2012. Estrogen receptor beta dependent attenuation of cytokine-induced cyclooxygenase-2 by androgens in human brain vascular smooth muscle cells and rat mesenteric arteries. Steroids 77, 8–9, 835–844.

Zuloaga, D.G., Zuloaga, K.L., Hinds, L.R., Carbone, D.L., Handa, R.J., 2014. Estrogen receptor beta expression in the mouse forebrain: age and sex differences. J. Comp. Neurol. 522 (2), 358–371.

# CHAPTER 14

# Sex Differences in Rodent Cognitive Processing and Responses to Chronic Stress

**Victoria N. Luine*, Juan L. Gomez****
*Hunter College of CUNY, Department of Psychology, New York, NY, USA
**Oregon Health & Sciences University, Department of Behavioral Neuroscience, Portland, OR, USA

## 1 INTRODUCTION

A growing literature shows that stress, either acute or chronic, does not elicit identical responses in the sexes (Bowman et al., 2003; Bowman, 2005; Luine et al., 2007; Luine and Dohanich, 2008; Luine, 2011). These sex differences range from physiological responses such as the amount of stress hormones released or amount of weight loss, effects on higher order neural functions like mood, anxiety, and cognition as well as neural substrates underlying these responses. In this review, we focus on sex differences in chronic stress effects on cognitive functions including learning and memory in adult rodents, but we also consider effects during the perinatal and adolescent period. We also discuss sex differences in strategies used to solve tasks and how these differences can impact responses to stress. In addition, interactive effects of stress and drugs in relation to sex on learning and memory are also considered. Results of these studies indicate that stress effects on cognition are not only different in the sexes, but that stress also appears to interact with other neural stimuli like drugs and alcohol, which results in different cognitive responses in males and females. Possible contributions of estradiol to sex differences in responses between the sexes are also discussed. Finally, these sexually differentiated responses are placed in the context that they may contribute to well-described sex differences in coping with chronic stress and disease incidence. Females show greater amounts of depression and posttraumatic stress disorder, which can be precipitated by stress, than males (Steiner and Young, 2008). Stress during adulthood can also precipitate drug and alcohol abuse, and males have a greater incidence of such abuse than females (Cooper et al., 1992). Developmental stresses may lead to greater sensitivity to mood disorders and drug abuse in adulthood (Romeo and Karatsoreos, 2011; Karatsoreos and McEwen, 2013). Thus, we consider whether better therapies for some stress-dependent psychological disorders might be dependent on different treatment courses for each sex depending on the nature of stress-induced changes.

*Sex Differences in the Central Nervous System.* http://dx.doi.org/10.1016/B978-0-12-802114-9.00014-7

## 2 CHRONIC STRESS EFFECTS ON LEARNING AND MEMORY ARE SEX DEPENDENT IN ADULTS

In male rodents, 1–3 weeks of stress elicited by daily restraint or different daily stressors (unpredictable chronic stress) generally results in impaired learning and memory. Impairments in male learning have been shown using spatial learning and memory tasks like the eight-arm radial task and the Morris water maze (Figure 14.1a, d). On the radial arm maze, stressed male rats make more mistakes per trial and take longer to reach a learning criterion than unstressed rats (Luine et al., 1994). In the water maze, a similar pattern of impaired learning is shown after chronic stressors (Kitraki et al., 2004). In tests assessing memory, chronic stress also generally impairs male performance (see Conrad, 2010 for review). Stressed male rats and mice are unable to significantly discriminate between known and new objects, object recognition memory (Beck and Luine, 1999, 2002), or a test of temporal order recognition memory (Wei et al., 2014). Spatial memory is also impaired by chronic stress in male rats in several tasks: object placement (Figure 14.1b; Beck and Luine, 1999; Bowman et al., 2009; Gomez et al., 2012) and the Y-maze task (Figure 14.1c; Conrad et al., 1996; Wright and Conrad, 2005; Gomez et al., 2013). Thus, chronic stress impairs learning and memory in male rodents in a variety of tasks.

When female rodents are given the same stress regimens and tested on the same tasks, a different pattern, compared with males, emerges (Figure 14.1). On the radial arm maze, stressed females perform better than unstressed females (Figure 14.1a). Stressed females make fewer mistakes/trial and more rapidly reach learning criteria than unstressed females (Bowman et al., 2001). Likewise, chronic stress enhances female rodent performance in the water maze (Figure 14.1d; Kitraki et al., 2004). In recognition memory tasks, stressed females show better performance than unstressed females in object placement (Figure 14.1b; Beck and Luine, 1999; Bowman et al., 2009) and memory is either not affected or enhanced on the Y-maze (Figure 14.1c; Conrad et al., 2003; Gomez and Luine, 2014), object recognition (Beck and Luine, 2002; Bisagnoet al., 2004; Bowman et al., 2009; Gomez and Luine, 2014), and temporal order recognition memory (Wei et al., 2014) following chronic stress. In addition, impairments in radial arm mazes, object recognition, and object placement were not found in female rats when stress was given for 28 or 35 days (Bowman et al., 2001; Bowman and Kelly, 2012). Thus, female rodents appear to show resilience to chronic stress, at least in terms of learning and memory, because female rodents generally show enhanced cognitive function after stress or are unaffected by the stress, depending on the specific task. However, it should be noted that much fewer experiments have been conducted in female rodents than in males and that most experiments in females utilized restraint stress. Thus, stress effects on female cognition need further investigation and confirmation.

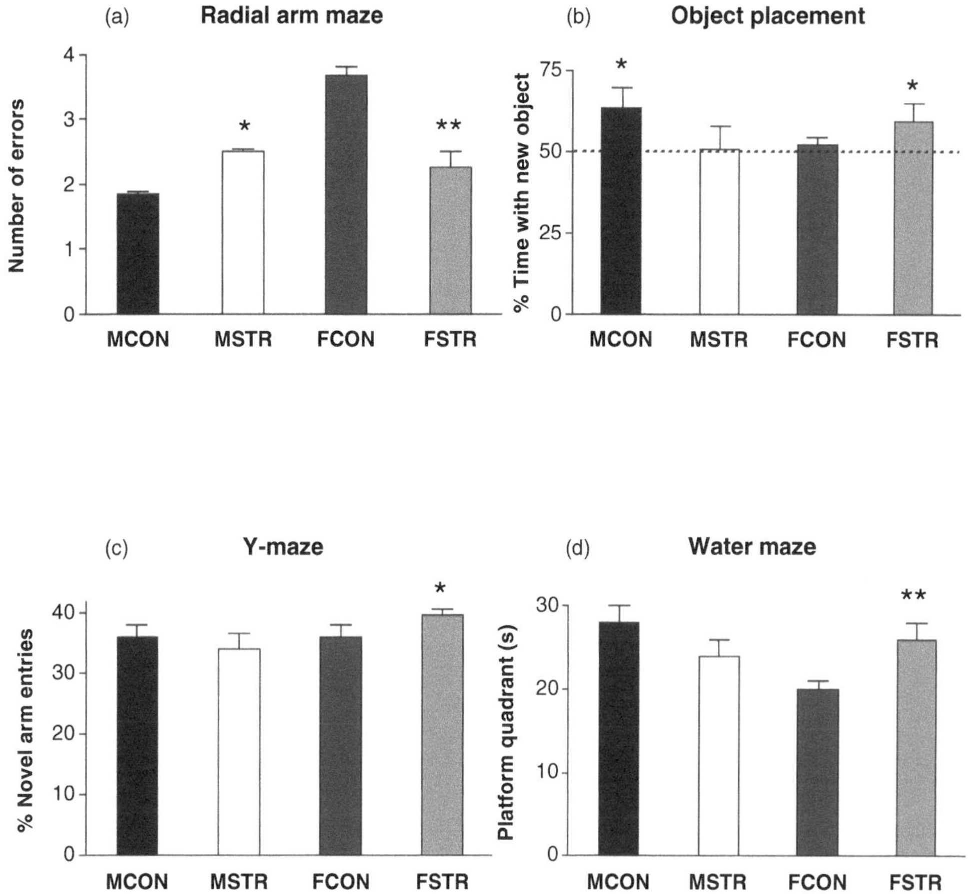


**Figure 14.1** ***Sex differences in chronic stress effects on spatial memory tasks.*** Adult male and female rats received 21 days of daily (6 h) restraint stress. (a) Radial arm maze. Data pooled from Luine et al. (1994) for males and from Bowman et al. (2001) for females. The number of errors to complete the task significantly increased in stress males and decreased in stress females suggesting that stress impairs male but enhances female performance of this task. (b) Object placement. Data from Beck and Luine (2002). The percentage time exploring objects in the new location is shown. Control males and stress females could significantly discriminate the old from new location but stress males and control females could not, suggesting that stress impairs male spatial memory but enhances female spatial memory. (c) Y-maze. Data from Conrad et al. (2003). The percentage of entries into the novel arms is shown. Stressed females entered the novel arm more than the start and other arms indicating enhanced memory. (d) Morris water maze. Data from Kitraki et al. (2004). Time spent in the target quadrant (platform location) during the probe trial. Stress enhanced female time in target quadrant. Not shown is that stress males spent significantly more time in the opposite quadrant than control males. Data suggest that chronic stress impairs male but enhances female memory. * $p < 0.05$; ** $p < 0.01$. *(Figure from Luine and Dohanich (2008). With permission from Oxford University Press.)*

## 3 MECHANISMS CONTRIBUTING TO SEX DIFFERENCES IN RESPONSE TO STRESS

The mechanisms underlying these sex differences in response to stress remain largely unexplored, but a few theories which are currently being investigated are discussed.

### 3.1 Role of sex differences in strategies

In completing spatial memory tasks like the radial arm and water maze, male and female rats often apply different strategies, which can lead to differences in the rate and accuracy of learning the tasks. Males generally reach criterion faster and are more accurate in performance than females in these tasks, but both sexes eventually reach the same level of performance (Luine and Dohanich, 2008). Experiments by Williams et al. (1990) showed that these differences are due to the organizing influence of estradiol (derived from testosterone by the enzyme aromatase) on male brains during the perinatal period. The nature of the sex differences was determined by removing cues surrounding the maze and by altering the geometry of the room with drapes. As depicted in Figure 14.2a, females make a detailed map of all landmarks and geometry, which is shown by the arrows emanating from the eyes and reaching all cues, whereas males use mainly the geometry of the room with a few landmarks, a "vectoring approach" to make a cognitive map (Figure 14.2b). Hence, males rapidly map and acquire the task, but male performance rapidly deteriorates when the environment is even minimally altered whereas female performance is well maintained with only a few landmarks (Williams et al., 1990).

The T or plus-maze is another spatial task where males outperform females, and it is useful because it can more directly assess strategies used by subjects (Figure 14.2c, d). When subjects reach criterion for performance (usually nine correct visits to the baited arm in 10 trials), the strategies can be assessed in a probe trial where the start arm is rotated 180°. When subjects continue to visit the same location on the probe trial, they are using a place or allocentric strategy while subjects that continue to turn left (or right) to solve the task are using a response or egocentric strategy. Neural processing through the hippocampus is associated with the place strategy while the striatum is associated with the response strategy (Packard and McGaugh, 1996; Chang and Gold, 2003). Approximately 90% of males (Figure 14.2c) initially adopt a place strategy but can change with continued training. No direct comparisons of male and female rats on the T-maze have been made, but Korol et al. (2004) found that females often utilize both strategies and which one depends on the stage of the estrous cycle. A place strategy predominates during proestrus while a response strategy predominates during estrous. The constant switching from a place to a response strategy most likely contributes to the poorer performance of females than males on this task. Some evidence for strategy differences by sex is also present in humans (Schwabe et al., 2007, 2008, 2009). For example, men and women exhibit different neural activity during navigation through a virtual maze, with

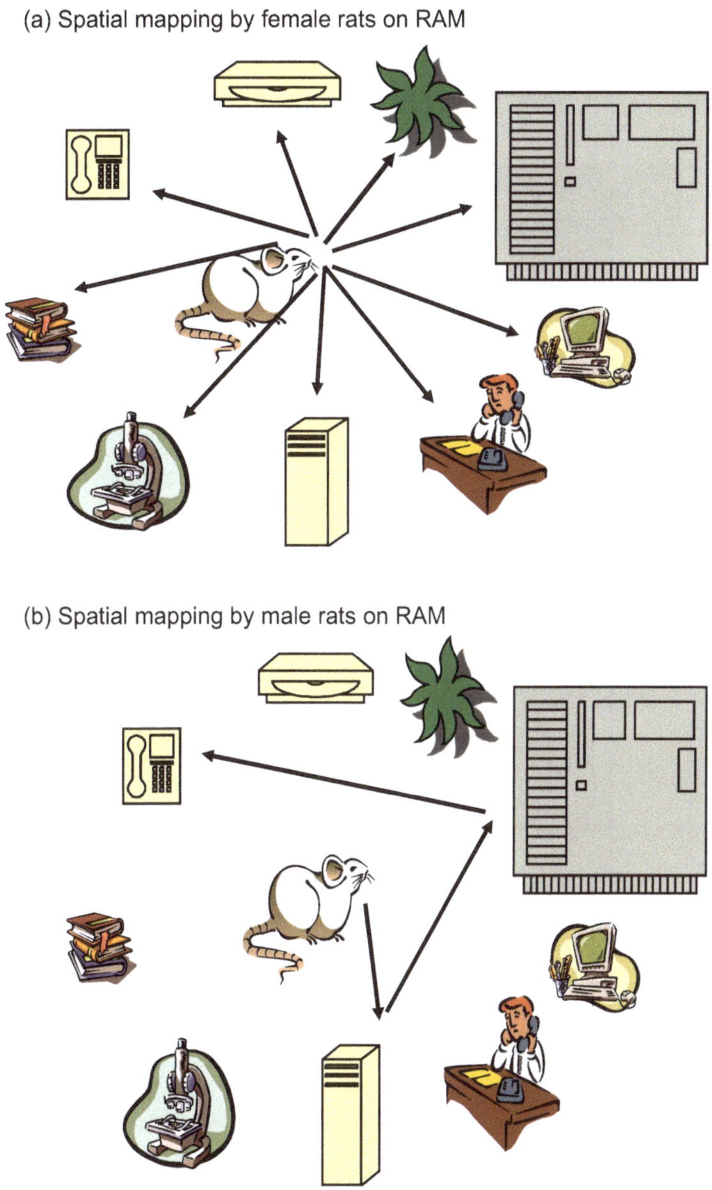


**Figure 14.2** ***Sex differences in strategies to solve spatial learning and memory tasks.*** (a) Learning strategy on radial arm maze (RAM): female rats. A female rat is shown in the center of a radial arm maze and looking around the room. Cues, both landmarks such as a plant, telephone, or microscope, and geometry of the room such as cupboards, which make angles, are used to make a cognitive map for exploration and remembrance of arms explored. Females make a detailed map of all landmarks and geometry, which are depicted by the arrows emanating from the eyes and going to all cues (see Williams et al. (1990) for further information and discussion). (b) Learning strategy on radial arm maze (RAM): male rats. A male rat is depicted in the same environment as the female in Figure 14.2a. Males do not make a detailed map of all landmarks and geometry, but use mainly geometry of the room coupled with a few landmarks to make a cognitive map. The detailed map made by females contributes to their slower acquisition of the task.

*(Continued)*

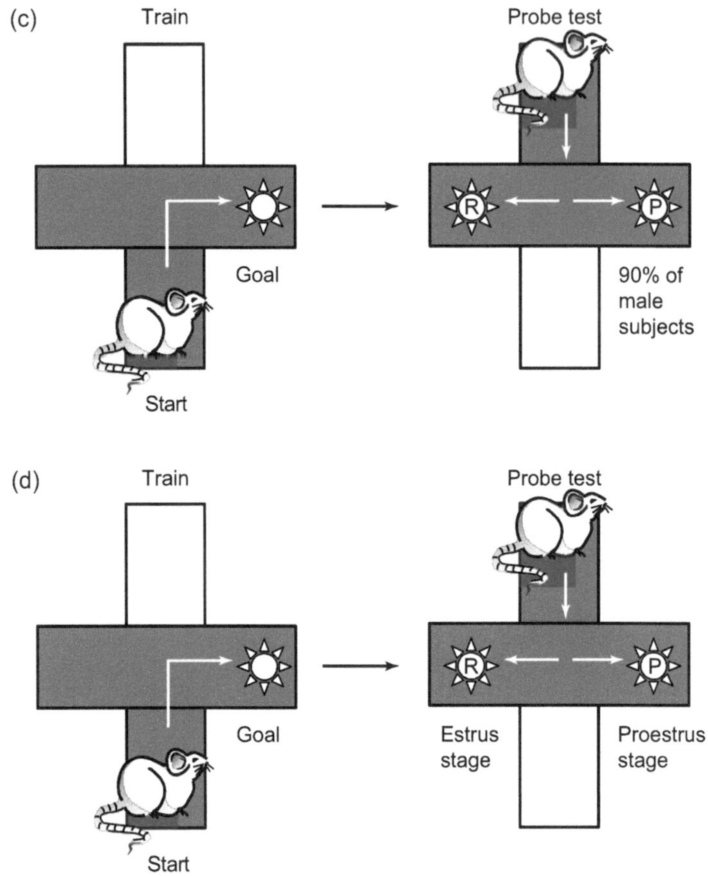


**Figure 14.2** ***(cont.)*** (c) Learning strategy on T-maze: male rats. Males were trained to find food reward (goal) on a T-maze. During training, the start and goal arms remain constant so that subjects could use either a place strategy (P, go to a specific place-arm) or a response strategy (R, turn right). During the probe trial, the maze was rotated 180°, and the strategy used during training assessed. As indicated in the probe figure, Packard and McGaugh (1996) reported that 90% of male rats used a place strategy. (d) Learning strategy on T-maze: female rats. Females were trained to find food reward (goal) on a T-maze. During training, the start and goal arms remain constant so that subjects could use either a place strategy (P, go to a specific place-arm) or a response strategy (R, turn right). During the probe trial, the maze was rotated 180°, and the strategy used during training assessed. As indicated in probe figure, Korol et al. (2004) reported that rats at proestrus were significantly more likely to use a place strategy (71%) while rats in estrus were significantly more likely to use a response strategy (71%). *(Adapted from Luine and Dohanich (2008). With permission from Oxford University Press, New York, NY.)*

males exhibiting greater activity in the left temporal lobe and females showing greater activity in the frontal and parietal cortices (Gron et al., 2000).

While these strategy differences between the sexes lead to a so-called "male advantage" in some learning and memory tasks, they may ultimately contribute to female resilience to stress. It is recognized that highly arousing stimuli (i.e., stress) cause changes in neural processing. These changes involve multiple neurotransmitter systems that shape processing in certain patterns across multiple brain regions presumably to focus neural resources to those stimuli that are important for survival. A recent theory posits that stress causes a shift in neural processing of associative relations from a spatial-contextual representation through the hippocampus to a stimulus-response procedural representation through the caudate nucleus–striatal area (see Beck and Luine, 2010 for further details and discussion). As indicated previously, males primarily use a spatial–contextual mode in learning and memory while females shift between modes depending on the estrus cycle. Thus, stress may impair male rat cognition because a new mode for cognitive function (response mode) must be learned whereas females simply switch from one already established mode to another one. Clearly, the contribution of strategy use by the sexes to differences in cognitive responses and resilience of females to stress requires extensive investigation, but this theory is gaining momentum (Cahill, 2014).

## 3.2 Role of estradiol in female resilience to stress

Some evidence supports the idea that estradiol, of both ovarian and neural origin, may be responsible for the cognitive resilience to chronic stress exhibited by females. While ovariectomized rats did not show enhanced radial-arm maze performance following 21 days of daily restraint stress, their performance was not impaired as in males (Bowman et al., 2002). In addition, McLaughlin et al. (2005) showed that chronically stressed, ovariectomized rats still showed some improvements on the Y-maze compared with nonstressed subjects. At the time that these experiments were completed, it was hypothesized that female cognitive resilience to stress might be due to both the enduring, organizing effects of estradiol during development and to activating influences at adulthood. However, it has now been shown that significant concentrations of estradiol are synthesized locally in discrete regions of the brain by the enzyme aromatase from androgen precursors such as testosterone and/or directly from cholesterol. In addition, estradiol levels in the hippocampus, hypothalamus, and prefrontal cortex are higher than are present in blood (Garcia-Segura et al., 2001). Moreover, estradiol is still present in the hippocampus following ovariectomy (Kato et al., 2013). Thus, intrahippocampal synthesis of estradiol appears to contribute to cognitive stress resilience in females such that estradiol is available to hippocampal and other sites even following ovariectomy. Fester and Rune (2014) reported that hippocampal neurons synthesize estradiol, which maintains LTP and synapses in females, but not males. In females, inhibition of estradiol synthesis results in impairment of LTP and synapse loss. These effects were not seen in males. Thus, the regulation of synaptic stability and plasticity by locally synthesized sexual steroids in males remains to be clarified.

Direct evidence that locally synthesized estradiol in the prefrontal cortex of females may protect against chronic stress effects on memory has been presented. Adult male rats that received 1 week of daily restraint stress were impaired on the temporal order recognition memory task, which is controlled by the prefrontal cortex, whereas female rats were not affected by the same stress (Wei et al., 2014). Impaired temporal order recognition memory in females following stress was shown when estrogen receptors were inhibited or knocked out in the prefrontal cortex, and impairments of performance in stressed males were prevented by estradiol administration. More importantly, blocking aromatase in the prefrontal cortex resulted in impaired temporal order recognition memory in stressed females (Wei et al., 2014). Thus, intraprefrontal cortex estradiol synthesis contributes to cognitive stress resilience in female rats. Another experiment also provides evidence for hippocampal-derived estradiol in memory. Local inhibition of aromatization impaired spatial memory in a food-finding task in zebra finches (Bailey et al., 2013). Thus, these results support the hypothesis that cortically and hippocampally derived estradiol may provide resilience against the detrimental effects of chronic stress on memory in females.

# 4 SEX-DEPENDENT EFFECTS OF STRESS ON LEARNING AND MEMORY ACROSS THE LIFESPAN

## 4.1 Prenatal stress effects

When experienced during adulthood, the effects of stress in both sexes are generally reversible and transitory in nature, but, during the prenatal and postnatal stage, chronic stress can result in long-lasting effects on some behaviors. A few experiments have examined whether prenatal stress exerts sexually dimorphic effects on later adult cognition and behavior (reviewed in Weinstock, 2011). Zagron and Weinstock (2006) found that once daily restraint of dams for 30 min on days 14–21 of gestation led to impaired water maze performance in males, but not females, at adulthood. Hosseini-Sharifabad and Hadinedoushan (2007) also reported that prenatal stress impaired Morris water maze performance at adulthood in males, but females were not tested. Markham et al. (2010) exposed dams to repeated variable prenatal stress during the third week of gestation and evaluated novel object recognition, cued and contextual fear conditioning, and the Morris water maze task at adulthood. Prenatally stressed males were impaired relative to unstressed males on every task, and the performance of females was less affected than the males. A similar pattern of effects was seen in mice prone to Alzheimer's disease (APPswe/PS1dE9 mice) when dams received restraint stress during the first week of gestation and offspring were behaviorally tested as adults (Sierksma et al., 2013). Stressed male mice showed spatial memory deficits while stressed females showed improved spatial memory performance. These results are consistent with cognitive effects of stress at adulthood, in which females appear more resilient than males to stress (see Section 2.1). However, we found a somewhat different pattern of effects when daily restraint stress was given from days 14 to 21 of pregnancy and cognition in offspring was tested at adulthood (Bowman et al., 2004). Regardless of sex, stressed subjects were not different from controls in object recognition or radial arm maze, suggesting that this prenatal stress regimen did not affect memory. However, stress did elicit a significant interactive effect of sex and prenatal treatment on radial arm maze performance such that stressed males and females performed comparably. Thus, for this spatial memory task, prenatal stress appeared to eliminate the normal sex difference in performance by masculinizing the females. Masculinization of females after prenatal stress may be due to activation of the hypothalamic–pituitary–adrenal (HPA) axis, which elevates both glucocorticoid and androgen secretion from the adrenal glands and masculinizes vulnerable brain regions (Sternberg, 1999).

In summary, prenatal stress, like adult stress, appears to be associated with sex-specific changes in learning and memory, but these results need further investigation. In some instances, prenatally stressed females, like adult females, appear resilient to stress. The role of estradiol in mediating/mitigating stress resilience during the perinatal period is difficult to assess and is complicated by contributions of adrenal androgens released during stress. While prenatally stressed males are often impaired as compared with controls at adulthood, females exposed to stress often perform no differently than control females at adulthood. Thus, lack of estrogen during stress periods does not appear to lead to cognitive impairments, but this lack may hinder the expression of enhancements by stress seen in adult females. Similarly, estradiol lack (ovariectomy) during adult stress exposure, while not detrimental to cognitive functioning, does not lead to behavioral enhancements (Bowman et al., 2002; McLaughlin et al., 2005). Thus, it is imperative to learn more about the effects of chronic stress and the extent to which it exerts sexually dimorphic effects during developmental periods because humans exposed to early life stresses have an increased incidence of mood and psychiatric disorders that are expressed differently between the sexes (McEwen, 2008; Romeo and Karatsoreos, 2011; Karatsoreos and McEwen, 2013).

## 4.2 Periadolescent/adolescent chronic stress effects

Many studies show increased vulnerability of the adolescent brain after stress or exposure to drugs and alcohol, but most research has focused on behaviors associated with anxiety, mood, and drug use/abuse (Romeo and Karatsoreos, 2011; Karatsoreos and McEwen, 2013) rather than cognition. However, Isgor et al. (2004) stressed 28-day-old males (no females) at the beginning of adolescence for 4 weeks and found that 3 weeks following the stress, Morris water maze performance was impaired. Similarly, adolescent male rats exposed to social defeat were impaired at adulthood on the delayed, alternating T-maze task compared with nonstressed rats (Novick et al., 2013). These persistent impairments in adolescent males should be viewed in the context that adults given comparable stresses are impaired immediately following, but not 10 days, poststress (Luine et al., 1994). Unfortunately, female rats were not examined in these studies. Thus, some evidence suggests that adolescent stress leads to enduring cognitive effects similar to when stress is given prenatally. While stress effects on female cognition have not been examined at adolescence, sex differences after adolescent restraint stress have been shown in other behaviors. For example, Ariza Traslavina et al. (2014) reported that when restraint stress was given for 1 h on P26, and anxiety was tested on the elevated plus-maze at P61, stressed females were anxiolytic compared with control females, but no changes were found in stressed males. Overall, it remains to be determined whether sex differences in the effects of chronic stress on cognitive and other neural functions are present during the adolescent period. Such information is important because data show that adolescent stress can lead to mood disorders and substance abuse at adulthood (McEwen, 2008).

## 4.3 Stress effects on neuronal function

Although beyond the scope of the current chapter, we will briefly review how stress is also associated with sex-specific alterations in areas important for learning and memory, like the prefrontal cortex and hippocampus, and how estradiol may protect against deleterious stress effects in both sexes. However, it should be noted that the vast majority of stress studies fail to include females. In morphological studies, chronic stress decreased short-term survival of new neurons in the dentate gyrus of adult male rats but increased survival in females (Westenbroek et al., 2004). Retraction/pruning of apical dendrites and spine loss in CA3 pyramidal neurons (Watanabe et al., 1992) and prefrontal cortex areas (Radley et al., 2004, 2006) was found in stressed males, but no retractions were seen in female apical CA3 dendrites whereas basal dendrites were pruned and CA1 spines were increased (Galea et al., 1997; McLaughlin et al., 2009). In support of the hypothesis that mitigations of stress effects in females are mediated by estrogen, Ortiz et al. (2013) found that 21 days of corticosterone treatment to gonadectomized rats caused retractions of CA3 apical dendrites in both male and female rats. However, estradiol treatment protected against the dendritic retractions. Prenatal stress also caused a sex-specific pattern of development in prefrontal cortex pyramidal cells wherein dendritic arborizations were disrupted during adolescence in prenatally stressed male, but not female, rats, but effects of the prenatal stress were evident by adulthood in females (Markham et al., 2013).

Monoaminergic systems are important responders to stress and norepinephrine, dopamine, and serotonin decreased in the hippocampus following 21 days of stress in males (Sunanda and Raju, 2000; Beck and Luine, 2002; Luine, 2002); however, opposite changes were found in females (Beck and Luine, 2002; Luine, 2002; Bowman et al., 2009). In addition, dopaminergic activity in the prefrontal cortex decreases in males and increases in females following chronic stress (Bowman et al., 2003). Prenatal stress also leads to sexually differentiated changes in monoamines at adulthood (Bowman et al., 2004). Glutamatergic neurotransmission and glutamate receptor surface expression was decreased in the prefrontal cortex of male rats following chronic stress but no changes were found in females (Wei et al., 2014). Future studies should target important neural systems in the hippocampus and prefrontal cortex in order to understand the mechanisms responsible for sex differences in neural and behavioral responses to stress.

### 4.4 Conclusion for stress effects on learning and memory

Mounting, but limited, evidence shows that chronic stress is associated with sexually differentiated effects on learning and memory in rats. Moreover, the nature of the sex differences may change over the lifecycle. Briefly, sexually differentiated responses to stress occur early in development (*in utero*), and some are permanent (present at adulthood). During adolescence, stress-dependent changes also appear permanent as some are maintained into adulthood, but since mainly males have been investigated, whether sex differences are present during this period remain to be investigated. At adulthood, sexually differentiated cognitive changes are robust, maybe even larger than in other periods, but do not generally appear permanent. The female hormone estradiol, of both gonadal and intraneural origin, may contribute to the sex differences in response to stress. In addition, use of different learning and memory strategies by the sexes and the influence of the estrus cycle on female strategies may impact stress responses. Thus, these observations raise the question of whether the classic formulation that stress responses change from adaptive to maladaptive at longer intervals and the concept of allostatic load may both need updating (Selye, 1976; Karatsoreos and McEwen, 2013). From the prospective of stress effects on learning and memory, updating should include information as to both the sex and age of the subject.

## 5 INTERACTIVE EFFECTS OF SEX, STRESS, AND ALCOHOL ON LEARNING AND MEMORY

Alcohol is often used or abused when people feel "stressed or burnt out," and surveys have validated that alcohol or cocaine use is increased (Sinha, 2008). Moreover, males use/abuse alcohol and drugs more frequently than females (see NIDA DrugFacts Sheet 2014 – http://www.drugabuse.gov/publications/drugfacts/nationwide-trends; Becker and Taylor, 2008; Weinberger et al., 2009). Thus, our laboratory became interested in exploring possible relationships between stress, sex, and alcohol in adult rats. First, we investigated whether rats could serve as a model, i.e., whether increased alcohol consumption would occur during chronic stress and whether such consumption might alter responses to stress.

Male rats were administered alcohol via a two-bottle choice (one bottle with water and the other with alcohol) and limited access paradigm for 1 h daily (Martinetti et al., 2000). When reliable drinking was established, the experiment began. Groups serving as controls (no stress, no alcohol) received alcohol, 6 h of daily restraint stress, or a combination of stress + alcohol for 10 days (Gomez et al., 2012). The bottles were available after the stress for 1 h. Figure 14.3a plots the regression lines for alcohol consumption in the alcohol and the stress + alcohol groups over time. The stressed rats showed greater alcohol consumption as compared with the alcohol only group, an effect seen in several (Pohorecky, 2008; Yaroslavsky and Tejani-Butt, 2010), but not all,

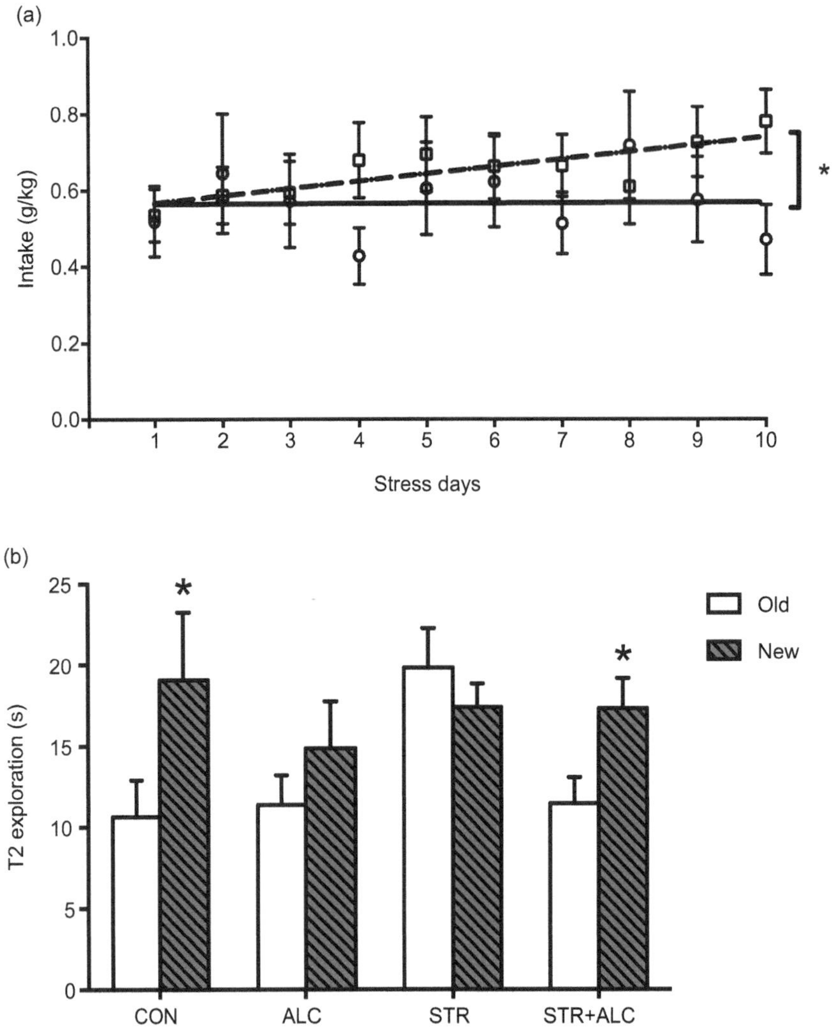


**Figure 14.3** ***Effects of chronic stress on alcohol consumption and spatial memory in male rats.*** (a) Mean alcohol consumption (g/kg) over days of restraint stress. Dashed regression line is plotted for stress + alcohol (STR + ALC) group and solid regression line is plotted for ALC group. * indicates significant difference ($R^2 = 0.25$, $p = 0.026$) between group regression lines. (b) Object placement. Shown is exploration at old and new locations 4 h after the initial sample (T1) trial. Only the control (CON) and STR + ALC groups spent more time exploring the object at the new location suggesting that STR or ALC impaired spatial memory but the combination did not. * indicates significant difference ($p < 0.05$). *(Reprinted from Gomez et al. (2012). With permission from Elsevier.)*

previous studies (Sillaber et al., 2002; Boyce-Rustay et al., 2008; Chester et al., 2004). Increased alcohol consumption is also seen in humans undergoing periods of stress and may be an attempt to cope with the negative effects of stress on mood and cognition (Sinha, 2008). This hypothesis received some support when the groups were tested for spatial memory on the object placement task as shown in Figure 14.3b. In the object placement task, rats explore two objects on an open field. Then an intertrial delay period is given, in this case 4 h. The rats are returned to the field where one object has been moved to a new location. Since rats seek novelty, they will explore an object in a new location more than an object in a known or old location, provided they remember the old location (Luine, 2015). Thus, the control rats spent significantly more time at the new as compared with the old location (approximately twice as much time); in contrast, the alcohol-consuming and the stressed rats spent approximately equal time exploring the objects at the new and old locations suggesting impaired memory. Interestingly, the stress + alcohol group performed similarly to the controls, spending more time at the new location. This result shows that the combination of alcohol and stress reverses the individual, deleterious effects of stress or alcohol. Whether alcohol blocks or mitigates the stress effect, or vice versa, is not known, but these results show that stressed rats increase alcohol consumption and suggest that the effects may be like human males who self-medicate with alcohol to relieve stress-related effects (Sinha, 2008).

We further investigated the interactive effects of stress and alcohol on cognition in male as well as female rats. As there was individual variability in the amount of consumption in the two-bottle paradigm, we utilized gavage of alcohol in order to deliver a constant dose to all rats. For 7 days male rats served as controls, were restrained for 6 h, were gavaged with alcohol (2.0 g/kg), or were restrained for 6 h and then received alcohol gavage (Gomez et al., 2013). Blood alcohol levels were approximately 125 mg/dL, an intoxicating level, and not different between the restrained and unrestrained alcohol treated groups, suggesting that stress had no effect on alcohol metabolism. Results of cognitive testing, object recognition, and Y-maze are shown in Figure 14.4. The Y-maze is a hippocampal-dependent spatial memory task similar to object placement. Memory is assessed by comparing entries into the novel arm (previously unexplored in training trial) with the start and other arms (previously explored arms in the training trials). Object recognition testing is conducted similar to object placement except that a new object replaces one of the identical objects in the T2-retention trial (Luine, 2015). No treatments affected exploration in the T1 trial for either object recognition (Figure 14.4a) or the Y-maze (Figure 14.4c). Consistent with previous stress (Beck and Luine, 1999; Bowman et al., 2009) or alcohol studies (Ryabinin et al., 2002), stress alone or alcohol alone impaired object recognition (Figure 14.4b). Rats in these groups spent equal time exploring at the old and new objects, but, surprisingly, the stress + alcohol group, like the control group, spent significantly more time exploring at the new, compared with the old (known) object. A similar pattern was seen for the Y-maze where the alcohol alone and stress alone groups showed no differences between exploration of the arms while the control and stress + alcohol groups explored the novel arm more than the other and start arms (Figure 14.4d). Thus, for both object recognition and Y-maze in male rats, there was an interactive effect between stress and alcohol that spared memory impairments caused by each treatment alone.

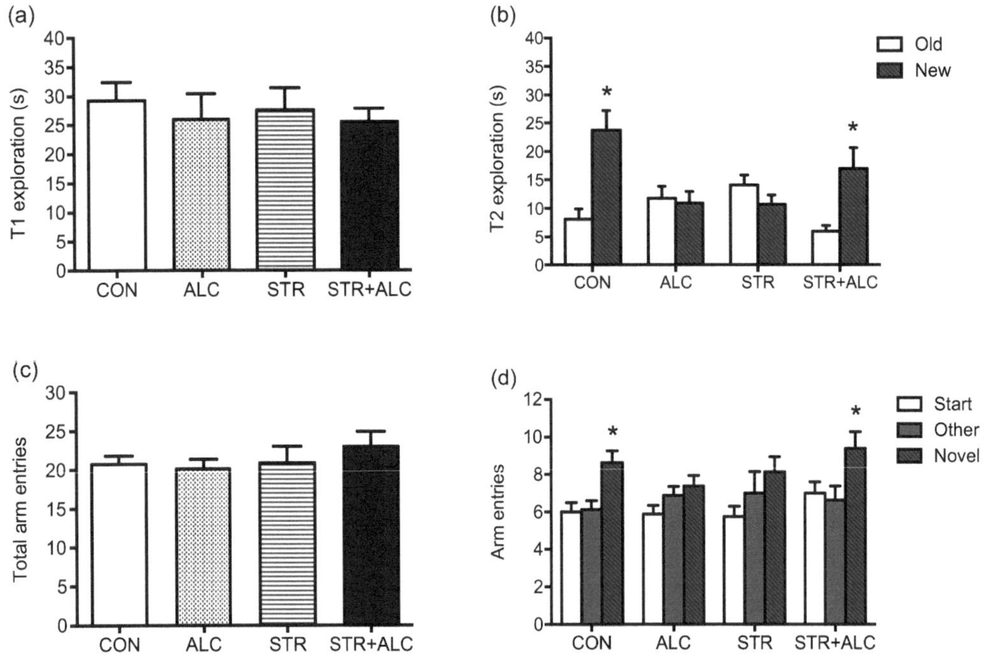


**Figure 14.4** ***Effects of chronic stress, alcohol, and stress + alcohol on memory in male rats.*** (a) Object recognition: exploration time (mean $\pm$ SEM) of identical objects during the sample (T1) trial. No significant differences between groups. (b) Object recognition: exploration time (mean $\pm$ SEM) of old and new objects during the retention (T2) trial conducted 4 h post-T1. Only the CON and STR + ALC groups spent more time exploring the new versus the old object suggesting that STR or ALC impaired memory but STR + ALC did not. * $p < 0.05$. (c) Y-maze: total entries into the start and other arms (mean $\pm$ SEM) in training trial. No significant differences between groups. (d) Y-maze: entries into each arm during the probe trial conducted 4 h posttraining trial. Only the CON and STR + ALC groups explored the novel arm more than the other arms (start and other) suggesting that ALC or STR impaired spatial memory but STR and ALC did not. * $p < 0.01$. *(Reprinted from Gomez et al. (2013). With permission from Elsevier.)*

An experiment in females, identical to the previous experiment in males, was conducted (Gomez and Luine, 2014; Figure 14.5). Object exploration (Figure 14.5a) and arm entries for the Y-maze (Figure 14.5c) in the T1 trial were not different among control, alcohol, stressed, or stressed + alcohol female groups. In the retention trial for object recognition, control and stressed females spent more time exploring the novel object, consistent with previous studies (Beck and Luine, 1999, 2002; Bowman et al., 2009) while the alcohol and stress + alcohol groups spent the same amount of time exploring old and new objects (Figure 14.5b). Thus, alcohol appears to impair object memory, and, unlike in males, the combination of stress + alcohol does not spare alcohol-dependent memory impairments in females. For Y-maze testing, only the control females showed greater exploration of the novel arm compared with the other and start arms (Figure 14.5d). Thus, alcohol, stress, or the combination impaired female memory on the Y-maze, a different

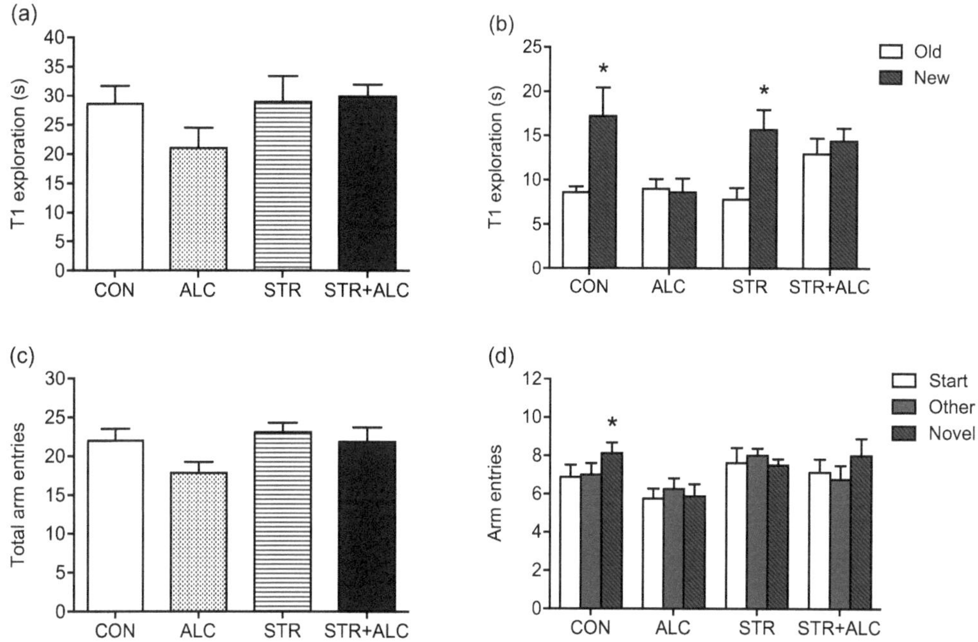


**Figure 14.5** ***Effects of chronic stress, alcohol, and stress + alcohol on memory in female rats.*** (a) Object recognition: exploration time (mean ± SEM) of identical objects during the sample (T1) trial. No significant differences between groups. (b) Object recognition: exploration time (mean ± SEM) of old and new objects during the retention (T2) trial conducted 4 h post-T1. The CON and STR group explored the new object more than the old object suggesting that ALC and STR + ALC impaired memory. (c) Y-maze: total entries into the start and other arms (mean ± SEM) during training trial. No significant differences between groups. (d) Y-maze: entries into each arm during the probe trial (mean ± SEM) conducted 4 h posttraining trial. Only the CON group explored the novel arm more than the other arms suggesting that ALC, STR, and ALC + STR impaired spatial memory. * $p < 0.03$. *(Reprinted from Gomez and Luine (2014). With permission from Elsevier.)*

pattern than in males where the combination treatment spared the impairing effects of alcohol or stress. Overall, a different pattern for the combination of stress + alcohol on cognitive function was seen in female versus male rats: stress + alcohol treatment impairs female, but not male, memory.

Taken together, these experiments examining the interactive effects of stress and alcohol on cognition indicate that robust sex differences are present not only in responses to stress alone but that the combination of stress with alcohol results in a further delineation of sex-dependent responses. For the cognitive realm, the combination of alcohol with stress appears to alleviate the stress induced impairments in memory, both spatial and nonspatial, in males. However, in females, stress normally has no effect on memory (object recognition) or enhances some spatial memory tasks, but combining alcohol with stress results in impairments in both types of memory. These results, as well as studies on mood and anxiety (see the subsequent section), may help explain some sex differences in mood disorders and drug use/abuse. The observation that stress-dependent cognitive deficits were reversed in males, but not females, when alcohol was given following the stress is consistent with the observation that males abuse drugs and alcohol more than females. This relationship suggests that drug therapies for depression or anxiety should be examined within the context that different effects may result depending on the sex of the subject.

# 6 CONCLUSIONS

## 6.1 Implications of sex differences in stress responses for mental health and disease

The mechanisms and neural circuits underlying cognition and emotion are linked and show overlap among brain areas that regulate them. Moreover, activation of the HPA axis as a result of stress causes hormone and neurotransmitter release throughout the brain, affecting the way we feel, remember, and behave. Thus, sex differences in cognitive responses to stress may contribute to known differences between the sexes in the incidence of depression and anxiety, which are higher in females, and in drug abuse, which is higher in males. We find that changes in mood and anxiety are present in the same rats whose cognition is altered by stress, and that sex differences are also present (Luine et al., 2007; Luine and Dohanich, 2008; Luine, 2011) but a discussion of these changes is beyond the scope of the current chapter (see also Karatsoreos and McEwen, 2013 for further discussion).

At first glance, the pattern of stress-dependent changes, a general resilience to stress for cognition function and anxiety in females, does not appear consistent with the general behavioral phenotype of females who show more depression, anxiety, and posttraumatic stress disorder than males (Weinberger et al., 2009). However, this review has focused on a memory circuit including mainly the hippocampus and prefrontal cortex, and stress effects on other, overlapping neural circuits may better reflect this behavioral phenotype (Shansky and Morrison, 2009). For example, males are more resilient than females to stress-induced remodeling of dendrites in a cortical–amygdalar pathway, and estradiol promotes sensitivity to the stress induced remodeling (Shansky et al., 2010). On the other hand, it might be envisioned that heightened cognition in females would facilitate remembrance of stressful events and/or that the experience would remain longer in memory. Thus, anxiety or depression might occur at an extended time, but not immediately, after the stressful event. In contrast, males, while more anxious following the stress, have dampened cognitive function, which might lead to forgetting of the stressful event and thereby promote wellbeing. Furthermore, the different interactions between stress and alcohol on cognition in the sexes – reversal of impairments by each alone in males but not females – may also contribute to greater anxiety and depression in females.

In conclusion, while sex differences in cognitive responses to stress have been demonstrated over the lifespan, the neural bases for these differences remain largely unknown. However, more information about sex differences in neural function maybe forthcoming because sex differences in stress-related diseases, as well responses to drug treatments, have initiated the gender medicine movement. Gender medicine posits that we need to do away with the assumption that men and women are similarly affected by diseases and treatments. Further emphasizing the importance of sex in biological systems, the National Institutes of Health (NIH) has unveiled policies to ensure that preclinical research funded by the United States NIH includes both females and males (Clayton and Collins, 2014). Clearly, most current research does not include both sexes, and we have only scratched the surface on how stress is processed in female versus male brains. This information is vital for understanding the link between stress and mental functioning, and, thus, it is imperative that future research and stress models include both sexes in their analyses in order to better understand and treat stress and its aftermath.

## REFERENCES


Ariza Traslavina, G.A., de Oliveira, F.L., Franci, C.R., 2014. Early adolescent stress alters behavior and the HPA axis response in male and female adult rats: the relevance of the natue and duration of the stressor. Physiol. Behav. 133, 178–189.

Bailey, D.J., Ma, C., Soma, K.K., Saldanha, C.J., 2013. Inhibition of hippocampal aromatization impairs spatial memory performance in a male songbird. Endocrinology 154, 4707–4714.

Beck, K.D., Luine, V.N., 1999. Food deprivation modulates chronic stress effects on object recognition in male rats: role of monoamines and amino acids. Brain Res. 830, 56–71.

Beck, K.D., Luine, V.N., 2002. Sex differences in behavioral and neurochemical profiles after chronic stress: role of housing conditions. Physiol. Behav. 75, 661–673.

Beck, K.D., Luine, V.N., 2010. Evidence for sex specific shifting of neural processes underlying learning and memory following stress. Physiol. Behav. 99, 204–211.

Becker, J.B., Taylor, J.R., 2008. Sex differences in motivation. In: Becker, J.B., Berkley, K.J., Geary, N., Hampson, E., Herman, J.P., Young, E.A. (Eds.), Sex Differences in the Brain: From Genes to Behavior. Oxford University Press, pp. 227–252.

Bisagno, V., Grillo, C.A., Piroli, G.G., Giraldo, P., McEwen, B., Luine, V.N., 2004. Chronic stress alters amphetamine effects on behavior and synaptophysin levels in female rats. Pharm. Biochem. Behav. 78, 541–550.

Bowman, R.E., 2005. Stress-induced changes in spatial memory are sexually differentiated and vary across the lifespan. J. Neuroendocrinol. 17, 526–535.

Bowman, R.E., Kelly, R., 2012. Chronically stressed female rats show increased anxiety but no behavioral alterations in object recognition or placement memory: a preliminary examination. Stress 15, 524–532.

Bowman, R.E., Zrull, M.C., Luine, V.N., 2001. Chronic restraint stress enhances radial arm maze performance in female rats. Brain Res. 904, 279–289.

Bowman, R.E., Ferguson, D., Luine, V.N., 2002. Effects of chronic restraint stress and estradiol on open field activity, spatial memory, and monoaminergic neurotransmitters in ovariectomized rats. Neuroscience 113, 401–410.

Bowman, R.E., Beck, K.D., Luine, V.N., 2003. Chronic stress effects on memory: sex differences in performance and monoamines. Horm. Behav. 43, 48–59.

Bowman, R.E., MacLusky, N.J., Sarmiento, Y., Frankfurt, M., Gordon, M., Luine, V.N., 2004. Sexually dimorphic effects of prenatal stress on cognition, hormonal responses and central neurotransmitters. Endocrinology 145, 3778–3787.

Bowman, R.E., Micik, R., Gautreaux, C., Fernandez, L., Luine, V.N., 2009. Sex dependent changes in anxiety, memory, and monoamines following one week of stress. Physiol. Behav. 97, 21–29.

Boyce-Rustay, J.M., Janos, A.L., Holmes, A., 2008. Effects of chronic swim stress on EtOH-related behaviors in C57BL/6J, DBA/2J and BALB/cByJ mice. Behav. Brain Res. 186, 133–137.
Cahill, L., 2014. Equal ≠ the same: sex differences in the human brain. Cerebrum. 2014, 5, eCollection.
Chang, Q., Gold, P.E., 2003. Switching memory systems during learning: changes in patterns of brain acetylcholine release in the hippocampus and striatum in rats. J. Neurosci. 23, 3001–3005.
Chester, J.A., Blose, A.M., Zweifel, M., Froehlich, J.C., 2004. Effects of stress on alcohol consumption in rats selectively bred for high or low alcohol drinking. Alcohol Clin. Exp. Res. 28, 385–393.
Clayton, J.A., Collins, F.S., 2014. Policy: NIH to balance sex in cell and animal studies. Nature 509, 282–283.
Conrad, C.D., 2010. A critical review of chronic stress effects on spatial learning and memory. Prog. Neuropsychopharmacol. Biol. Psychiat. 34, 742–755.
Conrad, C.D., Galea, L.A., Kuroda, Y., McEwen, B.S., 1996. Chronic stress impairs rat spatial memory on the Y maze, and this effect is blocked by tianeptine pretreatment. Behav. Neurosci. 110, 1321–1334.
Conrad, C.D., Grote, K.D., Hobbs, R.J., Ferayorni, A., 2003. Sex differences in spatial and non-spatial Y-maze performance after chronic stress. Neurobiol. Learn. Mem. 79, 32–40.
Cooper, M.L., Russell, M., Skinner, J.B., Frone, M.R., Mudar, P., 1992. Stress and alcohol use: moderating effects of gender, coping, and alcohol expectancies. J. Abnorm. Psychol. 101, 139–152.
Fester, L., Rune G.M., 2014. Sexual neurosteroids and synaptic plasticity in the hippocampus. Brain Res. Oct 28. pii: S0006-8993(14)01435-8. doi: 10.1016/j.brainres.2014.10.033.[Epub ahead of print].
Galea, L.A., McEwen, B.S., Tanapat, P., Deak, T., Spencer, R.L., Dhabhar, F.S., 1997. Sex differences in dendritic atrophy of CA3 pyramidal neurons in response to chronic restraint stress. Neuroscience 81, 689–697.
Garcia-Segura, L.M., Azcortia, I., DonCarlos, L.C., 2001. Neuroprotection by estrogen. Prog. Neurobiol. 63, 29–60.
Gomez, J.L., Luine, V., 2014. Female rats exposed to stress and alcohol show impaired memory and increased depressive-like behaviors. Physiol. Behav. 123, 47–54.
Gomez, J.L., Lewis, M., Luine, V., 2012. Alcohol access alleviates stress induced spatial memory impairments in male rats. Alcohol 46, 499–504.
Gomez, J.L., Lewis, M.J., Sebastian, V., Serrano, P., Luine, V., 2013. Alcohol administration blocks stress-induced impairments in memory and anxiety and alters hippocampal neurotransmitter receptor expression in male rats. Horm. Behav. 63, 659–661.
Gron, G., Wunderlich, A.P., Spitzer, M., Tomczak, R., Riepe, M.W., 2000. Brain activation during human navigation: gender-different neural networks as substrate of performance. Nat. Neurosci. 3, 404–408.
Hosseini-Sharifabad, M., Hadinedoushan, H., 2007. Prenatal stress induces learning deficits and is associated with a decrease in granules and CA3 cell dendritic tree size in rat hippocampus. Anat. Sci. Int. 82, 211–217.
Isgor, C., Kabbaj, M., Akil, H., Watson, S.J., 2004. Delayed effects of chronic variable stress during peripubertal-juvenile period on hippocampal morphology and on cognitive and stress axis functions in rats. Hippocampus 14, 636–648.
Karatsoreos, I.N., McEwen, B.S., 2013. Annual research review: the neurobiology and physiology of resilience and adaptation across the life course. J. Child. Psychol. Psychiat. 54, 337–347.
Kato, A., Hojo, Y., Higo, S., Komatsuzaki, Y., Murakami, G., Yoshino, H., Uebayashi, M., Kawato, S., 2013. Female hippocampal estrogens have a significant correlation with cyclic fluctuation of hippocampal spines. Front. Neural Circuits 7, 149.
Kitraki, E., Kremmyda, O., Youlatos, D., Alexis, M.N., Kittas, C., 2004. Gender-dependent alterations in corticosteroid receptor status and spatial performance following 21 days of restraint stress. Neuroscience 125, 47–55.
Korol, D.L., Malin, E.L., Borden, K.A., Busby, R.A., Couper-Leo, J., 2004. Shifts in preferred learning strategy across the estrous cycle in female rats. Horm. Behav. 45, 330–338.
Luine, V.N., 2002. Sex differences in chronic stress effects on memory in rats. Stress 5, 205–216.
Luine, V.N., 2011. Chronic stress and sex differences on cognition: animal studies. In: Conrad, C. (Ed.), Handbook of Stress: Neuropsychological Effects on the Brain. Wiley-Blackwell, Malden, MA.
Luine, V., 2015. Recognition memory tasks in neuroendocrine research. Behav. Brain Res. 285, 158–164.
Luine, V.N., Dohanich, G., 2008. Sex differences in cognitive function in rodents. In: Becker, J.B., Berkley, K.J. Geary, N., Hampson, E., Herman, J.P., Young, E.A. (Eds.), Sex Differences in the Brain: From Genes to Behavior. Oxford University Press, New York, pp. 227–252.

Luine, V., Villegas, M., Martinez, C., McEwen, B.S., 1994. Repeated stress causes reversible impairments of spatial memory performance. Brain Res. 639, 167–170.

Luine, V.N., Beck, K.D., Bowman, R.E., Frankfurt, M., MacLusky, N.J., 2007. Stress & neural function – accounting for sex and age. J. Neuroendocrinol. 19, 743–751.

Markham, J.A., Taylor, A.R., Taylor, S.B., Bell, D.B., Koenig, J.I., 2010. Characterization of the cognitive impairments induced by prenatal exposure to stress in the rat. Front. Behav. Neurosci. 25 (4), 173.

Markham, J.A., Mullins, S.E., Koenig, J.I., 2013. Periadolescent maturation of the prefrontal cortex is sex-specific and is disrupted by prenatal stress. J. Comp. Neurol. 521, 1828–1843.

Martinetti, M.P., Andrzejewski, M.E., Hineline, P.N., Lewis, M.J., 2000. Ethanol consumption and the matching law: a choice analysis using a limited-access paradigm. Exp. Clin. Psychopharmacol. 8, 395–403.

McEwen, B.S., 2008. Understanding the potency of early life experiences on brain and body function. Metabolism 57 (Suppl. 2), S11–S15.

McLaughlin, K.J., Baran, S.E., Wright, R.L., Conrad, C.D., 2005. Chronic stress enhances spatial memory in ovariectomized female rats despite CA3 dendritic retraction: possible involvement of CA1 neurons. Neuroscience 135, 1045–1054.

McLaughlin, K.J., Wilson, J.O., Harman, J., Wright, R.L., Wieczorek, L., Gomez, J., Korol, D.L., Conrad, D., 2009. Chronic 17β-estradiol or cholesterol prevents stress-induced hippocampal CA3 dendritic retraction in ovariectomized female rats: possible correspondence between CA1 spine properties and spatial acquisition. Hippocampus 20, 768–786.

Novick, A.M., Miller, L.C., Forster, G.L., Watt, M.J., 2013. Adolescent social defeat decreases spatial working memory performance in adulthood. Behav. Brain Funct., 39.

Ortiz, J.B., McLaughlin, K.J., Hamilton, G.F., Baran, S.E., Campbell, A.N., Conrad, C.D., 2013. Cholesterol and perhaps estradiol protect against corticosterone-induced hippocampal CA3 dendritic retraction in gonadectomized female and male rats. Neuroscience 246, 409–421.

Packard, M.G., McGaugh, J.L., 1966. Inactivation of hippocampus or caudate-putamen differentially strengthens place and response learning. Proc. Natl. Acad. Sci. USA 96, 12881–12886.

Pohorecky, L.A., 2008. Psychosocial stress and chronic ethanol ingestion in male rats: effects on elevated plus maze behavior and ultrasonic vocalizations. Physiol. Behav. 94, 432–447.

Radley, J.J., Sisti, H.M., Hao, J., Rocher, A.B., McCall, T., Hof, P.R., McEwen, B.S., Morrison, J.H., 2004. Chronic behavioral stress induces apical dendritic reorganization in pyramidal neurons of the medial prefrontal cortex. Neuroscience 125, 1–6.

Radley, J.J., Rocher, A.B., Miller, M., Janssen, W.G., Listen, C., Hof, P.R., McEwen, B.S., Morrison, J.H., 2006. Repeated stress induces dendritic spine loss in the rat medial prefrontal cortex. Cereb. Cortex 16, 313–320.

Romeo, R.D., Karatsoreos, I.N., 2011. Adolescence and stress: from hypothalamic-pituitary-adrenal function to brain development. In: Conrad C. (Ed.), Handbook of Stress: Neuropsychological Effects on the Brain. Wiley-Blackwell, Malden, MA.

Ryabinin, A.E., Miller, M.N., Durrant, S., 2002. Effects of acute alcohol administration on object recognition learning in C57BL/6J mice. Pharmacol. Biochem. Behav. 71, 307–312.

Schwabe, I., Oitzl, M.S., Philippsen, C., Richter, S., Bohringer, A., Wippich, W., et al., 2007. Stress modulates the use of spatial versus stimulus-response learning strategies in humans. Learn. Mem. 14, 109–116.

Schwabe, I., Dalm, S., Schachinger, H., Oitzel, M.S., 2008. Chronic stress modulates the use of spatial and stimulus-response learning strategies in mice and man. Neurobiol. Learn. Mem. 90, 495–503.

Schwabe, I., Oitzl, M.S., Richter, S., Schachinger, H., 2009. Modulation of spatial and stimulus-response learning strategies by exogenous cortisol in healthy young women. Psychoneuroendocrinology 34, 358–366.

Selye, H., 1976. The Stress of Life. McGraw Hill, New York.

Shansky, R.M., Morrison, J.H., 2009. Stress induced dendritic remodeling in the medial prefrontal cortex: effects of circuit, hormones and rest. Brain Res. 1293, 108–113.

Shansky, R.M., Hamo, C., Hof, P.R., Lou, W., McEwen, B.S., Morrison, J.H., 2010. Estrogen promotes stress sensitivity in a prefrontal cortex-amygdala pathway. Cereb. Cortex 20, 2560–2567.

Sierksma, A.S., Prickaerts, J., Chouliaras, L., Rostamian, S., Delbroek, L., Rutten, B.P., Steinbusch, H.W., van den Hove, D.L., 2013. Behavioral and neurobiological effects of prenatal stress exposure in male and female APPswe/PS1dE9 mice. Neurobiol. Aging 34, 319–337.

Sillaber, I., Rammes, G., Zimmermann, S., Mahal, B., Zieglgansberger, W., Wurst, W., Holsboer, F., Spanagel, R., 2002. Enhanced and delayed stress-induced alcohol drinking in mice lacking functional CRH1 receptors. Science 296, 931–933.

Sinha, R., 2008. Chronic stress, drug use, and vulnerability to addiction. Ann. NY Acad. Sci. 1141, 105–130.

Steiner, M., Young, E.A., 2008. Hormones and mood. In: Becker, J.B., Berkley, K.J., Geary, N., et al. (Eds.) Sex Differences in the Brain: From Genes to Behavior. Oxford University Press, New York, NY, pp. 405–426.

Sternberg, W.F., 1999. Sex differences in the effects of prenatal stress on stress-induced analgesia. Physiol. Behav. 68, 63–72.

Sunanda, R.B.S., Raju, T.R., 2000. Restraint stress-induced alterations in the levels of biogenic amines, amino acids, and AchE activity in the hippocampus. Neurochem. Res. 25, 1547–1552.

Watanabe, Y., Gould, E., McEwen, B.S., 1992. Stress induces atrophy of apical dendrites of hippocampal CA3 pyramidal neurons. Brain Res. 588, 341–345.

Wei, J., Yuen, E.Y., Liu, W., Li, X., Zhong, P., Karatsoreos, I.N., McEwen, B.S., Yan, Z., 2014. Estrogen protects against the detrimental effects of repeated stress on glutamatergic transmission and cognition. Mol. Psychiat. 588–598.

Weinberger, A.H., Maciejewski, P.K., McKee, S.A., Reutenauer, E.L., Mazure, C.M., 2009. Gender differences in associations between lifetime alcohol, depression, panic disorder, and posttraumatic stress disorder and tobacco withdrawal. Am. J. Addict. 18, 140–147.

Weinstock, M., 2011. Sex-dependent changes induced by prenatal stress in cortical and hippocampal morphology and behavior in rats: an update. Stress 14, 604–613.

Westenbroek, C., Den Boer, J.A., Veenhuis, M., Ter Horst, G.J., 2004. Chronic stress and social housing differentially affect neurogenesis in male and female rats. Brain Res. Bull. 64, 303–308.

Williams, C.L., Barnett, A.M., Meck, W.H., 1990. Organizational effects of early gonadal secretions on sexual differentiation in spatial memory. Behav. Neurosci. 104, 84–97.

Wright, R.L., Conrad, C.D., 2005. Chronic stress leaves novelty-seeking behavior intact while impairing spatial recognition memory in the Y-maze. Stress 8, 151–154.

Yaroslavsky, I., Tejani-Butt, S.M., 2010. Voluntary alcohol consumption alters stress-induced changes in dopamine-2 receptor binding in Wistar-Kyoto rat brain. Pharmacol. Biochem. Behav. 94, 471–476.

Zagron, G., Weinstock, M., 2006. Maternal adrenal hormone secretion mediates behavioural alterations induced by prenatal stress in male and female rats. Behav. Brain Res. 175, 323–328.

# CHAPTER 15

# Epigenetic Sex: Gene–Environment Contributions to Brain Sex Differences and their Impact on Mental Health Risk

**Anthony P. Auger**
University of Wisconsin–Madison, Neuroscience Training Program, Departments of Zoology and Psychology, Madison, WI, USA

## 1 INTRODUCTION

It is becoming clear that an individual's gene–environmental upbringing can influence their risk or resilience to mental health disorders. A more intriguing biological factor that can also alter one's resilience or risk to mental health disorder is biological sex. Indeed, biological sex is found to influence the occurrence, severity, and the age of onset of a number of mental health disorders. For example, sex differences in these factors have been reported in Alzheimer's disease, schizophrenia, Rett syndrome, and autism. While it is not fully clear what underlying factors in the brain result in these pathologies, less is known about why sex differences are observed in the incidence and severity of these disorders. Intriguingly, some childhood disorders, such as autism and attention deficit hyperactivity disorder, are diagnosed at higher rates in boys than girls. This suggests that biological sex differences in the male brain may put them more at risk for some types of developmental disorders. Likewise, the female brain may be somewhat resilient to these types of mental health disorders. Understanding the biological mechanisms governing sex differences within the developing brain may provide important clues as to why boys are at greater risk or why girls appear resistant to some mental health disorders. As such, gene–environment–biological sex interactions need to be considered when investigating the etiology or progression of mental health disorders.

## 2 SEXUAL DIFFERENTIATION OF THE BRAIN

### 2.1 Introduction

The development of sex differences in the brain is known to be regulated by a variety of factors, including genetic, hormonal, and environmental influences (Lonstein and Auger, 2009; McCarthy and Arnold, 2011; Auger and Auger, 2011); however, the most salient sex differences in the brain mainly result from altered gonadal steroid hormone exposure that is dictated by chromosomal sex. If biological sex is determined by

*Sex Differences in the Central Nervous System.* http://dx.doi.org/10.1016/B978-0-12-802114-9.00015-9

chromosomal sex, then sex differences start at fertilization by the presence or absence of the Y chromosome. The existence of the sex-determining region Y (Sry) gene located on the Y chromosome initiates gonadal differentiation. Once differentiated, the testes synthesize and release testosterone into the circulatory system differentiating male from female physiology. Within the brain, testosterone is converted into estradiol and dihydrotestosterone. Interestingly, while androgens can produce some brain sex differences, it is estradiol that is responsible for the most prominent sex differences identified within the brain. It is important to clarify that while exposure to a high level of estradiol is responsible for masculinizing the male brain, exposure to a low level of estradiol is necessary for active feminization of the female brain. Previous studies have reported that disrupting estradiol action within the developing female brain can further separate the female brain from that of males and disrupt typical female-like processes within the brain (Lonstein and Auger, 2009). Indeed, estradiol action within the developing female brain is necessary to produce typical levels of female sexual behavior and reproductive function. Further evidence that estradiol is important for female brain development comes from studies finding that the female brain can produce local levels of estradiol that equal male levels (Amateau et al., 2004). This indicates the presence of a sex difference that induces neurosteroid synthesis within the female brain in order to eliminate sex differences in estradiol exposure within some brain regions. A mechanism to ensure equal levels of estradiol exposure within some brain regions may act to reduce overt differences in function. Therefore, sometimes a sex difference functions in order to reduce or eliminate sex differences in brain physiology or function. The importance of active mechanisms in the female brain will be discussed later with regard to epigenetic programming, but it is important to note that the female brain is the product of an active process and not necessarily a passive default.

## 2.2 Hormonal contribution to sexual differentiation of the brain

While nonsteroidal mechanisms can contribute to the differentiation of male versus female brain (Olesen et al., 2005; De Vries et al., 2002), it is generally accepted that testosterone, and its metabolites estradiol and dihydrotestosterone, is important for organizing some of the most salient sex differences in brain and behavior (MacLusky and Naftolin, 1981; McCarthy, 2008). These differences mainly result from a postnatal surge in testosterone from the testes within hours after birth, as castration before this surge disrupts masculinization in males, and neonatal testosterone treatment can masculinize females (Baum, 1979). Gonadal steroid hormones produce sex differences within the brain mostly by binding to steroid/nuclear receptors located within the cytoplasm or the nuclear component of a cell. While gonadal steroid hormones bind to both nuclear and membrane receptors to elicit a variety of cellular actions, less is known about the functional consequences of membrane actions of steroids on sexual differentiation of the brain. With regard to epigenetics, nuclear receptors provide for a molecular pathway by which environmental signals can act upon the genome to elicit brief or lasting changes in neuronal or gene function to influence sexual differentiation of the brain.

A variety of steroid receptors have been implicated in sexually differentiating the brain, such as estrogen receptor (ER) α and β, androgen receptors (ARs), and progestin receptors. When gonadal hormones bind to steroid receptors, the hormone-bound steroid receptor forms a dimer complex with another like steroid bound receptor, and this complex binds to a response element located within the promoter region of a large number of target genes. Once bound, the steroid receptor complex recruits a variety of coregulatory proteins and other transcriptional factors that modifies the rate of gene transcription (Tetel et al., 2009). The recruited proteins that increase gene transcription rates are called coactivators and those that decrease gene transcription rates are called corepressors (Auger and Jessen, 2009; Tetel et al., 2009). The importance of these coregulatory proteins provide for a mechanism by which steroid hormones or other environmental signals that alter the levels of coregulatory protein expression can influence brain differentiation. Indeed, disrupting the expression of coregulatory proteins can have profound and lasting consequences on sexual differentiation of both juvenile and adult social and reproductive behaviors (Auger and Jessen, 2009). The presence and function of a large number of coregulatory proteins allows for diverse regulation of steroid hormone function that is dependent upon available levels of coregulatory proteins.

Perinatal hormone exposure is known to cause numerous functional and structural differences within the developing male and female brain. One of the most widely studied sex differences within the rodent brain is the sexually dimorphic nucleus of the preoptic area (SDN-POA). This nucleus has been reported to be around two to four times larger in males compared with females (Gorski et al., 1978). In humans, there is an analogous region that is also found to be larger in males contrasted to females. Although the exact subanatomical structure that resembles rodents remains somewhat in debate, analysis of four different cell groups referred to as the interstitial nuclei of the anterior hypothalamus (INAH 1-4) indicate that INAH-3 is found to be around two to three times larger in males. Therefore, it is suggested that INAH-3 is similar to that of the SDN-POA in rodents (Allen et al., 1989). Sex differences in volumetric size of the SDN-POA have been widely accepted and replicated by many. Functionally, this nucleus seems most likely to function to inhibit female sexual behavior. Additional brain regions that are larger in males include the bed nucleus of the stria terminalis and regions of the amygdala (Hines et al., 1992) that are involved in a wide variety of socio-emotional processes. In contrast to areas larger in males, the anteroventral periventricular nucleus is found to be two to three times larger in females (Bleier et al., 1982). This region is known to play an important role in regulating the luteinizing hormone surge in females; therefore, it is reasonable to predict that this area is more specialized in females. Additional differences are within the arcuate nucleus (Ciofi et al., 2006), ventromedial hypothalamus (Matsumoto and Arai, 1986), and substantia nigra pars compacta (Dewing et al., 2006). These data indicate that steroid receptor action within the developing brain can have lasting consequences on brain development. These effects can be measured as alterations of cell number, migration patterns, phenotypical differentiation, as well as morphological differentiation of brain cells (Lonstein and Auger, 2009). As differences in gene expression are regulated by epigenetic events, it is likely that these regions exhibit epigenetic differences between the sexes that are induced by steroid hormones.

The actions of steroid hormones not only shape differences in brain structure, but they also shape differences in social development. One of the first sex differences in social behaviors to emerge during development is juvenile social play behavior. Specifically, juvenile male rats engage in social play behavior at a higher frequency than do juvenile female rats (Olioff and Stewart, 1978). Although factors other than hormones have been found to alter the development of juvenile social play behavior (Auger and Olesen, 2009), masculinization of "rough-and-tumble" social play is known to be influenced by neonatal testosterone exposure (Meaney and Stewart, 1981; Beatty et al., 1981). Hormones shape lasting differences in social play via ARs (Casto et al., 2003; Meaney and Stewart, 1981); however, recent data also suggest a potential role for estrogen receptors (Olesen et al., 2005). Interestingly, maternal interactions can also alter the development of juvenile social play behavior (Chamove et al., 1973; Parent and Meaney, 2008). As variations in maternal care alter DNA methylation patterns of promoter regions for nuclear receptors, such as the ER$\alpha$ promoter region (Champagne et al., 2006), we will discuss the convergence of neonatal social experience and steroid receptor regulation in sculpting brain sex differences via epigenetic mechanisms.

## 3 EMERGENCE OF EPIGENETIC RESEARCH

### 3.1 Introduction

Conrad Waddington is credited with using the term "epigenetics" as a branch of biology that investigates how genes and their products produce a phenotype (Waddington, 1942). David Nanney later refined the term epigenetics to explain how cells with the same genotype can have different phenotypes (Nanney, 1958), which is perhaps the clearest definition to be used. Over the years, epigenetics has been conceptualized as a relatively stable phenomenon that persists throughout the lifespan. The relative stability of epigenetic mechanisms provides a potential explanation for how certain cellular "memories" may be maintained across cellular division, across a lifespan, and even passed on to future generations. A further refinement of the definition simplifies epigenetics to a modification to chromatin without changing the underlying code (Auger and Auger, 2011, 2013), which is similar to that redefined by Adrian Bird (Bird, 2007). Whether the marks are heritable or elicit an immediate change in gene transcription rates is not necessary for this definition. As such, an epigenetic change may not immediately result in a change in gene transcription rates, but it may alter a gene's response to future signaling events. For example, an epigenetic modification may be placed upon the promoter region of a gene near a response element. That epigenetic modification may not have an impact on gene expression unless a transcriptional factor is directed toward that response element. A difficultly in understanding the functional role of certain modifications to response elements may not become apparent until that pathway is activated across development. Therefore, it is conceivable that some epigenetic modifications may not impact gene expression until later in life.

While many epigenetic mechanisms have been identified, including modifications to cytosine residues, histone tail modifications, and posttranslational mechanisms, methylation of cytosine residues within DNA is perhaps the most widely studied epigenetic modification. DNA methyltransferases (DNMTs) appear to catalyze the addition of a methyl group to cytosine adjacent to guanine (CpG) dinucleotides on DNA (Grafstrom et al., 1985; Ramsahoye et al., 2000). While methylation of DNA occurs mostly at CpG sites, cytosines can also be methylated near thymine (CpT), adenine (CpA), as well as other cytosine (CpC) residues, and the functional role of methylation at other CpN sites remains to be determined. Methylation of DNA can also occur within the promoter region of a gene, upstream from the transcriptional start site, within coding and noncoding regions, and even downstream regions of a gene. As modifications can be proximal or distal to the gene, it is sometimes difficult to identify epigenetic markings that directly regulate gene expression. Nonetheless, methylation of DNA can have important consequences on gene expression.

While methylation of CpG sites is generally considered to be repressive to gene expression (Razin and Riggs, 1980), some data suggest it may participate in the cyclical activation of gene transcription (Metivier et al., 2008). Methylation may also repress silencers within the promoter, which could result in increased expression. Therefore, it remains surprising within the literature that most studies indicate that methylation has a repressive effect on gene expression. It is also important to note that while methylation of CpG sites may have some impact on gene transcription by itself, it more effectively serves as a signaling site for methyl-CpG-binding proteins (MBPs) to bind to methylated CpG sites and allow for the sequential recruitment of chromatin remodeling corepressor complexes to efficiently repress gene transcription (Klose and Bird, 2006; Auger and Auger, 2013) (Figure 15.1). These corepressor complexes contain histone deacetylase (HDAC), which removes acetyl groups from histone tails, resulting in chromatin condensation and gene repression (Yoon et al., 2003; Collingwood et al., 1999; Tetel et al., 2009). Importantly, emerging data suggest that there are sex differences in DNA methylation patterns, methyl binding proteins, and corepressor proteins that contribute to lasting sex differences in brain and juvenile behavior.

## 3.2 Epigenetic contribution to sexual differentiation of the brain

### *3.2.1 DNA methylation*

While there are various modifications to chromatin, one of the more intriguing epigenetic modifications to DNA is the methylation of CpG sites, as sometimes this

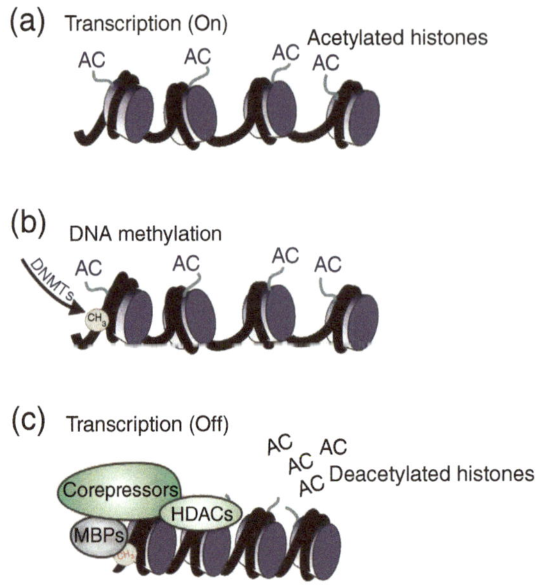


**Figure 15.1** ***Schematic representation of how epigenetic molecules can turn genes on or off.*** (a) In the active state, histone tails are acetylated and histones are apart, exposing DNA to the transcriptional machinery. (b) DNMTs facilitate the addition of a methyl group ($CH_3$) to DNA mainly at cytosine bases. (c) Methyl-binding proteins bind to the methylated DNA and recruit corepressor proteins and HDACs, which deacetylate histone tails, causing chromatin to tighten making it difficult for transcription factors to bind DNA.

modification can be duplicated across cellular division and even maintained throughout a lifespan. Methylation of DNA generally results from DNMTs conferring methyl groups to cytosine residues. While many DNMTs have been identified, DNMT1, DNMT3a, and DNMT3b have been the most studied. In general, DNMT1 is important in maintaining DNA methylation patterns (Pradhan et al., 1999; Yoder et al., 1997), whereas DNMT3a and DNMT3b are associated with *de novo* methylation (Okano et al., 1998; Hsieh, 1999). These functional classifications may prove to be an oversimplification in the future. With regard to sex differences in the brain, females have significantly more DNMT3a messenger (m)RNA than males within the developing amygdala (Kolodkin and Auger, 2011). Interestingly, while overall DNMT3a levels are higher in females during early brain development, DNMT1 activity appears higher in males (McCarthy et al., 2009). This suggests that the epigenome is undergoing sexual differentiation during a critical time point for brain development. If DNMT1 and DNMT3a are functionally different, then this would suggest a complicated interplay between these molecules within the developing male and female brain. While many sex differences within the developing brain are a result of differential hormonal exposure, it is not surprising to find that estradiol reduces DNMT3 levels within the developing brain. This suggests that hormone exposure may reduce the levels of methylation occurring on some genes within the developing male brain contrasted to female brain. However, to suggest that masculinization of the brain is due to reduced methylation is an oversimplification. Indeed, some genes need to be repressed for typical male brain development, which will be discussed in the subsequent section.

While hormonal contributions to brain sex differences have been well studied, other mechanisms, such as complicated interactions between the social environment, nonsteroidal activation of steroid receptors, and genetic and epigenetic mechanisms, are less well understood. Recent data suggest a convergence between these mechanisms within the developing brain. For example, maternal care has been shown to influence ERα promoter methylation within the developing brain (Champagne et al., 2006). As rat mothers typically groom the anogenital region of males more than females (Moore and Morelli, 1979; Moore, 1984), it is likely that differences in maternal care may contribute to sex differences in ERα expression and DNA promoter methylation patterns. In support, males have lower levels of ERα expression within the developing preoptic area (DonCarlos and Handa, 1994; Yamamoto et al., 2006; Yokosuka et al., 1997) and this difference appears to persist into adulthood (Brown et al., 1988; Lauber et al., 1991; Maerkel et al., 2007). Indeed, sex differences are found in ERα promoter CpG methylation within the developing rat POA. Specifically, males exhibit higher levels of methylation than females within the 5′ flanking region of ERα exon 1b promoter region, and this increase in methylation correlates with decreased ERα expression in males (Kurian et al., 2010). This sex difference in DNA methylation is partly due to differences in hormone exposure, as treatment with estradiol increased methylation and decreased ERα expression in neonatal females, suggesting that hormones contribute to the sex differences in ERα promoter methylation and expression. To determine the contributions of maternal care to the sex difference in ERα promoter methylation and expression, we modeled a component of maternal care using a paradigm called simulated maternal grooming (SMG). The tactile stimuli associated with maternal grooming were simulated with a soft nylon-bristled brush, which controls for other factors of maternal care and focuses on the tactile stimulation of licking and grooming. It was found that females receiving additional SMG had increased ERα promoter methylation to male-like levels within the developing POA (Kurian et al., 2010). These data suggest that differences in maternal grooming can masculinize the epigenome (Figure 15.2). As variations in maternal care can alter sex differences in juvenile social play (Edelmann et al., 2013) and adult sexual behavior (Moore, 1992; Rhees et al., 2001; Cameron et al., 2008), these data indicate that maternal interactions can guide brain sex differences and later behavioral differences by altering the epigenome.

This suggests that sex differences in the brain are due to both biological and environmental stimuli converging upon DNA, resulting in altered sex-specific DNA methylation patterns. These data are important in that they suggest that while differential

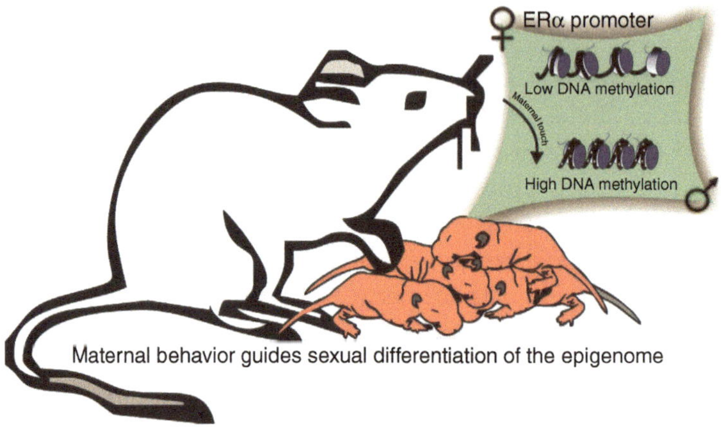


**Figure 15.2** ***Subtle differences in maternal behavior directed at male versus female offspring contribute to sexual differentiation of the epigenome.*** Mother rats tend to lick and groom neonatal males more than neonatal females. The increased maternal touch directed at males appears to contribute to increased methylation of the estrogen receptor α promoter region. Therefore, the mother rats' behavior is a contributing factor to further refining sexual differentiation of the epigenome, producing increased methylation of the estrogen receptor α promoter region in males.

hormone exposure is a critical factor in sexual differentiation of the brain, environmental influences, such as stress or parenting, can further modify sex differences in brain development. As stated earlier, methylation of DNA serves as a signal for methyl-CpG-binding proteins, which recruit additional proteins that modify chromatin structure and gene expression. Therefore, it is important to understand the functional role of methyl-binding proteins that interact with DNA methylation on modifying sex differences in the brain.

### *3.2.2 Methyl-binding proteins*

The methyl-CpG-binding family of proteins was first discovered after the characterization of a methyl-CpG-binding domain (MBD) that is responsible for binding of these classes of proteins to methylated CpG dinucleotides (Hendrich and Bird, 1998). Members within the family of MBP include Kaiso, MBD1, MBD2, MBD3, MBD4, and methyl-CpG-binding protein 2 (MeCP2). Upon binding to methylated DNA, MBPs recruit corepressor proteins and HDACs that ultimately modify chromatin to repress gene transcription. The functional importance of MBPs on typical gene function and their role on sex-biased neurological disorders is best illustrated with the findings of MeCP2.

MeCP2 was one of the first MBPs to be described (Meehan et al., 1989). Importantly, mutations in MECP2, an X-linked gene, are believed to be the major cause of Rett syndrome, a progressive neurodevelopmental disorder (Amir et al., 1999). Due to the presence of MECP2 on the X chromosome, and the phenomena of X inactivation, there is a sex difference in severity and diagnosis of Rett syndrome. While Rett syndrome occurs in approximately 10,000–22,000 female births and the severity of symptoms is variable in females, mutations in MECP2 in males are generally lethal. Interestingly, a mouse model of Rett syndrome has been developed through the deletion of the MECP2 gene that appears to recapitulate many of the symptoms exhibited in girls with Rett syndrome (Chen et al., 2001; Guy et al., 2001). Excitingly, recent studies have demonstrated that re-expressing MeCP2 later in life results in a reversal of many of the neurological symptoms of the Rett syndrome phenotype (Guy et al., 2007). These data are extremely exciting in that they suggest a neurodevelopmental disorder can be ameliorated later in life by altering the expression of a single molecule.

Interestingly, subtle reductions in MeCP2 levels are speculated to be involved in male dominated social and neurodevelopmental disorders, such as autism (Nagarajan et al., 2006). Furthermore, males may have reduced levels of MeCP2 expression within the developing amygdala around birth compared with females (Kurian et al., 2007). The lower levels of MeCP2 may put males at increased risk of developing symptoms inherent to MeCP2-related disorders. Indeed, transiently reducing MeCP2 levels with the developing male and female amygdala during the first few days of life using small interfering (si)RNA appears to disrupt later juvenile social play behavior in males, but not females (Kurian et al., 2008). As juvenile social play behavior is a highly rewarding driven behavior, it is intriguing to speculate that the reduction of MeCP2 may have induced social anhedonia in males. Furthermore, reduced expression of MeCP2 within the newborn male amygdala has lasting consequences on the expression of the socially relevant peptide arginine vasopressin (AVP; Figure 15.3). Specifically, transient reductions of MeCP2 using siRNA within the developing amygdala results in a lasting reduction of AVP expression in males, but not females (Forbes-Lorman et al., 2012). Together, these data suggest that the male brain is more sensitive to neonatal perturbations in MeCP2 expression within the developing amygdala. Likewise, females appear to be more resilient to perturbations in MeCP2 expression within the amygdala. These data suggest that subtle differences in epigenetic mechanisms between males and females appear to mediate risk or resilience to social disruptions in a sex-dependent manner.

### *3.2.3 Corepressor proteins: the brakes on the wheels*

As mentioned earlier, MBPs bind to methylated DNA and can initiate the recruitment of nuclear corepressor proteins and HDACs to form large corepressor complexes (Tetel et al., 2009; Auger and Jessen, 2009). These complexes remove acetyl groups from histones, restoring the positive charge on histones, leading to chromatin condensation and gene silencing (Figure 15.4). Numerous corepressor complexes have been identified, such as Sin3A, NuRD, CoREST, and the NCoR/SMRT repressor complexes, which contain rather unique combinations of proteins that can regulate gene transcription

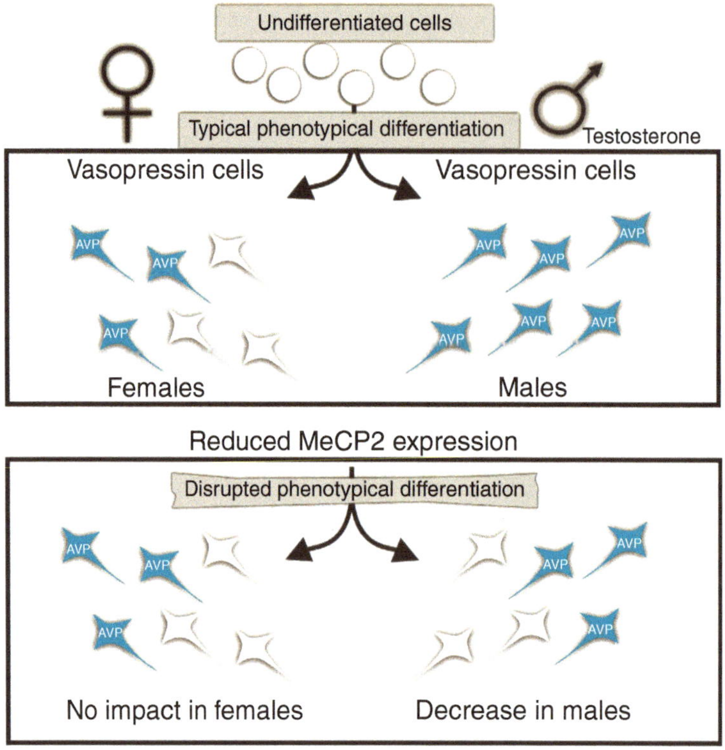


**Figure 15.3** ***During brain development, vasopressin cells are differentiated so that males express more vasopressin than females.*** Brief reductions in MeCP2 expression result in a lasting decrease in the number of vasopressin cells in males but not females, suggesting that the male brain may be less adept at compensating for MeCP2 loss of function.

(Schoch and Abel, 2014; Auger and Jessen, 2009). Different combinations of corepressor complexes allows for diverse gene regulation. One of the first corepressors to be identified was the nuclear corepressor (NCoR). NCoR has been found to interact with a variety of nuclear receptors, such as thyroid hormone receptors, ARs, ERs, and progestin receptors (Yoon et al., 2005; Chen and Evans, 1995; Horlein et al., 1995). Interestingly, NCoR has been shown to interact directly or indirectly with Kaiso and MeCP2 (Yoon et al., 2003; Cukier et al., 2008). Indeed, an NCoR/SMRT interaction domain was identified on the MeCP2 gene, and mutations of this site, which is seen in at least 5% of classical RTT cases, abolishes the interaction between NCoR and MeCP2 (Lyst et al., 2013). This suggests that some of the consequences of MeCP2 disruptions may result from abnormal interactions of MeCP2 with NCoR.

While the functional consequences of corepressor complexes within the developing brain are still being investigated, recent data indicate there is a sex difference in NCoR levels within the developing brain (Jessen et al., 2010). It was found that females express higher levels of NCoR within the developing amygdala compared to males, and these

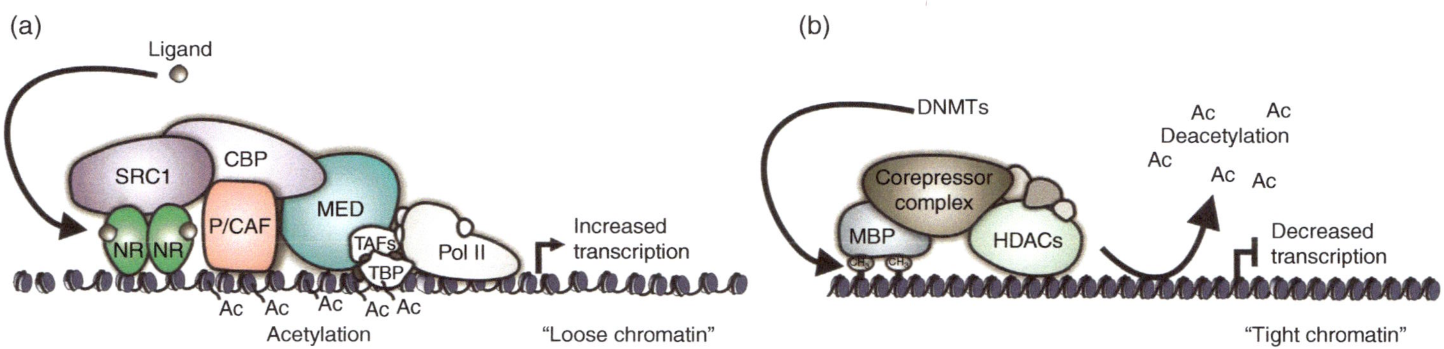


**Figure 15.4** ***Coactivator and corepressor complex dictate on and off states of gene transcription.*** (a) depicts ligand bound nuclear receptors bound to DNA. This binding recruits coactivator proteins (SRC-1 and CBP), which acetylate histone tails, causing histones to move apart and make DNA more accessible to the transcriptional complex. Included in the transcriptional complex are a host of molecules that mediate gene transcription (p300/CBP associated factor, p/CAF; mediator complex, MED; TATA box-binding protein, TBP; TBP-associated factors, TAFs; and DNA polymerase II, Pol II). (b) depicts methylation ($CH_3$) of DNA, facilitated by DNMT. MBPs bind to methyl groups ($CH_3$) and recruit a corepressor complex containing HDAC activity, which deacetylate histone tails. Deacetylation causes histones to move closer together making transcription factor binding more difficult. The functional implication of this entire complex is that transcription is decreased.

differences are partly due to differences in hormone exposure, as treatment of females with estradiol decreases the expression of NCoR to male-like levels. This suggests that NCoR is responsive to changes in the internal hormonal milieu. It is possible that other endogenous or exogenous perturbations, such as early life adversity, could alter NCoR levels to impact chromatin remodeling and have lasting consequences on behavior. Indeed, reducing NCoR levels within the neonatal amygdala produced lasting increases in juvenile social play behavior in males but not females. In contrast, the brief reductions of NCoR levels increased anxiety-like behavior in both juvenile males and females. This suggests that transient disruptions in NCoR within the developing brain can have sex-dependent consequences on juvenile social behavior. These data support the concept that targeting some epigenetic molecules may have a greater impact within the male versus the female brain, at least when considering juvenile social interactions. Interestingly, lowering NCoR expression levels within the developing amygdala increases juvenile play behavior in males; whereas, lowering MeCP2 levels within the developing amygdala decreases juvenile social play in males. Therefore, the organization of juvenile social interactions in males appears to be under the tight control of these epigenetic molecules. This further illustrates that combinations of these molecules are necessary for maintenance of typical juvenile behavioral patterns, and highlights the importance of the delicate balance necessary for epigenetic regulation. Furthermore, alterations in the availability, expression, or function of some epigenetic factors can lead to sex-specific outcomes on brain development and behavior.

### 3.2.4 Coactivator proteins: epigenetic accelerators

In contrast to corepressors, coactivator proteins enhance steroid receptor mediated gene expression by increasing the acetylation of histones, which relaxes chromatin structure and increases gene expression (Figure 15.4). Within the developing brain, steroid receptors have been found to interact with a variety of coactivator proteins, which include steroid receptor coactivator-1 (SRC-1) and the cAMP response element binding protein-binding protein (CPB), upon DNA (Tetel et al., 2009; Dasgupta and O'Malley, 2014; Auger et al., 2002). Coactivator complexes alter the acetylation of histones either through their own histone acetyltransferase (HAT) activity or association with other proteins that have HAT activity. As negatively charged DNA is tightly coiled around positively charged histones, acetylation of histones neutralizes the histone charge and thereby relaxes chromatin structure to allow more efficient binding of transcription factors that increase gene transcription. Due to the importance of coactivators in regulating steroid receptor activity, it is not surprising that both SRC-1 and CBP are critical for steroid mediated sexual differentiation of the brain. Specifically, reducing the expression of SRC-1 within the developing hypothalamus during the first few days of neonatal life interferes with sexual differentiation of adult male sexual behavior (Auger et al., 2000). Similarly, reductions in CBP within the developing hypothalamus also disrupt adult male sexual behavior. These data indicate that coactivators are critical for sexual differential of the male brain. Coactivators are also important in regulating hormone-sensitive behaviors in adulthood. Reducing SRC-1 or CBP levels in adult brain also disrupts steroid regulated sexual behavior in rats (Molenda et al., 2002), as well as in quail (Charlier et al., 2005). It is likely that coactivators play an important role in steroid induced differentiation of neuroendocrine systems early in development and perturbations to coactivators create lasting disruptions in neuroendocrine function, whereas coactivators participate in transient regulation of gene expression by steroid hormones later in adulthood. This would support the more complicated gene–environment–sex–time interaction.

## 4 CONCLUSIONS

### 4.1 Putting it together: chromatin remodeling and brain sex development

While a large amount of what we know about how epigenetic mechanisms influence behavior comes from examining epigenetic modifications occurring during perinatal brain development (Meaney, 2010; Auger et al., 2010), epigenetic modifications can also occur later in life (Zhao et al., 2010; Levenson and Sweatt, 2005; Auger and Auger, 2013). While it is the relative stability of methylation patterns that have led scientists to propose this as a mechanism for how early environmental influences can have lasting consequences on gene expression, some DNA methylation patterns in the adult brain are plastic and may only appear stable as they are actively being maintained by a hormonal signal (Auger and Auger, 2013; Auger et al., 2011). An example of this concept comes from studies indicating that DNA methylation patterns of the AVP gene within the bed nucleus of the stria terminalis of males, as well as its expression, require the presence of gonadal hormones (Figure 15.5). That is, the presence of testosterone maintains the expression of AVP mRNA perhaps through epigenetic means, as removal of testosterone, through castration, results in lower levels of AVP mRNA and increased methylation of specific CpG sites within the AVP promoter. Restoration of testosterone reduces methylation levels and restores AVP mRNA expression. These data indicate that some methylation marks may only appear to be stable because an endogenous signal is maintaining their patterns (Auger et al., 2011). As these data are obtained in mature, differentiated cells within the adult brain, they indicate that some methylation patterns can serve as homeostatic switches that are regulated by the environment in the adult. Methylation patterns need not be preprogrammed or static in the brain. The exciting notion is that methylation patterns can easily be rewritten in response to a changing environment regardless of whether that environment is internal or external. Therefore, understanding the mechanisms for these potentially homeostatic switches will be necessary to interpret the meaning of these marks in disease states.

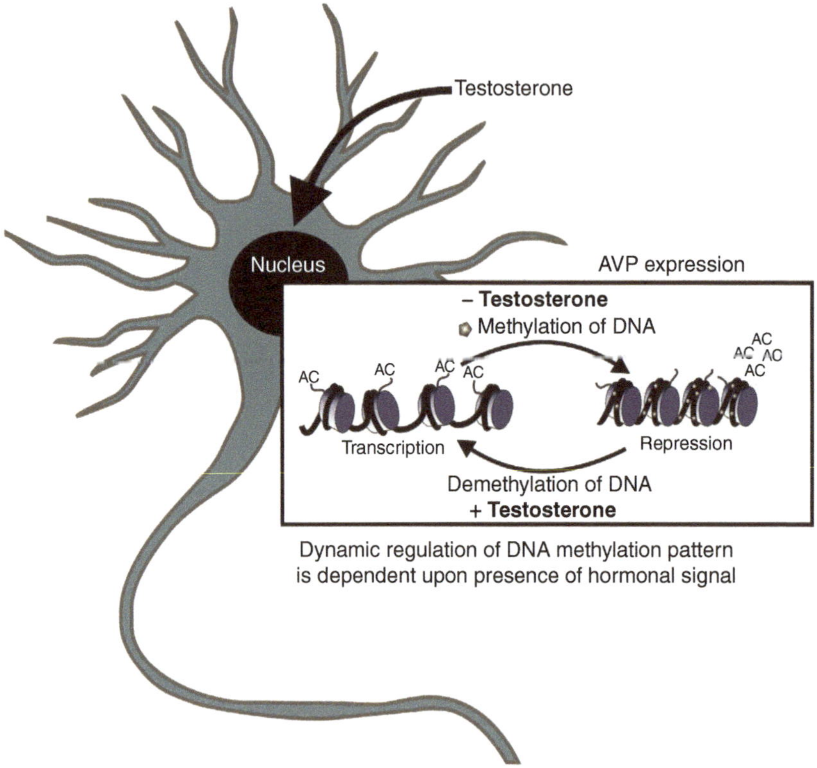


**Figure 15.5** ***DNA methylation is plastic within the adult brain.*** Within the adult male brain, circulating testosterone maintains DNA methylation patterns on the vasopressin gene. The loss of testosterone levels results in increased methylation of the vasopressin promoter region; whereas the restoration of testosterone levels quickly returns DNA methylation patterns back to baseline. Thus, a signal is needed to actively maintain some DNA methylation patterns in the adult brain.

## 4.2 Implications for mental health risk

Epigenetic differences are associated with a number of mental health disorders, including Alzheimer's disease, schizophrenia, and depression. Interestingly, sex differences in gene methylation patterns are found in individuals diagnosed with major psychosis (Mill et al., 2008; Connor and Akbarian, 2008). For example, *WNT1*, part of the Wnt signaling pathway, is found to be hypermethylated in females diagnosed with major psychosis. Additionally, *NURR1*, also known as *NR4A2*, is hypermethylated in females diagnosed with schizophrenia. It is important to note that it is not clear whether these differences occur as a result of major psychosis or are predictors of major psychosis. If the differences in these methylation patterns increase the risk for these disorders, then epigenetic mechanisms that are different between the sexes during brain development may be a contributing factor to vulnerability to mental health risk (Kigar and Auger, 2013; Jessen and Auger, 2011). As the developing female brain expresses higher levels of DNMT3a (Kolodkin and Auger, 2011), MeCP2 (Kurian et al., 2007), and NCoR (Jessen et al., 2010), as well as DNA methylation patterns (Kurian et al., 2010), cells are likely to respond differently to the same stimulus. As stated earlier, the differential response may be beneficial or may be harmful to cell functioning or to the individual. These data suggest that alterations in epigenetic mechanisms may underlie differences in juvenile or adult behavioral patterns.

As discussed earlier, Rett syndrome is a pervasive developmental disorder marked by impaired speech and social interactions, repetitive motor movements and seizures, due to mutations in MeCP2 expression or function (Amir et al., 1999). As MeCP2 is an X-linked gene, Rett syndrome is primarily diagnosed in females because MeCP2 mutations in males are usually embryonic lethal. As females have two X chromosomes, one of the X chromosomes is randomly inactivated by epigenetic mechanisms in females to correct for proper chromosomal dosage. Due to X inactivation, a mutated X chromosome may be randomly inactivated in some female brain cells, creating a female brain that is mosaic for the mutation. As males have no additional X chromosome, mutations on the X chromosome will impact all cells (Figure 15.6). Men are also more likely to be colorblind than women, as the genes responsible for red and green cone cells are located on the X chromosome. This is why females need to inherit two mutated copies for color blindness to occur. Therefore, the additional X chromosome confers resilience to some X-linked disorders. Understanding epigenetic inactivation of X chromosomes

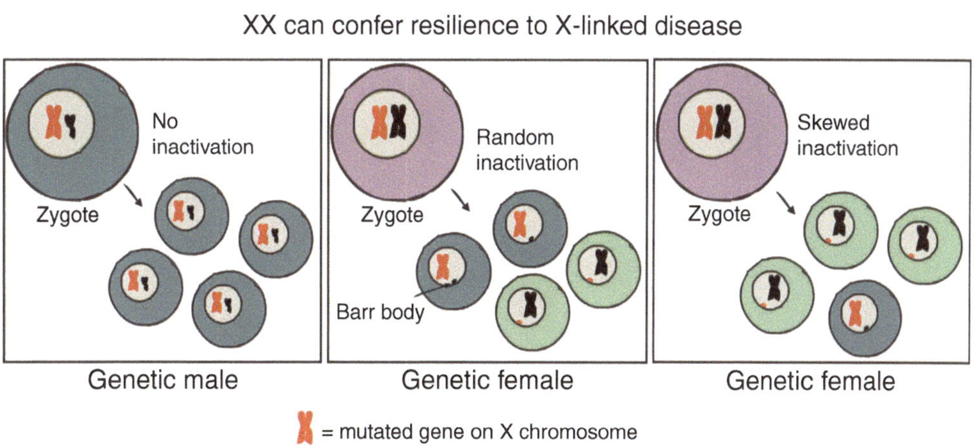


**Figure 15.6** ***Chromosomal sex and X inactivation as a contributing factor to female mental health resilience.*** In males, mutations on the X chromosome will have consequences for all cells. Females have two X chromosomes and one of the X chromosomes is randomly inactivated by epigenetic mechanisms to correct for proper chromosomal dosage. As X chromosomes with mutations in females may be inactivated in some brain cells, this creates a female brain that is mosaic for the mutation. Interestingly, skewing occurs when the X chromosome carrying the mutated gene is nonrandomly inactivated, favoring expression of the nonmutated allele.

may allow the silencing of mutated X chromosomes and increased expression of non-mutated X chromosomes.

As the field of neuroepigenetics matures, we will be able to further elucidate how the environment can reprogram lasting changes in gene expression that impact neuronal functioning. We will also have a better understanding of how environmentally induced modifications to chromatin structure as well as aberrant methylation patterns contribute to mental health disorders. While differences in some epigenetic signatures may result from environmental perturbations, others result from naturally occurring sexual differentiation of the brain. As discussed earlier, males and females exhibit different methylation patterns as a result of both testosterone exposure and maternal care within the promoter regions of important molecules influencing brain development (Ghahramani et al., 2014; Kurian et al., 2010). It is conceivable that epigenetic differences influence how a cell responds to a stimulus and this differing response may be beneficial or harmful to cell functioning or mental health (Figure 15.7). As epigenetic mechanisms are being implicated in numerous neurological or psychiatric disorders, it is not surprising that sex differences in epigenetic mechanisms may be a contributing factor to an individual's mental health risk. To fully understand the role of gene–environmental interactions in predicting disease, one must consider sex as a contributing factor. A further complication

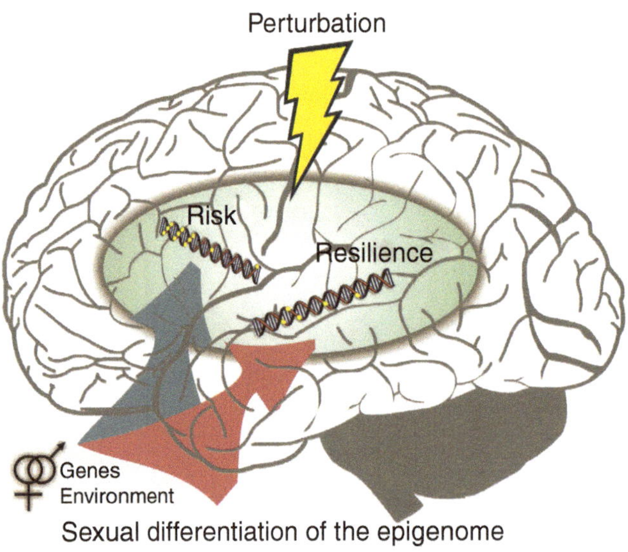


**Figure 15.7** ***Sexual differentiation of the epigenome may underlie sex differences in mental health risk or resilience.*** Converging mechanisms program sex differences in the epigenome that include differences in sex chromosomes, hormones, and environmental cues. Sex differences in the epigenome may change the way neurons respond to environmental stressors. Therefore, epigenomic sex differences may underlie sex differences in risk and resilience to neurological and psychiatric disorders that are induced by environmental or genetic perturbations.

to the model is that brain sex differences not only influence how one responds to environmental cues, but the environment can also produce sex differences in the epigenome. Specifically, sex differences in mother–offspring interactions can produce sex differences in the epigenome (Kurian et al., 2010). Therefore, sexual differentiation of the brain is a powerful model to study the contribution of gene–environment–sex interactions in potential mental health risk (Auger and Auger, 2013; Kigar and Auger, 2013).

## REFERENCES

Allen, L.S., Hines, M., Shryne, J.E., Gorski, R.A., 1989. Two sexually dimorphic cell groups in the human brain. J. Neurosci. 9, 497–506.

Amateau, S.K., Alt, J.J., Stamps, C.L., McCarthy, M.M., 2004. Brain estradiol content in newborn rats: sex differences, regional heterogeneity, and possible *de novo* synthesis by the female telencephalon. Endocrinology 145, 2906–2917.

Amir, R.E., Van den Veyver, I.B., Wan, M., Tran, C.Q., Francke, U., Zoghbi, H.Y., 1999. Rett syndrome is caused by mutations in X-linked MECP2, encoding methyl-CpG-binding protein 2. Nat. Genet. 23, 185–188.

Auger, A.P., Auger, C.J., 2011. Epigenetic turn ons and turn offs: chromatin reorganization and brain differentiation. Endocrinology 152, 349–353.

Auger, C.J., Auger, A.P., 2013. Permanent and plastic epigenesis in neuroendocrine systems. Front. Neuroendocrinol. 34 (3), 190–197.

Auger, A.P., Jessen, H.M., 2009. Corepressors, nuclear receptors, and epigenetic factors on DNA: a tail of repression. Psychoneuroendocrinology 34 (Suppl. 1), S39–S47.

Auger, A.P., Olesen, K.M., 2009. Brain sex differences and the organisation of juvenile social play behaviour. J. Neuroendocrinol. 21, 519–525.

Auger, A.P., Tetel, M.J., McCarthy, M.M., 2000. Steroid receptor coactivator-1 (SRC-1) mediates the development of sex-specific brain morphology and behavior. Proc. Natl. Acad. Sci. USA 97, 7551–7555.

Auger, A.P., Perrot-Sinal, T.S., Auger, C.J., Ekas, L.A., Tetel, M.J., McCarthy, M.M., 2002. Expression of the nuclear receptor coactivator, cAMP response element-binding protein, is sexually dimorphic and modulates sexual differentiation of neonatal rat brain. Endocrinology 143, 3009–3016.

Auger, A.P., Jessen, H.M., Edelmann, M.N., 2010. Epigenetic organization of brain sex differences and juvenile social play behavior. Horm. Behav. 3, 358–363.

Auger, C.J., Coss, D., Auger, A.P., Forbes-Lorman, R.M., 2011. Epigenetic control of vasopressin expression is maintained by steroid hormones in the adult male rat brain. Proc. Natl. Acad. Sci. USA 108, 4242–4247.

Baum, M.J., 1979. Differentiation of coital behavior in mammals: a comparative analysis. Neurosci. Biobehav. Rev. 3, 265–284.

Beatty, W.W., Dodge, A.M., Traylor, K.L., Meaney, M.J., 1981. Temporal boundary of the sensitive period for hormonal organization of social play in juvenile rats. Physiol. Behav. 26, 241–243.

Bird, A., 2007. Perceptions of epigenetics. Nature 447, 396–398.

Bleier, R., Byne, W., Siggelkow, I., 1982. Cytoarchitectonic sexual dimorphisms of the medial preoptic and anterior hypothalamic areas in guinea pig, rat, hamster, and mouse. J. Comp. Neurol. 212, 118–130.

Brown, T.J., Hochberg, R.B., Zielinski, J.E., MacLusky, N.J., 1988. Regional sex differences in cell nuclear estrogen-binding capacity in the rat hypothalamus and preoptic area. Endocrinology 123, 1761–1770.

Cameron, N.M., Fish, E.W., Meaney, M.J., 2008. Maternal influences on the sexual behavior and reproductive success of the female rat. Horm. Behav. 54, 178–184.

Casto, J.M., Ward, O.B., Bartke, A., 2003. Play, copulation, anatomy, and testosterone in gonadally intact male rats prenatally exposed to flutamide. Physiol. Behav. 79, 633–641.

Chamove, A.S., Rosenblum, L.A., Harlow, H.F., 1973. Monkeys (*Macaca mulatta*) raised only with peers. A pilot study. Anim. Behav. 21, 316–325.

Champagne, F.A., Weaver, I.C., Diorio, J., Dymov, S., Szyf, M., Meaney, M.J., 2006. Maternal care associated with methylation of the estrogen receptor-alpha1b promoter and estrogen receptor-alpha expression in the medial preoptic area of female offspring. Endocrinology 147, 2909–2915.

Charlier, T.D., Ball, G.F., Balthazart, J., 2005. Inhibition of steroid receptor coactivator-1 blocks estrogen and androgen action on male sex behavior and associated brain plasticity. J. Neurosci. 25, 906–913.

Chen, J.D., Evans, R.M., 1995. A transcriptional co-repressor that interacts with nuclear hormone receptors. Nature 377, 454–457.

Chen, R.Z., Akbarian, S., Tudor, M., Jaenisch, R., 2001. Deficiency of methyl-CpG binding protein-2 in CNS neurons results in a Rett-like phenotype in mice. Nat. Genet. 27, 327–331.

Ciofi, P., Leroy, D., Tramu, G., 2006. Sexual dimorphism in the organization of the rat hypothalamic infundibular area. Neuroscience 141, 1731–1745.

Collingwood, T.N., Urnov, F.D., Wolffe, A.P., 1999. Nuclear receptors: coactivators, corepressors and chromatin remodeling in the control of transcription. J. Mol. Endocrinol. 23, 255–275.

Connor, C.M., Akbarian, S., 2008. DNA methylation changes in schizophrenia and bipolar disorder. Epigenetics 3, 55–58.

Cukier, H.N., Perez, A.M., Collins, A.L., Zhou, Z., Zoghbi, H.Y., Botas, J., 2008. Genetic modifiers of MeCP2 function in Drosophila. PLoS Genet. 4, e1000179.

Dasgupta, S., O'Malley, B.W., 2014. Transcriptional coregulators: emerging roles of SRC family of coactivators in disease pathology. J. Mol. Endocrinol. 53, R47–R59.

De Vries, G.J., Rissman, E.F., Simerly, R.B., Yang, L.Y., Scordalakes, E.M., Auger, C.J., et al., 2002. A model system for study of sex chromosome effects on sexually dimorphic neural and behavioral traits. J. Neurosci. 22, 9005–9014.

Dewing, P., Chiang, C.W., Sinchak, K., Sim, H., Fernagut, P.O., Kelly, S., et al., 2006. Direct regulation of adult brain function by the male-specific factor SRY. Curr. Biol. 16, 415–420.

DonCarlos, L.L., Handa, R.J., 1994. Developmental profile of estrogen receptor mRNA in the preoptic area of male and female neonatal rats. Brain Res. Dev. Brain Res. 79, 283–289.

Edelmann, M.N., Demers, C.H., Auger, A.P., 2013. Maternal touch moderates sex differences in juvenile social play behavior. PLoS One 8, e57396.

Forbes-Lorman, R.M., Rautio, J.J., Kurian, J.R., Auger, A.P., Auger, C.J., 2012. Neonatal MeCP2 is important for the organization of sex differences in vasopressin expression. Epigenetics 7, 230–238.

Ghahramani, N.M., Ngun, T.C., Chen, P.Y., Tian, Y., Krishnan, S., Muir, S., et al., 2014. The effects of perinatal testosterone exposure on the DNA methylome of the mouse brain are late-emerging. Biol. Sex Differ. 5, 8.

Gorski, R.A., Gordon, J.H., Shryne, J.E., Southam, A.M., 1978. Evidence for a morphological sex difference within the medial preoptic area of the rat brain. Brain Res. 148, 333–346.

Grafstrom, R.H., Yuan, R., Hamilton, D.L., 1985. The characteristics of DNA methylation in an *in vitro* DNA synthesizing system from mouse fibroblasts. Nucleic Acid. Res. 13, 2827–2842.

Guy, J., Hendrich, B., Holmes, M., Martin, J.E., Bird, A., 2001. A mouse Mecp2-null mutation causes neurological symptoms that mimic Rett syndrome. Nat. Genet. 27, 322–326.

Guy, J., Gan, J., Selfridge, J., Cobb, S., Bird, A., 2007. Reversal of neurological defects in a mouse model of Rett syndrome. Science 315, 1143–1147.

Hendrich, B., Bird, A., 1998. Identification and characterization of a family of mammalian methyl-CpG binding proteins. Mol. Cell Biol. 18, 6538–6547.

Hines, M., Allen, L.S., Gorski, R.A., 1992. Sex differences in subregions of the medial nucleus of the amygdala and the bed nucleus of the stria terminalis of the rat. Brain Res. 579, 321–326.

Horlein, A.J., Naar, A.M., Heinzel, T., Torchia, J., Gloss, B., Kurokawa, R., et al., 1995. Ligand-independent repression by the thyroid hormone receptor mediated by a nuclear receptor co-repressor. Nature 377, 397–404.

Hsieh, C.L., 1999. *In vivo* activity of murine *de novo* methyltransferases, Dnmt3a and Dnmt3b. Mol. Cell Biol. 19, 8211–8218.

Jessen, H.M., Auger, A.P., 2011. Sex differences in epigenetic mechanisms may underlie risk and resilience for mental health disorders. Epigenetics 6, 857–861.

Jessen, H.M., Kolodkin, M.H., Bychowski, M.E., Auger, C.J., Auger, A.P., 2010. The nuclear receptor corepressor has organizational effects within the developing amygdala on juvenile social play and anxiety-like behavior. Endocrinology 151, 1212–1220.

Kigar, S.L., Auger, A.P., 2013. Epigenetic mechanisms may underlie the aetiology of sex differences in mental health risk and resilience. J. Neuroendocrinol. 25, 1141–1150.

Klose, R.J., Bird, A.P., 2006. Genomic DNA methylation: the mark and its mediators. Trends Biochem. Sci. 31, 89–97.

Kolodkin, M.H., Auger, A.P., 2011. Sex difference in the expression of DNA methyltransferase 3a (DNMT3a) in the rat amygdala during development. J. Neuroendocrinol. 7, 577–583.

Kurian, J.R., Forbes-Lorman, R.M., Auger, A.P., 2007. Sex difference in mecp2 expression during a critical period of rat brain development. Epigenetics 2, 173–178.

Kurian, J.R., Bychowski, M.E., Forbes-Lorman, R.M., Auger, C.J., Auger, A.P., 2008. Mecp2 organizes juvenile social behavior in a sex-specific manner. J. Neurosci. 28, 7137–7142.

Kurian, J.R., Olesen, K.M., Auger, A.P., 2010. Sex differences in epigenetic regulation of the estrogen receptor-alpha promoter within the developing preoptic area. Endocrinology 151, 2297–2305.

Lauber, A.H., Mobbs, C.V., Muramatsu, M., Pfaff, D.W., 1991. Estrogen receptor messenger RNA expression in rat hypothalamus as a function of genetic sex and estrogen dose. Endocrinology 129, 3180–3186.

Levenson, J.M., Sweatt, J.D., 2005. Epigenetic mechanisms in memory formation. Nat. Rev. Neurosci. 6, 108–118.

Lonstein, J.S., Auger, A.P., 2009. Perinatal gonadal hormone influences on neurobehavioral development. In: Blumberg, M., Freeman, J., Robinson, S. (Eds.), Handbook of Behavioral and Comparative Neuroscience. Oxford University Press, New York, pp. 424–453.

Lyst, M.J., Ekiert, R., Ebert, D.H., Merusi, C., Nowak, J., Selfridge, J., et al., 2013. Rett syndrome mutations abolish the interaction of MeCP2 with the NCoR/SMRT co-repressor. Nat. Neurosci. 16, 898–902.

MacLusky, N.J., Naftolin, F., 1981. Sexual differentiation of the central nervous system. Science 211, 1294–1302.

Maerkel, K., Durrer, S., Henseler, M., Schlumpf, M., Lichtensteiger, W., 2007. Sexually dimorphic gene regulation in brain as a target for endocrine disrupters: developmental exposure of rats to 4-methylbenzylidene camphor. Toxicol. Appl. Pharmacol. 218, 152–165.

Matsumoto, A., Arai, Y., 1986. Male-female difference in synaptic organization of the ventromedial nucleus of the hypothalamus in the rat. Neuroendocrinology 42, 232–236.

McCarthy, M.M., 2008. Estradiol and the developing brain. Physiol. Rev. 88, 91–124.

McCarthy, M.M., Arnold, A.P., 2011. Reframing sexual differentiation of the brain. Nat. Neurosci. 14, 677–683.

McCarthy, M.M., Auger, A.P., Bale, T.L., De Vries, G.J., Dunn, G.A., Forger, N.G., et al., 2009. The epigenetics of sex differences in the brain. J. Neurosci. 29, 12815–12823.

Meaney, M.J., 2010. Epigenetics and the biological definition of gene x environment interactions. Child Dev. 81, 41–79.

Meaney, M.J., Stewart, J., 1981. Neonatal-androgens influence the social play of prepubescent rats. Horm. Behav. 15, 197–213.

Meehan, R.R., Lewis, J.D., McKay, S., Kleiner, E.L., Bird, A.P., 1989. Identification of a mammalian protein that binds specifically to DNA containing methylated CpGs. Cell 58, 499–507.

Metivier, R., Gallais, R., Tiffoche, C., Le, P.C., Jurkowska, R.Z., Carmouche, R.P., et al., 2008. Cyclical DNA methylation of a transcriptionally active promoter. Nature 452, 45–50.

Mill, J., Tang, T., Kaminsky, Z., Khare, T., Yazdanpanah, S., Bouchard, L., 2008. Epigenomic profiling reveals DNA-methylation changes associated with major psychosis. Am. J. Hum. Genet. 82, 696–711.

Molenda, H.A., Griffen, A.L., Auger, A.P., McCarthy, M.M., Tetel, M.J., 2002. Nuclear receptor coactivators modulate hormone-dependent gene expression in brain and female reproductive behavior in rats. Endocrinology 143, 436–444.

Moore, C.L., 1984. Maternal contributions to the development of masculine sexual behavior in laboratory rats. Dev. Psychobiol. 17, 347–356.

Moore, C.L., 1992. The role of maternal stimulation in the development of sexual behavior and its neural basis. Ann. NY Acad. Sci. 662, 160–177.

Moore, C.L., Morelli, G.A., 1979. Mother rats interact differently with male and female offspring. J. Comp. Physiol. Psychol. 93, 677–684.

Nagarajan, R.P., Hogart, A.R., Gwye, Y., Martin, M.R., LaSalle, J.M., 2006. Reduced MeCP2 expression is frequent in autism frontal cortex and correlates with aberrant MECP2 promoter methylation. Epigenetics 1, e1–11.

Nanney, D.L., 1958. Epigenetic control systems. Proc. Natl. Acad. Sci. USA 44, 712–717.
Okano, M., Xie, S., Li, E., 1998. Cloning and characterization of a family of novel mammalian DNA (cytosine-5) methyltransferases. Nat. Genet. 19, 219–220.
Olesen, K.M., Jessen, H.M., Auger, C.J., Auger, A.P., 2005. Dopaminergic activation of estrogen receptors in neonatal brain alters progestin receptor expression and juvenile social play behavior. Endocrinology 146, 3705–3712.
Olioff, M., Stewart, J., 1978. Sex differences in the play behavior of prepubescent rats. Physiol. Behav. 20, 113–115.
Parent, C.I., Meaney, M.J., 2008. The influence of natural variations in maternal care on play fighting in the rat. Dev. Psychobiol. 50, 767–776.
Pradhan, S., Bacolla, A., Wells, R.D., Roberts, R.J., 1999. Recombinant human DNA (cytosine-5) methyltransferase. I. Expression, purification, and comparison of *de novo* and maintenance methylation. J. Biol. Chem. 274, 33002–33010.
Ramsahoye, B.H., Biniszkiewicz, D., Lyko, F., Clark, V., Bird, A.P., Jaenisch, R., 2000. Non-CpG methylation is prevalent in embryonic stem cells and may be mediated by DNA methyltransferase 3a. Proc. Natl. Acad. Sci. USA 97, 5237–5242.
Razin, A., Riggs, A.D., 1980. DNA methylation and gene function. Science 210, 604–610.
Rhees, R.W., Lephart, E.D., Eliason, D., 2001. Effects of maternal separation during early postnatal development on male sexual behavior and female reproductive function. Behav. Brain Res. 123, 1–10.
Schoch, H., Abel, T., 2014. Transcriptional co-repressors and memory storage. Neuropharmacology 80, 53–60.
Tetel, M.J., Auger, A.P., Charlier, T.D., 2009. Who's in charge? Nuclear receptor coactivator and corepressor function in brain and behavior. Front. Neuroendocrinol. 30, 328–342.
Waddington, C.H., 1942. The epigenotype. Endeavour 1, 18–20.
Yamamoto, Y., Carter, C.S., Cushing, B.S., 2006. Neonatal manipulation of oxytocin affects expression of estrogen receptor alpha. Neuroscience 137, 157–164.
Yoder, J.A., Soman, N.S., Verdine, G.L., Bestor, T.H., 1997. DNA (cytosine-5)-methyltransferases in mouse cells and tissues. Studies with a mechanism-based probe. J. Mol. Biol. 270, 385–395.
Yokosuka, M., Okamura, H., Hayashi, S., 1997. Postnatal development and sex difference in neurons containing estrogen receptor-alpha immunoreactivity in the preoptic brain, the diencephalon, and the amygdala in the rat. J. Comp Neurol. 389, 81–93.
Yoon, H.G., Chan, D.W., Reynolds, A.B., Qin, J., Wong, J., 2003. N-CoR mediates DNA methylation-dependent repression through a methyl CpG binding protein Kaiso. Mol. Cell 12, 723–734.
Yoon, H.G., Choi, Y., Cole, P.A., Wong, J., 2005. Reading and function of a histone code involved in targeting corepressor complexes for repression. Mol. Cell Biol. 25, 324–335.
Zhao, Z., Fan, L., Frick, K.M., 2010. Epigenetic alterations regulate estradiol-induced enhancement of memory consolidation. Proc. Natl. Acad. Sci. USA 107, 5605–5610.

# SUBJECT INDEX

## M



## N

Kolibri 360 GmbH
Ferdinand-Jühlke-Straße 7
99095 Erfurt, Deutschland
produktsicherheit@kolibri360.de